# Finite Simple Groups

Oxford Instructional Conference on.
"

# Finite Simple Groups

PROCEEDINGS OF AN INSTRUCTIONAL CONFERENCE
ORGANIZED BY THE LONDON MATHEMATICAL SOCIETY
(A NATO ADVANCED STUDY INSTITUTE)

*Edited by*

M. B. POWELL

*The Mathematical Institute and St. Peter's College,*
*University of Oxford, England*

and

G. HIGMAN

*The Mathematical Institute and Magdalen College,*
*University of Oxford, England*

1971
ACADEMIC PRESS
London and New York

ACADEMIC PRESS INC. (LONDON) LTD.
Berkeley Square House
Berkeley Square
London, W1X 6BA

*United States Edition published by*
ACADEMIC PRESS INC.
111 Fifth Avenue
New York, New York 10003





Library of Congress Catalog Card Number: 77–149703
ISBN: 0 12 563850 7

PRINTED IN GREAT BRITAIN BY THE WHITEFRIARS PRESS LTD.
LONDON AND TONBRIDGE

# Contributors

J. H. Conway, *Department of Pure Mathematics and Mathematical Statistics, 16 Mill Lane, Cambridge, England.*

C. W. Curtis, *Department of Mathematics, University of Oregon, Eugene, Oregon 97403, U.S.A.*

E. C. Dade, *Département de Mathématique, Université de Strasbourg, 7 Rue René Descartes, 67 Strasbourg, France.*

G. Glauberman, *Department of Mathematics, University of Chicago, Chicago, Illinois 60637, U.S.A.*

D. Gorenstein, *Department of Mathematics, The State University, New Brunswick, New Jersey 08903, U.S.A.*

G. Higman, *Mathematical Institute, University of Oxford, 24–29 St. Giles, Oxford OX1 3LB, England.*

M. Herzog, *Department of Mathematics, Tel-Aviv University, Ramat-Aviv, Israel.*

H. S. Leonard, Jr., *Department of Mathematics, Northern Illinois University, Dekalb, Illinois 60115, U.S.A.*

# Preface

After nearly fifty years of relative inactivity, considerable progress has been made in the last ten or fifteen years in the study of finite simple groups. At least two quite distinct factors have contributed to the renewed interest in the subject. First there was the systematic study of groups of Lie type, begun by Chevalley and continued by Steinberg, Tits and others. This not only drew attention to hitherto unnoticed series of simple groups but also, and more importantly, made available a great deal of information about a class of groups, that on any sensible measure includes most of the simple groups we know, and that in a uniform fashion. Secondly, there are the efforts that have been made by Brauer, and by those who, directly or indirectly, are of his school, to characterize classes of simple groups by internal properties, especially by the centralizers of involutions, and the structure of Sylow 2-subgroups. The earlier papers of this kind, for instance the seminal paper of Brauer, Suzuki and Wall characterizing the two-dimensional projective linear groups, are heavily character theoretical. However, the weightiest papers of the school, for instance the odd-order paper of Thompson and Feit, and Thompson's work on minimal simple groups, also use new and powerful techniques of combinatorial group theory, which later writers have made it their business to extend, systematize, and simplify. The two contributions are, of course, complementary. If we are ever going to know all finite simple groups, we shall have first to write down a long list of simple groups, and then prove a theorem which says that any finite simple group belongs to this list. Obviously, the better organized the list, the greater are our chances of proving the theorem. Much the most important contribution towards properly organizing the list of known simple groups has been Chevalley's. The contributions of the other school have been the first steps towards trying to prove the theorem.

The organizers of the Oxford Instructional Conference on Finite Simple Groups believed that the time was ripe for an exposition of some of this work, in the form of lecture courses, aimed primarily at young research workers, either in group theory, or in some subject sufficiently near it for them to want to know something about it. We did not feel that the subject was in any sense in a definitive state: clearly it is not. Rather we felt that the results so far obtained had generated the sort of interest that was best satisfied in this way. The response we received, the number of people wanting to take part suggests that we were right.

This volume contains the texts of four main courses of lectures given at the Conference, and of two short specialized series of lectures. Of the four main lectures, one, by Professor C. W. Curtis, deals with the contribution of Chevalley and his successors. The other three are an introduction to the study of characterization theorems. Professor E. C. Dade's lectures give the necessary character theory in its proper setting of linear algebra and ring theory. The lectures of Professor G. Glauberman and of Professor D. Gorenstein give different approaches to combinatorial methods in finite group theory. I would like here to thank all four lecturers for the insight they have given us.

*Mathematical Institute,* G. Higman
*Oxford*
*June*, 1971

## Contents

## CHAPTER III

## Chevalley Groups and Related Topics

C. W. CURTIS



## CHAPTER IV

## Finite Complex Linear Groups of Small Degree

H. S. LEONARD, JR.



## CHAPTER V

## Finite Groups with a Large Cyclic Sylow Subgroup

M. HERZOG

## CHAPTER VI

## Construction of Simple Groups from Character Tables

G. HIGMAN



## CHAPTER VII

## Three Lectures on Exceptional Groups

J. H. CONWAY



## CHAPTER VIII

## Character Theory Pertaining to Finite Simple Groups

E. C. DADE

CHAPTER I

# Global and Local Properties of Finite Groups

G. GLAUBERMAN*



## 1. Introduction

Let $G$ be a finite group, $p$ be a prime, and $S$ be a Sylow $p$-subgroup of $G$. We are concerned with the following problem: What is the relation between the local and global properties of $G$, that is, between the structure of various subgroups of $G$ and the structure of $G$ itself? We are particularly interested in properties involving the prime $p$, such as the existence of a factor group of order $p$.

We illustrate our problem in the following way. For every subgroup $H$ of $G$, let $\mathscr{L}(H)$ be a list of the elements and subgroups of $H$ and the equations of the multiplication table of $H$. For other subsets $H$ of $G$, let $\mathscr{L}(H)$ be a list of the elements of $H$. Imagine a file cabinet with four files, labelled "(A)",

* We wish to thank the London Mathematical Society for its kind invitation to participate in this conference. We are also grateful to the National Science Foundation and the Sloan Foundation for their support during the preparation of this work. We are indebted to W. Specht for a number of corrections to the original manuscript.

"(B)", "(C)", and "(D)". Assume that (A) contains $\mathscr{L}(G)$. Assume that (B) contains $\mathscr{L}(S)$ and every list of the form $\mathscr{L}(C \cap S)$, where $C$ is a conjugate class of $G$ that intefects $S$ non-trivially. Assume that file (C) contains every list of the form $\mathscr{L}(H)$, where $H$ is the normalizer in $G$ of a non-identity subgroup of $S$. Thus (C) is empty if $S = 1$. Assume that (D) is empty if $S = 1$. Otherwise, assume that (D) contains a single list $\mathscr{L}(H)$, where $H$ is the normalizer of some previously chosen non-identity normal (usually characteristic) subgroup $K$ of $S$.

We consider (A) to consist of "global" information, (C) and (D) to consist of "local" information, and (B) to consist of "fusion", which at first sight appears to be in between the local and the global realms. Clearly, if we are given (A), we may determine the contents of (B), (C), and (D). Similarly, (C) determines (D). We indicate these relations by solid arrows in the following diagram:

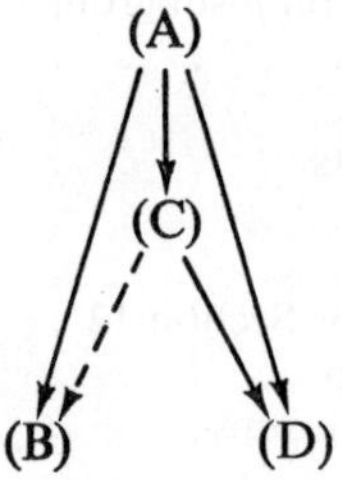


In Section 3, we will prove a theorem of Alperin that shows that (C) determines (B). We indicate this relation by a dashed arrow in the diagram. Our problem is to determine further relations between these files, in particular to determine partial information about (A) if we are given (B), (C), or (D).

In the past decade, many persons have contributed results toward the solution of this problem. Essential to the proof of many of these results have been the techniques of W. Burnside and J. Thompson, as illustrated in Theorems 4.2.5 and 14.3.1 in Hall (1959) and Hauptsatz IV.6.2 in Huppert (1967).

The material we present is divided into two parts. General results are discussed in Sections 2 to 8, and more restricted results in Sections 10 to 16. These are separated by a digression in Section 9. Until Section 14, the discussion is self-contained, with the following exceptions: initial assumed results in Section 2, two elementary applications of transfer in Section 4, three results on normal $p$-subgroups (Theorems 7.2, 10.1, and 10.2), and an isolated application of Burnside's $p^a q^b$-theorem (Corollary 12.6). Theorems 7.2 and 10.2 are proved in the appendix, and references for the others are given in the text. Only a few changes from the lectures given at the conference have been made, principally in Sections 7, 8, 14, 15, and 16.

## 2. Notation and Preliminary Results

All groups considered in this paper will be finite. We use standard notation; in particular, the following expressions concerning a finite group $G$ will be used most frequently:

| | |
|---|---|
| $H \subseteq G$, $G \supseteq H$ | $H$ is a subgroup of $G$ |
| $H \subset G$, $G \supset H$ | $H \subseteq G$ and $H \neq G$ |
| $H \trianglelefteq G$, $G \trianglerighteq H$ | $H$ is a normal subgroup of $G$ |
| $H \ntrianglelefteq G$, $G \ntrianglerighteq H$ | $H$ is not a normal subgroup of $G$ |
| $\mathbf{Z}(G)$ | center of $G$ |
| $\lvert G \rvert$ | order of $G$. |

For $H \subseteq G$, we define:

| | |
|---|---|
| $\mathbf{N}(H)$ | normalizer of $H$ |
| $\mathbf{C}(H)$ | centralizer of $H$ |
| $[G : H]$ | index of $H$ in $G$. |

For $H, K \subseteq G$, we define:

| | |
|---|---|
| $\mathbf{N}_H(K)$ | $\mathbf{N}(K) \cap H$ |
| $\mathbf{C}_H(K)$ | $\mathbf{C}(K) \cap H$ |
| $H \nsubseteq K$, $K \nsupseteq H$ | $H$ is not contained in $K$. |

We also define

| | |
|---|---|
| $A - B$ = set of all $x \in A$ such that $x \notin B$ | ($A$ and $B$ subsets of $G$) |
| $\langle A \rangle$ = subgroup of $G$ generated by $A$ | ($A$ a subset of $G$) |
| $[g, h] = g^{-1}h^{-1}gh$ | $(g, h \in G)$ |
| $[H, K] = \langle [h, k] : h \in H, k \in K \rangle$ | $(H, K \subseteq G)$ |
| $[H, K, L] = [[H, K], L]$ | $(H, K, L \subseteq G)$ |
| $G' = [G, G]$ | |
| $A^g = g^{-1}Ag$ | ($g \in G$, $A$ a subset of $G$) |
| $h^g = g^{-1}hg$ | $(g, h \in G)$ |
| $\mathbf{C}(g) = \mathbf{C}(\langle g \rangle)$ | $(g \in G)$ |
| $\mathbf{C}_H(g) = \mathbf{C}_H(\langle g \rangle)$ | $(g \in G; H \subseteq G)$. |

An automorphism of $G$ is said to *fix* an element or subset $A$ of $G$ if it maps $A$ to itself.

Let $G$ be a finite group and $p$ be a prime. Let $\lvert G \rvert_p$ be the highest power of $p$ that divides $\lvert G \rvert$. An element $g$ of $G$ is called a *p-element* if the order of $g$ is a power of $p$; it is called a *p'-element* if the order of $g$ is not divisible by $p$. We say that $G$ is a *p-group* if every element of $G$ is a $p$-element. If

$H \subseteq G$ and $H$ is a $p$-group, we say that $H$ is a *p-subgroup* of $G$. We define *p'-groups* and *p'-subgroups* similarly. Since $p^0 = 1$, the identity element of $G$ is both a $p$-element and a $p'$-element, and the identity subgroup of $G$ is both a $p$-subgroup and a $p'$-subgroup of $G$. If $H$ is an abelian $p$-subgroup of $G$, let $H^p$ be the group of all elements of the form $h^p$ for $h \in H$. A subgroup $S$ of $G$ is a *Sylow p-subgroup* of $G$ if $|S| = |G|_p$.

Henceforth we will adopt the convention that $G$ is an arbitrary finite group and that $p$ is an arbitrary prime.

Most of the results we discuss are based entirely on the elementary theory of groups and on the work of Sylow. We will assume as known the elementary properties of arbitrary groups and of finite groups (e.g., as given in Chapters 1 and 2 of Hall (1959) and in Sections 1.1 to 1.3 of Huppert (1967)). We will assume without proof the following two fundamental theorems of Sylow.

THEOREM 2.1. *If $|G|$ is a power of $p$ and $1 \subset N \trianglelefteq G$, then $N \cap \mathbf{Z}(G) \supset 1$.*

THEOREM 2.2.
(a) *There exists a Sylow p-subgroup $S$ of $G$.*
(b) *Any two Sylow p-subgroups of $G$ are conjugate.*
(c) *Suppose $H \subseteq G$, and $|H|$ is a power of $p$. Then $H$ is contained in some Sylow p-subgroup of $G$.*

Henceforth $S$ will denote an arbitrary Sylow $p$-subgroup of $G$. Theorems 2.1 and 2.2 and Lagrange's Theorem yield numerous corollaries:

COROLLARY 2.3 (Cauchy).
(a) *$G$ is a p-group if and only if $|G|$ is a power of $p$;*
(b) *$G$ is a p'-group if and only if $|G|$ is not divisible by $p$.*

COROLLARY 2.4. *If $G$ is a p-group and $G \neq 1$, then $\mathbf{Z}(G) \neq 1$.*

COROLLARY 2.5. *If $G$ is a p-group and $G \neq 1$, then there exists $N \trianglelefteq G$ such that $|G/N| = p$.*

*Proof.* Use Corollary 2.4 and induction.

COROLLARY 2.6. *Suppose $G$ is a p-group and $H \subset G$. Then $H \subset \mathbf{N}(H)$.*

*Proof.* Use induction on $|G|$. Let $Z = \mathbf{Z}(G)$. If $Z \subseteq H$, then by induction

$$\mathbf{N}(H)/Z = \mathbf{N}_{G/Z}(H/Z) \supset H/Z,$$

and consequently $\mathbf{N}(H) \supset H$. If $Z \nsubseteq H$, then $H \subset HZ \subseteq \mathbf{N}(H)$.

COROLLARY 2.7. *Suppose $T$ is a p-subgroup of $G$ and $T \trianglelefteq G$. Then $T \subseteq S$.*

*Proof.* Since $T \trianglelefteq G$, $TS \subseteq G$. Moreover, $|TS| = |T|[S : S \cap T]$, which is a power of $p$. Therefore, $|TS| \leq |G|_p = |S|$. Hence $TS = S$ and $T \subseteq S$.

COROLLARY 2.8. *Suppose* $T_1, \ldots, T_n$ *are normal p-subgroups of* $G$. *Then* $T_1 \ldots T_n$ *is a normal p-subgroup of* $G$.

*Proof.* By Corollary 2.7, $T_1 \ldots T_n \subseteq S$.

We now define $\mathbf{O}_p(G)$ to be the subgroup of $G$ generated by all the normal $p$-subgroups of $G$.

COROLLARY 2.9. $\mathbf{O}_p(G)$ *is a normal p-subgroup of* $G$.

LEMMA 2.10. *Suppose* $N \trianglelefteq G$. *Then*
(a) $N \cap S$ *is a Sylow p-subgroup of* $N$,
(b) $SN/N$ *is a Sylow p-subgroup of* $G/N$, *and*
(c) $G = SN$ *if and only if* $G/N$ *is a p-group.*

*Proof.* Clearly $N \cap S$ and $SN/N$ are $p$-groups. By Corollary 2.3, $|N \cap S|$ and $|SN/N|$ are powers of $p$. Therefore,

$$|N \cap S| \leq |N|_p \quad \text{and} \quad |SN/N| \leq |G/N|_p. \tag{2.1}$$

Since

$$|G|_p = |S| = |S/(S \cap N)|\,|S \cap N| = |SN/N|\,|S \cap N| \leq |G/N|_p |N|_p = |G|_p,$$

we must have equalities in (2.1). Thus we obtain (a) and (b). Clearly, (c) follows from (b).

LEMMA 2.11 (Frattini argument). *Suppose* $H \trianglelefteq G$ *and* $T$ *is a Sylow p-subgroup of* $H$. *Then*

$$G = \mathbf{N}(T)H = H\mathbf{N}(T).$$

*Proof.* Let $g \in G$. Then $T^g \subseteq H$. Therefore, $T^g$ is a Sylow $p$-subgroup of $H$. Take $h \in H$ such that $T^g = T^h$. Then $gh^{-1} \in \mathbf{N}(T)$, and $g = (gh^{-1})h$. Since $g$ is arbitrary, we are done.

We will frequently use two elementary results

LEMMA 2.12. *If* $H \subseteq G$, *then* $[H, \mathbf{N}(H)] \subseteq H$.

LEMMA 2.13. *If* $H \subseteq G$ *and* $g \in G$, *then* $\mathbf{N}(H^g) = \mathbf{N}(H)^g$ *and* $\mathbf{C}(H^g) = \mathbf{C}(H)^g$.

## 3. Alperin's Theorem

Our first step toward the goals described in the introduction is to prove that (C) determines (B), i.e., that the normalizers of non-identity subgroups of $S$ determine fusion. This important result was proved by Alperin in 1967.

*Definitions.* Let $\mathcal{F}$ be a set of subgroups of $S$.

Suppose that $A$ and $B$ are nonempty subsets of $S$ and $g \in G$. We say that $A$ is *$\mathcal{F}$-conjugate to $B$ via $g$* if there exist subgroups $T_1, \ldots, T_n$ in $\mathcal{F}$ and elements $g_1, \ldots, g_n$ of $G$ such that:

$$g_i \in \mathbf{N}(T_i), \quad (i = 1, \ldots, n); \tag{3.1}$$

$$\langle A \rangle \subseteq T_1 \text{ and } \langle A \rangle^{g_1 \cdots g_i} \subseteq T_{i+1}, \quad (i = 1, \ldots, n-1); \text{ and} \tag{3.2}$$

$$A^g = B, \quad \text{where } g = g_1 \ldots g_n. \tag{3.3}$$

We say that $\mathscr{F}$ is a *conjugation family* (for $S$ in $G$) if it has the following property. Whenever $A$ and $B$ are nonempty subsets of $S$ and $g \in G$ and $A^g = B$, then $A$ is $\mathscr{F}$-conjugate to $B$ via $g$.

LEMMA 3.1. *Suppose that $\mathscr{F}$ is a set of subgroups of $S$; $A_1, \ldots, A_{m+1}$ are subsets of $S$; and $h_1, \ldots, h_m \in G$. Assume that, for $i = 1, \ldots, m$, $A_i$ is $\mathscr{F}$-conjugate to $A_{i+1}$ via $h_i$. Then $A_1$ is $\mathscr{F}$-conjugate to $A_{m+1}$ via $h_1 \ldots h_m$.*

*Proof.* The cases $m = 1$ and $m = 2$ are easy. The general case follows by induction.

The following result is obvious.

LEMMA 3.2. *Suppose that $\mathscr{F}$ is a set of subgroups of $S$, $A$ and $B$ are subsets of $S$, $g \in G$, and $A$ is $\mathscr{F}$-conjugate to $B$ via $g$. Let $C$ be a non-empty subset of $A$. Then $C$ is $\mathscr{F}$-conjugate to $C^g$ via $g$.*

LEMMA 3.3. *Suppose that $T \subseteq S$. Then there exists $U \subseteq S$ such that $U$ is conjugate to $T$ in $G$ and $\mathbf{N}_S(U)$ is a Sylow $p$-subgroup of $\mathbf{N}_G(U)$.*

*Proof.* Let $V$ be a Sylow $p$-subgroup of $\mathbf{N}_G(T)$. By Corollary 2.7, $V \supseteq T$. Take $g \in G$ such that $V^g \subseteq S$. Let $U = T^g$. Since $T \trianglelefteq V$, $U = T^g \trianglelefteq V^g$. Moreover, $|V^g| = |V|$ and $|\mathbf{N}(U)| = |\mathbf{N}(T^g)| = |\mathbf{N}(T)^g| = |\mathbf{N}(T)|$; therefore, $V^g$ is a Sylow $p$-subgroup of $\mathbf{N}(U)$, as desired.

THEOREM 3.4. *Let $\mathscr{F}$ be a set of subgroups of $S$. Assume that, for every $T \subseteq S$, there exists $U \in \mathscr{F}$ such that*

(a) *$U$ is conjugate to $T$ in $G$, and*

(b) *$\mathbf{N}_S(U)$ is a Sylow $p$-subgroup of $\mathbf{N}_G(U)$.*

*Then $\mathscr{F}$ is a conjugation family for $S$ in $G$.*

*Proof.* Suppose that $A$ and $B$ are arbitrary nonempty subsets of $S$, $g \in G$, and $A^g = B$. We must prove that $A$ is $\mathscr{F}$-conjugate to $B$ via $g$. We use induction on the index of $\langle A \rangle$ in $S$. Let $T = \langle A \rangle$ and $V = \langle B \rangle$; then $T^g = V$.

First, suppose that $T = S$. Then $g \in \mathbf{N}(S)$. By (a), $\mathscr{F}$ contains a conjugate of $S$. Thus $S \in \mathscr{F}$. We now obtain (3.1) to (3.3) by putting $n = 1$, $T_1 = S$, and $g_1 = g$.

Suppose $T \neq S$. Then $T \subset \mathbf{N}_S(T)$ and $V \subset \mathbf{N}_S(V)$, by Corollary 2.6.

Take $U \in \mathscr{F}$ to satisfy (a) and (b). There exists $h_1 \in G$ such that $T^{gh_1} = V^{h_1} = U$. Then $(\mathbf{N}_S(T))^{gh_1} \subseteq \mathbf{N}_G(U)$. By (b), there exists $h_2 \in \mathbf{N}_G(U)$ such that

$$\mathbf{N}_S(T)^{gh_1h_2} = (\mathbf{N}_S(T)^{gh_1})^{h_2} \subseteq \mathbf{N}_S(U).$$

Let $h = h_1h_2$. Then $T^{gh} = U$ and $(\mathbf{N}_S(T))^{gh} \subseteq \mathbf{N}_S(U)$. Similarly, there exists $k \in G$ such that $V^k = U$ and $\mathbf{N}_S(V)^k \subseteq \mathbf{N}_S(U)$. Then $U^{h^{-1}k} = V^k = U$, and $h^{-1}k \in N_G(U)$. By induction, $\mathbf{N}_S(T)$ is $\mathscr{F}$-conjugate to $(\mathbf{N}_S(T))^{gh}$ via $gh$ and $(\mathbf{N}_S(V))^k$ is $\mathscr{F}$-conjugate to $\mathbf{N}_S(V)$ via $k^{-1}$. By Lemma 3.2, $A$ is $\mathscr{F}$-conjugate to $B^h$ via $gh$ and $B^k$ is $\mathscr{F}$-conjugate to $B$ via $k^{-1}$. Since $U \in \mathscr{F}$ and $h^{-1}k \in \mathbf{N}_G(U)$, $B^h$ is $\mathscr{F}$-conjugate to $B^k$ via $h^{-1}k$. Now, $g = (gh)(h^{-1}k)k^{-1}$. Hence $A$ is $\mathscr{F}$-conjugate to $B$ via $g$, by Lemma 3.1. This completes the proof of Theorem 3.4.

The above results yield the main goal of this section.

THEOREM 3.5 (Alperin). *Let $\mathscr{F}$ be the set of all subgroups $T$ of $S$ for which $\mathbf{N}_S(T)$ is a Sylow $p$-subgroup of $\mathbf{N}_G(T)$. Then $\mathscr{F}$ is a conjugation family for $S$ in $G$.*

REMARK. From Theorem 3.5, we see that (C) determines (B). We might express this by saying that two elements or subsets of $S$ are conjugate in $G$ if and only if they are "locally conjugate" in $G$.

Recall that $\mathbf{O}_p(G)$ is the maximal normal $p$-subgroup of $G$ (see Section 2).

LEMMA 3.6. *Suppose that $\mathscr{F}$ is a conjugation family for $S$ in $G$. Let $\mathscr{F}'$ be the set of all $T \in \mathscr{F}$ that contain $\mathbf{O}_p(G)$. Then $\mathscr{F}'$ is a conjugation family for $S$ in $G$.*

*Proof.* Suppose that $A$ and $B$ are nonempty subsets of $S$, $g \in G$, and $A^g = B$. Let $A^* = \langle A\rangle \mathbf{O}_p(G)$ and $B^* = \langle B\rangle \mathbf{O}_p(G)$. Then $(A^*)^g = B^*$ and, by Corollary 2.7, $S$ contains $A^*$ and $B^*$. Moreover, $\mathbf{O}_p(G) \subseteq (A^*)^h$ for all $h \in G$. Now use the definition of a conjugation family.

*Definitions.* Let $\mathscr{S}(G)$ be the set of all sequences $x = (x_1, \ldots, x_n)$ of distinct elements of $G$. For every $x = (x_1, \ldots, x_n) \in \mathscr{S}(G)$, let $\{x\}$ be the set $\{x_1, \ldots, x_n\}$ and $\langle x\rangle$ be the subgroup $\langle x_1, \ldots, x_n\rangle$ of $G$. For $x = (x_1, \ldots x_n)$ $\in \mathscr{S}(G)$ and $g \in G$, let

$$x^g = (x_1^g, \ldots, x_n^g).$$

We say that two elements $x,y$ of $\mathscr{S}(G)$ are *conjugate* in $G$ if $y = x^g$ for some $g \in G$.

Suppose that $\sim$ is an equivalence relation on $S$ (or on $\mathscr{S}(G)$). We will write

$$x \sim y \quad \text{to mean} \quad (x, y) \in \sim$$
$$x \nsim y \quad \text{to mean} \quad (x, y) \notin \sim$$

Let $H \subseteq G$. We say that $\sim$ *contains fusion* in $H$ if $x \sim y$ whenever $x,y \in S \cap H$ (or $x,y \in \mathscr{S}(S) \cap \mathscr{S}(H)$) and $x$ and $y$ are conjugate in $H$. (Equivalently, $\sim$ contains fusion if $\sim$ contains $(x,y)$ for every such pair $x,y$).

If $x,y \in S$ (or $x,y \in \mathscr{S}(S)$) and $x^g = y$ for some $g \in G$, then $\{x\}^g = \{y\}$. Therefore, Lemma 3.6 and the definition of a conjugation family yield

PROPOSITION 3.7. *Let $\sim$ be an equivalence relation on $S$ (or on $\mathscr{S}(S)$) and let $\mathscr{F}$ be a conjugation family for $S$ in $G$. Assume that $\sim$ does not contain fusion in $G$. Then there exist $T \in \mathscr{F}$ and $x,y \in T$ (or $x,y \in \mathscr{S}(T)$) such that $x$ and $y$ are conjugate in $\mathbf{N}(T)$, $x \nsim y$, and $T \supseteq \mathbf{O}_p(G)$.*

## 4. Abelian Factor Groups

In this section we investigate the question of whether $G$ has a factor group of order $p$. We first show that this "global" property of $G$ is determined by fusion and consequently by the "local" properties of $G$.

The main tool in investigating this property is the *transfer* homomorphism of $G$ into $S/S'$. Unfortunately, we do not have time to define this map. We will therefore assume without proof two elementary applications of transfer theory.

THEOREM 4.1 (Focal subgroup theorem). *We have*

$$S \cap G' = \langle x^{-1}y : x,y \in S \text{ and } x \text{ is conjugate to } y \text{ in } G\rangle.$$

LEMMA 4.2. *Let $n = [G : S]$. Then there exists a homomorphism $\phi$ of $G$ into $S/S'$ with the property that $\phi(g) = g^nS'$ for all $g \in S \cap \mathbf{Z}(G)$.*

Theorem 4.1 is called the focal subgroup theorem because $S \cap G'$ is called the *focal subgroup* of $S$ with respect to $G$. A proof of Theorem 4.1, including the relevant background in transfer theory, is given in pp. 245–251 of Gorenstein (1964). A similar proof of Lemma 4.2 is given in pp. 412–414 of Huppert (1967).

LEMMA 4.3. *Let $L$ be the subgroup of $G$ generated by all of the $p'$-elements of $G$, and let $M = G'L$. Then*

(a) *$G/L$ is a $p$-group and $G = SL$;*
(b) *$G = SM$ and $S \cap M = S \cap G'$;*
(c) *$G/M \cong S/(S \cap G')$; and*
(d) *$G$ has a factor group of order $p$ if and only if $S \cap G' \subset S$.*

*Proof.* (a) Let $g \in G$. By an elementary result, $g = hk$ for some $p$-element $h$ and some $p'$-element $k$. Since $k \in L$, we have $g \equiv h$ (modulo $L$). Hence the coset $gL$ is a $p$-element of $G/L$. Thus $G/L$ is a $p$-group. By Lemma 2.10, $G = SL$.

(b) We have $G = SL = S(G'L) = SM$. Since $M = G'L$, $M/G'$ is an abelian group generated by $p'$-elements. By an elementary result, $M/G'$ is a $p'$-group. Now, $S \cap G' \subseteq S \cap M$. Suppose that $g \in S \cap M$. Then $gG'$ is a $p$-element of $M/G'$; hence $gG' = G'$, and $g \in G'$. Thus $S \cap G' = S \cap M$.

(c) By (b),

$$S/(S \cap G') = S/(S \cap M) \cong SM/M = G/M.$$

(d) Suppose that $H \trianglelefteq G$ and $|G/H| = p$. Then $G/H$ is abelian and $H \supseteq G'$. Since $|G'|_p \leq |H|_p < |G|_p$, $S \nsubseteq H$.

Conversely, suppose that $S \cap G' \subset S$. By (c), $G/M$ is a non-identity $p$-group. By Corollary 2.5, $G/M$ has a factor group of order $p$. Hence $G$ has a factor group of order $p$.

REMARK. Theorem 4.1 and Lemma 4.3 show that fusion determines $S \cap G'$ and determines whether $G$ has a factor group of order $p$. Therefore, by Alperin's Theorem, the normalizer of non-identity subgroups of $G$ determine whether $G$ has a factor group of order $p$.

PROPOSITION 4.4. *We have* $S \cap \mathbf{Z}(G) \cap G' = S' \cap \mathbf{Z}(G)$.

*Proof.* Obviously, $S' \cap \mathbf{Z}(G) \subseteq S \cap \mathbf{Z}(G) \cap G'$. Conversely, suppose that $g \in S \cap \mathbf{Z}(G) \cap G'$. Take $n$ and $\phi$ as in Lemma 4.2. Then $\phi(g) = g^nS'$. Since $\phi$ maps $G$ into an abelian group, $G'$ is contained in the kernel of $\phi$. Thus $g^nS' = S'$, and $g^n \in S'$. Since $n = [G : S]$, $p$ does not divide $n$. As $g$ is a $p$-element and $g^n \in S'$, $g \in S'$, as desired.

DEFINITION. Let $A$ be an element or subset of $S$. Then $A$ is *weakly closed* in $S$ (with respect to $G$) if $A$ satisfies the following condition:

Whenever $g \in G$ and $A^g$ is contained in $S$, then $A^g = A$.

LEMMA 4.5. *Suppose that $T$ is a weakly closed subgroup of $S$ and $x,y \in \mathbf{C}_S(T)$. If $x$ and $y$ are conjugate in $G$, then $x$ and $y$ are conjugate in* $\mathbf{N}(T)$.

*Proof.* Suppose that $g \in G$ and $x^g = y$. Then $T \subseteq \mathbf{C}(y)$ and $T^g \subseteq \mathbf{C}(x)^g = \mathbf{C}(x^g) = \mathbf{C}(y)$. Let $U$ be a Sylow $p$-subgroup of $\mathbf{C}(y)$ that contains $T$. Take $h \in \mathbf{C}(y)$ and $k \in G$ such that $(T^g)^h \subseteq U$ and $U^k \subseteq S$. Then $T^{ghk} \subseteq S$ and $T^k \subseteq S$. By the weak closure of $T$ in $S$, $T^{ghk} = T = T^k$. Hence $k$, $ghk \in \mathbf{N}(T)$ and $gh \in \mathbf{N}(T)$. Since $x^{gh} = y^h = y$, we are done.

LEMMA 4.6. *Suppose that $S \subseteq H \subseteq G$. Assume that $H$ satisfies the following condition:*

*Whenever $x,y \in S$ and $x$ and $y$ are conjugate in $G$, then $x$ and $y$ are conjugate in $H$.*

*Then* $S \cap H' = S \cap G'$.

*Proof.* This is a direct application of the focal subgroup theorem.

THEOREM 4.7. (Grün). *Suppose that $T$ is a weakly closed subgroup of $S$ with respect to $G$ and $T \subseteq \mathbf{Z}(S)$. Let $H = \mathbf{N}(T)$. Then $S \cap G' = S \cap H'$.*

*Proof.* This follows immediately from Lemmas 4.5 and 4.6.

THEOREM 4.8. (Burnside). *Let $H = \mathbf{N}(S)$. Define $M$ as in Lemma* 4.3. *Then*
(a) *if $S$ is abelian, $S \cap G' = [S, H]$;*
(b) *if $S \subseteq \mathbf{Z}(H)$, $M$ is a characteristic $p'$-subgroup of $G$ and $G/M \cong S$.*

*Proof.* (a) Suppose that $S$ is abelian. By Grün's theorem, $S \cap G' = S \cap H'$. Since $S \trianglelefteq H$, we have $S \cap H' = [S, H]$ by the focal subgroup theorem.
(b) Suppose that $S \subseteq \mathbf{Z}(H)$. Then $S$ is abelian. By (a), $S \cap G' = 1$. By Lemma 4.3, $G = SM$ and $S \cap M = S \cap G' = 1$. Clearly $M$ is a characteristic subgroup of $G$. Moreover, $G/M = SM/M \cong S/(S \cap M) \cong S$. Thus $|M| = [G : S]$, which is not divisible by $p$.

PROPOSITION 4.9. *Let $S^* \subseteq S$. Let $\sim$ be the equivalence relation on $S$ given by*

$$x \sim y \quad \text{if} \quad S^*x = S^*y.$$

*Suppose that $H \subseteq G$, $T \subseteq S$, and $T$ is a Sylow $p$-subgroup of $H$. Then $\sim$ contains fusion in $H$ if and only if $H' \cap T \subseteq S^*$.*

*Proof.* By the focal subgroup theorem,
$H' \cap T = \langle xy^{-1} : x,y \in T$ and $x$ is conjugate to $y$ in $H\rangle$. The result follows.

PROPOSITION 4.10. *Let $\mathscr{F}$ be a conjugation family for $S$ in $G$. Let*

$$S^* = \langle [T, N(T)] : T \in \mathscr{F} \rangle$$

*Then $S^* = S \cap G'$.*

*Proof.* Clearly, $S^* \subseteq S \cap G'$. Define $\sim$ as in Proposition 4.9. If $T \in \mathscr{F}$ and $x,y \in T$ and $x$ and $y$ are conjugate in $N(T)$, then $xy^{-1} \in S^*$ and $x \sim y$. Therefore, $\sim$ contains fusion in $G$, by Proposition 3.7. By Proposition 4.9, $S \cap G' \subseteq S^*$.

## 5. Conjugacy Functors

The results of Section 3 have been extended by Alperin and Gorenstein. In this section, we prove some of their extensions, which require the following concepts.

*Definitions.* Let $\mathscr{C}_p(G)$ be the set of all $p$-subgroups of $G$. A *conjugacy functor* (for the prime $p$) on $G$ is a mapping $\mathbf{W}$ of $\mathscr{C}_p(G)$ into $\mathscr{C}_p(G)$ that satisfies the following three conditions for every $P \in \mathscr{C}_p(G)$:

(i) $\mathbf{W}(P) \subseteq P$;

(ii) $\mathbf{W}(P) \neq 1$ if $P \neq 1$; and

(iii) $\mathbf{W}(P^g) = (\mathbf{W}(P))^g$ for every $g \in G$.

Assume that $\mathbf{W}$ is a conjugacy functor on $G$. Let $T \subseteq S$. Define three sequences of subgroups of $G$ as follows:

$$W_1(T) = T, \qquad P_1(T) = \mathbf{N}_S(T), \qquad N_1(T) = \mathbf{N}_G(T),$$

and for $i = 1, 2, 3, \ldots$,

$$W_{i+1}(T) = \mathbf{W}(P_i(T)), \qquad P_{i+1}(T) = \mathbf{N}_S(W_{i+1}(T)),$$
$$N_{i+1}(T) = \mathbf{N}_G(W_{i+1}(T)).$$

We say that $T$ is a *well-placed* subgroup of $S$ (with respect to $\mathbf{W}$, $S$, and $G$) if $P_i(T)$ is a Sylow $p$-subgroup of $\mathbf{N}_i(T)$ ($i = 1, 2, 3, \ldots$).

These definitions easily yield

LEMMA 5.1. *Let* $\mathbf{W}$ *be a conjugacy functor on* $G$.

(a) *Suppose that* $T$ *is a well-placed subgroup of* $S$. *Then* $\mathbf{W}(\mathbf{N}_S(T))$ *is a well-placed subgroup of* $S$, *and* $\mathbf{N}_S(T)$ *is a Sylow* $p$-*subgroup of* $\mathbf{N}_G(T)$.

(b) *Suppose that* $P \in \mathscr{C}_p(G)$. *Then* $\mathbf{W}(P) \trianglelefteq \mathbf{N}(P)$.

(c) *Let* $T = \mathbf{O}_p(G)$. *Then* $T$ *is a well-placed subgroup of* $S$.

LEMMA 5.2. *Let* $\mathbf{W}$ *be a conjugacy functor on* $G$ *and let* $T \subseteq S$. *Then there exists a well-placed subgroup* $U$ *of* $S$ *such that* $U$ *is conjugate to* $T$ *in* $G$.

*Proof.* Define a sequence $T_1, T_2, \ldots$ of $p$-subgroups of $G$ as follows. Let $T_1$ be a Sylow $p$-subgroup of $\mathbf{N}_G(T)$. For $i = 1, 2, 3, \ldots$, let $T_{i+1}$ be a Sylow $p$-subgroup of $\mathbf{N}_G(\mathbf{W}(T_i))$ that contains $T_i$. Then

$$T \subseteq T_1 \subseteq T_2 \subseteq \ldots.$$

Let $S^*$ be the maximum of the subgroups $T_i$ ($i \geq 1$). Take $g \in G$ such that $S^{*g} \subseteq S$. Let $U = T^g$. Then

$$T_i^g \subseteq S \quad \text{for all } i \geq 1. \tag{5.1}$$

Define $W_i(U)$, $P_i(U)$, $N_i(U)$ ($i = 1, 2, 3, \ldots$) as in the definition of a well-placed subgroup. Then $T_1^g \subseteq (\mathbf{N}_G(T))^g = \mathbf{N}_G(U)$. Since $T_1$ is a Sylow $p$-subgroup of $\mathbf{N}_G(T)$, $T_1^g$ is a Sylow $p$-subgroup of $\mathbf{N}_G(U)$. By (5.1) $T_1^g = P_1(U)$. Suppose that $i \geq 1$ and $T_i^g$ is a Sylow $p$-subgroup of $\mathbf{N}_G(W_i(U))$. Then $T_i^g = P_i(U)$. Hence $W_{i+1}(U) = \mathbf{W}(P_i(U)) = \mathbf{W}(T_i^g) = (\mathbf{W}(T_i))^g$. Since $T_{i+1}$ is a Sylow $p$-subgroup of $\mathbf{N}_G(\mathbf{W}(T_i))$, $(T_{i+1})^g$ is a Sylow $p$-subgroup of $\mathbf{N}_G(W_{i+1}(U))$. By (5.1), $(T_{i+1})^g = P_{i+1}(U)$. By induction $P_i(U)$ is a Sylow $p$-subgroup of $\mathbf{N}_G(W_i(U))$ for all $i \geq 1$. Thus $U$ is a well-placed subgroup of $S$.

By Theorem 3.4 and Lemmas 5.1 and 5.2, we have

THEOREM 5.3. (Alperin–Gorenstein). *Let* $\mathbf{W}$ *be a conjugacy functor on* $G$, *and let* $\mathscr{F}$ *be the set of all well-placed subgroups of* $S$. *Then* $\mathscr{F}$ *is a conjugation family for* $S$ *in* $G$.

PROPOSITION 5.4. *Suppose that* $\sim$ *is an equivalence relation on* $S$ (*or on* $\mathscr{S}(S)$) *and* $\mathbf{W}$ *is a conjugacy functor on* $G$. *Let* $P = \mathbf{O}_p(G)$. *Assume that*

$$\sim \text{ contains fusion in } \mathbf{N}(\mathbf{W}(S)), \text{ and} \tag{5.2}$$

$$\sim \text{ does not contain fusion in } G. \tag{5.3}$$

*Then there exist non-identity subgroups* $T$ *and* $U$ *of* $S$ *such that* $T$ *is well placed in* $S$

(a) $\sim$ *contains fusion in* $\mathbf{N}(\mathbf{W}(\mathbf{N}_S(T)))$ *and in* $\mathbf{N}(\mathbf{W}(\mathbf{N}_S(T))) \cap \mathbf{N}(T)$,
(b) $\sim$ *does not contain fusion in* $\mathbf{N}(T)$,
(c) $U \supseteq P$, *and*
(d) *either*

(i) $T = U$ *and there exist* $x, y \in T$ (*or* $x, y \in \mathscr{S}(T)$) *for which* $x$ *is conjugate to* $y$ *in* $\mathbf{N}(T)$ *and* $x \nsim y$, *or*
(ii) $T = \mathbf{W}(U)$, $U \supset P$, *and* $U \supseteq \mathbf{C}_S(U)$.

*Proof.* Let $\mathscr{F}$ be the set of all well-placed subgroups of $S$. By Theorem 5.3, $\mathscr{F}$ is a conjugation family for $S$ in $G$. By Proposition 3.7, there exist some $T_1 \in \mathscr{F}$ and $x_1, y_1 \in T_1$ (or $x_1, y_1 \in \mathscr{S}(T_1)$) such that $T_1 \supseteq P$, $x_1$ and $y_1$ are conjugate in $\mathbf{N}(T_1)$, and $x_1 \nsim y_1$. If $T_1 = P$, we may let $T = U = T_1$.

Assume that $T_1 \supset P$. Let $\mathscr{F}'$ be the set of all $T \in \mathscr{F}$ that satisfy (b), (c) and (d) for some $U \subseteq S$ such that $U \supset P$. By taking $U = T_1$, we see that $T_1 \in \mathscr{F}'$. Hence $\mathscr{F}'$ is not empty. Choose $T \in \mathscr{F}'$ such that $|\mathbf{N}_S(T)|$ is maximal, that is, $|\mathbf{N}_S(T)| \geq |\mathbf{N}_S(T^*)|$ for all $T^* \in \mathscr{F}'$. Take some $U \subseteq S$ such that $T$ and $U$ satisfy (b), (c), and (d), and $U \supset P$. Let $S^* = \mathbf{N}_S(T)$ and $T^* = \mathbf{W}(S^*)$. By (d) and Lemma 5.1,

$$T^* \in \mathscr{F}', \qquad T^* \trianglelefteq \mathbf{N}(S^*), \qquad S^* \supseteq U \supset P, \quad \text{and } S^* \supseteq \mathbf{C}_S(T) \supseteq \mathbf{C}_S(S^*). \tag{5.4}$$

Suppose that $T^* \in \mathscr{F}'$. By (5.4) and by the choice of $T$,

$$|S^*| = |\mathbf{N}_S(T)| \geq |\mathbf{N}_S(T^*)| \geq |\mathbf{N}_S(S^*)|.$$

Since $S$ is a $p$-group and $S^* \subseteq S$, we obtain $S^* = S$ and $T^* = \mathbf{W}(S)$. But $\mathbf{W}(S)$ violates (b), by (5.2). This contradiction shows that $T^* \notin \mathscr{F}'$. Now, (c) and (d) are valid with $T$ and $U$ replaced by $T^*$ and $S^*$ respectively. Therefore, $T^*$ violates (b). Since $T^* = \mathbf{W}(S^*) = \mathbf{W}(\mathbf{N}_S(T))$, $\sim$ contains fusion in $\mathbf{N}(\mathbf{W}(S^*))$. Consequently, $\sim$ contains fusion in every subgroup of $\mathbf{N}(\mathbf{W}(S^*))$. This proves (a) and completes the proof of Proposition 5.4.

*Definitions*. Suppose that $\mathbf{W}$ is a conjugacy functor for $p$ on $G$. Let $H \subseteq G$.

(i) We say that $\mathbf{W}$ *controls strong fusion* in $H$ provided that there exists a Sylow $p$-subgroup $T$ of $H$ with the following property: whenever two elements of $\mathscr{S}(T)$ are conjugate in $H$, they are conjugate in $\mathbf{N}_H(\mathbf{W}(T))$.

(ii) We say that $\mathbf{W}$ *controls transfer* in $H$ if $T \cap H' = T \cap (\mathbf{N}_H(\mathbf{W}(T)))'$ for some Sylow $p$-subgroup $T$ of $H$.

(iii) We say that $\mathbf{W}$ *controls weak closure of elements* in $H$ provided that $\mathbf{W}(P) \supseteq \mathbf{Z}(P)$ for all $P \in \mathscr{C}_p(G)$ and that there exists a Sylow $p$-subgroup $T$ of $H$ with the following property: whenever $x \in T \cap \mathbf{Z}(\mathbf{N}(\mathbf{W}(T)))$, then $x$ is weakly closed in $T$ with respect to $H$.

Note that the above definitions do not depend on the choice of $T$. Moreover, in (iii), an element $x$ of $T$ is weakly closed in $T$ with respect to $\mathbf{N}(\mathbf{W}(T))$ if and only if $x \in \mathbf{Z}(\mathbf{N}(\mathbf{W}(T)))$.

THEOREM 5.5 (Alperin–Gorenstein). *Let* $\mathbf{W}$ *be a conjugacy functor on* $G$. *Let* $\mathscr{F}^*$ *be the set of all non-identity well-placed subgroups of* $S$.

(i) *If* $\mathbf{W}$ *controls strong fusion in* $\mathbf{N}(T)$ *for every* $T \in \mathscr{F}^*$, *then* $\mathbf{W}$ *controls strong fusion in* $G$.

(ii) *If* $\mathbf{W}$ *controls transfer in* $\mathbf{N}(T)$ *for every* $T \in \mathscr{F}^*$ *then* $\mathbf{W}$ *controls transfer in* $G$.

*Proof*. (i) Define an equivalence relation $\sim$ on $\mathscr{S}(S)$ as follows:

$$x \sim y \quad \text{if } x \text{ and } y \text{ are conjugate in } \mathbf{N}(\mathbf{W}(S)).$$

Obviously, $\sim$ contains fusion in $\mathbf{N}(\mathbf{W}(S))$. Moreover, $\sim$ contains fusion in $G$ if and only if $\mathbf{W}$ controls strong fusion in $G$. Suppose $\sim$ does not contain fusion in $G$. By Proposition 5.4, there exists $T \in \mathscr{F}^*$ such that $\sim$ contains fusion in $\mathbf{N}_{\mathbf{N}(T)}(\mathbf{W}(\mathbf{N}_S(T)))$ and $\sim$ does not contain fusion in $\mathbf{N}(T)$. By Lemma 5.1, $\mathbf{N}_S(T)$ is a Sylow $p$-subgroup of $\mathbf{N}(T)$. Hence $\mathbf{W}$ does not control strong fusion in $\mathbf{N}(T)$, contrary to hypothesis. Thus $\sim$ contains fusion in $G$, as desired.

(ii) Let $N = \mathbf{N}(\mathbf{W}(S))$ and $S^* = S \cap N'$. Define an equivalence relation $\sim$ on $S$ as in Proposition 4.9, namely,

$$x \sim y \quad \text{if } x \equiv y \text{ (modulo } S^*).$$

By Proposition 4.9, $\sim$ contains fusion in $N$, since $S \cap N' \subseteq S^*$. Suppose $\sim$ does not contain fusion in $G$. Take $T$ as in Proposition 5.4. Let $R = \mathbf{N}_S(T)$, $H = \mathbf{N}(T)$, and $V = \mathbf{W}(R)$. Then $T$ is well-placed in $S$, $\sim$ contains fusion in $\mathbf{N}_H(V)$, and $\sim$ does not contain fusion in $\mathbf{N}(T)$. Since $R$ is a Sylow $p$-subgroup of $H$ and $R \subseteq S$, Proposition 4.9 yields that

$$R \cap (\mathbf{N}_H(V))' \subseteq R \cap S^* \quad \text{and} \quad R \cap H' \nsubseteq R \cap S^*.$$

But, by hypothesis, $\mathbf{W}$ controls transfer in $H$, a contradiction. Thus $\sim$ contains fusion in $G$. By Proposition 4.9, $S \cap G' \subseteq S^*$. Since $S^* \subseteq S \cap G'$, we have $S \cap G' = S \cap N'$. This completes the proof of Theorem 5.5.

We require a similar theorem about weak closure of elements in $G$. We state it differently in order to obtain some applications when a conjugacy functor does not control weak closure of elements in $G$.

THEOREM 5.6. *Let* $\mathbf{W}$ *be a conjugacy functor for* $p$ *on* $G$. *Assume that* $\mathbf{W}(P) \supseteq \mathbf{Z}(P)$ *for every p-subgroup* $P$ *of* $G$. *Suppose that* $A$ *is an element (or subgroup) of* $\mathbf{Z}(S)$ *which is centralized (or normalized) by* $\mathbf{N}(\mathbf{W}(S))$, *and that* $A$ *is not weakly closed in* $S$ *with respect to* $G$. *Then there exists a well-placed subgroup* $T$ *of* $S$ *such that*

(a) $T$ *contains* $A$,
(b) $A$ *is centralized (or normalized) by* $\mathbf{N}(\mathbf{W}(\mathbf{N}_S(T)))$,
(c) $\mathbf{N}_S(T)$ *is a Sylow p-subgroup of* $\mathbf{N}(T)$,
(d) $\mathbf{Z}(\mathbf{N}_S(T))$ *contains* $A$, *and*
(e) $A$ *is not weakly closed in* $\mathbf{N}_S(T)$ *with respect to* $\mathbf{N}(T)$.

*Proof.* If $A \in \mathbf{Z}(S)$, define an equivalence relation $\sim$ on $S$ as follows:

$$y \sim z \quad \text{if } y = z = A \text{ or if } y,\, z \in S - \{A\}.$$

If $A \subseteq \mathbf{Z}(S)$, define an equivalence relation on $\mathscr{S}(S)$ as follows:

$$y \sim z \quad \text{if } \{y\} = \{z\} = A \text{ or if } \{y\},\, \{z\} \neq A.$$

By hypothesis, $\sim$ contains fusion in $\mathbf{N}(\mathbf{W}(S))$ but does not contain fusion in $G$. Choose $T$ and $U$ as in Proposition 5.4. If $T$ (or $\mathscr{S}(T)$) contains elements $y,z$ such that $y \nsim z$, then $y = A$ or $z = A$ (or $\{y\} = A$ or $\{z\} = A$). If $T = \mathbf{W}(U)$ and $U \supseteq \mathbf{C}_S(U)$, then $U \supseteq \mathbf{Z}(S)$ and $\mathbf{Z}(S) \subseteq \mathbf{Z}(U) \subseteq \mathbf{W}(U)$. Hence, both alternatives of part (d) of Proposition 5.4 yield that $T$ contains $A$. Now, Lemma 5.1 yields (c), and (d) is obvious.

Suppose $g \in \mathbf{N}(T)$. Then $A^g$ is contained in $T$. Hence (b) and (e) follow from parts (a) and (b) of Lemma 5.4. This completes the proof of Theorem 5.6.

The following easy result will be useful in the next section.

LEMMA 5.7. *Let* $\mathbf{W}$ *be a conjugacy functor on* $G$. *The following are equivalent:*

(a) $\mathbf{W}$ *controls strong fusion in* $G$;
(b) *whenever* $T \subseteq S$, $g \in G$, *and* $T^g \subseteq S$, *there exist* $c \in \mathbf{C}(T)$ *and* $n \in \mathbf{N}(\mathbf{W}(S))$ *such that* $cn = g$.

## 6. Section Conjugacy Functors

For any subgroups $H$ and $K$ of $G$ for which $K \trianglelefteq H$, the factor group $H/K$ is called a *section* of $G$. If $K = 1$, we will identify $H/K$ with $H$.

DEFINITION. Let $\mathscr{C}_p^*(G)$ be the set of all sections of $G$ that are $p$-groups. A *section conjugacy functor* (for $p$) on $G$ is a mapping $\mathbf{W}$ of $\mathscr{C}_p^*(G)$ into $\mathscr{C}_p^*(G)$ that satisfies the following four conditions for every section $H/K$ in $\mathscr{C}_p^*(G)$.

(i) The group $\mathbf{W}(H/K)$ is a subgroup of $H/K$.
(ii) If $H/K \neq 1$, then $\mathbf{W}(H/K) \neq 1$.
(iii) Let $\mathbf{W}(H/K) = L/K$. Then for all $g \in G$, $\mathbf{W}(H^g/K^g) = L^g/K^g$.
(iv) Suppose that $N \trianglelefteq H$ and $N \subseteq K$ and $K/N$ is a $p'$-group. Let $P/N$ be a Sylow $p$-subgroup of $H/N$, and let $\mathbf{W}(P/N) = L/N$. Then $\mathbf{W}(H/K) = LK/K$.

REMARK. In condition (iv), $H = KP$ and $H/K \cong P/(P \cap K) = P/N$; similarly,

$$\mathbf{W}(H/K) = LK/K \cong L/(L \cap K) = L/N = \mathbf{W}(P/N).$$

A sufficient condition for (iii) and (iv) is the following:

Whenever $Q$, $R \in \mathscr{C}_p^*(G)$ and $\phi$ is an isomorphism of $Q$ on to $R$, then $\phi(\mathbf{W}(Q)) = \mathbf{W}(R)$.

Note that the restriction of a section conjugacy functor on $G$ to $\mathscr{C}_p(G)$ is itself a conjugacy functor on $G$. We say that $\mathbf{W}$ *controls strong fusion* (or *controls transfer*) in a section $H/K$ of $G$ if the restriction of $\mathbf{W}$ to $\mathscr{C}_p(H/K)$ controls strong fusion (or controls transfer) in $H/K$. If $K, N \trianglelefteq H$ and $N \subseteq K$, we will identify $(H/N)/(K/N)$ with $H/K$. (By an elementary theorem, they are isomorphic.) A short argument yields

LEMMA 6.1. *Suppose* $\mathbf{W}$ *is a section conjugacy functor on* $G$ *and* $H/K$ *is a section of* $G$. *Then the restriction of* $\mathbf{W}$ *to* $\mathscr{C}_p^*(H/K)$ *is a section conjugacy functor on* $H/K$.

The following lemma is elementary (Hall, 1959; p. 124):

LEMMA 6.2 (Dedekind). *Let* $H, K, L \subseteq G$. *Assume that* $HL$ *is a subgroup of* $G$ *and* $L \subseteq K$. *Then* $HL \cap K = (H \cap K)L$.

LEMMA 6.3. *Let* $H, K \subseteq G$. *Assume that* $HS$ *is a subgroup of* $G$, $S \subseteq K$, *and* $G = HK$.

(a) *Suppose that* $T \subseteq S$, $g \in G$, *and* $T^g \subseteq S$. *Then there exist* $h \in H$, $k \in K$ *such that* $hk = g$ *and* $T^h \subseteq S$.

(b) *Suppose that* $\sim$ *is an equivalence relation on* $S$ *(or on* $\mathscr{S}(S)$*) and* $\sim$ *does not contain fusion in* $G$. *Then* $\sim$ *does not contain fusion in* $HS$ *or* $\sim$ *does not contain fusion in* $K$.

*Proof.* (a) Take $x \in H$, $y \in K$ such that $xy = g$. Then $T^x \subseteq \langle T, H \rangle \subseteq HS$

and $T^x = (T^g)^{y^{-1}} \subseteq S^{y^{-1}} \subseteq K$. Thus $T^x \subseteq HS \cap K$. Since $S \subseteq HS \cap K$, there exists $z \in HS \cap K$ such that $(T^x)^z \subseteq S$. Moreover, by Dedekind's Lemma, $HS \cap K = (H \cap K)S$. Take $u \in H \cap K$, $v \in S$ such that $uv = z$. Then $T^{xu} = T^{xzv^{-1}} \subseteq S^{v^{-1}}$. Let $h = xu$ and $k = u^{-1}y$.

(b) Take $x \in S$ (or $x \in \mathscr{S}(S)$) and $g \in G$ such that $x^g \in S$ (or $x^g \in \mathscr{S}(S)$) and $x \sim x^g$. Let $T = \langle x \rangle$. Take $h \in H$ and $k \in K$ as in (a). Then $x \sim x^h$ or $x^h \sim x^g = x^{hk}$.

LEMMA 6.4. *Let $T = \mathbf{O}_p(G)$. Suppose that $\mathbf{C}_S(T) \subseteq T$. Then there exists a $p'$-subgroup $H$ of $G$ having the following properties:*

(a) *$H$ is a characteristic subgroup of $G$,*
(b) *$\mathbf{C}(T) = \mathbf{Z}(T) \times H$, and*
(c) *for $\bar{G} = G/H$, $\mathbf{C}_{\bar{G}}(\mathbf{O}_p(\bar{G})) \subseteq \mathbf{O}_p(\bar{G})$.*

*Proof.* Let $C = \mathbf{C}(T)$. Since $T \trianglelefteq G$, $C \trianglelefteq G$. Hence $C \cap S$ is a Sylow $p$-subgroup of $C$. By hypothesis, $C \cap S = \mathbf{Z}(T)$. Thus $C \cap S$ is contained in the center of $C$. By Burnside's Theorem (Theorem 4.8), there exists a characteristic $p'$-subgroup $H$ of $C$ such that $C/H \cong C \cap S = \mathbf{Z}(T)$. Since $C$ is a characteristic subgroup of $G$, we obtain (a). Since $H \cap \mathbf{Z}(T) = 1$ and $\mathbf{Z}(T) \subseteq \mathbf{Z}(C)$, we obtain (b).

Let $M/H = \mathbf{O}_p(\bar{G})$ and $L/H = \mathbf{C}_{\bar{G}}(\mathbf{O}_p(\bar{G}))$. Then $HT \subseteq M$. Suppose $x \in T$ and $y \in L$. Then $[x,y] \in T$ because $T \trianglelefteq G$, and $[x,y] \equiv 1$ (modulo $H$) because $y \in L$. Thus $[x,y] \in T \cap H = 1$. Since $x$ and $y$ are arbitrary, $L \subseteq \mathbf{C}(T) \subseteq TH \subseteq M$, and this yields (c).

LEMMA 6.5. *Let $Q$ be a normal subgroup of a finite $p$-group $P$. Let $A$ be the set of all automorphisms of $P$ that fix every element of $Q$ and every coset of $Q$. Then $A$ is a $p$-group.*

*Proof.* Clearly, $A$ is a group. Let $\alpha \in A$ and $n = |P|$. We claim that $\alpha^n = 1$. Suppose that $g \in P$ and $g^\alpha = gh$. Then $h \in Q$ and

$$g^\alpha = gh, \qquad g^{\alpha^2} = (gh)^\alpha = gh^2, \ldots, g^{\alpha^n} = gh^n = g.$$

THEOREM 6.6. *Let $\mathbf{W}$ be a section conjugacy functor on $G$. Assume that $\mathbf{W}$ satisfies the following condition:*
*Whenever $G^*$ is a section of $G$, $S^*$ is a Sylow $p$-subgroup of $G^*$, and*

$$\mathbf{C}_{G^*}(\mathbf{O}_p(G^*)) \subseteq \mathbf{O}_p(G^*), \text{ then } \mathbf{W}(S^*) \trianglelefteq G^*. \tag{6.1}$$

*Then $\mathbf{W}$ controls strong fusion in every section of $G$. In particular, $\mathbf{W}$ controls strong fusion in $G$.*

REMARK. Let $T = \mathbf{O}_p(G)$. If $\mathbf{C}(T) \subseteq T$ and $\mathbf{W}$ controls strong fusion in $G$, then $\mathbf{W}(S) \trianglelefteq G$, by Lemma 5.7. Thus (6.1) is a necessary condition for $\mathbf{W}$ to control strong fusion in every section of $G$.

*Proof.* We use induction on $|G|$. Every section of $G$ satisfies (6.1); by Lemma 6.1 and induction, $\mathbf{W}$ controls strong fusion in every section of $G$ except possibly $G$. Therefore, it suffices to prove that $\mathbf{W}$ controls strong fusion in $G$ itself. By induction and the theorem of Alperin and Gorenstein, we may assume that $\mathbf{O}_p(G) \neq 1$. Let $\sim$ be the equivalence relation on $\mathscr{S}(S)$ defined as follows:

$$x \sim y \quad \text{if } x \text{ and } y \text{ are conjugate in } \mathbf{N}(\mathbf{W}(S)).$$

By induction,

$$\text{whenever } S \subseteq H \subset G, \sim \text{ contains fusion in } H. \tag{6.2}$$

Moreover,

$$\text{if } \sim \text{ contains fusion in } G, \text{ then } \mathbf{W} \text{ controls strong fusion in } G. \tag{6.3}$$

Let $P = \mathbf{O}_p(G)$, $C = \mathbf{C}(P)$, $N = \mathbf{N}(\mathbf{W}(S))$, and $R = PC \cap S$. Since $PC \trianglelefteq G$, the Frattini argument yields that $G = PC\mathbf{N}(R) = CP\mathbf{N}(R) = C\mathbf{N}(R)$. As $CS$ is a subgroup of $G$, we may assume that $G = CS$ or $G = \mathbf{N}(R)$, by (6.2), (6.3), and Lemma 6.3.

Assume that $G = \mathbf{N}(R)$. Then $R \subseteq \mathbf{O}_p(G) = P$ and $\mathbf{C}_S(P) \subseteq R \subseteq P$. By Lemma 6.4, there exists a normal $p'$-subgroup $L$ of $G$ such that

$$\mathbf{C}_{G/L}(\mathbf{O}_p(G/L)) \subseteq \mathbf{O}_p(G/L).$$

By (6.1), $\mathbf{W}(SL/L) \trianglelefteq G/L$. By part (iv) of the definition of a section conjugacy functor, $\mathbf{W}(SL/L) = \mathbf{W}(S)L/L$. Thus $\mathbf{W}(S)L \trianglelefteq G$. Since $\mathbf{W}(S)$ is a Sylow $p$-subgroup of $\mathbf{W}(S)L$, we have

$$G = \mathbf{N}(\mathbf{W}(S))(\mathbf{W}(S)L) = NL. \tag{6.4}$$

As $LS$ is a subgroup of $G$, we may assume that $G = N$ or $G = LS$, by (6.2), (6.3), and Lemma 6.3. Clearly, we may assume that $G = LS$. Since $[P, L] \subseteq P \cap L = 1$, we have $L \subseteq C$ and $G = CS$.

Thus, we will assume that $G = CS$ for the remainder of the proof. Suppose $T \subseteq S$, $g \in G$, and $T^g \subseteq S$. By Lemma 5.7, it is sufficient to find $c \in \mathbf{C}(T)$ and $n \in \mathbf{N}(\mathbf{W}(S))$ such that $cn = g$. Take $c_0 \in \mathbf{C}(P)$ and $n_0 \in S$ such that $c_0 n_0 = g$. Then $n_0 \in N$ and, if $T \subseteq P$, $c_0 \in \mathbf{C}(T)$. We will therefore assume that $T \nsubseteq P$. Since $(TP)^g \subseteq S$, we will even assume that $P \subset T$. Then $1 \subset P \subset S$. Let $\bar{G} = G/P$, $\bar{S} = S/P$, and $\bar{T} = T/P$, and let $\bar{g}$ be the coset $gP$. By induction,

$$\bar{g} = \bar{c}\bar{n} \quad \text{for some } \bar{c} \in \mathbf{C}_{\bar{G}}(\bar{T}) \text{ and } \bar{n} \in \mathbf{N}_{\bar{G}}(\mathbf{W}(\bar{S})). \tag{6.5}$$

Take $U \subseteq G$ such that $P \subseteq U$ and $U/P = \mathbf{W}(\bar{S})$. Since $P \subset S$, we have

$$\bar{S} \neq 1, \mathbf{W}(\bar{S}) \neq 1, P \subset U, \text{ and } S \subseteq \mathbf{N}(U) \subset G. \tag{6.6}$$

Moreover,

$$\bar{T}^{\bar{n}} = \bar{T}^{\bar{c}\bar{n}} = \bar{T}^{\bar{g}} \subseteq \bar{S}.$$

Let $n$ be an element of the coset $\bar{n}$. Then $n \in \mathbf{N}(U)$ and $T^n \subseteq S$. By (6.6) and induction, there exist $d \in \mathbf{C}(T)$ and $m \in N$ such that $dm = n$. Obviously,

$d \in \mathbf{C}_G(\overline{T})$ (the subgroup of $G$ fixing every coset of $P$ in $T$ by conjugation). By (6.5), $gn^{-1} \in \mathbf{C}_G(\overline{T})$. Thus

$$g \in \mathbf{C}_G(\overline{T})N \quad \text{and} \quad T^g = T^n. \tag{6.7}$$

By Lemma 3.3, there exists $T^* \subseteq S$ such that $T^*$ is conjugate to $T$ in $G$ and $\mathbf{N}_S(T^*)$ is a Sylow $p$-subgroup of $\mathbf{N}_G(T^*)$. By the argument that yields (6.7),

$$T = (T^*)^m \quad \text{for some } m \in N. \tag{6.8}$$

Let $D = \mathbf{C}_G(\overline{T^*})$. Note that $P \subseteq T^*$ and $\mathbf{C}(T^*) \subseteq D$. Clearly, $D \trianglelefteq \mathbf{N}_G(T^*)$. Hence $D \cap S$ is a Sylow $p$-subgroup of $D$. By Lemma 6.5, $\mathbf{C}_D(P)/\mathbf{C}(T^*)$ is a $p$-group. Since

$$D/\mathbf{C}_D(P) = D/(D \cap C) \cong DC/C \subseteq G/C = CS/C \cong S/(C \cap S),$$

$D/\mathbf{C}_D(P)$ is a $p$-group. Therefore, $D/\mathbf{C}(T^*)$ is a $p$-group and

$$D = \mathbf{C}(T^*)\,(D \cap S) = \mathbf{C}(T^*)\,(D \cap N). \text{ By (6.8),}$$

$$\mathbf{C}_G(\overline{T}) = (\mathbf{C}_G(\overline{T^*}))^m = D^m = (\mathbf{C}(T^*))^m(D \cap N)^m = \mathbf{C}(T)\,(D^m \cap N). \tag{6.9}$$

By (6.7) and (6.9), $g = dk$ for some $d \in \mathbf{C}(T)$ and $k \in N$. This completes the proof of Theorem 6.6.

LEMMA 6.7. *Let $P$ be a normal $p$-subgroup of $G$. Assume that $S \subseteq H \subseteq G$ and that $[P, H] = [P, G]$. Then $P \cap H' = P \cap G'$.*

*Proof.* Let $Q = [P, G]$ and $R = P \cap G'$. Then $Q \trianglelefteq G$, $Q \subseteq R$, and $R/Q \subseteq \mathbf{Z}(G/Q)$. By Proposition 4.4,

$$R/Q \subseteq \mathbf{Z}(G/Q) \cap (S/Q) \cap (G/Q)' \subseteq (S/Q)' = S'Q/Q.$$

Thus $R \subseteq S'Q \subseteq S'H' \subseteq H'$. Hence $R \subseteq P \cap H'$. Therefore, $R = P \cap H'$.

THEOREM 6.8. *Let $\mathbf{W}$ be a section conjugacy functor on $G$. Assume that $\mathbf{W}$ satisfies the following condition:*
*Whenever $G^*$ is a section of $G$, $S^*$ is a Sylow $p$-subgroup of $G^*$, and*

$$\mathbf{C}_{G^*}(\mathbf{O}_p(G^*)) \subseteq \mathbf{O}_p(G^*),$$

*then*

$$\mathbf{O}_p(G^*) \cap (G^*)' = \mathbf{O}_p(G^*) \cap (\mathbf{N}_{G^*}(\mathbf{W}(S^*)))'. \tag{6.10}$$

*Then $\mathbf{W}$ controls transfer in every section of $G$.*

REMARK. Note that the conclusion of (6.10) is valid if $\mathbf{W}$ controls transfer in $G^*$.

*Proof.* We use induction on $|G|$. Let $P = \mathbf{O}_p(G)$, $C = \mathbf{C}(P)$, and $N = \mathbf{N}(\mathbf{W}(S))$. By Lemma 6.1 and induction, it suffices to prove that $\mathbf{W}$ controls transfer in $G$. We may define an equivalence relation on $S$ as in the proof of part (ii) of Theorem 5.5; by the method used in the proof of Theorem

6.6, we may assume that $P \neq 1$ and that

$$G = CS \quad \text{or} \quad PC \cap S \trianglelefteq G. \tag{6.11}$$

Suppose that $1 \subset K \trianglelefteq G$. For every subgroup $H$ of $G$, let $\bar{H} = HK/K$. By induction,

$$\bar{G}' \cap \bar{S} = (\mathbf{N}_{\bar{G}}(\mathbf{W}(\bar{S})))' \cap \bar{S}. \tag{6.12}$$

Take $L$ to be the subgroup of $G$ that contains $K$ and satisfies $\bar{L} = \mathbf{N}_{\bar{G}}(\mathbf{W}(\bar{S}))$. By (6.12),

$$(G' \cap S)K/K \subseteq \bar{G}' \cap \bar{S} = \bar{L}' \cap \bar{S} \subseteq \bar{L}' = L'K/K.$$

Hence

$$G' \cap S \subseteq L'K \cap S. \tag{6.13}$$

Assume first that $K$ is a $p'$-group. Let $W = \mathbf{W}(S)$. By the definition of a section conjugacy functor, $\mathbf{W}(\bar{S}) = \bar{W}$. Hence $L = \mathbf{N}_G(WK)$. Since $W$ is a Sylow $p$-subgroup of $WK$ and $N \subseteq L$, the Frattini argument yields that $L = \mathbf{N}_G(W)K = NK$. By (6.13),

$$|G' \cap S| \leq |L'K|_p \leq |N'K|_p = |N'|_p = |N' \cap S|.$$

Since $N' \cap S \subseteq G' \cap S$, $N' \cap S = G' \cap S$. Thus we may assume that

$$G \text{ has no non-identity normal } p'\text{-subgroups.} \tag{6.14}$$

Now assume above that $K = P$. Suppose first that $P \subset S$. Then $\bar{S} \neq 1$. Therefore, $\mathbf{W}(\bar{S}) \neq 1$. As $\mathbf{O}_p(\bar{G}) = 1$, $\bar{L} \subset \bar{G}$. Hence $S \subseteq L \subset G$. By induction,

$$L' \cap S = (\mathbf{N}_L(\mathbf{W}(S)))' \cap S \subseteq N' \cap S.$$

Consequently, by (6.13) and Dedekind's Lemma,

$$G' \cap S \subseteq L'P \cap S = (L' \cap S)P \subseteq (N' \cap S)P.$$

Let $T_1 = G' \cap S$ and $T_2 = N' \cap S$. By a second application of Dedekind's Lemma,

$$G' \cap S = T_1 = T_1 \cap T_2P = (T_1 \cap P)T_2 = (G' \cap P)(N' \cap S). \tag{6.15}$$

Note that (6.15) is also valid when $P = S$.

We claim that

$$P \cap N' = P \cap G'. \tag{6.16}$$

By (6.11), $G = CS$ or $PC \cap S \trianglelefteq G$. Suppose first that $G = CS$. Then $[P, G] = [P, S] = [P, N]$, and (6.16) follows from Lemma 6.7. Suppose that $PC \cap S \trianglelefteq G$. Then $\mathbf{C}_S(P) \subseteq PC \cap S \subseteq \mathbf{O}_p(G) = P$. By (6.14) and Lemma 6.4, $\mathbf{C}(P) \subseteq P$. Then (6.10) yields (6.16). Thus (6.16) occurs in both cases. By (6.15) and (6.16),

$$G' \cap S = (G' \cap P)(N' \cap S) = (N' \cap P)(N' \cap S) = N' \cap S.$$

This completes the proof of Theorem 6.8.

The following observation was made by J. Thompson.

PROPOSITION 6.9. *Suppose* $\mathbf{W}$ *is a section conjugacy function on G that controls transfer in every section of G. Assume that* $S \neq 1$ *and that* $\mathbf{N}(S)/\mathbf{C}(S)$ *is a p-group. Then G has a factor group of order p.*

*Proof.* By Lemma 4.3, $G$ has a factor group of order $p$ if and only if $G' \cap S \subset S$. Let $H = \mathbf{N}(\mathbf{W}(S))$, $N = \mathbf{N}(S)$, and $C = \mathbf{C}(S)$. As $S \subseteq H$, we have $H' \cap S \subset S$ by induction if $H \subset G$. Since $H' \cap S = G' \cap S$, we may assume that $H = G$. Then $\mathbf{W}(S) \trianglelefteq G$. Let $Z = \mathbf{Z}(\mathbf{W}(S))$. For every subgroup $K$ of $G$ that contains $Z$, let $\bar{K} = K/Z$.

Suppose that $Z \subset S$. Then $\bar{S} \neq 1$, $\mathbf{N}_{\bar{G}}(\bar{S}) \neq \bar{N}$, and $\mathbf{C}_{\bar{G}}(\bar{S}) \supseteq \bar{C}$. Hence $\mathbf{N}_{\bar{G}}(\bar{S})/\mathbf{C}_{\bar{G}}(\bar{S})$ is a $p$-group. By induction, $\bar{G}$ has a factor group of order $p$. Thus $G$ also has a factor group of order $p$.

Suppose $Z = S$. Then $S \subseteq C$. Since $N/C$ is a $p$-group, $N/C = 1$. Therefore, $S \subseteq \mathbf{Z}(N)$. By Burnside's theorem (Theorem 4.8), $S \cap G' = [S, N] = 1 \subset S$. This completes the proof of Proposition 6.9.

## 7. Commutator Conditions on *p*-subgroups

DEFINITION. Suppose that $H \subseteq G$ and $g \in G$. Define

$$[H, g] = \langle [h, g] : h \in H \rangle$$

Let $[H, g, g] = [[H, g], g]$. For $n = 0, 1, \ldots$, define $[H, g; n]$ inductively by

$$[H, g; 0] = H, \text{ and}$$

$$[H, g; i+1] = [[H, g; i], g], \quad (i = 0, 1, \ldots).$$

We will generally consider the groups $[H, g; n]$ in cases where $H$ is a $p$-group and $g$ is a $p$-element that normalizes $H$. The importance of these groups was originally demonstrated in the work of Wielandt (1940), Hall and Higman (1956), and Thompson (1960). Gorenstein and Walter (1964) introduced the related concept of $p$-stability, which we will define later in this section.

DEFINITION. Suppose $G$ is abelian. Then $G$ is an *elementary abelian* group if there exists a prime $q$ such that $x^q = 1$ for all $x \in G$.

DEFINITION. Suppose that $H, K \trianglelefteq G$ and $K \subset H$. Then $H/K$ is a *chief factor* of $G$ if $H$ and $K$ are the only normal subgroups $N$ of $G$ for which $K \subseteq N \subseteq H$.

REMARK. Note that $H/K$ is a chief factor of $G$ if and only if $H/K$ is a minimal normal subgroup of $G/K$.

LEMMA 7.1. *Suppose that* $H/K$ *is a chief factor of G and H is a p-group. Then* $H/K$ *is an elementary abelian group.*

*Proof*. Clearly, $H/K$ is a chief factor of $G/K$. We may therefore assume that $K = 1$. Let $L$ be the set of all $x \in Z(H)$ such that $x^p = 1$. Then $L$ is an elementary abelian characteristic subgroup of $H$ and $1 = K \subset L \subseteq H$. Hence $L \trianglelefteq G$ and $L = H$.

The following two theorems (obtained independently by the author and L. Scott) extend slightly a result of Feit (1969). A proof of Theorem 7.2 is given in the appendix to this paper. Theorem 7.3 follows from Theorem 7.2 and Lemma 6.7.

THEOREM 7.2. *Suppose that P is a normal p-subgroup of G and $E_0$ is a non-empty subset of S. Assume that:*

(a) *$<E_0>$ is weakly closed in S with respect to G; and*

(b) *$[X, g; p-1] \subseteq Y$ for every $g \in E_0$ and every chief factor $X/Y$ of G such that $X \subseteq P$.*

*Let $E = \langle E_0 \rangle$ and $L = \mathbf{N}(E)$. Then $S \subseteq L$, $P \cap \mathbf{Z}(G) = P \cap \mathbf{Z}(L)$, and $[P, G] = [P, L]$.*

THEOREM 7.3. *Assume the hypothesis and notation of Theorem 7.2. Then $S \subseteq L$ and $P \cap G' = P \cap L'$.*

THEOREM 7.4. *Let* **W** *be a section conjugacy functor on G. Assume that* **W** *satisfies the following condition:*

*Whenever $G^*$ is a section of G, $S^*$ is a Sylow p-subgroup of $G^*$, $\mathbf{C}_{G^*}(\mathbf{O}_p(G^*)) \subseteq \mathbf{O}_p(G^*)$, and $\mathbf{W}(S^*) \ntrianglelefteq G^*$, then there exists $g \in S^* - \mathbf{O}_p(G^*)$ such that $[X, g; p-1] \subseteq Y$ for every chief factor $X/Y$ of $G^*$ for which $X \subseteq \mathbf{O}_p(G^*)$.* (7.1)

*Then* **W** *controls transfer in every section of G.*

*Proof*. We use induction on $|G|$. Thus we may assume that **W** controls transfer in every section of $G$ except possibly $G$ itself. Let $P = \mathbf{O}_p(G)$ and $N = \mathbf{N}(\mathbf{W}(S))$. By Theorem 6.8 and the remark following, it suffices to prove that $P \cap G' = P \cap N'$ if $\mathbf{C}(P) \subseteq P$.

Assume that $\mathbf{C}(P) \subseteq P$. If $\mathbf{W}(S) \trianglelefteq G$, then $G = N$. Assume that $\mathbf{W}(S) \ntrianglelefteq G$. Let $E_0$ be the set of all $g \in S$ enjoying the property that $[X, g; p-1] \subseteq Y$ for every chief factor $X/Y$ of $G$ for which $X \subseteq \mathbf{O}_p(G)$. Put $E = \langle E_0 \rangle$ and $L = \mathbf{N}(E)$. Clearly, $E_0$ satisfies conditions (a) and (b) of Theorem 7.2. Therefore, by Theorem 7.3,

$$S \subseteq L \quad \text{and} \quad P \cap G' = P \cap L'. \tag{7.2}$$

By (7.1), $E \nsubseteq P$, Hence $L = \mathbf{N}(E) \subset G$. By (7.2) and induction,

$$P \cap G' = P \cap L' = P \cap (\mathbf{N}_L(\mathbf{W}(S)))' \subseteq P \cap N' \subseteq P \cap G'.$$

Thus $P \cap G' = P \cap N'$, as desired. This completes the proof of Theorem 7.4.

THEOREM 7.5. *Let* $\mathbf{W}$ *be a conjugacy functor on* $G$. *Assume that* $\mathbf{W}(P) \supseteq \mathbf{Z}(P)$ *for every p-subgroup* $P$ *of* $G$ *and that* $\mathbf{W}$ *satisfies the following condition:*

*Whenever* $G^* \subseteq G$, $S^*$ *is a Sylow p-subgroup of* $G^*$, *and* $\mathbf{W}(S^*) \ntrianglelefteq G^*$, *then there exists* $g \in S^* - \mathbf{O}_p(G^*)$ *such that* $[X, g; p-1] \subseteq Y$ *for every chief factor* $X/Y$ *of* $G^*$ *for which* $X \subseteq \mathbf{Z}(\mathbf{O}_p(G^*))$. (7.3)

*Then* $\mathbf{W}$ *controls weak closure of elements in every subgroup of* $G$.

*Proof.* We imitate the proof of Theorem 7.4. Use induction on $|G|$. Suppose $x \in S \cap \mathbf{Z}(\mathbf{N}(\mathbf{W}(S)))$. We must prove that $x$ is weakly closed in $S$ with respect to $G$. By Theorem 5.6 and induction, we may assume that $x \in \mathbf{O}_p(G)$.

Let $Z = \mathbf{Z}(\mathbf{O}_p(G))$ and $N = \mathbf{N}(\mathbf{W}(S))$. Then $x \in S \cap \mathbf{Z}(N) \subseteq \mathbf{Z}(S)$. Hence $x \in Z$. Since $x \notin \mathbf{Z}(G)$, $\mathbf{W}(S) \ntrianglelefteq G$. By (7.3) and an argument similar to one used in the proof of Theorem 7.4, there exists $L \subset G$ such that

$$L \supseteq S \quad \text{and} \quad Z \cap \mathbf{Z}(G) = Z \cap \mathbf{Z}(L). \tag{7.4}$$

By induction, $x$ is weakly closed in $S$ with respect to $L$. Since $x \in \mathbf{O}_p(L)$, $x \in \mathbf{Z}(L)$. By (7.4), $x \in \mathbf{Z}(G)$. This completes the proof of Theorem 7.5.

DEFINITION. For $H \subseteq G$ and $g \in G$, let $[H, g, g] = [[H, g], g]$.

The following important concept was introduced by Gorenstein and Walter (1964). Our definition is the same as that of Gorenstein (1968), as we will prove in Proposition 7.8.

DEFINITION. We say that $G$ is a *p-stable* group if it satisfies the following condition:

Whenever $P$ is a $p$-subgroup of $G$, $g \in N(P)$, and $[P, g, g] = 1$, then the coset $g\mathbf{C}(P)$ lies in $\mathbf{O}_p(\mathbf{N}(P)/\mathbf{C}(P))$.

PROPOSITION 7.6. *Suppose that* $H/K$ *is a chief factor of* $G$ *and* $H$ *is a p-group. Let* $\bar{G} = G/K$ *and* $\bar{H} = H/K$. *Then*

(a) $\mathbf{O}_p(\bar{G}/\mathbf{C}_{\bar{G}}(\bar{H})) = 1$, *and*

(b) *if* $\bar{G}$ *is p-stable,* $g \in G$, *and* $[H, g, g] \subseteq K$, *then* $[H, g] \subseteq K$.

*Proof.* Statement (b) is equivalent to the statement:

(b′) if $\bar{G}$ is $p$-stable, $\bar{g} \in \bar{G}$, and $[\bar{H}, \bar{g}, \bar{g}] = 1$, then $\bar{g} \in \mathbf{C}_{\bar{G}}(\bar{H})$.

Since $\bar{H}$ is a chief factor of $\bar{G}$, we may assume that $K = 1$. We will prove (a) and (b′). Let $C = \mathbf{C}(H)$, and take $N \trianglelefteq G$ such that $N \supseteq C$ and $N/C = \mathbf{O}_p(G/C)$.

(a) By Corollary 2.10, $S \cap N$ is a Sylow $p$-subgroup of $N$ and $N = (S \cap N)C$. Let $Z = H \cap \mathbf{Z}(S \cap N)$. Then

$$Z = H \cap \mathbf{Z}(N) \trianglelefteq G \quad \text{and} \quad 1 \subset Z \subseteq H,$$

by Theorem 2.1. Hence $Z = H$, and $H \subseteq \mathbf{Z}(N)$. Thus $N = C$.

(b′) Assume that $G$ is $p$-stable and $[H, g, g] = 1$. By the definition of a $p$-stable group and by (a), $gC \in N/C = C/C$. Thus $g \in C$.

LEMMA 7.7.
(a) *Let* $x,y \in G$. *Then* $[y,x] = [x,y]^{-1}$.
(b) *Let* $x,y,z \in G$. *Then* $[xy,z] = [x,z]^y [y,z]$.
(c) *Let* $z \in G$ *and* $H \subseteq G$. *Then* $H$ *normalizes* $[H,z]$.
(d) *Suppose that* $x,y \in G$ *and* $G' \subseteq \mathbf{Z}(G)$. *Then* $[x^n,y] = [x,y^n] = [x,y]^n$ *for every positive integer* $n$.

*Proof.* We may verify (a) and (b) by checking that

$$y^{-1}x^{-1}yx = (x^{-1}y^{-1}xy)^{-1} \text{ and}$$

$$(xy)^{-1}z^{-1}(xy)z = y^{-1}(x^{-1}z^{-1}xz)y(y^{-1}z^{-1}yz).$$

To prove (c), let $g, h \in H$. By (b)

$$[g, z]^h = [gh, z][h, z]^{-1} \in [H, z].$$

Finally, (d) follows from (b) by induction.

PROPOSITION 7.8. *The following condition is a necessary and sufficient condition for* $G$ *to be* $p$*-stable.*
*Let* $P$ *be an arbitrary* $p$*-subgroup of* $G$ *such that* $G = K\mathbf{N}(P)$ *for some* $K \trianglelefteq G$. *Let* $A$ *be an arbitrary* $p$*-subgroup of* $\mathbf{N}(P)$. *Then, if* $[P, A, A] = 1$, *we have*

$$A\mathbf{C}(P)/\mathbf{C}(P) \subseteq \mathbf{O}_p(\mathbf{N}(P)/\mathbf{C}(P)). \tag{7.5}$$

REMARK. Condition (7.5) is the definition of a $p$-stable group (Gorenstein, 1968).

*Proof.* Suppose that $G$ is $p$-stable and that $P$ and $A$ satisfy the hypothesis of (7.5). Assume that $[P, A, A] = 1$. Then $[P, g, g] = 1$ for every $g \in A$. Hence

$$g\mathbf{C}(P) \in \mathbf{O}_p(\mathbf{N}(P)/\mathbf{C}(P)) \quad \text{for every } g \in A,$$

which proves (7.5).

Conversely, assume $G$ satisfies (7.5). Suppose that $P$ is a $p$-subgroup of $G$, $g \in N(P)$, and $[P, g, g] = 1$. Let $N = \mathbf{N}(P)$, $C = \mathbf{C}(P)$, $A = \langle g \rangle$, and $Q = [P, g]$. Then $g$ centralizes $Q$, and $P$ normalizes $Q$, by Lemma 7.7. Hence $Q \trianglelefteq PA$. Consequently, $[P, A] = Q$ and $[P, A, A] = 1$. Taking $K = G$ in (7.5), we obtain $AC/C \subseteq \mathbf{O}_p(N/C)$. Therefore, $gC \in \mathbf{O}_p(N/C)$, which proves that $G$ is $p$-stable.

Recall that $H^p = \{h^p : h \in H\}$ for every abelian $p$-subgroup $H$ of $G$. In discussing properties of $p$-stable groups, we will sometimes verify condition (7.6) of the following result.

THEOREM 7.9. *Let* $\mathbf{W}$ *be a conjugacy functor on* $G$. *Assume that* $\mathbf{W}(P) \supseteq \mathbf{Z}(P)$ *for every p-subgroup* $P$ *of* $G$ *and that* $\mathbf{W}$ *satisfies the following condition: Whenever* $G^* \subseteq G$, $S^*$ *is a Sylow p-subgroup of* $G^*$, *and* $\mathbf{W}(S^*) \ntrianglelefteq G^*$, *then there exists* $g \in S^* - \mathbf{O}_p(G^*)$ *such that* $[\mathbf{Z}(\mathbf{O}_p(G^*)), g, g] = 1$. (7.6)

*Let* $Z = \mathbf{Z}(S)^p$. *Suppose that* $A$ *is an element of* $Z \cap \mathbf{Z}(\mathbf{N}(\mathbf{W}(S)))$ *or* $A$ *is a normal subgroup of* $\mathbf{N}(\mathbf{W}(S))$ *contained in* $Z$. *Then* $A$ *is weakly closed in* $S$ *with respect to* $G$.

*Proof.* We use induction on $|G|$. If $A$ is not weakly closed in $S$, we may take $T$ as in Theorem 5.6; then $A \subseteq Z \subseteq (\mathbf{Z}(\mathbf{N}_S(T)))^p$. Therefore, by induction, we may assume that $\mathbf{O}_p(G)$ contains $A$.

Let $P = \mathbf{O}_p(G)$, $Q = \mathbf{Z}(P)$, and $R = \mathbf{C}_S(P)$. By the Frattini argument, $G = \mathbf{C}(P)\mathbf{N}(R)$. Since $S \subseteq \mathbf{N}(R)$, we may assume by induction tha $G = \mathbf{N}(R)$. Hence $R \subseteq P$ and $\mathbf{Z}(S) \subseteq R \subseteq P$. Therefore, $\mathbf{Z}(S) \subseteq Q$ and $Z \subseteq Q^p$. Applying the Frattini argument again, we obtain $G = \mathbf{C}(Q_p)\mathbf{N}(\mathbf{C}_S(Q_p))$. Now we may assume that $G = \mathbf{N}(\mathbf{C}_S(Q^p))$. Then

$$\mathbf{C}_S(Q^p) \subseteq \mathbf{O}_p(G). \tag{7.7}$$

Suppose that $\mathbf{W}(S) \ntrianglelefteq G$. Take $g$ as in (7.6), and let $T = [Q, g]$ and $H = \langle Q, g \rangle$. Then $Q$ and $g$ centralize $T$. Consequently,

$$T \subseteq \mathbf{Z}(H) \quad \text{and} \quad H' \subseteq T. \tag{7.8}$$

Let $h$ be a power of $g$ such that $h \notin P$ and $h^p \in P$. By (7.8) and Lemma 7.7,

$$[x^p, h] = [x, h^p] \in [Q, P] = 1 \quad \text{for all } x \in Q.$$

Hence $h \in \mathbf{C}_S(Q^p) \subseteq \mathbf{O}_p(G)$ by (7.7), contrary to the choice of $h$. This contradiction shows that $\mathbf{W}(S) \trianglelefteq G$ and therefore that $A^g = A$ for all $g \in G$. This completes the proof of Theorem 7.9.

## 8. Normal *p*-complements

DEFINITION. Suppose that $H \trianglelefteq G$. We say that $H$ is a *normal p-complement* in $G$ if $G = HT$ and $H \cap T = 1$ for every Sylow $p$-subgroup $T$ of $G$.

Let $\mathbf{O}^p(G)$ be the subgroup of $G$ generated by all the $p'$-elements of $G$.

LEMMA 8.1. *The following conditions are equivalent:*

(a) *$G$ has a normal p-complement;*

(b) *$G$ has a normal subgroup $H$ and a Sylow p-subgroup $T$ for which $HT = G$ and $H \cap T = 1$;*

(c) *$\mathbf{O}^p(G)$ is a normal p-complement in $G$ and is the only normal p-complement in $G$;*

(d) *the set of all p'-elements of $G$ is closed under multiplication;*

(e) *$\mathbf{O}^p(G)$ is a p'-group.*

*Proof.* We give a cyclic proof by showing that any of the conditions (a), (b), (c) and (d) implies the one following it and that (e) implies (a).

It is obvious that (a) implies (b). Next, suppose $H$ and $T$ satisfy (b). Then $|G| = |H||T| = |H||G|_p$. Hence $H$ is a $p'$-subgroup of $G$, and $H \subseteq \mathbf{O}^p(G)$. Since $G/H \cong T$, every $p'$-element of $G$ lies in $H$. Therefore, $\mathbf{O}^p(G) \subseteq H$. Thus $\mathbf{O}^p(G) = H$, which proves (c).

Assume (c). Then $|\mathbf{O}^p(G)| = [G : S]$, and $\mathbf{O}^p(G)$ is a $p'$-group. This yields (d).

Assume (d). Since $G$ is finite, the set of all $p'$-elements of $G$ is a subgroup of $G$. Clearly, this subgroup is $\mathbf{O}^p(G)$, which proves (e).

Assume (e). Then $\mathbf{O}^p(G) \cap S = 1$. By Lemma 4.3, $G = \mathbf{O}^p(G)S$. Now, for all $g \in G$, we have

$$1 = 1^g = \mathbf{O}^p(G) \cap S^g \quad \text{and} \quad G = G^g = \mathbf{O}^p(G)S^g.$$

Thus $\mathbf{O}^p(G)$ is a normal $p$-complement in $G$.

LEMMA 8.2. *Suppose $N \trianglelefteq G$ and $G/N$ is a $p$-group. Then $\mathbf{O}^p(G) = \mathbf{O}^p(N)$.*

*Proof.* Clearly, every $p'$-element of $G$ is a $p'$-element of $N$, and vice versa.

Using the concept of a normal $p$-complement, we obtain another statement of part (b) of Theorem 4.8; namely

THEOREM 8.3 (Burnside). *If $S \subseteq \mathbf{Z}(\mathbf{N}(S))$, then $G$ has a normal $p$-complement.*

LEMMA 8.4. *Suppose that $G$ is a finite $p$-group and $1 \subset H \subseteq G$. Then there exists a normal subgroup $N$ of $G$ such that*

$$1 \subset HN/N \subseteq \mathbf{Z}(G/N).$$

*Proof.* Let $N$ be a normal subgroup of maximal order in $G$ subject to the restriction that $H \nsubseteq N$. Let $M/N = \mathbf{Z}(G/N)$. By Theorem 2.1, $M \supset N$. Since $M \trianglelefteq G$, $M \supseteq H$. Hence $M \supseteq HN$.

PROPOSITION 8.5. *Suppose that $H \subseteq G$, $T \subseteq S \cap H$, and $T$ is a Sylow $p$-subgroup of $H$. Assume that whenever two elements of $T$ are conjugate in $H$, they are conjugate in $S$. Then $H$ has a normal $p$-complement.*

*Proof.* We use induction on $|T|$. If $T = 1$, then $H$ is a $p'$-group. Therefore, we will assume that $T \neq 1$. By Lemma 8.4, there exists $N \trianglelefteq S$ such that

$$1 \subset TN/N \subseteq \mathbf{Z}(S/N). \tag{8.1}$$

Suppose that $x, y \in T$ and $x$ and $y$ are conjugate in $H$. By hypothesis, $x$ and $y$ are conjugate in $S$. Therefore, $x \equiv y \pmod N$ by (8.1). Hence $xy^{-1} \in T \cap N$. By the focal subgroup theorem and (8.1), $T \cap H' \subseteq T \cap N \subset T$. By Lemma 4.3, $H$ has a normal subgroup $K$ of index $p$.

Now, $K \cap S = K \cap (H \cap S) = K \cap T$. By Corollary 2.10, $K \cap T$ is a Sylow $p$-subgroup of $K$. Therefore, $K$ has a normal $p$-complement, by induction. Lemmas 8.1 and 8.2 yield that $\mathbf{O}^p(K)$ is a $p'$-group and that $\mathbf{O}^p(H) = \mathbf{O}^p(K)$. Consequently, $H$ has a normal $p$-complement, by Lemma 8.1.

THEOREM 8.6 (Frobenius). *The following conditions are equivalent:*

(a) *$G$ has a normal $p$-complement;*

(b) *whenever $P$ is a non-identity $p$-subgroup of $G$, then $\mathbf{N}(P)$ has a normal $p$-complement;*

(c) *whenever $P$ is a non-identity $p$-subgroup of $G$, then $\mathbf{N}(P)/\mathbf{C}(P)$ is a $p$-group;*

(d) *whenever two elements of $S$ are conjugate in $G$, they are conjugate in $S$.*

*Proof.* We give a cyclic proof. We note first that (a) implies (b) by Lemma 8.1.

Assume (b). Suppose that $P$ is a non-identity $p$-subgroup of $G$. Let $K = \mathbf{O}^p(\mathbf{N}(P))$. By Lemma 8.1, $K$ is a $p'$-group. Since $K$ and $P$ normalize each other,

$$[K, P] \subseteq K \cap P = 1.$$

Hence $K \subseteq \mathbf{C}(P)$. Since $\mathbf{N}(P)/K$ is a $p$-group, $\mathbf{N}(P)/\mathbf{C}(P)$ is a $p$-group.

Assume (c). Let $\mathscr{F}$ be the set of all subgroups $T$ of $S$ for which $\mathbf{N}_S(T)$ is a Sylow $p$-subgroup of $\mathbf{N}(T)$. By Alperin's theorem (Theorem 3.5), $\mathscr{F}$ is a conjugation family for $S$ in $G$. Let $\sim$ be the equivalence relation on $S$ given by

$$x \sim y \quad \text{if } x \text{ and } y \text{ are conjugate in } S.$$

Suppose that $T \in \mathscr{F}$, that $x, y \in T$, and that $x$ and $y$ are conjugate in $\mathbf{N}(T)$. By (c) and Corollary 2.10, $\mathbf{N}(T) = \mathbf{C}(T)\mathbf{N}_S(T)$. Hence $x$ and $y$ are conjugate in $\mathbf{N}_S(T)$. Therefore, $x \sim y$. By Proposition 3.7, $\sim$ contains fusion in $G$. This proves (d).

By Proposition 8.5, (d) implies (a). This completes the proof of Theorem 8.6.

THEOREM 8.7 (Thompson). *Let $\mathbf{W}$ be a section conjugacy functor on $G$. Suppose that $\mathbf{N}(\mathbf{W}(S))$ has a normal $p$-complement and that $G$ satisfies the following condition:*

*Let $G^*$ be a section of $G$ for which $G^*/\mathbf{O}_p(G^*)$ has a normal $p$-complement and $\mathbf{C}_{G^*}(\mathbf{O}_p(G^*)) \subseteq \mathbf{O}_p(G^*)$. Let $S^*$ be a Sylow $p$-subgroup of $G^*$. Assume that $S^*$ is a maximal subgroup of $G^*$. Then $\mathbf{W}(S^*) \trianglelefteq G^*$.* (8.2)

*Then $G$ has a normal $p$-complement.*

*Proof.* We use induction on $|G|$. Assume that the theorem is false and that

$G$ is a counterexample of minimal order. Let $\mathscr{P}$ be the set of all non-identity $p$-subgroups $P$ of $G$ for which $\mathbf{N}(P)$ does not have a normal $p$-complement. Define a partial ordering $\leq$ on $\mathscr{P}$ in the following way: $P_1 \leq P_2$ if

$$\begin{aligned} &\text{(i) } |\mathbf{N}(P_1)|_p < |\mathbf{N}(P_2)|_p \text{ or} \\ &\text{(ii) } |\mathbf{N}(P_1)|_p = |\mathbf{N}(P_2)|_p \quad \text{and} \quad |P_1| < |P_2|, \text{ or} \\ &\text{(iii) } P_1 = P_2. \end{aligned} \tag{8.3}$$

By Theorem 8.6, $\mathscr{P}$ is not empty. Let $P$ be a maximal element of $\mathscr{P}$ with respect to $\leq$, and let $G_1 = \mathbf{N}(P)$.

Suppose that $G_1 \subset G$. Let $S_1$ be a Sylow $p$-subgroup of $G_1$. By induction $\mathbf{N}_{G_1}(\mathbf{W}(S_1))$ does not have a normal $p$-complement. So $\mathbf{W}(S_1) \in \mathscr{P}$, by Lemma 8.1. Because of the maximal choice of $P$,

$$|S_1| = |\mathbf{N}(P)|_p \geq |\mathbf{N}(\mathbf{W}(S_1))|_p \geq |\mathbf{N}(S_1)|_p \geq |S_1|. \tag{8.4}$$

Let $S_2$ be a Sylow $p$-subgroup of $G$ that contains $S_1$. By (8.4), $\mathbf{N}_{S_2}(S_1) = S_1$. By Corollary 2.6, $S_1 = S_2$. Hence $S_1$ is conjugate to $S$. Consequently, $\mathbf{N}(\mathbf{W}(S_1))$ is conjugate and, therefore, isomorphic to $\mathbf{N}(\mathbf{W}(S))$. Thus $\mathbf{N}(\mathbf{W}(S_1))$ has a normal $p$-complement, a contradiction.

Thus $G_1 = G$, and $P \trianglelefteq G$. Then $P \subseteq \mathbf{O}_p(G)$. By the maximal choice of $P, P = \mathbf{O}_p(G)$. Since $\mathbf{W}(S) \trianglelefteq \mathbf{N}(S), P \subset S$. Hence $S/P \neq 1$ and $\mathbf{W}(S/P) \neq 1$. Take $H \subseteq G$ such that

$$H \supseteq P \quad \text{and} \quad H/P = \mathbf{N}_{G/P}(\mathbf{W}(S/P)).$$

Since $\mathbf{O}_p(G/P) = 1$, $H \subset G$. Moreover, $S \subseteq H$. Since $\mathbf{N}(\mathbf{W}(S))$ has a normal $p$-complement, $\mathbf{N}_H(\mathbf{W}(S))$ has a normal $p$-complement, by Lemma 8.1. Therefore, $H$ has a normal $p$-complement, by induction. Applying induction again, we find that

$$\text{there exists a normal } p\text{-complement } L/P \text{ in } G/P. \tag{8.5}$$

Suppose $K$ is a non-identity normal $p'$-subgroup of $G$. Let $W = \mathbf{W}(S)$. By the definition of a section conjugacy functor, $\mathbf{W}(SK/K) = WK/K$. Clearly, $W$ is a Sylow $p$-subgroup of $WK$ and $\mathbf{N}(W) \subseteq \mathbf{N}(WK)$. By the Frattini argument, $\mathbf{N}(WK) = \mathbf{N}(W)(WK) = \mathbf{N}(W)K$. Hence $\mathbf{N}(WK)/K$ has a normal $p$-complement. However,

$$\mathbf{N}(WK)/K = \mathbf{N}_{G/K}(WK/K) = \mathbf{N}_{G/K}(\mathbf{W}(SK/K)).$$

By induction, $G/K$ has a normal $p$-complement, say, $M/K$. Then $M$ is a normal $p'$-subgroup of $G$ and $G/M$ is a $p$-group. By Corollary 2.10, $M$ is a normal $p$-complement in $G$, a contradiction. Thus

$$G \text{ has no non-identity normal } p'\text{-subgroups.} \tag{8.6}$$

Take $L$ as in (8.5). Obviously, $P$ is a Sylow $p$-subgroup of $L$. By (8.6), $\mathbf{C}_L(P)$ has no non-identity characteristic $p'$-subgroups. Hence, by Lemma 6.4 applied to $L$ instead of $G$, $\mathbf{C}_L(P) \subseteq P$. Since $\mathbf{C}(P)/\mathbf{C}_L(P)$ is isomorphic to a subgroup of $G/L$, $\mathbf{C}(P)/\mathbf{C}_L(P)$ is a $p$-group. Therefore, $\mathbf{C}(P)$ is a $p$-group. Since $\mathbf{C}(P) \trianglelefteq G$,

$$\mathbf{C}(P) \subseteq \mathbf{O}_p(G) = P. \tag{8.7}$$

Suppose that $S \subseteq H \subset G$. By induction, $H$ has a normal $p$-complement, say, $K$. Then

$$[P, K] \subseteq P \cap K = 1.$$

Thus $K \subseteq \mathbf{O}^p(C(P)) \subseteq \mathbf{O}^p(P) = 1$, by (8.7). Hence $H = SK = S$. This proves that $S$ is a maximal subgroup of $G$. Now $\mathbf{W}(S) \trianglelefteq G$, by (8.2). Therefore, $G = \mathbf{N}(\mathbf{W}(S))$, and $G$ has a normal $p$-complement. This contradiction completes the proof of Theorem 8.7.

The following result is useful in discussing groups that satisfy condition (8.2) of Theorem 8.7.

LEMMA 8.8. *Let $P = \mathbf{O}_p(G)$. Suppose that $S$ is a maximal subgroup of $G$ and $G/P$ has a normal $p$-complement. Then there exist a prime $q \neq p$ and an elementary abelian Sylow $q$-subgroup $Q$ of $G$ such that $QP/P$ is a minimal normal subgroup of $G/P$ and $G = SQ = S(PQ)$.*

*Proof.* Since the hypothesis is inherited by $G/P$, it suffices to consider the case where $P = 1$. Let $L$ be the normal $p$-complement of $G$. Let $q$ be a prime divisor of $|L|$ and $Q$ be a Sylow $q$-subgroup of $L$. By the Frattini argument, $G = L\mathbf{N}(Q)$. Since $L$ is a $p'$-group, $|G|_p = |\mathbf{N}(Q)|_p$. Therefore, $\mathbf{N}(Q)$ contains some conjugate of $S$. Replacing $Q$ by a conjugate of $Q$ if necessary, we may assume that $S$ normalizes $Q$. Since $S$ is a maximal subgroup of $G$, $G = SQ$.

Let $N$ be a minimal normal subgroup of $G$ contained in $Q$. By Lemma 7.1, $N$ is elementary abelian. Since $S$ is a maximal subgroup of $G$, $G = SN$, as desired.

THEOREM 8.9. *Let $\mathbf{W}$ be a section conjugacy functor on $G$. Suppose that $\mathbf{N}(\mathbf{W}(S))$ has a normal $p$-complement and that $G$ satisfies the following condition.*

*Whenever $G^*$ is a section of $G$, $S^*$ is a Sylow $p$-subgroup of $G^*$, $\mathbf{C}_{G^*}(\mathbf{O}_p(G^*)) \subseteq \mathbf{O}_p(G^*)$, and $\mathbf{W}(S^*) \ntrianglelefteq G^*$, then there exist $g \in S^* - \mathbf{O}_p(G^*)$ and a chief factor $X/Y$ of $G^*$ for which*

$$X \subseteq \mathbf{O}_p(G^*), \qquad [X, g; p-1] \subseteq Y, \qquad \text{and } [X, g] \nsubseteq Y. \tag{8.8}$$

*Then $G$ has a normal $p$-complement.*

*Proof.* Assume the result is false. Then $G$ violates (8.2). Choose $G^*$, $S^*$, $X$, $Y$, and $g$ to satisfy the hypothesis of (8.2) and violate its conclusion. Then $\mathbf{N}(\mathbf{W}(S^*)) = S^*$, since $S^*$ is a maximal subgroup of $G^*$. Thus $\mathbf{N}(\mathbf{W}(S^*))$

has a normal $p$-complement. If $G^*$ has a normal $p$-complement, then $\mathbf{O}^p(G^*)$ is a $p'$-group and

$$[\mathbf{O}_p(G^*), \mathbf{O}^p(G^*)] \subseteq \mathbf{O}_p(G^*) \cap \mathbf{O}^p(G^*) = 1,$$

contrary to $\mathbf{C}(\mathbf{O}_p(G^*)) \subseteq \mathbf{O}_p(G^*)$. Hence $G^*$ does not have a normal $p$-complement. Thus it suffices to assume that $G^* = G$ and that $S^* = S$.

For every subgroup $H$ of $G$ that contains $Y$, let $\bar{H} = H/Y$. Let $E_0$ be the set of all $h \in \bar{S}$ for which $[\bar{X}, h; p-1] = 1$. Set $E = \langle E_0 \rangle$ and $L = \mathbf{N}_{\bar{G}}(E)$. By Theorem 7.2 (with $\bar{G}$ for $G$ and $\bar{X}$ for $P$), we obtain

$$\bar{S} \subseteq L \quad \text{and} \quad \bar{X} \cap \mathbf{Z}(\bar{G}) = \bar{X} \cap \mathbf{Z}(\bar{L}). \tag{8.9}$$

Since $g \in S - \mathbf{O}_p(G)$ and $\mathbf{O}_p(\bar{G}) = \mathbf{O}_p(G)/Y$, $E \nsubseteq \mathbf{O}_p(\bar{G})$. Hence $\bar{L} \subset \bar{G}$ Since $\bar{S}$ is a maximal subgroup of $\bar{G}$, $\bar{S} = \bar{L}$. By (8.9) and Theorem 2.1, $\bar{X} \cap \mathbf{Z}(\bar{G}) = \bar{X} \cap \mathbf{Z}(\bar{S}) \neq 1$. Since $\bar{X}$ is a minimal normal subgroup of $\bar{G}$, $\bar{X} \subseteq \mathbf{Z}(\bar{G})$. Then $[X,g] \subseteq Y$, a contradiction. This completes the proof of Theorem 8.9.

LEMMA 8.10. *Suppose that $T \trianglelefteq S$, $T \supseteq \mathbf{C}_S(T)$, and $\mathbf{N}(T)/\mathbf{C}(T)$ is a $p$-group. Then $\mathbf{N}(T)$ has a normal $p$-complement.*

*Proof.* Since $S \subseteq \mathbf{N}(T)$, we may assume that $G = \mathbf{N}(T)$. Let $C = \mathbf{C}(T)$. By Corollary 2.10, $C \cap S$ is a Sylow $p$-subgroup of $C$. Since $C \cap S \subseteq T$,

$$C \cap S = \mathbf{Z}(T) \subseteq \mathbf{Z}(C).$$

By Burnside's Theorem (Theorem 8.3), $C$ has a normal $p$-complement, By Lemmas 8.1 and 8.2, $\mathbf{O}^p(C)$ is a $p'$-group and $\mathbf{O}^p(G) = \mathbf{O}^p(C)$. Hence $G$ has a normal $p$-complement, by Lemma 8.1.

## 9. Digression

Section 8 marks the end of the first part of this work. In the previous sections. we have discussed rather formal, general properties of finite groups. We have relied on only a few facts beyond the elementary theory of groups. In particular, from the vast theory of finite $p$-groups, which requires almost one hundred and fifty pages in Huppert (1967), we have used only Theorem 2.1 and some of its easy consequences, proved in Section 2.

Let us recall the information files (A), (B), (C), and (D) mentioned in the introduction. In Section 3, we have shown by Alperin's Theorem that (C) determines (B). In Section 4, we have shown that (B), and therefore (C), determines some partial information about (A); for example, whether $G$ has a factor group of order $p$. Suppose we are given a single non-identity characteristic (or merely normal) subgroup of $T$ of $S$ and we wish to see how much (D) determines (A), (B), and (C). If there exists a conjugacy functor or section conjugacy functor $\mathbf{W}$ for which $T = \mathbf{W}(S)$, then the results of Sections 5 to 8 may be helpful. These results, however, are only reductions.

To use them, we must prove that their conditions on subgroups or on sections are satisfied.

To do this we must move in several ways from the general to the particular. Unfortunately, this will entail the loss of much of the symmetry of the previous results. Some functors satisfy our conditions and some do not; we must choose particular functors and work with them. This will require further investigation into the structure and embedding of $p$-subgroups in $G$. We must also choose the properties we wish to prove. By Lemma 4.6, a conjugacy functor that controls strong fusion must control transfer; the converse is false. Thus in some cases we will try to prove the easier result. Finally, we will see that some theorems are true for certain primes, but not for others.

## 10. Further Necessary Results

In this section we state a number of results that we will use in later sections. Our first result is an important result of Thompson. It is proved as Theorem 5.3.11 in Gorenstein (1968).

THEOREM 10.1. *Let $P$ be a finite $p$-group. Then $P$ contains a characteristic subgroup $Q$ that satisfies the following conditions:*

(a) *$Q/\mathbf{Z}(Q)$ is an elementary abelian group;*
(b) *$[P, Q] \subseteq \mathbf{Z}(Q)$;*
(c) *$\mathbf{C}_P(Q) = \mathbf{Z}(Q)$; and*
(d) *every non-identity $p'$-element of the automorphism group of $P$ determines a non-identity automorphism of $Q$.*

We will denote the set of characteristic subgroups $Q$ of $P$ that satisfy Theorem 10.1 by $\mathcal{Q}(P)$.

The following result will be useful in discussing transfer. It is proved in the Appendix.

THEOREM 10.2. *Let $P$ be a normal $p$-subgroup of $G$. Suppose that $Q \in \mathcal{Q}(P)$, $g \in G$, and $i$ and $j$ are positive integers. Assume that for every chief factor $X/Y$ of $G$, we have*

$$[X, g; i] \subseteq Y \quad \textit{if } X \subseteq Q, \textit{ and}$$
$$[X, g; j] \subseteq Y \quad \textit{if } X \subseteq \mathbf{Z}(Q).$$

*Then for every chief factor $X/Y$ of $G$ such that $X \subseteq P$, we have*

$$[X, g; i+j-1] \subseteq Y.$$

DEFINITION. Let $H \subseteq G$. Then $H$ is a *subnormal* subgroup of $G$ if there exists a chain of subgroups of the form

$$H = H_0 \subseteq H_1 \subseteq \ldots \subseteq H_n = G$$

such that $H_{i-1} \trianglelefteq H_i$ $(i = 1, \ldots, n)$.

THEOREM 10.3. *Suppose that $M$ is a minimal normal subgroup of $G$ and $H$ is a subnormal subgroup of $G$ that contains $M$. Then $M$ is a direct product of minimal normal subgroups of $H$.*

*Proof.* We use induction on $[G : H]$. We may assume that $H \subset G$. By the definition of a subnormal subgroup, there exists $K \subseteq G$ such that $K$ is subnormal in $G$, $H \subset K$, and $H \trianglelefteq K$. Suppose $K \subset G$. By induction, $M$ is a direct product of minimal normal subgroups of $K$. Similarly, every minimal normal subgroup of $K$ contained in $M$ is a direct product of minimal normal subgroups of $H$. Thus $M$ is a direct product of minimal normal subgroups of $H$.

Suppose $K = G$. Then $H \trianglelefteq G$. If $M$ is a $p$-group, then by considering $M$ to be a $G$-module over the field of $p$ elements, we may obtain the conclusion by Clifford's Theorem on representation theory (Theorem 3.4.1, Gorenstein, 1968). However, we may give a short direct proof. Let $N$ be a subgroup of maximal order in $M$ subject to the condition that $N = N_1 \times N_2 \times \ldots \times N_r$ for some $r \geq 1$ and some minimal normal subgroups $N_1, N_2, \ldots, N_r$ of $H$. For any $N_i$ and any $g \in G$, $(N_i)^g$ is a minimal normal subgroup of $H$ because $H \trianglelefteq G$. Let $L = (N_i)^g$. Then $N \cap L$ is 1 or $L$. In the former case, $NL = N \times L$, contrary to the maximal choice of $N$; thus $N \cap L = L$ and $N \supseteq L$. Since $N_i$ is arbitrary, $N \supseteq N^g$. As $g$ is arbitrary, $N \trianglelefteq G$. Since $1 \subset N \subseteq M$ and $M$ is a minimal normal subgroup of $G$, $N = M$. This completes the proof of Theorem 10.3.

LEMMA 10.4. *Suppose that $P$ is a finite $p$-group and*

$$P = P_0 \supseteq P_1 \supseteq \ldots \supseteq P_n = 1$$

*is a series of normal subgroups of $P$. Let $A$ be the group of all automorphisms of $P$ that fix $P_0, P_1, \ldots,$ and $P_n$. Let $B$ be the subgroup of $A$ consisting of the elements of $A$ that fix every coset of $P_{i-1}$ in $P_i$ $(i = 1, \ldots, n)$. Then $B$ is a normal $p$-subgroup of $A$.*

*Proof.* It is easy to see that $B \trianglelefteq A$. We use induction on $n$ to show that $B$ is a $p$-group. This is obvious if $n = 0$ or $n = 1$. Suppose that $n \geq 2$. Let $C$ be the subgroup of $B$ consisting of the elements of $B$ that fix every element of $P_1$. Then $C \trianglelefteq B$. By considering $B/C$ as a group of automorphisms of $P_1$, we see that $B/C$ is a $p$-group, by induction. By Lemma 6.5, $C$ is a $p$-group. Therefore, $B$ is a $p$-group.

LEMMA 10.5. *Suppose that $P$ is a finite $p$-group and $Q \subseteq P$. Then $Q$ is a subnormal subgroup of $P$.*

*Proof.* We use induction on $[P : Q]$. We may assume that $[P : Q] > 1$. By Corollary 2.6, $Q \subset \mathbf{N}_P(Q)$. By induction, $\mathbf{N}_P(Q)$ is subnormal in $P$.

From the definition of a subnormal subgroup it easily follows that a normal subgroup of a subnormal subgroup of $P$ is also a subnormal subgroup of $P$.

DEFINITION. Let $X$ be a group. Then $X$ is *involved* in $G$ if $X$ is isomorphic to a section of $G$.

LEMMA 10.6. *Let $X$ be a group. Suppose that $X$ is involved in a section $H/K$ of $G$ for which $\mathbf{O}_p(H/K) \neq 1$. Then there exists a subgroup $L$ of $H$ for which $X$ is involved in $L$ and $\mathbf{O}_p(L) \neq 1$.*

*Proof.* Let $\mathbf{O}_p(H/K) = M/K$, and let $P$ be a Sylow $p$-subgroup of $M$. Then $P \neq 1$. Define $L = \mathbf{N}_H(P)$. By Corollary 2.10, $M = KP$. By the Frattini argument,

$$H = M\mathbf{N}_H(P) = ML = KPL = KL.$$

Hence $L/(L \cap K) \cong LK/K = H/K$.

## 11. Some Counterexamples

DEFINITION. Suppose that $e$ is a positive integer. The finite field of $p^e$ elements is denoted by $GF(p^e)$.

DEFINITION. Suppose that $e$ and $n$ are positive integers. Let $GL(n, p^e)$ be the *general linear group* of degree $n$ over $GF(p^e)$; that is, the group of all non-singular matrices of degree $n$ over $GF(p^e)$. Let $SL(n, p^e)$ be the *special linear group* of degree $n$ over $GF(p^e)$; that is, the group of all elements of determinant one in $GL(n, p^e)$.

Let $Qd(p)$ be the subgroup of $SL(3, p)$ consisting of all matrices of the form

$$\begin{bmatrix} a & b & c \\ d & e & f \\ 0 & 0 & 1 \end{bmatrix} \quad (ae - bd = 1).$$

Let $F(p)$ be the subgroup of $Qd(p)$ consisting of all matrices of the form

$$\begin{bmatrix} a & b & c \\ 0 & a^{-1} & d \\ 0 & 0 & 1 \end{bmatrix} \quad (a \neq 0).$$

Note that $|Qd(p)| = p^3(p^2-1)$, $|F(p)| = p^3(p-1)$. Moreover, $\mathbf{O}_p(F(p))$ is a Sylow $p$-subgroup of $Qd(p)$, and $F(p)$ is the normalizer of $\mathbf{O}_p(F(p))$ in $Qd(p)$.

LEMMA 11.1. *Assume that $|S| > |\mathbf{Z}(S)| = p$ and that $S$ is generated by some conjugates of $\mathbf{Z}(S)$ in $G$.*

(a) *Suppose that $1 \subset T \trianglelefteq S$. Then there exists $U \subseteq S$ such that $\mathbf{Z}(S)$ and $U$ are conjugate in $G$ but not in $\mathbf{N}(T)$.*

(b) *No conjugacy functor on $G$ controls strong fusion in $G$.*

*Proof.* Clearly, (b) follows from (a). We prove (a). By Theorem 2.1, $T \cap \mathbf{Z}(S) \neq 1$. Hence $T \supseteq \mathbf{Z}(S)$. Let $U_1, \ldots, U_n$ be conjugates of $\mathbf{Z}(S)$ that generate $S$. If $T \subset S$, then some $U_i$ is not contained in $T$ and is therefore not conjugate to $\mathbf{Z}(S)$ in $\mathrm{N}(T)$. Suppose that $T = S$. Take $U_i$ such that $U_i \neq \mathbf{Z}(S)$. Then $\mathbf{Z}(S) \trianglelefteq \mathrm{N}(T)$, so $\mathbf{Z}(S)$ is not conjugate to $U_i$ in $\mathrm{N}(T)$.

EXAMPLE 11.2. Suppose that $G = GL(3, p)$. We may assume that $S$ is the group of all matrices of the form

$$\sigma(a, b, c) = \begin{bmatrix} 1 & a & b \\ 0 & 1 & c \\ 0 & 0 & 1 \end{bmatrix},$$

for $a, b, c \in GF(p)$. Calculations show that

$$\mathbf{Z}(S) = \langle \sigma(0, 1, 0) \rangle \quad \text{and} \quad S = \langle \sigma(1, 0, 0), \sigma(0, 0, 1) \rangle.$$

Moreover, $\sigma(0, 1, 0)$, $\sigma(1, 0, 0)$, and $\sigma(0, 0, 1)$ are conjugate in $G$, by linear algebra (Herstein, 1964; pp. 248–255). Hence, by Lemma 11.1, no conjugacy functor on $G$ controls strong fusion in $G$.

EXAMPLE 11.3. Let $H = SL(2, 17)$. Suppose that $G = H/\mathbf{Z}(H)$. (Then $G$ is usually denoted by $PSL(2, 17)$.) Assume that $p = 2$. Then (Section II.8, Huppert, 1967).

(a) $S$ is a dihedral group of order 16;
(b) $S = \mathrm{N}(T)$ whenever $1 \subset T \trianglelefteq S$;
(c) any two elements of order two in $G$ are conjugate; and
(d) $G$ is a simple group.

By (a), $|\mathbf{Z}(S)| = 2$. Let $x \in \mathbf{Z}(S) - \{1\}$. By (a) and (c), $S$ is generated by two conjugates of $x$. Consequently, for any conjugacy functor $\mathbf{W}$ on $G$, we have

(i) $\mathbf{W}$ does not control transfer in $G$;
(ii) $\mathbf{W}$ does not control weak closure of elements in $G$; and
(iii) $\mathrm{N}(\mathbf{W}(S))$ has a normal 2-complement, but $G$ has no normal 2-complement.

EXAMPLE 11.4. Suppose that $G = Qd(p)$ and $S = \mathbf{O}_p(F(p))$. Then $|S| = p^3$. If $p = 2$, $S$ is a dihedral group; if $p$ is odd, $S$ is a non-abelian group and every non-identity element of $S$ has order $p$. These conditions determine $S$ up to isomorphism (Section 4.4, Hall, 1959). Moreover, the only characteristic subgroups of $S$ are 1, $\mathbf{Z}(S)$, $S$, and, if $p = 2$, the unique cyclic subgroup of order four in $S$.

Let $H = \mathbf{O}_p(G)$. Then $H$ is an elementary abelian subgroup of order $p^2$, and

$$G/H \cong SL(2, p) \quad \text{and} \quad \mathrm{C}(H) = H.$$

Therefore, $S/H \nsubseteq \mathbf{O}_p(G/H)$. Since $[H, S, S] = 1$, $G$ is not $p$-stable.

Suppose **W** is a conjugacy functor on $G$ that controls strong fusion in $G$. By the remark after Theorem 6.6, $\mathbf{W}(S) \trianglelefteq G$. Since $\mathbf{Z}(S)$ is not normal in $G$, $\mathbf{W}(S)$ cannot be a characteristic subgroup of $S$.

EXAMPLE 11.5. Suppose that $p = 2$. Let $q$ be an arbitrary odd prime, and let $r = 2^{q-1}$. Then $r \equiv 1$ (modulo $q$). Hence $GF(r)$ contains a nonzero element $\theta$ of multiplicative order $q$. Let $x$ and $y$ be the elements of $SL(3, r)$ given by

$$x = \begin{bmatrix} 0 & 1 & 0 \\ 1 & 0 & 0 \\ 0 & 0 & 1 \end{bmatrix}, \quad y = \begin{bmatrix} \theta & 0 & 0 \\ 0 & \theta^{-1} & 0 \\ 0 & 0 & 1 \end{bmatrix}.$$

Let $H$ be the subgroup of $SL(3, r)$ consisting of all elements of the form

$$\begin{bmatrix} 1 & 0 & a \\ 0 & 1 & b \\ 0 & 0 & 1 \end{bmatrix}, \quad (a, b \in GF(r)).$$

Suppose that $G = \langle x, y, H \rangle$. Calculations show that $H$ is an elementary abelian normal 2-subgroup of $G$ and that

$$x^2 = y^q = 1, \qquad x^{-1}yx = y^{-1}, \qquad \text{and } \langle x, y \rangle \cap H = 1.$$

Therefore, $\langle x, y \rangle$ is a dihedral group of order $2q$ and $\mathbf{C}(H) = H = \mathbf{O}_2(G)$. For every $z \in H$,

$$[z, x, x] = [z^{-1}, z^x, x] = [zz^x, x] = zz^x z^x z^{x^2} = (zz^x)^2 = 1.$$

Hence $[H, x, x] = 1$. However, $\mathbf{O}_2(G/\mathbf{C}(H)) = \mathbf{O}_2(G/H) = 1$. Since $x \notin H$, $G$ is not 2-stable.

Let **W** be a conjugacy functor on $G$. Suppose that $\mathbf{W}(S) = S$ or $\mathbf{W}(S) = \mathbf{Z}(S)$. Then $\mathbf{W}(S) \ntrianglelefteq G$. Since $[G : S]$ is a prime, $\mathbf{N}(\mathbf{W}(S)) = S$. It is easy to see that $S' \subset H = S \cap G'$ and that $H \cap \mathbf{Z}(G) = 1$. Therefore, **W** satisfies conditions (i), (ii), and (iii) of Example 11.3.

## 12. The K-subgroups

Section 11 illustrated some cases in which conjugacy functors are inadequate. In this section and the following section, we will prove the existence of a section conjugacy functor on $G$ that enjoys the following properties:

(i) $\mathbf{W}(P)$ is a characteristic subgroup of $P$ and $\mathbf{W}(P) \supseteq \mathbf{Z}(P)$, for every section $P$ of $G$ that is a $p$-group;
(ii) if $p \geq 5$, then **W** controls transfer in $G$;
(iii) if $p$ is odd, then **W** controls weak closure of elements in $G$; and
(iv) if $p$ is odd and every section of $G$ is $p$-stable, then **W** controls strong fusion in $G$.

Because of the length of the definitions and proofs, we will defer them to Section 13. There, for every finite $p$-group $P$ we will define two characteristic subgroups $\mathbf{K}_\infty(P)$ and $\mathbf{K}^\infty(P)$. Both of these subgroups will contain $\mathbf{Z}(P)$ (by Proposition 12.1, below) and will therefore be non-trivial when $P$ is not trivial. To prove the above statements, we will consider $\mathbf{K}_\infty$ to be a function on $\mathscr{C}_p^*(G)$ by restriction. By the remark following the definition of a section conjugacy functor, it will be obvious that $\mathbf{K}_\infty$ is a section conjugacy functor on $G$. By the results below, $\mathbf{K}_\infty$ has the properties (i)–(iv). It will also be clear that $\mathbf{K}^\infty$ has these properties as well.

In order to define the groups $\mathbf{K}_\infty(P)$ and $\mathbf{K}^\infty(P)$, we will define characteristic subgroups $\mathbf{K}_i(P)$ $(i = -1, 0, 1, 2, \ldots)$. The following two results will be proved in Section 13.

PROPOSITION 12.1. *Let $P$ be a finite $p$-group. Then*
(a) $\mathbf{K}_\infty(P) \supseteq \mathbf{C}_P(\mathbf{K}_\infty(P)) \supseteq \mathbf{Z}(P)$ *and* $\mathbf{K}^\infty(P) \supseteq \mathbf{C}_P(\mathbf{K}^\infty(P)) \supseteq \mathbf{Z}(P)$; *and*
(b) *if* $Q \subseteq P$ *and* $Q$ *contains* $\mathbf{K}_i(P)$ *for every non-negative* $i$, *then* $\mathbf{K}_\infty(Q) = \mathbf{K}_\infty(P)$ *and* $\mathbf{K}^\infty(Q) = \mathbf{K}^\infty(P)$.

THEOREM 12.2. *Let $T = \mathbf{O}_p(G)$ and let $A \in \mathscr{Q}(T)$. Suppose that $\mathbf{K}_\infty(S)$ or $\mathbf{K}^\infty(S)$ is not a normal subgroup of $G$. Then there exists an element $g$ of $S - T$ that has the following properties:*

(a) $[A, g; 3] \subseteq \mathbf{Z}(A)$;
(b) $[Z(A), g, g] = 1$; *and*
(c) *if* $\mathbf{C}(T) \subseteq T$, *then there exists a chief factor $X/Y$ of $G$ for which* $X \subseteq A$ *and* $[X, g, g] \subseteq Y$ *and* $[X, g] \nsubseteq Y$.

Proposition 12.1 is a useful result about the $K$-subgroups. Theorem 12.2 is the main theorem of this section. We use it to obtain Theorem 12.3, from which the other results of this section follow.

THEOREM 12.3. *Let $T = \mathbf{O}_p(G)$. Suppose that $\mathbf{K}_\infty(S)$ or $\mathbf{K}^\infty(S)$ is not a normal sub-group of $G$. Then there exists an element $g$ of $S - T$ that has the following properties:*

(a) $[X, g; 4] \subseteq Y$ *for every chief factor $X/Y$ of $G$ such that* $X \subseteq T$;
(b) $[\mathbf{Z}(T), g, g] = 1$; *and*
(c) *if* $\mathbf{C}(T) = T$, *then some factor group of $G$ is not $p$-stable.*

*Proof.* By Theorem 10.1, there exists $A \in \mathscr{Q}(T)$. We obtain (a) from Theorems 10.2 and 12.2. Since $\mathbf{Z}(T) \subseteq \mathbf{C}_T(A) = \mathbf{Z}(A)$, Theorem 12.2 yields that

$$[\mathbf{Z}(T), g, g] \subseteq [\mathbf{Z}(A), g, g] = 1.$$

Finally, (c) follows from Proposition 7.6 and Theorem 12.2.

For some of the following results, we regard $\mathbf{K}_\infty$ as a section conjugacy functor on $G$. Note that all of the results below about $\mathbf{K}_\infty$ are valid for $\mathbf{K}^\infty$ as well.

Theorems 7.4 and 12.3 yield

THEOREM 12.4. *If $p \geq 5$, then $\mathbf{K}_\infty$ controls transfer in $G$.*

Theorem 12.4 and Proposition 6.9 yield

COROLLARY 12.5. *If $p \geq 5$, $S \neq 1$, and $\mathbf{N}(S)/\mathbf{C}(S)$ is a $p$-group, then $G$ has a factor group of order $p$.*

For the following result we require a theorem of Burnside that asserts that all groups of order $p^a q^b$ are solvable (for $a, b \geq 0$ and distinct primes $p$, $q$). This result is proved by character theory (Hall, 1959). Part (b) of the following result proves a conjecture in Huppert (1967). After (b) was presented at the Conference, Professor B. Klaiber raised the question of whether (a) was valid.

COROLLARY 12.6. *Suppose that $\mathbf{N}(Q) = Q$ for every prime divisor $q$ of $|G|$ and every Sylow $q$-subgroup $Q$ of $G$. Then*

(a) *$G$ is a $q$-group for some prime $q$, and*

(b) *$G$ is not a non-abelian simple group.*

*Proof.* We may assume that $G \neq 1$. We claim that $G$ has a normal subgroup $H$ of prime index. If $|G| = 2^a 3^b$ for some $a, b \geq 0$, then $G$ is solvable by the result of Burnside, and $H$ exists. Otherwise, $H$ exists by Corollary 12.5.

Let $q = |G/H|$. Consider a prime $r$ other than $q$ and a Sylow $r$-subgroup $R$ of $G$. Then $R \subseteq H$. By the Frattini argument, $G = H\mathbf{N}(R)$. Thus $\mathbf{N}(R) \not\subseteq R$. Hence $r$ does not divide $|G|$. Since $r$ is an arbitrary prime other than $q$, $G$ is a $q$-group. Consequently, $G$ is not a non-abelian simple group.

Propositions 12.1 and Theorems 7.5 and 12.3 yield

THEOREM 12.7. *If $p$ is odd, then $\mathbf{K}_\infty$ controls weak closure of elements in $G$.*

Proposition 12.1 and Theorems 7.9 and 12.3 yield

THEOREM 12.8. *Let $Z = \mathbf{Z}(S)^p$. Suppose that $A$ is an element of $Z \cap \mathbf{Z}(\mathbf{N}(\mathbf{K}_\infty(S)))$, or $A$ is a normal subgroup of $\mathbf{N}(\mathbf{K}_\infty(S))$ contained in $Z$. Then $A$ is weakly closed in $S$ with respect to $G$.*

Theorems 6.6. and 12.3 yield

THEOREM 12.9. *If $p$ is odd and every section of $G$ is $p$-stable, then $\mathbf{K}_\infty$ controls strong fusion in $G$.*

Theorems 8.9 and 12.9, Lemma 8.10, and Proposition 12.1 yield

THEOREM 12.10. *Suppose that p is odd. Assume that*

(a) $\mathbf{N}(\mathbf{K}_\infty(S))$ *has a normal p-complement, or*
(b) $\mathbf{N}(\mathbf{K}_\infty(S))/\mathbf{C}(\mathbf{K}_\infty(S))$ *is a p-group.*

*Then G has a normal p-complement.*

The following result follows directly from the definitions of the sets $\mathscr{K}_i(P)$ $(i \geq -1)$ in Section 13:

LEMMA 12.11. *Suppose that P is a finite p-group.*

(a) *If A is a normal abelian subgroup of P and* $i \geq -1$, *then* $A \in \mathscr{K}_i(P)$.
(b) *If P is generated by its normal abelian subgroups, then* $\mathbf{K}_\infty(P) = \mathbf{K}^\infty(P) = P$.

## 13. Definitions and Proofs for Section 12

In this section, we define the groups $\mathbf{K}_\infty(P)$ and $\mathbf{K}^\infty(P)$ and prove Proposition 12.1 and Theorem 12.2.

DEFINITION. Suppose that $P$ is a finite $p$-group. Let $\mathscr{B}(P)$ be the set of all subgroups $Q$ of $P$ for which $Q' \subseteq \mathbf{Z}(Q)$.

*Definitions.* Suppose that $P$ is a finite $p$-group and $Q \subseteq P$.
Let $\mathscr{K}^*(P; Q)$ be the set of all $B \in \mathscr{B}(P)$ for which $[B, Q] \subseteq \mathbf{Z}(B)$.
Let $\mathscr{K}_*(P; Q)$ be the set of all $B \in \mathscr{B}(P)$ that satisfy the following conditions:

(i) $Q \subseteq \mathbf{N}(B)$, and
(ii) whenever $A \in \mathscr{B}(P)$, $A \subseteq Q$, $B \subseteq \mathbf{N}(A)$, and $[B, A'] = [\mathbf{Z}(B), A, A] = 1$, then $[B,A] \subseteq \mathbf{Z}(B)$.

*Definitions.* Suppose $P$ is a finite $p$-group. Define inductively subsets $\mathscr{K}_i(P)$ of $\mathscr{B}(P)$ and subgroups $\mathbf{K}_i(P)$ for $i = 1, 0, 1, 2, \ldots$ as follows:

(a) $\mathscr{K}_{-1}(P) = \mathscr{B}(P)$;
(b) $\mathscr{K}_i(P) = \mathscr{K}_*(P; \mathbf{K}_{i-1}(P))$ if $i$ is even and $i \geq 0$, and $\mathscr{K}_i(P) = \mathscr{K}^*(P; \mathbf{K}_{i-1}(P))$ if $i$ is odd and $i \geq -1$; and
(c) $\mathbf{K}_i(P) = \langle \mathscr{K}_i(P) \rangle$ for every $i \geq -1$.

*Definitions.* Let $\mathbf{K}_\infty(P)$ be the subgroup of $G$ generated by the groups $\mathbf{K}_i(P)$ for even values of $i$. Let $\mathbf{K}^\infty(P)$ be the intersection of the groups $\mathbf{K}_i(P)$ for all odd $i$.

Note that $\mathbf{K}_{-1}(P) = P$ and that $\mathbf{K}_{i-1}(P) \subseteq \mathbf{N}(B)$ for all $i \geq 0$ and all $B \in \mathscr{K}_i(P)$.

Proposition 12.1 is included in the following result.

LEMMA 13.1. *Suppose that $P$ is a finite $p$-group.*

(a) *If $R \subseteq Q \subseteq P$, then $\mathscr{K}_*(P; Q) \subseteq \mathscr{K}_*(P; R)$ and $\mathscr{K}^*(P; Q) \subseteq \mathscr{K}^*(P; R)$.*

(b) *We have*

$$\mathbf{K}_{-1}(P) \supseteq \mathbf{K}_1(P) \supseteq \mathbf{K}_3(P) \supseteq \ldots, \text{ and } \mathbf{K}_0(P) \subseteq \mathbf{K}_2(P) \subseteq \mathbf{K}_4(P) \subseteq \ldots.$$

(c) *For some $n \geq 0$,*

$$\mathbf{K}_i(P) = \mathbf{K}_\infty(P) \quad \text{for all even } i \geq n, \text{ and}$$
$$\mathbf{K}_i(P) = \mathbf{K}^\infty(P) \quad \text{for all odd } i \geq n.$$

(d) *Suppose that $Q \subseteq P$ and $Q$ contains $\mathbf{K}_i(P)$ for all $i \geq 0$. Then $\mathbf{K}_\infty(Q) = \mathbf{K}_\infty(P)$ and $\mathbf{K}^\infty(Q) = \mathbf{K}^\infty(P)$.*

(e) *Suppose $i \geq -1$ and $B$ is a maximal element of $\mathscr{K}_i(P)$ under inclusion. Then $B \supseteq \mathbf{C}_P(B)$.*

(f) *We have $\mathbf{K}_\infty(P) \supseteq \mathbf{C}_P(\mathbf{K}_\infty(P)) \supseteq \mathbf{Z}(P)$ and $\mathbf{K}^\infty(P) \supseteq \mathbf{C}_P(\mathbf{K}^\infty(P)) \supseteq \mathbf{Z}(P)$.*

*Proof.* Part (a) follows from the definitions. We prove parts (b), (c), and (d) together. Suppose that $Q \subseteq P$ and $Q$ contains $\mathbf{K}_i(P)$ for all $i \geq 0$. We claim that

$$\mathbf{K}_i(P) \supseteq \mathbf{K}_i(Q) \supseteq \mathbf{K}_{i+2}(P) \quad \text{for all odd } i \geq -1, \text{ and} \tag{13.1}$$
$$\mathbf{K}_i(P) \subseteq \mathbf{K}_i(Q) \subseteq \mathbf{K}_{i+2}(P) \quad \text{for all even } i \geq 0. \tag{13.2}$$

We will prove (13.1) and (13.2) by induction on $i$.

Now, (13.1) is obvious for $i = -1$. We give the argument from $i$ to $i+1$ for odd $i$; the proof for even $i$ is similar. From (a) and from (13.1) for a particular odd $i$, we obtain

$$\mathscr{K}_{i+1}(Q) = \mathscr{K}_*(Q; \mathbf{K}_i(Q)) \subseteq \mathscr{K}_*(P; \mathbf{K}_i(Q)) \subseteq \mathscr{K}_*(P; \mathbf{K}_{i+2}(P)) = \mathscr{K}_{i+3}(P) \tag{13.3}$$

and

$$\mathscr{K}_{i+1}(P) = \mathscr{K}_*(P; \mathbf{K}_i(P)) \subseteq \mathscr{K}_*(P; \mathbf{K}_i(Q)). \tag{13.4}$$

If $A \in \mathscr{K}_{i+1}(P)$, then $A \subseteq \mathbf{K}_{i+1}(P) \subseteq Q$. Thus (13.4) yields

$$\mathscr{K}_{i+1}(P) \subseteq \mathscr{K}_*(P; \mathbf{K}_i(Q)) \cap \mathscr{B}(Q) = \mathscr{K}_*(Q; \mathbf{K}_i(Q)) = \mathscr{K}_{i+1}(Q). \tag{13.5}$$

From (13.3) and (13.5), we obtain (13.2) with $i$ replaced by $i+1$ (for our particular choice of $i$).

From (13.1) and (13.2),

$$\mathbf{K}_{-1}(P) \supseteq \mathbf{K}_{-1}(Q) \supseteq \mathbf{K}_1(P) \supseteq \mathbf{K}_1(Q) \supseteq \ldots,$$

and

$$\mathbf{K}_0(P) \subseteq \mathbf{K}_0(Q) \subseteq \mathbf{K}_2(P) \subseteq \mathbf{K}_2(Q) \subseteq \ldots.$$

By considering the case where $Q = P$, we obtain (b) and (c). Returning to the general case, we obtain (d).

Suppose $i \geq -1$ and $B \in \mathscr{K}_i(P)$. Let $C = \mathbf{C}_P(B)$ and $N = \mathbf{N}_P(B)$. Then $C \cap B = \mathbf{Z}(B)$. Suppose that $C \nsubseteq B$. Now, $1 \subset C/\mathbf{Z}(B) \trianglelefteq N/\mathbf{Z}(B)$. Therefore, $C/\mathbf{Z}(B)$ contains a subgroup $C_0$ of order $p$ in the centre of $N/\mathbf{Z}(B)$. Let $B^* = BC_0$. Then $B \subset B^*$. We claim that $B^* \in \mathscr{K}_i(P)$.

Take $x \in C_0 - B$. Then $C_0 = \langle x, \mathbf{Z}(B)\rangle$ and $B^* = \langle x, B\rangle$, and $x \in \mathbf{Z}(B^*)$. Hence

$$(B^*)' = B' \subseteq \mathbf{Z}(B) \subseteq \mathbf{Z}(B^*) = C_0,$$

and $B^* \in \mathscr{B}(P)$. If $i = -1$, then $B^* \in \mathscr{K}_i(P)$. Assume that $i \geq 0$. Suppose $i$ is odd. Then $\mathbf{K}_{i-1}(P) \subseteq N \subseteq \mathbf{N}_P(C_0)$, and

$$[B^*, \mathbf{K}_{i-1}(P)] \subseteq [B, \mathbf{K}_{i-1}(P)]C_0 \subseteq \mathbf{Z}(B)C_0 = C_0 = \mathbf{Z}(B^*).$$

Thus $B^* \in \mathscr{K}_i(P)$.

Assume that $i$ is even. Then $\mathbf{K}_{i-1}(P) \subseteq N \subseteq \mathbf{N}_P(B^*)$. Suppose $A \in \mathscr{B}(P)$, $A \subseteq \mathbf{K}_{i-1}(P)$, $B \subseteq \mathbf{N}(A^*)$, and

$$[B^*, A'] = [Z(B^*), A, A] = 1.$$

Since $B \subseteq B^*$ and $\mathbf{Z}(B) \subseteq \mathbf{Z}(B^*)$, we obtain $[B, A] \subseteq \mathbf{Z}(B)$ because $B \in \mathscr{K}_i(P)$. As in the previous paragraph, we have $[B^*, A] \subseteq Z(B^*)$. Hence $B^* \in \mathscr{K}_i(P)$. Thus we obtain (e). Part (f) follows from (c) and (e).

LEMMA 13.2. *Let $T = \mathbf{O}_p(G)$ and let $A \in \mathscr{Q}(T)$. Suppose that $1 = H_0 \subset H_1 \subset \ldots \subset H_n = A$ is a series of normal subgroups of $G$. Assume that, for $i = 1, \ldots, n$, there is no normal subgroup $N$ of $G$ such that $H_{i-1} \subset N \subset H_i$. If $g \in S - T$ and $[H_i, g, g] \subseteq H_{i-1}$ for $i = 1, \ldots, n$, then $g$ satisfies condition* (c) *of Theorem* 12.2.

*Proof.* By the definition of a chief factor, $H_i/H_{i-1}$ is a chief factor of $G$ for $i = 1, \ldots, n$. Suppose $\mathbf{C}(T) \subseteq T$. Let $K$ be the set of all $x \in G$ such that $[H_i, x] \subseteq H_{i-1}$ for $i = 1, \ldots, n$. By Lemma 10.4, $K \trianglelefteq G$ and $K/\mathbf{C}_K(A)$ is a $p$-group. By the definition of $\mathscr{Q}(T)$, $\mathbf{C}_K(A)/\mathbf{C}_K(T)$ is a $p$-group. Hence $K/\mathbf{C}_K(T)$ is a $p$-group. Since $\mathbf{C}_K(T) \subseteq \mathbf{C}(T) \subseteq T$, $K$ is a $p$-group. Therefore, $K \subseteq \mathbf{O}_p(G) = T$. Thus, if $g \in S - T$, then $g \notin K$. This completes the proof of Lemma 13.2.

LEMMA 13.3. *Let $T = \mathbf{O}_p(G)$ and let $A \in \mathscr{Q}(T)$. Suppose $g \in S - T$, $A_1 \subseteq A$, and $A_1 \trianglelefteq G$. Assume that $[A, g, g] \subseteq A_1$ and $[A_1, g, g] = 1$. Then $g$ satisfies condition* (c) *of Theorem* 12.2.

*Proof.* If $A = 1$, then $T \subseteq \mathbf{C}_T(A) = \mathbf{Z}(A) = 1$. Assume that $A \neq 1$ and suppose that $1 \subset A_1 \subset A$. Expand the series $1 \subset A_1 \subset A$ to a series that satisfies the hypothesis of Lemma 13.2 and apply Lemma 13.2. If $A_1 = 1$ or $A_1 = A$, consider the series $1 \subset A$ and argue similarly.

We may now prove Theorem 12.2. We first restate it.

THEOREM 12.2. *Let* $T = \mathbf{O}_p(G)$ *and let* $A \in \mathscr{Q}(T)$. *Suppose that* $\mathbf{K}_\infty(S)$ *or* $\mathbf{K}^\infty(S)$ *is not a normal subgroup of G. Then there exists an element g of* $S-T$ *that has the following properties:*

(a) $[A, g; 3] \subseteq \mathbf{Z}(A)$;

(b) $[Z(A), g, g] = 1$; *and*

(c) *if* $\mathbf{C}(T) \subseteq T$, *then there exists a chief factor* $X/Y$ *of G for which* $X \subseteq A$ *and* $[X, g, g] \subseteq Y$ *and* $[X, g] \nsubseteq Y$.

*Proof of Theorem* 12.2: By Proposition 12.1, $\mathbf{K}_i(S) \nsubseteq T$ for some nonnegative $i$. Take $i$ minimal with respect to this property. We first assume that $A \notin \mathscr{K}_{i-1}(S)$. Then $i > 0$. If $i$ is even, then $i-2 \geq 0$ and $\mathbf{K}_{i-2}(S) \subseteq T$. But then $[\mathbf{K}_{i-2}(S), A] \subseteq \mathbf{Z}(A)$, contrary to the assumption that $A \notin \mathscr{K}_{i-1}(S)$. Thus $i$ is odd. Obviously, $\mathbf{K}_{i-2}(S) \subseteq G = \mathbf{N}(A)$. Hence there exists $A^* \subseteq \mathbf{K}_{i-2}(S)$ such that

$$A \subseteq \mathbf{N}(A^*), \tag{13.6}$$

$$[A, (A^*)'] = [\mathbf{Z}(A), A^*, A^*] = 1, \tag{13.7}$$

and

$$[A, A^*] \nsubseteq \mathbf{Z}(A). \tag{13.8}$$

By (13.8), there exists $g \in A^* - T$. By (13.6) and (13.7), $g$ satisfies (b) and

$$[A, g, g] \subseteq A \cap [A^*, g] \subseteq A \cap (A^*)' \subseteq \mathbf{Z}(A).$$

This proves (a). We obtain (c) by taking $A_1 = \mathbf{Z}(A)$ in Lemma 13.3.

Thus we may assume that $A \in \mathscr{K}_{i-1}(S)$. Take $B \in \mathscr{K}_i(S)$ such that $B \nsubseteq T$. Take $g \in B-T$. Suppose first that $i$ is odd. Then

$$[B, A] \subseteq [B, \mathbf{K}_{i-1}(S)] \subseteq \mathbf{Z}(B),$$

and consequently $[A, B, B] = 1$. This proves (a) and (b). By putting $A_1 = A$ in Lemma 13.3, we obtain (c).

Now assume that $i$ is even. Then $A$ normalizes $B$. Hence

$$[[A, B], B, B] \subseteq [B, B, B] \subseteq [\mathbf{Z}(B), B] = 1,$$

which yields (a). Since $\mathbf{Z}(A)$ is abelian and is normal in $G$, $\mathbf{Z}(A) \in \mathscr{K}_{i-1}(S)$. Thus $\mathbf{Z}(A) \subseteq \mathbf{K}_{i-1}(S)$. As

$$(\mathbf{Z}(A))' = [G, \mathbf{Z}(A), \mathbf{Z}(A)] = 1,$$

we obtain $[\mathbf{Z}(A), B] \subseteq \mathbf{Z}(B)$, by the definition of $\mathscr{K}_i(S)$. So $[\mathbf{Z}(A), B, B] = 1$, which yields (b).

We must now verify (c). Assume first that $\mathbf{Z}(B) \nsubseteq T$. Then we may suppose that $g \in \mathbf{Z}(B)$. Since $A$ normalizes $B$, we have $[A, \mathbf{Z}(B), \mathbf{Z}(B)] \subseteq (\mathbf{Z}(B))' = 1$. By letting $A_1 = A$ in Lemma 13.3, we obtain (c).

Assume that $\mathbf{Z}(B) \subseteq T$. Since $[\mathbf{Z}(A), B, B] = 1$, we may assume that $[X, g] \subseteq Y$ for every chief factor $X/Y$ of $G$ such that $X \subseteq \mathbf{Z}(A)$. Define

$C = \mathbf{C}(\mathbf{Z}(A))$, and let $N$ be the normal subgroup of $G$ that contains $C$ and satisfies $N/C = \mathbf{O}_p(G/C)$. By Lemma 10.4, $g \in N$.

Let $W = A' \cap B'$ and let $\tilde{A}/W = \mathbf{Z}(A/W)$. Then $\tilde{A} \subseteq A$ and $B \subseteq \mathbf{N}(\tilde{A})$. Moreover,

$$[B, \tilde{A}'] \subseteq [B, W] = 1 \text{ and } [\mathbf{Z}(B), \tilde{A}, \tilde{A}] \subseteq [T, A, A] \subseteq [\mathbf{Z}(A), A] = 1.$$

By the definition of $\mathscr{K}_i(S)$, we have $[B, \tilde{A}] \subseteq \mathbf{Z}(B)$ and

$$[\tilde{A}, B, B] = 1. \tag{13.9}$$

Since $[A, B, B] \subseteq B' \cap A$ and $[B' \cap A, A] \subseteq B' \cap A' \subseteq W$,

$$[A, B, B] \subseteq \tilde{A}. \tag{13.10}$$

Let $L = \langle g, C \rangle$. Then $W \subseteq \mathbf{Z}(L)$, so $W \trianglelefteq L$ and $\tilde{A} \trianglelefteq L$. Suppose that $X/Y$ is a chief factor of $L$ and $X \subseteq A$. We claim that

$$[X, g, g] \subseteq Y. \tag{13.11}$$

Let $Z = (X \cap \tilde{A})Y$. Then $Y \subseteq Z \subseteq X$ and $Z \trianglelefteq L$. Therefore, $Z = Y$ or $Z = X$. If $Z = Y$, then (13.11) follows from (13.10). If $Z = X$, then (13.9) yields that

$$[X, g, g] = [(X \cap \tilde{A})Y, g, g] \subseteq [X \cap \tilde{A}, g, g]Y = Y,$$

as claimed.

Since $N/C$ is a $p$-group and $C \subseteq L \subseteq N$, $L$ is a subnormal subgroup of $N$, by Lemma 10.5. Hence $L$ is a subnormal subgroup of $G$. By Theorem 10.3, we obtain (13.11) if $X/Y$ is a chief factor of $G$ and $X \subseteq A$. Take a series of subgroups of $A$ as in Lemma 13.2; we obtain (c).

## 14. The Thompson Subgroup

*Definitions*. Suppose that $P$ is a finite $p$-group. Let $d(P)$ be the maximum of the orders of the abelian subgroups of $P$. Define $\mathscr{A}(P)$ to be the set of all abelian subgroups of order $d(P)$ in $P$, and let $\mathbf{J}(P)$ be the subgroup of $P$ generated by the elements of $\mathscr{A}(P)$. The group $\mathbf{J}(P)$ is called the *Thompson subgroup* of $P$.

We may regard $\mathbf{J}$ and $\mathbf{ZJ}$ (given by $\mathbf{ZJ}(P) = \mathbf{Z}(\mathbf{J}(P))$) as section conjugacy functors on $G$. Many of the results stated in Section 12 are analogues of propositions originally proved for $\mathbf{J}$ or for $\mathbf{ZJ}$. For example, the following results are proved in Gorenstein (1968, Sections 8.2 and 8.3):

Suppose that $p$ is odd. If $G$ is $p$-stable and $\mathbf{C}(\mathbf{O}_p(G)) \subseteq \mathbf{O}_p(G)$, then

$$\mathbf{Z}(\mathbf{J}(S)) \trianglelefteq G. \tag{14.1}$$

If $p$ is odd and $\mathbf{N}(\mathbf{Z}(\mathbf{J}(S)))$ has a normal $p$-complement, then $G$ has a normal $p$-complement. (14.2)

Professor Gorenstein (1968) has used (14.1) in his lectures. We show later that (14.2) follows from (14.1). In this section, we discuss some properties of $\mathbf{J}$ and $\mathbf{ZJ}$ for which analogues have not been proved for $\mathbf{K}_\infty$ and $\mathbf{K}^\infty$.

The following two results are proved in pp. 274–278 of Gorenstein (1968).

PROPOSITION 14.1. *Let P be a finite p-group.*

(a) *We have* $A = \mathbf{C}_P(A)$ *for all* $A \in \mathscr{A}(P)$ *and also* $\mathbf{Z}(P) \subseteq \mathbf{J}(P)$.

(b) *If* $Q \subseteq P$, $A \in \mathscr{A}(Q)$, *and* $A \subseteq Q$, *then* $A \in \mathscr{A}(P)$ *and* $A \subseteq \mathbf{J}(Q) \subseteq \mathbf{J}(P)$.

(c) *If* $Q \subseteq P$ *and* $\mathbf{J}(P) \subseteq Q$, *then* $\mathbf{J}(P) = \mathbf{J}(Q)$.

(d) *If* $A \in \mathscr{A}(P)$ *and B is an abelian subgroup of P normalized by A, then* $|A/\mathbf{C}_A(B)| \geq |B/\mathbf{C}_B(A)|$.

THEOREM 14.2 (Thompson). *Let P be a finite p-group. Suppose that B is an abelian subgroup of P and* $A \in \mathscr{A}(P)$. *Assume that A normalizes B and that B does not normalize A (e.g.,* $[B, A, A] \neq 1$). *Then there exists* $A^* \in \mathscr{A}(P)$ *such that*

$$A^* \cap B \supset A \cap B, \text{ and} \tag{14.3}$$

$$A^* \text{ normalizes } A \text{ (and thus } [A^*, A, A] = 1). \tag{14.4}$$

Proposition 14.1 follows directly from the definitions, except that (d) requires (a) and the inequality $|B\mathbf{C}_A(B)| \leq |A|$. Theorem 14.2 is called the Replacement Theorem. Its proof is based on an elementary but ingenious calculation, and it is the fundamental tool for dealing with $\mathbf{J}(P)$ and $\mathbf{ZJ}(P)$. We now give an example of its applications.

THEOREM 14.3 (Thompson). *Let* $T = \mathbf{O}_p(G)$. *Suppose that* $\mathbf{J}(S) \ntrianglelefteq G$. *Then there exists* $A \in \mathscr{A}(S)$ *such that* $A \nsubseteq T$ *and* $[\mathbf{Z}(T), A, A] = 1$.

*Proof.* Let $B = \mathbf{Z}(T)$. Since $\mathbf{J}(S) \ntrianglelefteq G$, $\mathbf{J}(S) \neq \mathbf{J}(T)$. By Proposition 14.1, $\mathbf{J}(S) \nsubseteq T$. So there exists $A \in \mathscr{A}(S)$ for which $A \nsubseteq T$. Of all such subgroups $A$, choose one for which $|A \cap B|$ is maximal. Denote it by $A$.

We claim that $[B, A, A] = 1$. Suppose otherwise. Take $A^* \in \mathscr{A}(S)$ as in the Replacement Theorem. Then $|A \cap B| < |A^* \cap B|$. By the maximal choice of $A$, we have $A^* \subseteq T$. Hence $A^*B$ is abelian. Since $|A^*B| \leq d(S) = |A^*|$, $A^*B = A^*$ and $B \subseteq A^*$. Thus

$$[B, A, A] \subseteq [A^*, A, A] = 1,$$

a contradiction.

Note that the conclusion of Theorem 14.3 is valid if $\mathbf{Z}(\mathbf{J}(S)) \ntrianglelefteq G$, because $\mathbf{Z}(\mathbf{J}(S))$ is a characteristic subgroup of $\mathbf{J}(S)$. By Proposition 14.1 and Theorems 7.5, 7.9, and 14.3, we obtain the following analogues of Theorems 12.7 and 12.8.

THEOREM 14.4. *If p is odd, then* $\mathbf{J}$ *controls weak closure of elements in G.*

THEOREM 14.5. *Let* $Z = \mathbf{Z}(S)^p$. *Suppose A is an element of* $Z \cap \mathbf{Z}(\mathbf{N}(\mathbf{J}(S)))$, *or A is a normal subgroup of* $\mathbf{N}(\mathbf{J}(S))$ *contained in Z. Then A is weakly closed in S with respect to G.*

Suppose $p$ is odd. An important theorem of Alperin, Gorenstein and Walter (Theorem 8.1.2, Gorenstein, 1968) asserts that $G$ is $p$-stable if $SL(2, p)$ is not involved in $G$. Because of Lemma 10.6 and the definition of $p$-stability, we may state their theorem as follows.

THEOREM 14.6. *Assume that p is odd. Suppose that, for every non-identity p-subgroup P of G, the group SL(2, p) is not involved in* $\mathbf{N}(P)$. *Then every section of G is p-stable.*

We have seen in Example 11.4 that $Qd(p)$ is not a $p$-stable group. The proof of Theorem 14.6 also yields the following result (Lemma 6.3, Glauberman, 1968a).

PROPOSITION 14.7. *Assume that p is odd. Then the following conditions on G are equivalent.*

(a) *The group Qd(p) is not involved in G.*
(b) *Every section of G is p-stable.*

By Theorems 6.6 and 12.9, Example 11.4, Proposition 14.7, and result (14.1), we obtain the following result.

THEOREM 14.8. *Assume that p is odd. Then the following conditions on G are equivalent.*

(a) *The group Qd(p) is not involved in G.*
(b) *The section conjugacy functor* $\mathbf{ZJ}$ *controls strong fusion in every section of G.*
(c) *The section conjugacy functor* $\mathbf{K}_\infty$ *controls strong fusion in every section of G.*

The next result concerns arbitrary section conjugacy functors.

THEOREM 14.9. *Let* $\mathbf{W}$ *be a section conjugacy functor on G. Suppose p is odd and* $\mathbf{N}(\mathbf{W}(S))$ *has a normal p-complement. Assume that* $\mathbf{W}$ *controls strong fusion in* $G^*$ *whenever* $G^*$ *is a section of G and every section of* $G^*$ *is p-stable. Then G has a normal p-complement.*

*Proof.* We will verify condition (8.2) of Theorem 8.7. Suppose that $G^*$ and $S^*$ satisfy the hypothesis of (8.2). Then $\mathbf{C}_{G^*}(\mathbf{O}_p(G^*) \subseteq \mathbf{O}_p(G^*)$. By Lemma 8.8, $[G^* : S^*]$ is divisible by a unique prime $q$, and $G^*$ has an abelian Sylow $q$-subgroup. Since $p$ is odd, $G^*$ has an abelian Sylow 2-subgroup. As $SL(2, p)$ has a non-abelian Sylow 2-subgroup (Section II.8, Huppert, 1967), $SL(2, p)$ is not involved in $G^*$. By Theorem 14.6, every section of $G^*$ is $p$-stable. Therefore, $\mathbf{W}$ controls strong fusion in $G^*$. By the remark after Theorem 6.6, $\mathbf{W}(S^*) \trianglelefteq G^*$. This proves (8.2). By Theorem 8.7, $G$ has a normal $p$-complement.

Now (14.2) follows from (14.1) and Theorem 14.9.

DEFINITION. Denote the symmetric group of degree four by $S^4$. We say that $G$ is *$S^4$-free* if $S^4$ is not involved in $G$.

Suppose that $p = 2$ and $G = Qd(p)$. Then $G \cong S^4$, and $G/\mathbf{O}_2(G)$ is a dihedral group of order 6. As mentioned above, $G$ is not 2-stable. Thus condition (a) of Proposition 14.7 follows from condition (b). However, the converse is false. Example 11.5 shows that for every odd prime $q > 3$ there exists a group $H$ for which $H/\mathbf{O}_2(H)$ is a dihedral group of order $2q$ and $H$ is not 2-stable. In fact, by Theorem 8.7, Proposition 7.6, and Lemma 10.4, it is easy to prove the following statement.

A necessary and sufficient condition for every section of $G$ to be 2-stable is that $G$ have a normal 2-complement. (14.5)

The contrast between (14.5) and Proposition 14.7 illustrates some of the difficulties in the case where $p = 2$. Here the condition of $p$-stability is too strong to be useful. Similarly, the condition, "$[X, g; p-1] \subseteq Y$" of (7.3) (in Theorem 7.5) states that $gY$ actually centralizes $X/Y$. Thus, with the conspicuous exception of Theorem 7.9, the results on commutator conditions discussed in Section 7 do not appear to be useful for the case $p = 2$. Because of this situation, we discuss two lines of thought relevant to this case.

1. As an analogue to the condition of $p$-stability for odd primes $p$, we will consider condition (a) of Proposition 14.7. For $p = 2$, this condition asserts that $G$ is $S^4$-free. Of course, this is violated in many classes of groups, including some in which every section is $p$-stable for every odd prime $p$. Even in these classes, however, information about $S^4$-free subgroups can help to pinpoint the subgroups that cause difficulties. Example 11.3 shows that some restrictions on $G$ are necessary if we wish to prove theorems for the prime 2 comparable to those we have proved for odd primes. No similar examples are known in groups that are $S^4$-free.

2. To prove a result about $p$-stable groups, we usually apply induction. Equivalently, we may assume that $G$ is a minimal counterexample to the theorem and then derive a contradiction by obtaining a commutator relation that shows that $G$ (or some section of $G$) is not $p$-stable. Usually, the existence of this commutator relation for a particular functor $\mathbf{W}$ depends mainly on the relation between $\mathbf{W}(S)$ and $S$ and thus on the internal properties of $p$-groups. Hence it is a "local" property of $G$. Suppose, however, that $p = 2$ and that $G$ is a minimal counterexample to some theorem. The condition of being $S^4$-free is not equivalent to a commutator condition or,

as far as we know, to any other "local" condition on $S$. Therefore, we will obtain a contradiction by investigating the "global" structure of $G$.

These two ideas were actually suggested by the proof of the following two results.

THEOREM 14.10. *Suppose that* $x \in S \cap \mathbf{Z}(\mathbf{N}(\mathbf{J}(S)))$. *Assume that*

(a) *$p$ is odd, or*
(b) $x \in (\mathbf{Z}(S))^p$, *or*
(c) *$G$ is $S^4$-free.*

*Then $x$ is weakly closed in $S$ with respect to $G$.*

THEOREM 14.11 (Thompson). *Suppose that $p$ is odd or $S^4$ is not involved in $G$. If* $\mathbf{C}(\mathbf{Z}(S))$ *and* $\mathbf{N}(\mathbf{J}(S))$ *both have normal $p$-complements, then $G$ has a normal $p$-complement.*

Theorem 14.10 is the main result of Glauberman (1969). Except for some details, it is proved by considering a minimal counterexample $G$ to the proposition that $\mathbf{J}$ controls weak closure of elements in all finite groups. In part of the proof the structure of $G/\mathbf{C}(\mathbf{Z}(\mathbf{O}_p(G)))$ is explicitly determined by identifying the group of automorphisms of $\mathbf{Z}(\mathbf{O}_p(G))$ induced by conjugation by elements of $G$. At this point, it is clear that $p = 2$. Condition (c) was chosen for the hypothesis of Theorem 14.10 because the only obvious common property of the counterexamples was the involvement of $S^4$.

We now see that this proof was not necessary for cases (a) and (b); they are proved by simpler means as Theorems 14.5 and 14.6 above. However, we know of no similar simplification of case (c).

Thompson's proof of Theorem 14.11 is remarkably short. It is given, for odd $p$ and for a slightly different definition of $\mathbf{J}$, in Section IV. 6 of Huppert (1967). The first part of the proof is incorporated in the proofs of Theorem 8.7 and Lemma 8.8. The last part of the proof in Huppert (1967) is a determination of $G/\mathbf{O}_p(G)$ in a minimal counterexample, and it is easy to obtain the result for $p = 2$ from this step.

Example 11.5 and Lemma 12.11 show that Theorem 14.11 and part (c) of Theorem 14.10 are false if we replace $\mathbf{J}(S)$ by $\mathbf{K}_\infty(S)$ or $\mathbf{Z}(\mathbf{K}_\infty(S))$.

Under the hypothesis of Theorem 14.11, we obtain $\mathbf{Z}(S) \subseteq \mathbf{Z}(\mathbf{N}(\mathbf{J}(S)))$. Therefore, Theorem 14.11 can be obtained from Lemma 4.5, Theorem 8.6, and Theorem 14.10. This is not surprising, since many of the concepts and techniques used in the proof of Theorem 14.10 and in the results of the previous sections were introduced by Thompson in the proofs of Theorem 14.11 and his earlier result on normal $p$-complements (Thompson, 1960a).

Theorem 14.10 is used in the proof of the following joint result of Thompson (1968) and the author:

THEOREM 14.12. *Suppose $Q$ and $R$ are subgroups of $S$, and $S = Q \times R$. Assume that*

(a) *$p$ is odd or $G$ is $S^4$-free, and*
(b) *whenever $R = R_1 \times R_2$ and $R_1 \neq 1$, then $R_1$ is not isomorphic to a subgroup of $Q$.*

*Then $S = Q \times R^*$ for some weakly closed subgroup $R^*$ of $S$ in $G$.*

Suppose that $p$ is odd. By result (14.2), **ZJ** controls strong fusion in $G$ if $\mathbf{N}(\mathbf{ZJ}(S))$ has a normal $p$-complement. By an extension of the methods used to prove Theorem 14.10, we may obtain weaker conditions on $\mathbf{N}(\mathbf{ZJ}(S))$ to guarantee that **ZJ** controls strong fusion in $G$.

THEOREM 14.13. *Let $H = \mathbf{N}(\mathbf{Z}(\mathbf{J}(S)))$. Suppose that $p$ is odd and* **ZJ** *does not control strong fusion in $G$. Then*

(a) *$F(p)$ is involved in $H$, and*
(b) *$\mathbf{Z}(\mathbf{J}(S))$ contains a cyclic subgroup $C$ of order $p$ such that $|\mathbf{N}_H(C)/\mathbf{C}_H(C)| = p-1$.*

Theorem 14.13 is proved by the author in work to be published. By Lemma 4.5 and Theorem 8.6, (14.2) follows from Theorem 14.13. The next result is also proved in the same work; however, it was obtained independently by M. J. Collins (1969), with a shorter proof.

THEOREM 14.14. *Suppose that $p$ is odd, $T \subseteq \mathbf{Z}(S)$, and $T \trianglelefteq \mathbf{N}(\mathbf{J}(S))$. Let $H = \mathbf{N}(S)$. If $T$ is not weakly closed in $S$ with respect to $G$, then $T$ contains a cyclic subgroup $C$ of order $p$ such that $|\mathbf{N}_H(C)/\mathbf{C}_H(C)| = p-1$.*

Note that Theorem 14.14 generalizes case (a) of Theorem 14.10, which is also Theorem 14.4.

## 15. Applications to Simple Groups

For the purposes of this section, we will say that $G$ is *simple* if $G$ is not abelian and if $G$ has no normal subgroup except the identity subgroup and $G$ itself. The preceding sections yield various classes of applications to simple groups. For example, suppose that $G$ is a simple group and $p$ divides $|G|$. Then $G$ has no normal $p$-complement and $G$ has no factor group of order $p$. Therefore, our results on normal $p$-complements and on control of transfer yield a number of restrictions on $G$. Among them are the following.

If $p \geq 5$, then $\mathbf{N}(S)/\mathbf{C}(S)$ is not a $p$-group and $\mathbf{S} \subseteq (\mathbf{N}(\mathbf{K}_\infty(S)))'$. (Corollary 12.5 and Theorem 12.4). (15.1)

If $p = 3$, $\mathbf{N}(\mathbf{K}_\infty(S))/\mathbf{C}(\mathbf{K}_\infty(S))$ is not a $p$-group (Theorem 12.10). (15.2)

If $p = 2$ and $G$ is $S^4$-free, then $\mathbf{C}(\mathbf{Z}(S))$ or $\mathbf{N}(\mathbf{J}(S))$ does not have a normal $p$-complement (Theorem 14.11). (15.3)

Let $\mathbf{W}$ be a conjugacy functor on $G$. If $\mathbf{W}$ controls strong fusion in $G$, then $\mathbf{W}$ controls transfer in $G$, by Lemma 4.6. In this case, we obtain the restriction that

$$S \subseteq (\mathbf{N}(\mathbf{W}(S)))' \quad \text{if } G \text{ is simple.} \tag{15.4}$$

For this reason, and for its own sake, it is useful to have sufficient conditions for $\mathbf{W}$ to control strong fusion in $G$. Theorems 14.8 and 14.13 give us two such conditions for the case where $\mathbf{W}$ is $\mathbf{ZJ}$. As a corollary of Theorem 14.8, we obtain

COROLLARY 15.1. *Assume that $p$ is odd and that $G$ satisfies one of the following conditions:*

(a) *$G$ has an abelian or dihedral Sylow 2-subgroup; or*
(b) *$p \geq 5$, and $\mathbf{N}(P)$ is solvable for every non-identity $p$-subgroup $p$ of $G$.*

*Then every section of $G$ is $p$-stable.*

The derivation of Corollary 15.1 from Theorem 14.8 is similar to the proof of Theorem 3.8.4 in Gorenstein (1968). The hypothesis of Corollary 15.1 is satisfied in several classes of "small" simple groups and also in several classes of "small" subgroups of "large" simple groups. By our previous results, $\mathbf{ZJ}$, $\mathbf{K}_\infty$, and $\mathbf{K}^\infty$ control strong fusion in all of the groups in these classes if $p$ is odd. In particular, they control strong fusion in subgroups of odd order; Professor Gorenstein has discussed some applications of this observation in his lectures.

The following direct corollary of (14.1) has also been useful.

COROLLARY 15.2. *Suppose that $p$ is odd, $p$ divides $|G|$, and $G$ is $p$-stable. Let $\mathscr{H}$ be the class of all subgroups of $G$ of the form $\mathbf{N}(T)$ for some non-identity normal subgroup $T$ of $S$. Let $H_0 = \mathbf{N}(\mathbf{Z}(\mathbf{J}(S)))$. Suppose $\mathbf{C}_H(\mathbf{O}_p(H)) \subseteq \mathbf{O}_p(H)$ for every $H \in \mathscr{H}$. Then $H_0$ is the unique maximal element of $\mathscr{H}$ under inclusion, that is, $H_0 \in \mathscr{H}$ and $H \subseteq H_0$ for all $H \in \mathscr{H}$.*

In this section and in the following section, we will denote the largest normal subgroup of odd order in $G$ by $\mathbf{O}_{2'}(G)$. Grün's Theorem (Theorem 4.7) shows that if $G$ is simple and $T$ is a subgroup of $\mathbf{Z}(S)$ that is weakly closed in $S$, then $S \subseteq (\mathbf{N}(T))'$. Thus our previous results on weak closure yield several restrictions on $G$. Some additional restrictions can be obtained from the following two results, which were proved by the theory of blocks of characters.

THEOREM 15.3. *Suppose that $p = 2$, $x \in S$, and $x$ is weakly closed in $S$ with respect to $G$. Let $\mathbf{Z}^*(G)$ be the subgroup of $G$ that contains $\mathbf{O}_{2'}(G)$ and satisfies*

$$\mathbf{Z}^*(G)/\mathbf{O}_{2'}(G) = \mathbf{Z}(G/\mathbf{O}_{2'}(G))$$

*Then $x \in \mathbf{Z}^*(G)$.*

THEOREM 15.4 (D. Goldschmidt). *Suppose* $p = 2$, $1 \subset T \subseteq (\mathbf{Z}(S))^p$, *and* $T$ *is weakly closed in* $S$ *with respect to* $G$. *Then*

$$T\mathbf{O}_{2'}(G)/\mathbf{O}_{2'}(G) \subseteq O_2(G/\mathbf{O}_{2'}(G))$$

Theorem 15.3 and some applications are proved in Glauberman (1966). In his lectures, Professor Dade has also proved the main part of Theorem 15.3, namely, the case in which $x^2 = 1$. Theorem 15.4 and some related results are proved by Goldschmidt in work to be published. Because of Theorem 14.5, we obtain the following corollary of Theorem 15.4.

COROLLARY 15.5. *Suppose that* $p = 2$, $1 \subset T \subseteq (\mathbf{Z}(S))^p$, *and* $T \trianglelefteq \mathbf{N}(\mathbf{J}(S))$. *Then* $G$ *is not a simple group.*

Theorem 15.3 may be used to show that some groups cannot occur as Sylow 2-subgroups of simple groups. Part (a) of the following result is proved as Corollary 7 of Glauberman (1966); part (b), which requires Theorem 14.10, is part of Corollary 7 of Glauberman (1969).

THEOREM 15.6. *Suppose that* $p = 2$ *and* $S = Q \times R$ *for some generalized quaternion subgroup* $Q$. *Assume that*

(a) *every element of order two in* $R$ *lies in* $\mathbf{Z}(R)$, *or*
(b) $G$ *is* $S^4$*-free.*

*Then* $G$ *is not simple.*

The following result is obtained from Theorem 15.3 and the Feit–Thompson Theorem on groups of odd order.

THEOREM 15.7. *Suppose that* $G$ *is simple and* $p = 2$. *Assume either that* $S$ *is generated by two elements or that* $S$ *is generated by three elements and* $\mathbf{N}(S)/\mathbf{C}(S)$ *is not a 2-group. Let* $A$ *be the factor group given by the group of all automorphisms of* $G$ *taken modulo the group of all inner automorphisms of* $G$. *Then* $A$ *is solvable.*

Theorem 15.4 is used to obtain the following result

THEOREM 15.8 (Goldschmidt). *Suppose that* $p = 2$, $S$ *has nilpotence class* $n$, *and* $\mathbf{O}_2(G/\mathbf{O}_{2'}(G)) = 1$. *Let* $m = 2$ *if* $n = 1$, *and* $m = 2^{n-1}$ *if* $n > 1$. *Let* $k = 2$ *if* $n = 1$, *and* $k = 2^{n(n-1)}$ *if* $n > 1$. *Then* $x^m = 1$ *for all* $x \in Z(S)$, *and* $x^k = 1$ *for all* $x \in S$.

## 16. Some Open Questions

Between the counterexamples of Section 11 and the results of the later sections lie a number of questions not yet resolved. The most serious gap appears to occur in the case where $p = 2$.

*Question* 16.1. Suppose that $P$ is a finite 2-group. Does there exist a characteristic subgroup $\mathbf{L}(P)$ of $P$ such that $\mathbf{L}(P) \trianglelefteq H$ for every finite group $H$ that satisfies the following conditions:

(a) $P$ is a Sylow 2-subgroup of $H$,
(b) $H$ is $S^4$-free, and
(c) $\mathbf{C}_H(\mathbf{O}_2(H)) \subseteq \mathbf{O}_2(H)$?

By Theorem 6.6 an affirmative answer to Question 16.1 for every $P$ would tell us that $\mathbf{L}$ controls strong fusion in every $S^4$-free group. Several weaker results would also be of interest. We may restrict $H$ to be solvable in Question 16.1. We may ask only that $\mathbf{O}_2(H) \cap (N(\mathbf{L}(P)))' = \mathbf{O}_2(H) \cap H'$, for then $L$ would control transfer in every $S^4$-free group, by Theorem 6.8. Lemma 12.11 and Example 11.5 show that we cannot take $\mathbf{K}_\infty(P)$, $\mathbf{K}^\infty(P)$, $\mathbf{Z}(\mathbf{K}_\infty(P))$, or $\mathbf{Z}(\mathbf{K}^\infty(P))$ for $\mathbf{L}(P)$. Similar examples (in Section 10 of Glauberman, 1968a) show that we cannot take $\mathbf{J}(P)$ or $\mathbf{Z}(\mathbf{J}(P))$ for $\mathbf{L}(P)$. A candidate for $\mathbf{L}(P)$ has been suggested by Thompson (1969).

By Theorem 14.10, $\mathbf{J}$ controls weak closure of elements in every $S^4$-free group if $p = 2$. The case in which $G$ is not $S^4$-free was investigated by Glauberman (1968b, 1969). The following question is related to these investigations and to Question 16.1.

*Question* 16.2. Suppose that $p = 2$.

(a) What are the possibilities for $G$ under the restrictions that $S$ is a maximal subgroup of $G$ and $\mathbf{O}_2(G) = \mathbf{O}_{2'}(G) = 1$? In particular, what are the possibilities if $G$ is also assumed to be simple?

(b) Suppose $S = Q \times R$, where $Q$ is a generalized quaternion group. Can $G$ be a simple group?

Some special cases of (a) are proved by Thompson (1960b). Theorem 15.6 gives some special cases of (b).

Another obvious gap in our knowledge is the following.

*Question* 16.3. Suppose that $p = 3$. Does there exist a conjugacy functor $\mathbf{W}$ that controls transfer in $G$?

For every section conjugacy functor $\mathbf{W}$, define the degree of $\mathbf{W}$ to be the smallest positive integer $n$ that satisfies the following condition:

Whenever $G^*$ is a section of $G$, $S^*$ is a Sylow $p$-subgroup of $G^*$, and $\mathbf{W}(S^*) \ntrianglelefteq G^*$, then there exists $g \in S^* - \mathbf{O}_p(G^*)$ with the property that $[X, g; n] \subseteq Y$ for every chief factor $X/Y$ of $G^*$ such that $X \subseteq \mathbf{O}_p(G^*)$.

Let $n$ be the degree of $W$. Using Lemma 7.1 and Lemma A1.7 (in the Appendix), we may show that $n \leq p$. By Theorem 12.3, $n \leq 4$ if $\mathbf{W}(P) = \mathbf{K}_\infty(P)$ for all $P \in \mathscr{C}_p^*(G)$. By Theorem 3 of Glauberman (1968a) we have $n \leq 6$ if $\mathbf{W}(P) = \mathbf{J}(P)$ for all $P \in \mathscr{C}_p^*(G)$.

*Question* 16.4. Does there exist a section conjugacy functor on $G$ of degree at most 2 if $p \geq 3$? At most 3, if $p \geq 5$?

By Theorem 6.8, an affirmative answer to the first part of Question 16.4 would also provide an affirmative answer to Question 16.3. An affirmative answer to the second part would also be interesting and suggestive.

*Question* 16.5. Let $n$ be a positive integer. Suppose $G \subseteq GL(n, p)$ and $A$ is an abelian $p$-subgroup of $G$. Consider $G$ to be a group of linear transformations on a vector space $V$ of dimension $n$ over $GF(p)$. Let $W$ be the subspace of $V$ consisting of all the vectors fixed by every element of $A$. Assume that $(g-1)(h-1) = 0$ for all $g, h \in A$ and that $G$ is generated by the conjugates of $A$ in $G$. What are the possibilities for $G$ and $V$ if $\mathbf{O}_p(G) = 1$ and if either $p$ is odd or $|V/W| \leq |A|$?

An important partial answer to this question was obtained by McLaughlin (1967, 1969). Other partial answers were used in the proofs of Theorems 14.10 and 14.13. Another question related to Theorem 14.13 is:

*Question* 16.6. What are the consequences for $G$ if $p$ is odd and $\mathbf{N}(S)/\mathbf{C}(S)$ is a $p$-group? For example, if $p = 3$, can $G$ be simple? Can this occur if $\mathbf{ZJ}$, $\mathbf{K}_\infty$, or $\mathbf{K}^\infty$ does not control strong fusion in $G$?

*Question* 16.7. Is there a valid analogue to Theorem 15.3 if $p$ is odd?

An affirmative partial answer to Question 16.7 was proved by Shult (1966). The analogue to Goldschmidt's Theorem is false if $p$ is odd. Let $q = p(p-1)$ and let $H = SL(2, 2^q)$. Then $H$ is a simple group with a cyclic Sylow $p$-subgroup. By elementary number theory, $|H|_p \geq p^2$.

Let $A$ be a cyclic group of order $p$. Consider $G$ to be a trivial operator group on $A$. Let $N = \mathbf{N}(\mathbf{K}_\infty(S))$. Then Theorem 12.4 is equivalent to a statement about cohomology groups (Hall, 1959) namely,

$$H^1(G, A) \cong H^1(N, A) \quad \text{if } p \geq 5. \tag{16.1}$$

This suggests the following problem related to Professor Alperin's lecture.

*Question* 16.8. Does there exist a function $f$ from the positive integers to the positive integers such that

$$H^i(G, A) \cong H^i(N, A) \quad \text{whenever } p \geq f(i)?$$

## Appendix A1. Proof of Theorem 7.2

DEFINITION. Suppose that $A$, $B$, and $C$ are non-empty subsets of $G$. Define

$$AB = \{ab : a \in A, b \in B\}.$$

Put $ABC = (AB)C$. We say that $B$ is a *transversal* to $A$ in $G$ provided that, for every $g \in G$, there exist unique elements $a \in A$ and $b \in B$ such that $ab = g$.

DEFINITION. Let $V$ be a group. Suppose that for every $g \in G$ we are given an automorphism $a_g$ of $V$. For $v \in G$ and $g \in G$, denote the image of $v$ under $a_g$ by $v^g$. Then $G$ is an *operator group* on $V$ if

$$(v^g)^h = v^{gh}, \quad \text{for all } v \in V \text{ and } g, h \in G.$$

Suppose $G$ is an operator group on $V$. If $H \subseteq G$, let $\mathbf{C}_V(H) = \{v \in V : v^h = v \text{ for all } h \in H\}$. If $g \in G$ and $\mathbf{C}_V(\langle g \rangle) = V$, we say that $g$ *centralizes* $V$. If $H \subseteq G$ and $W \subseteq V$, then $W$ is *$H$-invariant* if $w^h \in W$ for all $w \in W$ and $h \in H$. If $X \subset W \subseteq V$ and $W, X \trianglelefteq V$, then $W/X$ is a *composition factor* of $V$ under $G$ if $X$ and $W$ are $G$-invariant and if no $G$-invariant normal subgroup $Y$ of $V$ satisfies $X \subset Y \subset W$. We say that $V$ is *irreducible* under $G$, or that $G$ is *irreducible* on $V$, if $V$ is abelian and if 1 and $V$ are the only $G$-invariant subgroups of $V$.

Let $g \in G$ and $W \subseteq V$. Define

$$[v, g; 1] = [v, g] = v^{-1}v^g,$$

and

$$[v, g; i+1] = [[v, g; i], g], \quad (v \in V \text{ and } i \geq 1).$$

Define

$$[W, g] = \langle [w, g] : w \in W \rangle.$$

For $H \subseteq G$, define $[W, H] = \langle [W, h] : h \in H \rangle$. Define

$$[V, g; 0] = V \quad \text{and} \quad [V, g; i+1] = [[V, g; i], g] \quad (i \geq 0).$$

REMARK. Note that these definitions of $\mathbf{C}_V(H)$ and commutator elements and subgroups agree with the previous definitions in the case where $V$ and $G$ are subgroups of a larger group $G^*$, $V \trianglelefteq G^*$, and $v^g$ is given by $g^{-1}vg$ for $v \in V$, $g \in G$.

Our first lemma is elementary.

LEMMA A1.1. (a) *Let $H$ and $K$ be subgroups of a group $G$. Suppose that $T_1$ is a transversal to $H \cap K$ in $K$ and $T_2$ is a transversal to $HK$ in $G$. Then $\{xy : x \in T_1, y \in T_2\}$ is a transversal to $H$ in $G$.*

(b) *Suppose $H$ and $T$ are subsets of a finite group $G$, $HT = G$, and $|T| \leq |G|/|H|$. Then $T$ is a transversal to $H$ in $G$.*

LEMMA A1.2. *Suppose that $Q$ and $R$ are subgroups of a finite $p$-group $P$. Then there exists a transversal to $QR$ in $P$.*

*Proof.* We use induction on $|P|+[P : R]$. We may assume that $QR \neq P$. Then $P \neq 1$, and $\mathbf{Z}(P)$ contains a subgroup $S$ of order $p$.

Suppose $S \subseteq R$. By induction, there exists a transversal $T_0$ to $(QS/S)(R/S)$ in $P/S$. Let $T$ be a set that contains one element from each element of $T_0$. Then $QRT = P$ and $|T| = |T_0| \leq |P|/|QR|$. By Lemma A1.1(b), $T$ is a transversal to $QR$ in $P$.

Suppose that $S \nsubseteq R$. Then $RS \subseteq P$ and $[RS : R] = p$. By induction, there exists a transversal $T_1$ to $Q(RS)$ in $P$. If $Q(RS) = QR$, we are done. Assume that $QR \neq Q(RS)$. Let $m = |Q(RS)|/|QR|$. Since $|Q(RS)| = |(QR)S| \leq |QR||S|$, we have $1 < m \leq p$. However, $|QR| = |Q||R|/|Q \cap R|$; therefore $|QR|$, and likewise $|Q(RS)|$, are powers of $p$. Consequently, $m$ is an integer and $m = p$. Let $T = \{xy : x \in S,\ y \in T_1\}$. Then $P = Q(RS)T_1 = QRT$ and $|T| \leq |S||T_1| = p|T_1| = p|P|/|Q(RS)| = |P|/|QR|$. By Lemma A1.1(b), $T$ is a transversal to $QR$ in $P$.

LEMMA A1.3. *Suppose that G is an operator group on a group V. Then the group $[V, G]$ is a G-invariant normal subgroup of V.*

*Proof.* Let $W = [V, G]$. Since $(v^{-1}v^g)^h = (v^{-1}v^h)^{-1}(v^{-1}v^{gh})$, $W$ is $G$-invariant. Likewise $W \trianglelefteq V$ because, for $v, w \in V$,

$$w^{-1}(v^{-1}v^g)w = (vw)^{-1}(vw)^g(w^{-1}w^g)^{-1}.$$

THEOREM A1.4. *Suppose that V is a finite p-group and G is an operator group on V. Let $\mathscr{A}$ be a non-empty set of subgroups of S and B be the subgroup of S generated by the elements of $\mathscr{A}$. Assume that $\mathscr{A}$ satisfies the following conditions.*

*B is weakly closed in S with respect to G.* (A1.1)

*Whenever $A \in \mathscr{A}$, $g \in G$, $A \nsubseteq S^g$, W is a composition factor of V under G, and $v \in \mathbf{C}_W(A \cap S^g)$, then* (A1.2)

$$\prod_{h \in R} v^h = 1 \qquad \text{(A1.3)}$$

*for every transversal R of $A \cap S^g$ in A.*

*Then $B \trianglelefteq S$ and $\mathbf{C}_V(G) = \mathbf{C}_V(\mathbf{N}_G(B))$.*

*Proof.* Note that (A1.3) is independent of the choice of the transversal $R$, since $v^{ah} = v^h$ whenever $a \in A \cap S^g$, $h \in A$, and $v \in \mathbf{C}_V(A \cap S^g)$. By (A1.1), $B \trianglelefteq S$.

Let $M = \mathbf{N}_G(B)$. We prove that $\mathbf{C}_V(G) = \mathbf{C}_V(M)$ by induction on $|V|$. We may assume that $|V| > 1$. Let $V_0 = \mathbf{C}_V(M)$. Clearly, $\mathbf{C}_V(G) \subseteq V_0$. We must prove that $V_0 \subseteq \mathbf{C}_V(G)$.

Let $Q$ be a transversal to $M$ in $G$. Define a mapping $\phi : V_0 \cap \mathbf{Z}(V) \to \mathbf{Z}(V)$ by

$$\phi(v) = \prod_{g \in Q} v^g, \quad (v \in V_0 \cap \mathbf{Z}(V)). \qquad \text{(A1.4)}$$

Clearly, $\phi$ is a homomorphism. Since $v^{mg} = v^g$ for all $v \in V_0 \cap \mathbf{Z}(V)$, $m \in M$, and $g \in G$, the definition of $\phi$ does not depend on the choice of the transversal $Q$. Suppose that $h \in G$. Then the set $\{gh : g \in Q\}$ is also a transversal to $M$ in $G$. Therefore, for $v \in V_0 \subseteq \mathbf{Z}(V)$,

$$\phi(v) = \prod_{g \in Q} v^{gh} = \Big(\prod_{g \in Q} v^g\Big)^h = \phi(v)^h.$$

Thus

$$\phi \text{ maps } V_0 \cap \mathbf{Z}(V) \text{ into } \mathbf{C}_{\mathbf{Z}(V)}(G). \qquad \text{(A1.5)}$$

Let $X$ be a non-identity subgroup of $\mathbf{Z}(V)$ that is minimal with respect to the property of being $G$-invariant. We first assume that $X \subset V$. Suppose that $v \in V_0$. We must show that $v \in \mathbf{C}_V(G)$. Now, $M$ centralizes the coset $vX$. By induction, G centralizes $vX$. Consequently, the group $\langle X, v\rangle$ is $G$-invariant. If $\langle X, v\rangle \subset V$, then $v \in \mathbf{C}_V(G)$, by induction. Suppose $\langle X, v\rangle = V$. Since $v$ and $X$ centralize each other, $V$ is abelian. As $X \subset V$, $v \notin X$. Let $n = [G : M]$. Then

$$\phi(v) \equiv \prod_{g \in Q} v^g \equiv v^n, \quad \text{modulo } X.$$

Since $S \subseteq M$, $p$ does not divide $n$. Hence $v^n \notin X$. By (A1.5), $\phi(v) \in \mathbf{C}_V(G)$. Therefore, by induction,

$$|\mathbf{C}_V(G)/(X \cap V_0)| = |\mathbf{C}_V(G)/\mathbf{C}_X(G)| = p = |V_0/(X \cap V_0)|.$$

Since $\mathbf{C}_V(G) \subseteq V_0$, $\mathbf{C}_V(G) = V_0$. Thus $v \in \mathbf{C}_V(G)$.

Now assume that $X = V$. Then (A1.3) is satisfied for $W = V$. Let

$$G = \bigcup_{i \in I} Mg_iB$$

be a double coset decomposition of $G$. We may assume that $0 \in I$ and that $g_0 = 1$. For each $i \in I$, let $B_i = B \cap M^{g_i}$, and let $R_i$ be a transversal to $B_i$ in $B$. We may assume that $R_0 = \{1\}$. By an easy argument, $Mg_iB$ is the disjoint union of the sets $Mg_ib$, $b \in B_i$. Therefore, the set $\{g_ib : i \in I, b \in R_i\}$ is a transversal to $M$ in $G$. Thus, for $v \in V_0$.

$$\phi(v) = \prod_{i \in I,\, b \in R_i} v^{g_ib} = v \prod_{i \in I - \{0\}} \Big(\prod_{b \in R_i} v^{g_ib}\Big). \tag{A1.6}$$

Take $i \in I - \{0\}$. There exists $m \in M$ such that $(B_i^{g_i^{-1}})^m \subseteq S$. Let $h = g_i^{-1}m$ and $E = B_i^h$. If $B_i = B$, then $E = B$, by (A1.1). But then

$$g_i^{-1}m \in \mathbf{N}_G(E) = M,$$

$g_i \in M$, and $i = 0$, a contradiction. Hence $B_i \subset B$. Consequently, there exists $A \in \mathscr{A}$ such that $A \nsubseteq B_i$. Now

$$A \cap B_i = A \cap B_i \cap S^{h^{-1}} = A \cap B \cap M^{g_i} \cap S^{h^{-1}} = A \cap S^{h^{-1}}.$$

By Lemmas A1.1 and A1.2, there exists a transversal $R$ to $A \cap B_i$ in $A$ and a subset $T$ of $B$ such that the set $\{xy : x \in R, y \in T\}$ is a transversal to $B_i$ in $B$. We may assume that $R_i$ is this set.

Let $v \in V_0$. Take $i$ and $R_i$ as above, and let $w = v^{g_i}$. Then

$$w \in \mathbf{C}_V(M)^{g_i} = \mathbf{C}_V(M^{g_i}) \subseteq \mathbf{C}_V(A \cap B_i).$$

Hence

$$\prod_{b \in R_i} v^{g_ib} = \prod_{y \in T} \Big(\prod_{x \in R} w^x\Big)^y = 1,$$

by (A1.2). Since $i$ is arbitrary, (A1.6) yields that $\phi(v) = v$. Thus $v \in \mathbf{C}_V(G)$. This completes the proof of Theorem A1.4.

In order to prove our next result, we introduce some further concepts. Let $C_p$ be a fixed multiplicative group of order $p$.

DEFINITION. Suppose that $V$ is an elementary abelian $p$-group and $G$ is an operator group on $V$. The group *dual* to $V$ is the group $V^*$ of all homomorphisms of $V$ into $C_p$, in which multiplication is defined by

$$(f_1f_2)(v) = f_1(v)f_2(v) \qquad (f_1, f_2 \in V^* \text{ and } v \in V).$$

We define $G$ to be an operator group on $V^*$ by

$$f^g(v) = f(v^{g^{-1}}) \qquad (f \in V^*, g \in G, v \in V).$$

For every subgroup $W$ of $V$, let

$$W^\perp = \{f \in V^* : f(v) = 1 \quad \text{for all } v \in W\}.$$

For every subgroup $W$ of $V^*$, let

$$W^\perp = \{v \in V : f(v) = 1 \quad \text{for all } f \in W\}.$$

LEMMA A1.5. *Suppose that $V$ is an elementary p-group and $G$ is an operator group on $V$.*

(a) *For every subgroup $W$ of $V$ or $V^*$, $(W^\perp)^\perp = W$.*
(b) *If $W_1$ and $W_2$ are subgroups of $V$ (or $V^*$) and $W_1 \subset W_2$, then $W_2^\perp \subset W_1^\perp$.*
(c) *If $W$ is a $G$-invariant subgroup of $V$ (or $V^*$), then $W^\perp$ is a $G$-invariant subgroup of $V^*$ (or $V$).*
(d) *Suppose that $W_1$ and $W_2$ are $G$-invariant subgroups of $V$ (or $V^*$) and $W_1 \subset W_2$. Then $W_2/W_1$ is irreducible under $G$ if and only if $W_1^\perp/W_2^\perp$ is irreducible under $G$.*
(e) *Suppose $H \subseteq G$. Then $\mathbf{C}_{V^*}(H) = [V, H]^\perp$.*

*Proof.* By using additive notation instead of multiplicative, and by considering $C_p$ to be $GF(p)$ and $V$ to be a vector space over $GF(p)$, we obtain (a) and (b) by elementary linear algebra (Herstein, 1964). Part (c) is trivial. Part (d) follows from part (c). To prove (e), let $v \in V$, $f \in V^*$, and $h \in H$. If $f \in \mathbf{C}_{V^*}(H)$, then

$$f(v^hv^{-1}) = f(v^h)f(v^{-1}) = f^{h^{-1}}(v)f(v^{-1}) = f(v)f(v^{-1}) = 1.$$

If $f \in [V, H]^\perp$, then

$$f^hf^{-1}(v) = f^h(v)f(v^{-1}) = f(v^{h^{-1}})f(v^{-1}) = f(v^{h^{-1}}v^{-1}) = 1.$$

THEOREM A1.6. *Suppose that $V$ is a finite p-group and $G$ is an operator group on $V$. Let $\mathscr{A}$ be a non-empty set of subgroups of $S$ and $B$ be the subgroup of $S$ generated by the elements of $\mathscr{A}$. Assume that $\mathscr{A}$ satisfies the following conditions.*

*$B$ is weakly closed in $S$ with respect to $G$.* (A1.1)

*Whenever $A \in \mathscr{A}$, $g \in G$, $A \nsubseteq S^g$, $W$ is a composition factor of $V$ under $G$, and $v \in W$, then* (A1.7)

$$\prod_{h \in R} v^{h^{-1}} \equiv 1, \quad \textit{modulo } [W, A \cap S^g], \qquad \text{(A1.8)}$$

*for every transversal $R$ of $A \cap S^g$ in $A$.*

*Then $B \trianglelefteq S$ and $[V, G] = [V, \mathbf{N}_G(B)]$.*

*Proof.* We imitate the proof of Theorem A1.4. Let $M = \mathbf{N}_G(B)$. Clearly, $S \subseteq M$. We prove that $[V, G] = [V, M]$ by induction on $|V|$. Let $X = [V, M]$, $Y = [V, G]$ and

$$Z = \{v \in \mathbf{Z}(V) : v^p = 1\}.$$

Clearly, $X \subseteq Y$. We must prove that $Y \subseteq X$.

Assume first that $Z \subset V$. Let $Z_1 = [Z, G]$. By induction, $Z_1 = [Z, M]$. By Lemma A1.3, $Z_1$ is $G$-invariant. If $Z_1 \neq 1$, we may apply induction to $V/Z_1$ to obtain $X = Y$. If $Z_1 = 1$ and $Z \cap X \neq 1$, we may similarly apply induction to $V/(Z \cap X)$.

Suppose that $Z_1 = Z \cap X = 1$. Then $\mathbf{Z}(V) \cap X = 1$. However, $X \trianglelefteq V$ by Lemma A1.3. Hence, $X = 1$ by Theorem 2.1. By applying induction to $V/Z$, we obtain

$$Y = [V, G] \subseteq Z.$$

Thus every element of $G$ centralizes $V/Z$ and $Z$. By Lemma 6.5, the automorphisms of $V$ determined by the elements of $G$ form a $p$-group. Since $S \subseteq M$ and $[V, M] = 1$, we obtain $[V, G] = 1 = [V, M]$.

Thus we may assume that $Z = V$, i.e., that $V$ is an elementary abelian $p$-group. Let $V^*$ be the group dual to $V$, and consider $G$ to be an operator group on $V^*$, as defined previously. We claim that $G$, $\mathscr{A}$, and $V^*$ satisfy the hypothesis of Theorem A1.4. Since (A1.1) is part of our hypothesis, we need only verify (A1.2). Let $A \in \mathscr{A}$ and $g \in G$. Suppose $W$ is a composition factor of $V^*$ under $G$ and $f \in \mathbf{C}_W(A \cap S^g)$. Assume that $A \nsubseteq S^g$ and that $R$ is a transversal to $A \cap S^g$ in $A$. Let $W = W_1/W_2$. Put $V_i = W_i^{\perp}$ for $i = 1, 2$. By Lemma A1.5, $V_2/V_1$ is irreducible and $W_i = V_i^{\perp}$ for $i = 1, 2$. Let $e$ be an element of $W_1$ in the coset $f$. Let $e' = \Pi\, e^h$ and $f' = \Pi\, f^h$, where $h$ ranges over $R$.

Suppose that $v \in V_2$ and $x \in A \cap S^g$. Since $x^{-1}$ centralizes $f$, $e^{x^{-1}}e^{-1} \in W_2$. Hence

$$1 = e^{x^{-1}}e^{-1}(v) = e(v^x)e(v)^{-1} = e(v^x v^{-1}).$$

Thus

$$[V_2, A \cap S^g] \text{ is contained in the kernel of } e. \tag{A1.9}$$

Let $u \in V_2$. Then

$$e'(u) = \prod_{h \in R} e^h(u) = \prod_{h \in R} e(u^{h^{-1}}) = e\Big(\prod_{h \in R} u^{h^{-1}}\Big) = 1,$$

by (A1.8) and (A1.9). Therefore, $e' \in V_2^{\perp} = W_2$. So $f' = 1$. This proves (A1.2) for $G$, $\mathscr{A}$, and $V^*$. By Theorem A1.4, $\mathbf{C}_{V^*}(G) = \mathbf{C}_{V^*}(M)$. By Lemma A1.5,

$$[V, G] = \mathbf{C}_{V^*}(G)^{\perp} = \mathbf{C}_{V^*}(M)^{\perp} = [V, M].$$

This completes the proof of Theorem A1.6.

LEMMA A1.7. *Suppose $V$ is an elementary abelian p-group and $G$ is an operator group on $V$. Let $g \in G$. Then*

(a) $[V, g; p] = 1$ *if and only if* $g^p$ *centralizes* $V$;

(b) $\prod_{0 \le i \le p-1} v^{g^i} = [v, g; p-1]$, *for all* $v \in V$; *and*

(c) *if* $g^p$ *centralizes* $V$ *and* $n \ge 2$, *then*

$$\prod_{0 \le i \le p^n-1} v^{g^i} = 1, \quad \textit{for all } v \in V.$$

*Proof.* Let $E$ be the set of all endomorphisms of $V$. It is well known that $E$ forms a ring under the operations of addition and multiplication, as given by

$$v^{\alpha+\beta} = v^\alpha v^\beta, \qquad v^{\alpha\beta} = (v^\alpha)^\beta \qquad (\text{for } v \in V \text{ and } \alpha, \beta \in E).$$

Moreover, $E$ has a zero element and a unity element given by $v^0 = 1$, $v^1 = v$ ($v \in V$). For every $\alpha \in E$, the sum $\alpha+\alpha+\ldots+\alpha$ (with $p$ terms) is zero. Consequently, for every $\alpha \in E$,

$$(\alpha-1)^p = \alpha^p - 1,$$
$$(\alpha-1)^{p-1} = \alpha^{p-1}+\alpha^{p-2}+\ldots+\alpha+1.$$

In addition, if $\alpha^p = 1$ and $n \ge 2$, then

$$\begin{aligned} 1+\alpha+\alpha^2+\ldots+\alpha^{p^n-1} &= (1+\alpha+\ldots+\alpha^{p-1})(1+\alpha^p+\ldots+\alpha^{p^n-p}) \\ &= (1+\alpha+\ldots+\alpha^{p-1})(1+1+\ldots+1) = 0. \end{aligned}$$

By taking $\alpha$ to be the automorphism of $V$ given by $v^{\dot{\alpha}} = v^g$ ($v \in V$), we obtain the desired results.

LEMMA A1.8. *Suppose that $V$ is an elementary abelian p-group, $G$ is an operator group on $V$, $g \in S$, and $A = \langle g \rangle$. Assume that $[V, g; p-1] = 1$. Then for every proper subgroup $A_1$ of $A$ and every transversal $R$ to $A_1$ in $A$,*

$$\prod_{h \in R} v^h = \prod_{h \in R} v^{h^{-1}} = 1 \quad \textit{for all } v \in V. \tag{A1.10}$$

*Proof.* Take $A_1 \subset A$.

Clearly, $[V, g; p] = 1$. By Lemma A1.7(a), $g^p$ centralizes $V$. So every element of $A_1$ centralizes $V$. Therefore, (A1.10) does not depend on the choice of $R$. Let $p^n = |A/A_1|$. We may assume that $R = \{g^i : 0 \le i \le p^n-1\}$, or we may assume that $R = \{g^{-i} : 0 \le i \le p^n-1\}$. Thus (A1.10) follows from Lemma A1.7.

We may now prove Theorem 7.2. Take $P$, $E_0$, $E$, and $L$ as in the theorem and let $\mathscr{A} = \{\langle g \rangle : g \in E_0\}$. Then $E$ is the group generated by the elements of $\mathscr{A}$, and $L = \mathrm{N}(E)$. We may consider $G$ to be an operator group on $P$ by conjugation. Now the theorem follows directly from Lemma A1.8 and Theorems A1.4 and A1.6.

## Appendix A2. Proof of Theorem 10.2

DEFINITION. Suppose that $V$ and $W$ are abelian groups and $G$ is an operator group on $V$ and on $W$. Let $\text{Hom}(V, W)$ be the group of all homomorphisms from $V$ into $W$, where multiplication is defined by

$$(f_1 f_2)(v) = f_1(v) f_2(v) \quad \text{for } v \in V.$$

We define $G$ to be an operator group on $\text{Hom}(V, W)$ by

$$f^g(v) = (f(v^{g^{-1}}))^g$$

for $v \in V$, $g \in G$, and $f \in \text{Hom}(V, W)$. An element $f$ of $\text{Hom}(V, W)$ is said to be a *G-homomorphism* if $f^g = f$ for all $g \in G$, that is,

$$f(v^g) = (f(v))^g \quad \text{for all } v \in V,\ g \in G.$$

We say that $V$ and $W$ are *G-isomorphic* if there exists an isomorphism of $V$ onto $W$ which is a $G$-homomorphism.

LEMMA A2.1. *Suppose that $G$ is an operator group on a finite p-group $P$. Let $X/Y$ be a composition factor of $P$ under $G$. Then $X/Y$ is an elementary abelian p-group and $[X, P] \subseteq Y$.*

*Proof.* Since $G$ is an operator group on $P/Y$, we may assume that $Y = 1$. Since $1 \subset X \trianglelefteq P$, $X \cap \mathbf{Z}(P) \neq 1$. As $X \cap \mathbf{Z}(P)$ is $G$-invariant, $X \subseteq \mathbf{Z}(P)$. So $[X, P] \subseteq Y$. Similarly, $X = \{x \in X : x^p = 1\}$. Thus $X$ is an elementary abelian group.

The following result is proved in p. 19 of Gorenstein (1968).

LEMMA A2.2. (Three subgroups lemma). *Suppose that $H$, $K$, $L \subseteq G$ and that $[[H, K], L] = [[K, L], H] = 1$. Then $[[L, H], K] = 1$.*

LEMMA A2.3. *Suppose that $V$ and $W$ are abelian groups and $G$ is an operator group on $V$ and on $W$. Let $H = \text{Hom}(V, W)$. Assume that $[V, G; i] = 1$ and $[W, G; j] = 1$ for some positive integers $i$ and $j$. Then $[H, G; i+j-1] = 1$.*

*Proof.* For all $f \in H$, $g \in G$, and $v \in V$, we have

$$f^g f^{-1}(v) = (f(v^{g^{-1}}))^g (f(v^{g^{-1}}))^{-1} f(v^{g^{-1}} v^{-1}). \tag{A2.1}$$

Let $V_k = [V, G; k]$ $(k \geq 0)$. By (A2.1) and induction on $r$ we obtain

$$f_r(V) \subseteq \langle [f(V_k), G; r-k] : 0 \leq k \leq r \rangle \quad \text{for all } f_r \in [H, G; r].$$

By hypothesis, $V_i = 1$ and $[W, G; j] = 1$. Hence, for $r \geq i+j-1$, $f_r(V) = 1$ and $f_r = 1$. Thus $[H, G; i+j-1] = 1$.

Theorem 10.2 follows from Lemma A2.3 and the following result, which is implicitly contained in the proof of Glauberman (1968c).

THEOREM A2.4. *Suppose that $G$ is an operator group on a finite p-group $P$. Let $Q \in \mathscr{Q}(P)$. Suppose that $X/Y$ is a composition factor of $P$ under $G$. Then* *either $X/Y$ is $G$-isomorphic to a composition factor of $\mathbf{Z}(Q)$ under $G$, or there exist composition factors $U/U_1$ and $V/V_1$ of $P$ under $G$ such that*

(a) $U \subseteq Q$,

(b) $V \subseteq [U, P] \subseteq \mathbf{Z}(Q)$, *and*

(c) *$X/Y$ is $G$-isomorphic to a composition factor of* $\mathrm{Hom}(U/U_1, V/V_1)$ *under $G$.*

*Proof.* Choose $X_1$ to be minimal subject to the condition that there exists a composition factor $X_1/Y_1$ of $P$ under $G$ which is $G$-isomorphic to $X/Y$. If $X_1 \subseteq \mathbf{Z}(Q)$, we are done. Assume that $X_1 \nsubseteq \mathbf{Z}(Q)$. For convenience in notation, we may assume that $X_1 = X$.

Suppose that $N \subset X$ and $N$ is a $G$-invariant normal subgroup of $P$. If $N \nsubseteq Y$, then $NY = X$ and $N/(N \cap Y)$ is $G$-isomorphic to $X/Y$. But this contradicts the minimal choice of $X$. Therefore,

$$\text{if } N \subset X,\ N \trianglelefteq P, \text{ and } N \text{ is } G\text{-invariant, then } N \subseteq Y. \qquad \text{(A2.2)}$$

Since $\mathbf{C}_P(Q) = \mathbf{Z}(Q)$, $X \nsubseteq \mathbf{C}_P(Q)$. Let $U$ be a normal $G$-invariant subgroup of $P$ that is minimal subject to the conditions that $U \subseteq Q$ and $U \nsubseteq \mathbf{C}_Q(X)$. Let $V = [U, X]$. Then $V$ is a normal $G$-invariant subgroup of $P$. Take $U_1, V_1 \trianglelefteq P$ such that $U_1 \subset U$, $V_1 \subset V$, and the factor groups $U/U_1$ and $V/V_1$ are composition factors of $P$ under $G$. By Lemma A2.1, these factor groups and $X/Y$ are elementary abelian $p$-groups and

$$[X, P] \subseteq Y, \qquad [U, P] \subseteq U_1, \quad \text{and} \quad [V, P] \subseteq V_1. \qquad \text{(A2.3)}$$

By our choice of $U$,

$$U_1 \text{ centralizes } X. \qquad \text{(A2.4)}$$

By (A2.3) and (A2.4),

$$[[U, X], P] = [V, P] \subseteq V_1, \quad \text{and} \quad [[P, U], X] \subseteq [U_1, X] = 1.$$

Applying the three subgroups lemma to $P/V_1$, we see that

$$[X, P]V_1/V_1 \text{ centralizes } U/V_1. \qquad \text{(A2.5)}$$

Let $M = U/U_1$, $N = V/V_1$, and $W = X/[X, P]$. Define $H = \mathrm{Hom}(M, N)$. Then each of these groups is abelian, and $G$ may be considered to be an operator group on each of them in the natural way. Define a mapping $\phi : W \to H$ by

$$\phi(x[X, P])(uU_1) = [u, x]V_1.$$

That $\phi$ is well defined and is a homomorphism follows from (A2.3), (A2.4), (A2.5) and the identities

$$[u_1 u_2, x] = [u_1, x][u_1, x, u_2][u_2, x]$$

and

$$[u, x_1 x_2] = [u, x_2][u, x_1][u, x_1, x_2].$$

Furthermore, $\phi$ is a $G$-homomorphism, since

$$\phi(\bar{x}^g)(\bar{u}) = [u, x^g]V_1 = ([u^{g^{-1}}, x]V_1)^g = (\phi(\bar{x}))^g(\bar{u})$$

for $\bar{u} = uU$ and $\bar{x} = x[X, P]$. Now, the kernel of $\phi$ has the form $K/[X, P]$ for some subgroup $K$ of $X$ that contains $[X, P]$. Clearly, $K \trianglelefteq P$. Since $\phi$ is a $G$-homomorphism $K$ is $G$-invariant and $X/K$ is $G$-isomorphic to a subgroup of $H$. Moreover, $K \subset X$ because $V = [X, U] \supset V_1$. By (A2.2), $K \subseteq Y$. Thus $X/Y$ is $G$-isomorphic to a composition factor of $H$ under $G$. This completes the proof of Theorem A2.4.

## Appendix A3. Notes on the Previous Sections

*Section 2*

Theorems 2.1 and 2.2 were published by Sylow in 1872 and Corollary 2.3 by Cauchy in 1844.

*Section 3*

Alperin (1967) has produced several generalizations of Theorem 3.5 and several applications of them. In particular, let $\mathscr{F}$ be a conjugation family for $S$ in $G$ such that $N_S(T)$ is a Sylow $p$-subgroup of $N(T)$ for every $T \in \mathscr{F}$. Let $\mathscr{F}^*$ be the set of all $T \in \mathscr{F}$ such that $C_S(T) \subseteq T$. Alperin proves that whenever $A$ and $B$ are non-empty conjugate subsets of $S$ there exists $g \in G$ such that $A$ is $\mathscr{F}^*$-conjugate to $B$ via $g$. Note that this result yields an improvement of Proposition 3.7, namely, that $C_S(T) \subseteq T$ may be required in the conclusion.

Goldschmidt (1970) has proved the following generalization of Theorem 3.5: Take $\mathscr{F}$ as above. For each $T \in \mathscr{F}$, let

$$M(T) = \langle N(T) \cap N(T^*) : T \subset T^* \subseteq S \rangle.$$

Let $\mathscr{F}_1$ be the set of $T \in \mathscr{F}$ such that $M(T) \subset N(T)$. Let $\mathscr{F}_2 = \mathscr{F}_1 \cup \{S\}$. Then $\mathscr{F}_2$ is a conjugation family for $S$ in $G$.

Note that if $T \in \mathscr{F}_1$ and $p = 2$ then $N(T)/T$ has a strongly embedded subgroup and is therefore of known type, by the work of Bender (see Professor Gorenstein's lectures).

*Section 5*

Theorem 5.5 and several generalizations and related results were proved by Alperin and Gorenstein (1967).

Because of the improvement of Proposition 3.7 mentioned in our notes on Section 3, alternative (i) of part (d) of Proposition 5.4 may be improved by requiring that $U \supseteq C_S(U)$.

*Section 6*

Theorem 6.6 is a slight extension of a result stated without proof in Remark 5.1 of Glauberman (1968a).

Suppose that **W** is an arbitrary section conjugacy functor on $G$ that controls strong fusion in all sections of $G$. If $\mathbf{W}(P)$ is abelian for all $P \in \mathscr{C}_p^*(G)$, then there exists a section conjugacy functor $\mathbf{W}^*$ on $G$ that controls strong fusion in all sections of $G$ and satisfies $\mathbf{W}^*(P) \supseteq \mathbf{C}_P(\mathbf{W}^*(P))$ for all $P \in \mathscr{C}_p^*(G)$. This was proved for the functor **ZJ** (Section 9, Glauberman, 1968a) but the method of proof is valid in general. (The functor $\mathbf{ZJ}^*$ is used in some of Professor Gorenstein's lectures.) If $\mathbf{W}(P)$ is not abelian for some $P$, we may replace **W** by the functor **V** that takes each $P \in \mathscr{C}_p^*(G)$ to $\mathbf{Z}(\mathbf{W}(P))$, and then define $\mathbf{V}^*$ as above.

Suppose **W** is an arbitrary conjugacy functor on $G$ and $\mathbf{W}(P) \supseteq \mathbf{Z}(P)$ for every $p$-subgroup $P$ of $G$. For every $p$-subgroup $P$ and every normal $p'$-subgroup $K$ of $G$, define $\mathbf{W}^*(PK/K) = \mathbf{W}(P)K/K$; then $\mathbf{W}^*(PK/K) \supseteq \mathbf{Z}(PK/K)$. By using the Frattini argument and Lemma 6.4, we may prove that $\mathbf{W}^*$ controls weak closure of elements in $G$ if it controls weak closure of elements in $G/K$ whenever $K$ is a $p'$-subgroup of $G$ and $\mathbf{C}_{G/K}(\mathbf{O}_p(G/K)) \subseteq \mathbf{O}_p(G/K)$.

*Section 7*

Theorem 7.3 was suggested by the work of Wielandt (1940).

Theorems 6.8 and 7.4 generalize parts of Theorem 6 in Glauberman (1970). Moreover, suppose we weaken condition (7.1) in Theorem 7.4 by allowing $g$ to vary, according to the choice of $X/Y$. Then the conclusion of Theorem 7.4 is still valid, as is evident from the proof of part (e) of Theorem 6, Glauberman, (1970).

Let (7.6′) be the condition obtained from (7.6) by replacing the condition "$[\mathbf{Z}(\mathbf{O}_p(G^*)), g, g] = 1$" by "$[\mathbf{Z}(\mathbf{O}_p(G^*)), g; p] = 1$". Put $Z = (\mathbf{Z}(S))^p$. Let $\mathscr{F}$ be the set of all $T \subseteq S$ for which $\mathbf{N}_S(T)$ is a Sylow $p$-subgroup of $\mathbf{N}(T)$. Suppose $G$ satisfies (7.6′). By Alperin's Theorem, $\mathscr{F}$ is a conjugation family for $G$. Goldschmidt (to be published) has proved that $\mathscr{F}$ satisfies the definition of a conjugation family if (3.1) is replaced by the stronger condition

$$g_i \in \mathbf{N}(T_i) \cap \mathbf{N}(\mathbf{W}(S)) \quad \text{or} \quad g_i \in \mathbf{N}(T_i) \cap \mathbf{C}(Z), \quad \text{for } i = 1, \ldots, n. \qquad (3.1')$$

Note that this result generalizes Theorem 7.6 and shows that it is still valid if condition (7.6) is replaced by (7.6′).

*Section 8*

The main ideas of the proofs of Theorem 8.7 and Lemma 8.8 are given in the original proof of Theorem 14.11.

Let us say that a section conjugacy functor **W** on $G$ *controls normal $p$-complements* in $G$ if $\mathbf{N}(\mathbf{W}(S))$ has no normal $p$-complement or if $G$ has a normal $p$-complement. By Theorem 8.7, **W** controls normal $p$-complements in $G$ if it controls normal $p$-complements in every section $G^*$ of $G$ that satisfies the hypothesis of Lemma 8.8. This observation and the Hall–Higman Theorem (Chapter 11, Gorenstein, 1964) immediately yield Theorem 8.9.

Let $H$ be an arbitrary subgroup of $G$ that contains $S$. An important result of Tate (1964) states that if $H'\mathbf{O}^p(H) = H \cap G'\mathbf{O}^p(G)$, then $\mathbf{O}^p(H) = H \cap \mathbf{O}^p(G)$. An easy argument (Step (I), Section 4, Glauberman, 1968) then yields that if $S \cap H' = S \cap G'$, then $S \cap \mathbf{O}^p(H) = S \cap \mathbf{O}^p(G)$. Now suppose $H = \mathbf{N}(\mathbf{W}(S))$ for some conjugacy functor $\mathbf{W}$ on $G$. Tate's result shows that if $\mathbf{W}$ controls transfer in $G$, then $\mathbf{O}^p(\mathbf{N}(\mathbf{W}(S))) = \mathbf{N}(\mathbf{W}(S)) \cap \mathbf{O}^p(G)$ and $\mathbf{W}$ controls normal $p$-complements in $G$.

*Section 9*

Suppose $T \subseteq S$ and $T \trianglelefteq \mathbf{N}(S)$. Then we can define a conjugacy functor $\mathbf{W}$ on $G$ as follows:

$$\mathbf{W}(S^g) = T^g \quad \text{for all } g \in G;$$

$$\mathbf{W}(U) = U \quad \text{if } U \text{ is a } p\text{-subgroup of } G \text{ not conjugate to } S.$$

Thus $\mathbf{W}(S) = T$.

*Section 12*

Theorem 1 of Glauberman (1970) extends part (b) of Proposition 12.1 by asserting that if $Q$ contains $\mathbf{K}_\infty(P)$ and $\mathbf{K}^\infty(P)$, then $\mathbf{K}_\infty(P) = \mathbf{K}_\infty(Q)$ and $\mathbf{K}^\infty(P) = \mathbf{K}^\infty(Q)$. Theorems 10.2, 12.2, 12.3, and 12.4 are proved in Glauberman (1970).

The hypothesis of Corollary 12.6 occurred in Feit's investigation of Frobenius groups (Feit, 1957).

Let $\mathbf{W}$ be a conjugacy functor on $G$, and let $H = \mathbf{N}(\mathbf{W}(S))$. In our notes on Section 8, we mentioned that Tate's work gives the following result.

If $\mathbf{W}$ controls transfer in $G$, then $\mathbf{O}^p(H) = H \cap \mathbf{O}^p(G)$.

Consider the following condition on $\mathbf{W}$.

Let $T$ be the largest normal $p$-subgroup of $H$ such that $H/\mathbf{C}_H(T)$ is a $p$-group. If $x \in T$, $g \in G$, and $x^g \in S$, then $x^g = x^h$ for some $h \in S$. (A3.1)

A "dual" statement to Tate's result might assert that if $\mathbf{W}$ controls weak closure of elements in $G$, then $\mathbf{W}$ satisfies (A3.1). Unfortunately, this statement if easily seen to be false (e.g., when $p = 2$, $G = SL(2, 7)$, and $\mathbf{W}(P) = P$ for every $p$-subgroup $P$ of $G$). However, the results of Section 12 yield that $\mathbf{K}_\infty$ and $\mathbf{K}^\infty$ both satisfy (A3.1) if $p \geq 5$.

*Section 14*

An extension of (14.1) is proved in Section II.7 of Alperin, Brauer, and Gorenstein (1970).

Suppose that $\mathbf{C}(\mathbf{O}_p(G)) \subseteq \mathbf{O}_p(G)$. Let $R$ be a subgroup of $S$ that contains $\mathbf{O}_p(G)$. Assume either that $R \trianglelefteq S$ or that $RK/K$ contains $\mathbf{O}_p(G/K)$ for every normal subgroup $K$ of $G$. Assume further that whenever $A, P \subseteq R$ and $P \trianglelefteq G$, and $[P, A, A] = 1$, then

$$AC(P)/P \subseteq \mathbf{O}_p(\mathbf{N}(P)/\mathbf{C}(P)).$$

Then $\mathbf{Z}(\mathbf{J}(R)) \trianglelefteq G$.

The functor $\mathbf{ZJ}^*$ mentioned in our notes on Section 6 satisfies those results of Section 14 that are stated for $\mathbf{ZJ}$ except Theorem 14.13.

An extension of Theorem 14.10 for $p = 2$ has been obtained (Glauberman, 1969).

*Section 15*

Denote by $\mathbf{O}_{2',2}(G)$ the normal subgroup $N$ of $G$ that contains $\mathbf{O}_{2'}(G)$ and satisfies $N/\mathbf{O}_{2'}(G) = \mathbf{O}_2(G/\mathbf{O}_{2'}(G))$. Theorem 15.4 is proved by induction and the following result of work to be published by Goldschmidt, which generalizes the main part of the proof of Theorem 15.3.

Suppose that $p = 2$, $W \subseteq \mathbf{Z}(S)$, and $W$ is a weakly closed subgroup of $S$ with respect to $G$. Suppose $t \in S - W$ and $t^2 = 1$. Assume that $W \subseteq \mathbf{O}_{2',2}(\mathbf{C}(tw))$ for all $w \in W$. Then $W \subseteq \mathbf{O}_{2',2}(G)$.

In some recent unpublished work, Shult has proved the following extensive generalization of Theorem 15.3.

Suppose $p = 2$, $x \in S$, and $x$ has order two. Let $T$ be the subgroup of $\mathbf{C}(x)$ generated by all conjugates of $x$ that are contained in $\mathbf{C}(x)$. Assume that $T$ is abelian and $T^g \cap \mathbf{N}(T) \subseteq T$ for all $g \in G$. Then $x$ is contained in a normal subgroup $N$ of $G$ such that

$$N/Z(N) \cong N_1 \times \ldots \times N_r \times M,$$

where $M$ has an elementary abelian Sylow 2-subgroup and a normal 2-complement, and each $N_i$ is isomorphic to $SL(2, 2^{n_i})$, $Sz(2^{n_i})$, or $PSU(3, 2^{n_i})$ for some $n_i$.

Here $Sz(2^n)$ denotes one of the infinite family of Suzuki groups (Gorenstein, 1968) and $PSU(3, 2^n)$ is a unitary group (pp. 236–237, Huppert, 1967).

*Section 16*

As indicated in Sections 12, 14, and 16, we have few results for $p = 2$; most of these results involve $\mathbf{J}$ rather than $\mathbf{K}_\infty$ (or $\mathbf{K}^\infty$). In considering Questions 16.1 and 16.2, we may find it useful to examine the similarities and differences between $\mathbf{J}$ and $\mathbf{K}_\infty$. Theorem 12.2 and the Replacement Theorem enable us to use commutator conditions with both functors in roughly the same way; the discussion preceding Question 16.4 suggests that $\mathbf{K}_\infty$ may be slightly more useful in this respect for $p \geq 5$. The main advantages of $\mathbf{J}$ for $p = 2$ seem to be the "hereditary" and numerical properties that yield parts

(b) and (d) of Proposition 14.1. These properties are used in the proofs of Theorems 14.10 and 14.11 and appear to be absent from $\mathbf{K}_\infty$.

The main part of Question 16.5 concerns the case in which $G$ is irreducible on $V$ (as defined in Section A1). In some recent unpublished work, Thompson has completely answered Question 16.5 in this case when $p \geq 5$.

## Bibliography


Alperin, J. L. (1967). Sylow intersections and fusion. *J. Algebra* **6,** 222.

Alperin, J. L., Brauer, R., and Gorenstein, D. (1970). Finite groups with quasi-dihedral and wreathed Sylow 2-subgroups. *Trans. Amer. Math. Soc.* **151,** 1.

Alperin, J. L. and Gorenstein, D. (1967). Transfer and fusion in finite groups. *J. Algebra* **6,** 242.

Collins, M. (1969). "Some problems in the theory of finite insoluble groups". Ph.D. thesis, University of Oxford, Oxford, England.

Feit, W. (1957). On the structure of Frobenius groups. *Canad. J. Math.* **9,** 587.

Feit, W. (1969). Some properties of the Green correspondence. *In* "Theory of Finite Groups, A Symposium". W. A. Benjamin, New York.

Glauberman, G. (1966). Central elements of core-free groups. *J. Algebra* **4,** 403.

Glauberman, G. (1968a). A characteristic subgroup of a $p$-stable group. *Canad. J. Math.* **20,** 1101.

Glauberman, G. (1968b). Weakly closed elements of Sylow subgroups I, II. *Math. Zeit.* **107,** 1; (1969). **112,** 89.

Glauberman, G. (1968c). Prime-power factor groups of finite groups I, II. *Math. Zeit.* **107,** 159; (1970). **117,** 46.

Glauberman, G. A sufficient condition for $p$-stability. To be published.

Glauberman, G. and Thompson, J. G. (1968). Weakly closed direct factors of Sylow subgroups. *Pacific J. Math.* **26,** 73.

Goldschmidt, D. (1970). A conjugation family for finite groups.

Goldschmidt, D. An application of Brauer's second main theorem. To be published.

Goldschmidt, D. On the 2-exponent of a finite group. To be published.

Gorenstein, D. (1968). "Finite Groups". Harper and Row, New York.

Gorenstein, D. and Walter, J. (1964). On the maximal subgroups of finite simple groups. *J. Algebra* **1,** 168.

Hall, M. (1959). "The Theory of Groups". Macmillan, New York.

Hall, P. and Higman, G. (1956). On the $p$-length of $p$-solvable groups and reduction theorems for Burnside's problem. *Proc. London Math. Soc.* **(3)6,** 1.

Herstein, I. N. (1964). "Topics in Algebra". Blaisdell, New York.

Huppert, B. (1967). "Endliche Gruppen I". Springer-Verlag, New York.

McLaughlin, J. (1967). Some groups generated by transvections. *Archiv der Math.* **18,** 364.

McLaughlin, J. (1969). Some subgroups of $SL_n(F_2)$. *Ill. J. Math.* **13,** 108.

Shult, E. (1966). Some analogues of Glauberman's $Z^*$-Theorem. *Proc. Amer. Math Soc.* **17,** 1186.

Shult, E. On the fusion of an involution in its centralizer. To be published.

Tate, J. (1964). Nilpotent quotient groups. *Topology* **3,** 109.

Thompson, J. G. (1960a). Normal $p$-complements for finite groups. *Math. Zeit.* **72** 332.

Thompson, J. G. (1960b). A special class of non-solvable groups. *Math. Zeit.* **72,** 458.

Thompson, J. G. (1969). A replacement theorem for finite groups and a conjecture. *J. Algebra* **13,** 149.

Thompson, J. G. (1970). Normal *p*-complements and irreducible characters. *J. Algebra.* **14,** 129.

Thompson, J. G. Quadratic pairs. To be published.

Wielandt, H. (1940). *p*-Sylowgruppen und *p*-Faktorgruppen. *J. reine angew. Math.* **182,** 180.

CHAPTER II

# Centralizers of Involutions in Finite Simple Groups

D. GORENSTEIN

## 1. General Introduction

### 1.1. Centralizers of Involutions

In the study of finite simple groups, the centralizers of involutions (that is, elements of order 2) have come to play a fundamental role. There exist a number of results which would lead one to anticipate the importance of involutions. The celebrated Feit–Thompson theorem that groups of odd order are solvable implies that every nonabelian simple group has even order. Then there is an older result of Brauer and Fowler which asserts that there are at most a finite number of simple groups in which the centralizer of an involution has a given structure.

More recently a considerable number of results of the following general type have been established. Let $G^*$ be some one of the presently known simple groups, let $z^*$ be an involution in the center of a Sylow 2-subgroup of $G^*$ (for brevity, a *central* involution), and let $N^* = C_{G^*}(z^*)$ be the centralizer of $z^*$ in $G^*$. Now let $G$ be an abstract simple group, let $z$ be a central involution of $G$, and put $N = C_G(z)$. Suppose that $N \cong N^*$. Then $G \cong G^*$.

In other words, the simple group $G^*$ is completely *characterized* by the structure of the centralizer of one of its central involutions. In some situations there is more than one involution in the center of a given Sylow 2-subgroup and one may hypothesize a knowledge of each of these centralizers. Still other variations have been considered. The central point, however, is the following. Sufficient information concerning the centralizers of certain of their involutions is enough to characterize many of the known simple groups.

Of course, one can equally well prove non-existence theorems of this nature—namely, no simple group exists in which the centralizer of some involution or involutions has a specified structure. On the other hand, there are instances in which there exists more than one simple group with a given centralizer. For example, the centralizer of a central involution in the alternating group $A_8$ is isomorphic to that in $A_9$. A similar situation occurs in Janko's two recently discovered simple groups $J_2$ and $J_3$. An even more striking illustration occurs in the *three* groups $GL(5, 2)$, the five-dimensional linear group over $GF(2)$, the field with 2 elements, $M_{24}$, the largest of the five simple groups of Mathieu, and the new simple group of Held. In fact, Held's group was discovered in the process of attempting to characterize $GL(5, 2)$ and $M_{24}$ by the structure of the centralizer of a central involution. However, this ambiguous situation is the exception rather than the rule. The more typical situation can be exemplified by the following theorem of Brauer, representing one of the pioneering results in this direction.

Theorem. *Let $G$ be a simple group with an involution whose centralizer is isomorphic to $GL(2, q)/X$, where $X$ is a subgroup of odd order in the center of*

*$GL(2, q)$ and where $q$ is an odd prime power congruent to* $-1 \pmod 4$. *Then either*

*$G \cong PSL(3, q)$, the three-dimensional projective linear group; or*
*$q = 3$ and $G \cong M_{11}$, the smallest Mathieu group.*

Actually, $X$ has order 1 or 3 according as $q-1$ is not or is divisible by 3 However, this fact is proved in the course of the argument and so it need not be assumed. To see that these conditions, in fact, hold in $PSL(3, q)$, one has only to compute the centralizer $N$ of the involution $z = \begin{pmatrix} -1 & 0 & 0 \\ 0 & -1 & 0 \\ 0 & 0 & 1 \end{pmatrix}$ in $SL(3, q)$, which is clearly all matrices of the form $\begin{pmatrix} A & & 0 \\ & & 0 \\ 0 & 0 & a \end{pmatrix}$, where $a \cdot \det A = 1$. Thus, $N \cong GL(2, q)$. The centralizer of the image of $z$ in $PSL(3, q)$ is the image of $N$ in $PSL(3, q)$, which is $N/N \cap Z(SL(3, q))$. ($Z(H)$ will always denote the center of the group $H$.) But $N \cap Z(SL(3, q))$ is of order 3 or 1 according as 3 does or does not divide $q-1$. Thus the given conditions do hold in $PSL(3, q)$.

This particular area of finite group theory is presently in considerable flux and the subject of much effort. It is perhaps premature, but certainly not unreasonable, to think that ultimately each of the presently known simple groups will be characterized in terms of the structure of the centralizers of their central involutions.

It is not my object in these talks to discuss this fundamental portion of the study of finite simple groups but, rather, to use it as motivation for our central topic. Let us then accept, for the purpose of these talks, the following general principle.

If, in an abstract simple group $G$, the centralizer of a central involution is isomorphic to that in one of the known simple groups $G^*$, then (allowing for ambiguities) methods exist for proving that $G$ is isomorphic to $G^*$.

## 1.2. The Core of the Centralizer of an Involution

The effect of the above principle is to shift the focus in the problem of classifying the finite simple groups to the following basic question.

If $G$ is an abstract simple group and $z$ is a central involution of $G$, is it possible to prove that $N = C_G(z)$ is isomorphic to the centralizer of a central involution in one of the presently known simple groups?

It is this question to which I intend to address myself in these talks. There is gradually developing a body of techniques for successfully attacking problems of this type and I hope to describe some of these to you in detail.

Unfortunately in the form that I have posed the question, the problem is so general that it will not be possible for me to do justice to it in a short series of lectures. Instead, I shall concentrate on a single, but very important, aspect of the problem; namely, the (2-regular) *core* of $N$; that is, the largest normal subgroup of $N$ of odd order, which we denote by $O(N)$. Thus, we ask.

What can we say about the structure of $O(N)$ in our abstract simple group $G$?

This might seem to be only a very minor aspect of the overall problem posed above. However, every general classification problem solved to date has dealt in very large part with this specific question. This includes the classification of

(a) groups with a dihedral Sylow 2-subgroup $S$
(here $S = \langle a, b : a^{2^n} = b^2 = 1 \quad \text{and} \quad b^{-1}ab = a^{-1}, \quad n \geq 1\rangle$);
(b) groups with a quasi-dihedral Sylow 2-subgroup $S$
(here $S = \langle a, b : a^{2^n} = b^2 = 1 \quad \text{and} \quad b^{-1}ab = a^{-1+2^{n-1}}, \quad n \geq 3\rangle$);
(c) groups with an abelian Sylow 2-subgroup;
(d) minimal simple groups (i.e., simple groups whose proper subgroups are solvable) and, more generally, $N$-groups (groups whose nontrivial solvable subgroups have solvable normalizers).

It has been proved that the only simple groups with dihedral Sylow 2-subgroups are the groups $PSL(2, q)$ ($q$ odd) and $A_7$; and those with quasi-dihedral Sylow 2-subgroups are the groups $PSL(3, q)$ ($q \equiv -1 (\text{mod } 4)$), $M_{11}$, and $PSU(3, q)$ ($q \equiv 1 (\text{mod } 4)$). (The latter group is the projective special unitary group, the general unitary group $GU(3, q)$ being the group of matrices $X$ over $GF(q^2)$ such that $((X^\sigma)^t)^{-1} = X$, where $t$ denotes transpose and $X^\sigma$ is the automorphism of $X$ induced by raising each entry of $X$ to the $q$th-power). In both of these problems, as well as in the abelian Sylow 2-subgroup problem, the bulk of the argument involves the analysis of $O(N)$; while in the $N$-group problem a large portion of the work involves the study of $O(N)$.

For example, in the quasi-dihedral case, one knows very early in the argument (using the dihedral classification and the fact that $N/\langle z\rangle$ has dihedral Sylow 2-subgroups) that $N$ possesses a normal subgroup $L$ containing $O(N)$ such that

(a) $N/L$ is cyclic of odd order; and
(b) $L/O(N) \cong SL^{\pm}(2, q)$ for some $q \equiv -1$ (mod 4) or $SU^{\pm}(2, q)$ for some $q \equiv 1$ (mod 4) (here the $\pm$ refers to matrices of determinant $+1$ or $-1$).

Almost the entire paper consists in establishing the following results.

(1) $O(N)$ is cyclic.
(2) $O(N) \subseteq Z(L)$.

(3) $|O(N)|$ divides $q-1$ or $q+1$ according as $q \equiv -1 \pmod 4$ or $q \equiv 1 \pmod 4$.

These three conditions assert precisely that $L$ is isomorphic respectively to $GL(2, q)/X$ or $GU(2, q)/X$, $X$ being a central subgroup of odd order. In the case that $q \equiv -1 \pmod 4$, we see then that to complete the classification all that remains is to prove that $N = L$, for once this is established $G$ satisfies the hypotheses of Brauer's theorem and so $G \cong PSL(3, q)$ or $M_{11}$.

Whatever may be required to pin down the structure of $N$ in more general classification problems, one can assert without reservation that the first step in the analysis will involve the determination of the structure of $O(N)$. I hope this brief discussion will convince you of the essential importance of the study of $O(N)$ for the ultimate determination of the structure of $N$.

### 1.3. Cores in Known Simple Groups

To know what kind of results we would like to prove, it is natural to inquire first about the structure of $O(N)$ when $G$ is actually one of the presently known simple groups. Remarkably enough, there seems to be a very simple general answer, not only when $z$ is a central involution, but when $z$ is an arbitrary involution. We have

(1) $O(N)$ is cyclic;
(2) $C_N(O(N))$ has index 1 or 2 in $N$.

Note that if the index is 1 (equivalently $N = C_N(O(N))$), then $O(N) \subseteq Z(N)$. There are many cases, however, where the index is 2; for example, in $PSL(2, q)$ ($q$ odd, $q$ not a Mersenne or Fermat prime or 9), and in $A_{4n+3}$.

In general then, we see that the structure and embedding of $O(N)$ in $N$ appears to be very similar to that which we encounter in the quasi-dihedral case. Frequently, however, $O(N)$ actually turns out to be trival for every involution $z$ of $G$. To describe the occurrence of this phenomenon, we must say a few words about the presently known simple groups (of composite order).

The bulk of these are the so-called groups of *Lie type* over a finite field. They include the classical groups $PSL(n, q)$, $PSU(n, q)$, symplectic, and orthogonal groups as well as those associated with the exceptional Lie algebras. It would take too long to describe them in any detail. Let me just say that these families of groups are now well understood. Their existence and properties have been systematically investigated by a number of people, notably, Chevalley, Steinberg, and Tits.

Apart from these and the alternating groups, there exist only the so-called *sporadic* simple groups. These are the groups which do not appear to be part of an infinite family of simple groups—for example, the five Mathieu groups, the three Janko groups, and Held's group. There have been a number of such recently discovered groups and at present the list is approaching twenty in number. One of the most fascinating questions of finite group theory is whether the number of sporadic groups is finite or infinite. If the latter were true, we would be in serious difficulties, for it would then seem to be impossible ever to complete the classification of all finite simple groups.

For which of these types is $O(N)$ always trivial? The answer seems to be roughly as follows:

(1) groups of Lie type over fields of characteristic 2;
(2) $A_m$ $(m = 4n, 4n+1, 4n+2)$; and
(3) most (and perhaps all) of the sporadic groups.

In general, in the groups of Lie type over fields of odd characteristic, $O(N)$ is nontrivial. However, exceptions occur for special characteristics (for example, $PSL(3, 3)$ or $PSU(3, 5)$) and also for certain families (for example, $G_2(q)$ and $PS_p(4, q)$ ($q$ odd), $G_2(q)$ being the group associated with the exceptional Lie algebra $G_2$ and $PS_p(4, q)$ being the four dimensional projective symplectic group. In the latter case, $O(N)$ is trivial if $z$ is a central involution, but is, in general, non-trivial if $z$ is non-central.)

Primarily, but not entirely, the triviality of $O(N)$ is a property associated with the groups defined over fields of characteristic 2. Thus, depending upon the particular classification problem under consideration, one's objective will be to establish one of the following two assertions:

(1) $O(N)$ is trivial, or
(2) $O(N)$ is a nontrivial cyclic group.

It turns out that it is technically much easier to handle those situations in which the goal is to prove that $O(N) = 1$.

One can give a plausible explanation of this quite easily. Indeed, turned around, the problem invariably comes down to deriving a contradiction from the existence of a suitable *nontrivial* normal subgroup $K$ of $O(N)$ such that $O(N)/K$ is cyclic, for some involution $z$ of $G$. In case (1), we know which subgroup to take for $K$ in advance—namely, $O(N)$; whereas in case (2), the determination of the proper choice of this normal subgroup is itself a part of the problem. Construction of the "right" subgroup $K$ in the general situation is quite complicated to describe. For this reason, it will be preferable for me to limit myself to problems in which the task is to prove that $O(N)$ is trivial. In doing this, none of the general features of the over-all problem will be lost; in fact, they will be much easier to discern.

## 1.4. Groups with 2-Constrained Centralizers

I should like to lead us now to a quite general set of assumptions in which it is possible to establish the triviality of $O(N)$. As indicated above, this condition is somewhat related to the groups of Lie type over fields of characteristic 2. I wish to describe an additional property of these groups that will help to distinguish them much more precisely.

For any group $X$, we use the standard terminology $O_p(X)$, $O_{p'}(X)$, $O_{p',p}(X)$, $p$ a prime. Thus $O_p(X)$, $O_{p'}(X)$ denote the unique largest normal $p$-subgroup and subgroup of order prime to $p$ respectively, of $X$, while $O_{p',p}(X)$ denotes the inverse image in $X$ of $O_p(X/O_{p'}(X))$. In particular, $O_{2'}(X)$ is simply the core $O(X)$ of $X$. We shall also use the bar convention: if $\bar{X}$ is a homomorphic image of $X$, $\bar{Y}$ will always denote the image in $\bar{X}$ of an element, subset, or subgroup $Y$ of $X$.

If $G$ is any simple group of Lie type over a field of characteristic 2, it appears to be always the case that $C_N(O_2(N)) \subseteq O_2(N)$ for any involution $z$ of $G$. To see that this is not a general property of some of the other families, take $G = A_n$ with $n \geq 9$ and $z = (12)(34)$. One sees that

$$N = C_G(z) = (T \times L)A,$$

where $T = \langle(12)(34),\ (13)(24)\rangle$ is a four group (that is, abelian of type (2, 2)), $L \cong A_{n-4}$, the alternating group on the letters $\{5, 6, \ldots, n\}$ and $A = \langle(15)(67)\rangle$ is of order 2. Thus, $N/T \cong S_{n-4}$. Since $n-4 \geq 5$, $L$ is simple and consequently $O(N) = 1$ and $O_2(N) = T$. But $C_N(T) = T \times L \nsubseteq T$, so $C_N(O_2(N)) \nsubseteq O_2(N)$.

Similarly, return to the group $G = PSL(3, q)$, where $q \equiv -1 \pmod 4$ $(q > 3)$, and, for definiteness, with 3 not dividing $q-1$, in which case $N \cong GL(2, q)$. Of course, $O(N)$ is now nontrivial and is in the center of $N$, so we really cannot expect to have $C_N(O_2(N)) \subseteq O_2(N)$. To be fair, we must really ask the question in $\bar{N} = N/O(N)$, which is isomorphic to $SL^{\pm}(2, q)$. Since $q > 3$, $O_2(\bar{N})$ is of order 2 and is in the center of $\bar{N}$ (under the isomorphism of $\bar{N}$ with $SL^{\pm}(2, q)$, $O_2(\bar{N})$ corresponds to $\left\langle\begin{pmatrix} -1 & 0 \\ 0 & -1 \end{pmatrix}\right\rangle$). Hence $C_{\bar{N}}(O_2(\bar{N})) = \bar{N} \neq O_2(\bar{N})$.

Likewise this condition is not satisfied in the groups $PSp(4, q)$ or $G_2(q)$ when $q > 3$.

Let us formalize this property as follows.

Definition If $X$ is a group in which $O_{p'}(X) = 1$ ($p$ prime), we say that $X$ is *p-constrained* if

$$C_X(O_p(X)) \subseteq O_p(X).$$

More generally, if $O_{p'}(X) \neq 1$, we say that $X$ is *p-constrained* if $X/O_{p'}(X)$ is $p$-constrained.

(Note that $O_{p'}(X/O_{p'}(X)) = 1$.)

The concept of constraint is actually a direct generalization of a well-known lemma of P. Hall and G. Higman concerning solvable groups. In the present terminology, it can be phrased as follows.

THEOREM. *A solvable group is p-constrained for every prime p.*

Thus in the groups of Lie type over fields of characteristic 2 the centralizer of every involution is 2-constrained, whereas in those of odd characteristic and in the alternating groups this is not the case with the exception of a few groups of low orders. On the other hand, remarkably enough, the centralizer of every involution is 2-constrained in many, but not all, of the sporadic groups. It is in simple groups with this property—the centralizer of every involution 2-constrained—that we shall be primarily interested in these talks. At the end, I shall consider a somewhat larger class of groups which will encompass all the sporadic groups along with the groups of Lie type of characteristic 2.

## 1.5. GROUPS OF LOW 2-RANK

In order to reach a precise theorem, we shall need to discuss briefly one additional topic. We require a preliminary definition.

DEFINITION. For any group $X$ and any prime $p$, the *p-rank* of $X$ is the maximum rank of an abelian $p$-subgroup of $X$.

(The rank of an abelian $p$-group $A$, denoted by $m(A)$, is, as customary, the number of elements in a basis of $A$). Clearly the $p$-rank of $X$ is the same as the $p$-rank of a Sylow $p$-subgroup of $X$.

One may also define the *normal* $p$-rank of $X$ as the maximum rank of an abelian $p$-subgroup $A$ of $X$ with $A$ normal in a Sylow $p$-subgroup $P$ of $X$.

This is closely related to the Feit–Thompson concept of $SCN(p)$, which is the collection of abelian subgroups $A$ which are maximal subject to being normal in a Sylow $p$-subgroup $P$ of $X$. Such a subgroup $A$ is self-centralizing in $P$ and this accounts for the notation $SCN$. One writes $A \in SCN_r(p)$ if $A \in SCN(p)$ and $m(A) \geq r$. Thus it is clear that the normal $p$-rank of $X$ is precisely the maximum integer $r$ for which $SCN_r(p)$ is non-empty.

There is actually some connection between the $p$-rank and the normal $p$-rank of a group. For example, the combined results of Blackburn and MacWilliams tell us that if the normal $p$-rank of $X$ is less than 3 (equivalently, if $SCN_3(p)$ is empty), then the $p$-rank of $X$ is less than 3 if $p$ is odd and is less than 5 if $p = 2$; moreover, it is well known that if $X$ has $p$-rank 1, then its Sylow $p$-subgroups are either cyclic or generalized quaternion.

Our concern here, naturally, is primarily with the 2-rank. It has been evident for some time that special problems and special methods exist in connection with simple groups of low 2-rank. In Thompson's classification of $N$-groups, it was necessary for him to give a completely separate analysis in the case that $SCN_3(2)$ was empty. Fortunately the methods of modular character theory seem to be especially suited to dealing with these low rank groups, particularly those of 2-rank less than 3.

Moreover, in these cases one can pin down in advance the possible structure of a Sylow 2-subgroup of such a simple group $G$. For example, Alperin has shown that if $G$ has 2-rank less than 3, then a Sylow 2-subgroup of $G$ is of one of the following types:

(1) dihedral;
(2) quasi-dihedral;
(3) wreathed (that is, $Z_{2^n} \int Z_2$ ($n \geq 2$); such groups occur as the Sylow 2-subgroups of $PSL(3, q)$ ($q \equiv 1 \pmod 4$)) and $PSU(3, q)$ ($q \equiv -1 \pmod 4$);
(4) isomorphic to a Sylow 2-subgroup of $PSU(3, 4)$.

We note that in each case $G$ is actually of 2-rank 2 because a 2-group of rank 1 is either cyclic or generalized quaternion. Hence, if $G$ has 2-rank 1, this result together with either Burnside's transfer theorem or a theorem of Brauer and Suzuki implies that $G$ is not simple.

Simple groups of the first two types have been completely classified. A very large portion of the classification of those with wreathed Sylow 2-subgroups is included in the quasi-dihedral analysis. Furthermore, Lyons has recently shown, using modular character theory, that $PSU(3, 4)$ is the only simple group of type 4. Thus we can say that the simple groups of 2-rank less than 3 are essentially completely classified (see Note Added in Proof 2).

Likewise MacWilliams and Thompson have almost completed the determination of the possible structure of a Sylow 2-subgroup of a simple group in which $SCN_3(2)$ is empty. Perhaps it will be possible to treat similarly the more general problem of groups of 2-rank at most 4. The classification of simple groups with specific Sylow 2-subgroups of rank 3 or 4 is just beginning and would seem to depend upon modifications of the techniques that I shall be discussing in these talks. I might mention that Harada and I have recently shown that $J_2$ and $J_3$ are the only simple groups having a Sylow 2-subgroup $S$ isomorphic to that of $J_2$ or $J_3$. Such a group $S$ is of order $2^7$ and 2-rank 4, $SCN_3(2)$ being empty.

## 1.6. Statement of the Main Theorem

Let us return then to our central topic—groups in which the centralizer of every involution is 2-constrained. The discussion of the preceding section was designed to justify our making some assumption that the 2-rank of our group $G$ be not too low. Our objective in the balance of these talks will be to give a proof of the following theorem, which was obtained jointly with John H. Walter.

THEOREM. *Let G be a group of 2-rank at least 5 in which $O(G) = 1$ and assume that the centralizer of every involution of G is 2-constrained. Then*

$$O(C_G(x)) = 1$$

*for every involution x of G.*†

We do not require the simplicity of $G$ for our result, but only the weaker assumption that $O(G) = 1$. An analogous result is obtained by Thompson in the $N$-group situation and our theorem can be considered to be a generalization of his result. In particular, the theorem holds if the centralizer of every involution of $G$ is solvable. A special case of this gives a new classification theorem as a corollary.

A group $X$ is said to be 2-*closed* if it has a normal Sylow 2-subgroup; and it is said to be of 2-*length* 1 if $X/O(X)$ is 2-*closed.* Since groups of odd order are solvable, every group of 2-length 1 is solvable. Moreover, if $X$ is of 2-length 1 with $O(X) = 1$, then $X$ is 2-closed. In a long sequence of papers Suzuki has classified all groups in which the centralizer of every involution is 2-closed. Combining his results with the preceding theorem, we obtain the following corollary:

COROLLARY. *If G is a simple group of 2-rank at least 5 in which the centralizer of every involution has 2-length 1, then G is isomorphic to one of the groups* $PSL(2, 2^n)$, $Sz(2^n)$, $PSU(3, 2^n)$ $(n \geq 5)$, *or* $PSL(3, 2^n)$ $(n \geq 3)$.

The restriction on $n$ is a consequence of our assumption on the 2-rank of $G$. Here the groups $Sz(2^n)$ denote the simple groups of Lie type discovered by Suzuki in the course of his investigation of simple groups in which the centralizer of every involution is a 2-group.

The proof of our main theorem divides into two parts. The first part involves some general results of mine about centralizers of involutions, obtained in three papers (Gorenstein, 1969a, 1969b, 1970). We shall refer to these as C.I.1, 2, 3 respectively. The second part is a particular case of a more general theorem obtained jointly with John Walter. The bulk of our talks will be devoted to the discussion of these various results.

I have tried in these introductory remarks to place the main theorem in its proper context within the broader problem of classifying the simple groups and, in particular, the problem of determining the structure of the centralizers of involutions in abstract simple groups.

† Because of recent improvements in the signalizer functor theorem made by Goldschmidt, this theorem is now valid under the weaker assumption that $G$ has 2-rank 3 and $SCN_3(2)$ is nonempty (or, more generally, connected). See the final comment.

## 2. Local Group-theoretic Analysis

### 2.1. Preliminary Remarks

The proof of our main theorem lies at the heart of what has come to be called *local group-theoretic analysis*. The normalizers of the nontrivial subgroups of prime power order in a group $G$ are called the *local* subgroups of $G$ (*p-local* if we wish to distinguish the particular prime). We note that the centralizer of an involution $x$ of $G$ is the same as the normalizer of the subgroup $\langle x \rangle$ and so is an example of a 2-local subgroup.

Any theorem whose proof involves solely a study of the structure, embedding, and interrelationships of various local subgroups is said to be proved by local group-theoretic analysis. Such a proof can involve no character theory, for one regards the characters of $G$ from this point of view as global properties of $G$ (I do not wish to involve myself here with the question of whether the characters are determined from the local structure of $G$). As we proceed, we shall come to see what is meant by the term local group-theoretic analysis.

Sometimes it is referred to as *Sylow-type arguments*. A lovely example is Philip Hall's classical characterization of solvable groups (allowing for the use of Burnside's $p^a q^b$-theorem, which, of course, is proved by character theory). It was Thompson who first revealed the true power of this technique, when pushed relentlessly; beginning with his original proof of the Frobenius conjecture in his thesis and reaching full flower in Chapter IV of the odd order paper, it is then extended to unimagined lengths in the $N$-group paper. Thus, despite the fact that our main theorem concerns involutions, its proof is profoundly influenced by results and ideas of Chapter IV of the odd order paper.

In the first part of my book on finite groups I have tried to develop systematically the main tools needed to carry out local group-theoretic analysis. I think it will be valuable for you if I begin by summarizing some of the essential results—to the extent that we shall need them for the main theorem. In this way you will have before you an outline of the basic techniques that are used. No attempt will be made to prove any of these results; almost all of them can be found in my book and it will be better to allow the maximum time for seeing how they are applied.

I must, of course, assume a general familiarity with the essentials of finite group theory and so I make no pretense of a systematic development. Rather, my aim will be to list a number of salient results and then to explain their significance for local group-theoretic analysis.

### 2.2. Groups of Prime Power Order

There is a great variety of results about $p$-groups that are used in local analysis—some concern their automorphisms, others their action on $p'$-

groups, still others concern their embedding in larger groups. As an example of the third type, we may mention the simple, but fundamental, consequence of Sylow's theorem, known as the Frattini argument.

THEOREM. *If $K \lhd H$ and $P$ is a Sylow $p$-subgroup of $K$, then $H = KN_H(P)$.*

Suppose that $H$ above is contained in our group $G$ under investigation. Then $H = K(N_G(P) \cap H)$. But $N_G(P)$ is a $p$-local subgroup of $G$ if $P \neq 1$. Thus knowledge of $H$ is obtained from that of $K$ and the local subgroup $N_G(P)$.

The following, equally basic result, on the action of $p$-groups on $p'$-groups, can often be used to obtain knowledge of $K$ as well from a knowledge of $p$-local subgroups.

THEOREM. *If $P$ is a noncyclic abelian $p$-group acting on the $p'$-group $L$, then $L$ is generated by its subgroups $C_L(P_0)$, where $P_0$ ranges over the nontrivial subgroups of $P$.*

This result reduces ultimately to the important fact that a faithful, irreducible representation of an abelian group on any vector space is necessarily cyclic.

As an example, if $P$ is a *four group* (that is, $P$ is abelian of type (2, 2)) with involutions $t_1, t_2, t_3$, the theorem asserts that $L = \langle C_L(t_i) : 1 \leq i \leq 3 \rangle$. We shall have more to say about this situation later.

Consider again a subgroup $H$ of our group $G$, and take $K$ to be $O_{p',p}(H)$. Set $L = O_{p'}(H)$ and let $P$ be a Sylow $p$-subgroup of $O_{p',p}(H)$, so that $K = LP$. By the Frattini argument

$$H = K(N_G(P) \cap H) = LP(N_G(P) \cap H) = L(N_G(P) \cap H).$$

Suppose now that $P$ contains a noncyclic abelian subgroup. Then the preceding theorem implies that

$$L = \langle C_L(P_0) : 1 \subset P_0 \subseteq P \rangle.$$

We see then that in this case $H$ is generated by its intersections with the $p$-local subgroups $N_G(P_0)$ as $P_0$ ranges over the nontrivial subgroups of $P$.

Next we state a well-known theorem of Burnside on $p'$-automorphisms of $p$-groups.

THEOREM. *If $\alpha$ is a $p'$-automorphism of the $p$-group $P$ and $\alpha$ induces the trivial automorphism on the Frattini factor group $P/\Phi(P)$, then $\alpha = 1$.*

To see how this theorem can be used, consider a $p$-constrained group $H$ in which $O_{p'}(H) = 1$. If we put $P = O_p(H)$, it follows by definition that $C_H(P) \subseteq P$. Now $\Phi(P)$, being characteristic in $P$, is normal in $H$. Setting

$\bar{H} = H/\Phi(P)$, we have $\bar{P} = O_p(\bar{H})$ (here by our convention $\bar{P}$ is the image of $P$ in $\bar{H}$). We claim that $C_{\bar{H}}(\bar{P}) \subseteq \bar{P}$. Indeed, we need only show that $C_{\bar{H}}(\bar{P})$ is a $p$-group, since then, being normal in $\bar{H}$, it must be contained in $O_p(\bar{H}) = \bar{P}$, the unique largest normal $p$-subgroup of $\bar{H}$. However, if the assertion is false, $C_{\bar{H}}(\bar{P})$ contains a nontrivial $p'$-element $\bar{a}$. Then $H$ contains a $p'$-element $a$ which maps on $\bar{a}$. If $\alpha$ denotes the automorphism of $P$ determined by conjugation by $a$, then $\alpha$ induces the trivial automorphism of $\bar{P}$ as $\bar{a}$ centralizes $\bar{P}$. Thus $\alpha$ is trivial on $P$ by Burnside's theorem—that is, $a \in C_H(P)$. But $C_H(P) \subseteq P$, whence $a \in P$. Since $a$ is a $p'$-element, this forces $a = 1$, contrary to the fact that $\bar{a} \neq 1$.

What is the implication of this result? We know that $\bar{P} = P/\Phi(P)$ is an elementary abelian $p$-group and so can be identified with a vector space over $GF(p)$. Moreover, under this identification, each automorphism of $\bar{P}$ becomes a linear transformation of $\bar{P}$. Thus $\bar{H}/\bar{P}$ is represented faithfully as a group of linear transformations of $\bar{P}$. But $\bar{H}/\bar{P} \cong H/P$. We conclude therefore that $H/O_p(H)$ is *faithfully* represented as a linear group on the Frattini factor group of $O_p(H)$. As a consequence of this, many questions concerning the structure of $H$ (under the given assumptions) can be translated into problems concerning linear groups over $GF(p)$. This is a very basic technique.

Of course, even when $H$ is not $p$-constrained, one can still obtain a faithful representation of $H/C_H(P)$ on $\bar{P}$. Unfortunately the critical portion of the structure of $H$ may lie in $C_H(P)$ and this is lost in this representation. In the constrained case, $C_H(P) \subseteq P$ and so the full structure of $H/P$ is preserved in the representation. Clearly then other techniques must be used to study the structure of $H$ in the non-constrained case. In these talks we shall only be able to touch very briefly on them.

In the next section we shall discuss in detail one of the basic concepts whose analysis utilizes this linear representation—the concept of *p-stability*. We prefer to continue our present discussion by mentioning another frequently used procedure.

Suppose in our group $G$ there is an abelian $p$-group $A$ acting on a $q$-subgroup $Q$ with $q \neq p$ and with $A$ not centralizing $Q$. One has the following theorem.

THEOREM. *$Q$ possesses an $A$-invariant subgroup $Q_0$ such that*

(1) *$A$ acts irreducibly on $Q_0/\Phi(Q_0)$ and trivially on $\Phi(Q_0)$;*
(2) *$A$ does not centralize $Q_0$.*

If $Q_0$ is chosen to be an $A$-invariant subgroup of $Q$ of minimal order which is not centralized by $A$, then $Q_0$ has these properties. Actually $Q_0$ will be a special $q$-group and, in particular, will have class at most 2. The significance of this result can be seen when $A$ is noncyclic. Indeed, if $\bar{Q}_0 = Q_0/\Phi(Q_0)$ and $A_0 = C_A(\bar{Q}_0)$, Burnside's theorem tells us that $A_0 = C_A(Q_0)$ as $Q_0$ is a $p'$-group. Moreover, since $A/A_0$ is faithfully and irreducibly represented on $\bar{Q}_0$, $A/A_0$ must be cyclic and consequently $A_0 \neq 1$. Hence if we consider $K = N_G(A_0)$, we see that $AQ_0$ is contained in the $p$-local subgroup $K$.

There is an elementary lemma of Thompson, which is often used in conjunction with the preceding discussion.

LEMMA. *Let the group $X \times Y$ act on the p-group $P$, where $X$ is a p-group and $Y$ is a $p'$-group. If $Y$ acts trivially on $C_P(X)$, then $Y$ acts trivially on $P$.*

Turned about, it says that any element of $Y$ which induces a nontrivial automorphism of $P$ induces a nontrivial automorphism of $C_P(X)$.

In the situation considered above, suppose that our subgroup $AQ$ was contained in a $p$-constrained $p$-local subgroup $H$ of $G$, and, for simplicity, suppose also that $O_{p'}(H) = 1$. Setting $P = O_p(H)$, we have $C_H(P) \subseteq P$ and so both $Q_0$ and $Q$ induce faithful groups of automorphisms of $P$. Now we have seen above that $A_0$ centralizes $Q_0$, so $A_0Q_0 = A_0 \times Q_0$. Thompson's result thus tells us that if $P_0 = C_P(A_0)$, then $Q_0$ induces a faithful group of automorphisms of $P_0$; that is, $C_{p_0}(P_0) = 1$. Furthermore, our $p$-local subgroup $K = N_G(A_0)$, in fact, contains the group $P_0Q_0A_0$ with $Q_0$ acting on $P_0$ and $A_0$ acting on $Q_0$ in a prescribed way. Now in practice, one has hypotheses on the structure of the $p$-local subgroups of $G$ which enables one to draw strong implications from the existence of subgroups of this form inside a $p$-local subgroup $K$. In particular, one can often make assertions about the *weak closure* of $A$ in a Sylow $p$-subgroup $\tilde{P}$ of $G$—that is, the subgroup generated by all conjugates of $A$ in $G$ which lie in $\tilde{P}$. As we shall see in the next section, such information is extremely important in carrying out local analysis.

We wish to mention briefly some other results. The various parts of the following theorem are used repeatedly.

THEOREM. *Let the p-group $P$ act on the $p'$-group $L$. Then we have*

(i) *for each prime $q$, $P$ leaves invariant some Sylow $q$-subgroup of $L$;*
(ii) *any two $P$-invariant Sylow $q$-subgroups of $L$ are conjugate by an element of $C_L(P)$;*
(iii) *$[L, P]$ is a $P$-invariant normal subgroup of $L$;*
(iv) *$[L, P, P] = [L, P]$;*
(v) *$L = [L, P]C_L(P)$.*

A result of a similar nature is the following.

THEOREM. *Let $P$ be a $p$-subgroup of $H$ and let $L$ be a normal $p'$-subgroup of $H$. If $\overline{H} = H/L$, then*

$$\overline{C_H(P)} = C_{\overline{H}}(\overline{P}) \quad \text{and} \quad \overline{N_H(P)} = N_{\overline{H}}(\overline{P}).$$

We have already mentioned the fact that if a four group $T$ with involutions $t_1, t_2, t_3$ acts on the group $L$ of odd order, then $L = \langle C_L(t_i) : 1 \leq i \leq 3\rangle$. This result can be sharpened as a consequence of the following so-called order formula of Brauer and Wielandt.

THEOREM. *If $L_i = C_L(t_i)$, $1 \leq i \leq 3$, and $L_0 = C_L(T)$, then*

$$|L| = \frac{|L_1||L_2||L_3|}{|L_0|^2}$$

As a consequence, one obtains

COROLLARY. $L = L_1L_2L_3$.

The corollary thus asserts that every element $x$ of $L$ can be written in the form $x = x_1x_2x_3$ with $x_i$ centralizing $t_i$, $1 \leq i \leq 3$. One can sharpen the corollary slightly on the basis of the following.

LEMMA. *Let $\alpha$ be an automorphism of order 2 of the group $X$ of odd order. Set $X_0 = C_X(\alpha)$ and let $X'$ denote the subset of $X$ inverted by $\alpha$ (that is, the set of elements transformed into their inverses by $\alpha$). Then*

$$X = X_0X' = X'X_0 \quad \text{and} \quad X' \cap X_0 = 1.$$

In the case of our above group $L$, $t_j$ and $t_it_j$ induce the same automorphism of $L_i$ for $i \neq j$ and $C_{L_i}(t_j) = C_L(T) = L_0$. Hence using the lemma, one obtains the further

COROLLARY. *If $L'_i$ denotes the subset of $L_i$ inverted by $t_j$ ($i \neq j$, $1 \leq i, j \leq 3$) then*

$$L = L_0L'_1L'_2L'_3.$$

*Moreover, every element $x$ of $L$ has a unique representation in the form $x = x_0x_1x_2x_3$ with $x_0 \in L_0$ and $x_i \in L'_i$ ($1 \leq i \leq 3$).*

The representations of $L$ in the form $L_1L_2L_3$ and $L_0L'_1L'_2L'_3$ are known as the *T-decompositions* of $L$. Clearly these factorizations are basic for analyzing the action of 2-groups on groups of odd order.

The Brauer–Wielandt formula can be extended to noncyclic elementary abelian groups of arbitrary order. In the proof of our main theorem, we shall need the following consequence of this result.

THEOREM. *Let A be a noncyclic elementary abelian 2-group acting on the group L of odd order and let $\mathscr{B}$ denote the set of subgroups B of index at most 2 in A. Then the order of L is completely determined by the orders of the subgroups $C_L(B)$ as B ranges over $\mathscr{B}$.*

In carrying out local analysis on a simple group $G$, transfer is an essential tool. One makes strong use of Burnside's theorem and Grün's theorems as well as Alperin's fusion theorem and the so-called "focal subgroup theorem", the main objective being to obtain contradictions to the existence of particular configurations by forcing $G$ to have a normal subgroup of index $p$.

I should like to give an application of transfer of a somewhat different nature.

THEOREM. *If $A \in SCN(p)$ in the group G, then*

$$C_G(A) = A \times D,$$

*where D is a p'-group.*

*Proof.* By definition, $A$ is a maximal abelian normal subgroup of a Sylow $p$-subgroup $P$ of $G$, so $C_P(A) = A$. Putting $H = C_G(A)$, we see that $P$ normalizes $H$ and that $P \cap H = A$. Hence $K = PH$ is a group and $P$ is a Sylow $p$-subgroup of $K$. However, in general, if $P$ is a Sylow $p$-subgroup of a group $K$ and $H \lhd K$, then $P \cap H$ is a Sylow $p$-subgroup of $H$. Thus $A$ is a Sylow $p$-subgroup of $H$. But $A$ is abelian and $A$ is in the center of its normalizer in $H$, this normalizer being $H$ itself. Burnside's transfer theorem now tells us that $H$ has a normal $p$-complement $D$. Therefore, $H = AD$, with $D$ a $p'$-group. Since $A$ centralizes $D$, $AD = A \times D$, whence $H = C_G(A) = A \times D$, as asserted.

I should also like to mention a result which sheds some light on the nature of the hypothesis of our main theorem. Its proof, which is not difficult, utilizes many of the general results we have quoted above. It appears in my paper in I.H.E.S.

THEOREM. *For any group G and any prime p, the following three statements are equivalent.*

(i) *every p-local subgroup of G is p-constrained;*
(ii) *every maximal p-local subgroup of G (under inclusion) is p-constrained;*
(iii) *the centralizer of every element of order p of G is p-constrained.*

Hence our assumption that the centralizer of every involution of $G$ is 2-constrained implies, in fact, that every 2-local subgroup of $G$ is 2-constrained.

We conclude with the statement of Glauberman's fundamental $Z^*$-theorem, which holds only for the prime 2. In spirit, it is used in much the same way as transfer; namely, to obtain contradictions to the existence of particular configurations by forcing $O_{2',2}(G)$ to be nontrivial.

THEOREM. *If S is a Sylow 2-subgroup of the group G and the involution z of S is conjugate in G to no other involution of S, then the image of z is in the center of G/O(G).*

## 2.3. p-STABILITY AND GLAUBERMAN'S ZJ-THEOREM

Once again consider a $p$-constrained group $H$ with $O_{p'}(H) = 1$ and put $P = O_p(H)$, so that, as we know, $\bar{H} = H/P$ is faithfully represented on $\bar{P} = P/\Phi(P)$. Let $A$ be an abelian $p$-subgroup of $H$ with $A \nsubseteq P$ and suppose

$$[P, A, A] = 1.$$

Let me illustrate how we might come to encounter such a situation. Suppose $H$ happens to equal $N_G(P)$ for some $p$-subgroup $P$ of a group $G$. Let $Z$ be a nontrivial subgroup of $Z(P)$, set $M = N_G(Z)$, and consider $A = Z(O_p(M))$. Then $P \subseteq M$ and as $A \lhd M$, $[P, A] \subseteq A$. Since $A$ is abelian, this implies that $[[P, A], A] = [P, A, A] = 1$. Of course, in general $A$ need not normalize $P$ and so need not lie in $H$. But there are circumstances of this type in which we do know that $A$ normalizes $P$.

Commutator identities of the type $[P, A, A, \ldots, A] = 1$ constantly arise in local analysis, particularly in connection with questions of weak closure.

In the above situation, if we pass to $\bar{H}$, we have that $\bar{A} \neq 1$ and that $[\bar{P}, \bar{A}, \bar{A}] = 1$. What does this condition say if we view $\bar{P}$ as a vector space and $\bar{H}$ as a linear group acting on $\bar{P}$? Writing $\bar{P}$ additively, our identity reads that for all $\bar{v}$ in $\bar{P}$ and all $\bar{a}$ in $\bar{A}$,

$$\bar{v}(1-\bar{a})^2 = 0.$$

Hence every element of $\bar{A}$ has a minimal polynomial dividing $(1-x)^2$. If $\bar{a} \neq 1$, its minimal polynomial is not $1-x$. Consequently, each element of $\bar{A}^{\#}$ (i.e., $\bar{A}-\{1\}$) has a *quadratic* minimal polynomial on $\bar{P}$. Note also that $O_p(\bar{H}) = 1$ since $\bar{H} = H/P = H/O_p(H)$. We are thus led to the concept of $p$-stability for groups having no nontrivial normal $p$-subgroups.

DEFINITION. If $L$ is a group in which $O_p(L) = 1$, we say that $L$ is *p-stable* provided that in each faithful representation of $L$ on a vector space over $GF(p)$, no $p$-element of $L$ has a quadratic minimal polynomial. In the contrary case, $L$ is not $p$-stable. Likewise a particular representation is either $p$-stable or not.

This is a meaningful concept only in the case of *odd p*, for if $p = 2$, then clearly in any faithful representation of $L$ on a vector space over $GF(2)$ every involution $z$ of $L$ necessarily has a quadratic minimal polynomial since $z^2 = 1$. The condition that $O_p(L) = 1$ is essential here, otherwise elements of $O_p(L)$ could have quadratic minimal polynomials and it would not be possible to establish general results about $p$-stability. Shortly, we shall extend the concept to groups in which $O_p(L) \neq 1$.

As an example, the natural representation of $SL(2, p)$, $p$ odd, is obviously not $p$-stable, since *every* nonidentity element has a quadratic characteristic and minimal polynomial. On the other hand, all the other faithful irreducible representations of $SL(2, p)$ are $p$-stable. However, according to the definition, $SL(2, p)$ is therefore not a $p$-stable group.

The following is the basic result in this connection.

THEOREM. *If $L$ is a group in which $O_p(L) = 1$ and $L$ is not $p$-stable, then $SL(2, p)$ is involved in $L$ (i.e., for suitable subgroups $X$, $Y$ of $L$ with $Y \lhd X$, $X/Y \cong SL(2, p)$).*

A Sylow 2-subgroup of $SL(2, p)$, is generalized quaternion, so $L$ must, in particular, involve a quaternion group. Thus we have the following.

COROLLARY. *If $O_p(L) = 1$, $p$ odd, and either $L$ has odd order, $L$ has abelian Sylow 2-subgroups, or $L$ has dihedral Sylow 2-subgroups, then $L$ is $p$-stable.*

We shall now extend the concept of $p$-stability to arbitrary groups and, in particular, to $p$-local subgroups of a group $G$. Consider once again our group $H$ with $O_{p'}(H) = 1$ and $P = O_p(H)$. Clearly the simplest definition would be to say that $H$ is $p$-stable provided $H/P$ is $p$-stable. However, in the arguments it is often necessary to consider other subgroups $Q$ of $P$ that are normal in $H$ and to study the action of $H/C_H(Q)$ on $\bar{Q} = Q/\Phi(Q)$. Unfortunately if we now put $\bar{H} = H/C_H(Q)$, we do not necessarily have $O_p(\bar{H}) = 1$. On the other hand, observe that if we consider some composition factor $\tilde{V}$ of $\bar{Q}$, under the action of $\bar{H}$, then we obtain a faithful irreducible representation of $\tilde{H} = \bar{H}/C_{\bar{H}}(\tilde{V})$ on $\tilde{V}$. But now by a general elementary fact concerning faithful irreducible representations of groups on vector space over $GF(p)$, we have $O_p(\tilde{H}) = 1$. So the question of $p$-stability really comes down to whether each of the groups $\tilde{H}$ that arise in this way is $p$-stable. Indeed, the relation $[P, A, A] = 1$ implies that $[Q, A, A] = 1$, whence $[\bar{Q}, \bar{A}, \bar{A}] = 1$ and consequently also $[\tilde{V}, \tilde{A}, \tilde{A}] = 1$. Now if $\tilde{H}$ is, in fact, $p$-stable, this last relation can occur only if $\tilde{A} = 1$—that is, if $\bar{A} \subseteq C_{\bar{H}}(\tilde{V})$. But if this holds for each composition factor $\tilde{V}$ of $\bar{Q}$, it follows easily that $\bar{A}$ must, in fact, lie in $O_p(\bar{H}) = O_p(H/C_H(Q))$. We can therefore formulate the concept of $p$-stability for groups $H$ in which $O_{p'}(H) = 1$ and $O_p(H) \neq 1$ as follows.

DEFINITION. $H$ is $p$-stable if, for every non-trivial normal $p$-subgroup $Q$ of $H$ and every abelian $p$-subgroup $A$ of $H$ such that $[Q, A, A] = 1$, we have

$$AC_H(P)/C_H(P) \subseteq O_p(H/C_H(P)).$$

More generally, if $O_{p'}(H) \neq 1$, we pass to $H/O_{p'}(H)$ and apply the definition to that group.

The discussion preceding the definition shows that the question of $p$-stability in the present sense always reduces to questions about $p$-stability in the previous sense for suitable homomorphic images of $H$.

We come now to Glauberman's fundamental $ZJ$-theorem which concerns $p$-stable groups. For any $p$-group $P$, the *Thompson* subgroup $J(P)$ of $P$ is the subgroup of $P$ generated by all abelian subgroups of $P$ of *maximum order*. Clearly $J(P)$ is characteristic in $P$. More than that, it is weakly closed in $P$ if $P$ is a Sylow $p$-subgroup of $G$. Indeed, suppose $J(P)^x \subseteq P$ for some $x$ in $G$. If $A$ is one of the generating abelian subgroups of $P$, $A \subseteq J(P)$, so $A^x \subseteq P$. But $|A^x| = |A|$, so $A^x$ is also abelian of maximum order, whence $A^x \subseteq J(P)$. Hence $J(P)^x \subseteq J(P)$ and the result follows.

We can now state Glauberman's theorem.

THEOREM *Let $H$ be a group in which $O_{p'}(H) = 1$ and $O_p(H) \neq 1$, $p$ an odd prime, and let $P$ be a Sylow $p$-subgroup of $H$. If $H$ is both $p$-constrained and $p$-stable, then $Z(J(P))$ is normal in $H$.*

If, under the above conditions, $H$ is a $p$-local subgroup of our group $G$, the theorem asserts that $H$ is contained in the $p$-local subgroup $M = N_G(Z(J(P)))$. What is the significance of this? Let $R$ be a Sylow $p$-subgroup of $G$ containing $P$. If $P = R$, then, of course, $H$ contains a Sylow $p$-subgroup of $G$. On the other hand, if $P \subset R$, then by a basic property of $p$-groups, $Q = N_R(P) \supset P$. But $Q$ normalizes $Z(J(P))$ since it is characteristic in $P$. Thus $Q \subseteq M$ and consequently a Sylow $p$-subgroup of $M$ contains $P$ *properly*.

If $M$ is $p$-constrained and $p$-stable, and $O_{p'}(M) = 1$, one can repeat this argument again. In fact, if every $p$-local subgroup of $G$ is both $p$-constrained and $p$-stable and has no nontrivial normal $p'$-subgroups, we are led to a very striking conclusion: Namely, every maximal $p$-local subgroup of $G$ is conjugate to $N_G(Z(J(R)))$, where $R$ is a Sylow $p$-subgroup of $G$. Hence under these very strong hypotheses Glauberman's theorem gives us a complete picture of the maximal $p$-local subgroups of $G$.

In general, the aim of local analysis is primarily involved with determining the maximal local subgroups of $G$ and their interrelationships. Clearly then Glauberman's theorem is a central tool in this endeavor. Its principal use is that it enables one to "push up" from a $p$-local subgroup to one having a Sylow $p$-subgroup of higher order.

Glauberman has obtained a slight extension of this result which we shall need for our main theorem. $ZJ(P)$ does not necessarily contain its own centralizer in $P$, a fact which is often technically inconvenient. However, he has constructed a characteristic subgroup of $P$, denoted by $ZJ^*(P)$, which contains it own centralizer in $P$ and at the same time is normal in $H$ under the hypotheses of the above theorem.

### 2.4. Solvable Groups

The theory of solvable groups is a very elaborate and beautiful one. However, for local analysis very few properties of solvable groups are required. Of course, one uses constantly the fact that groups of odd order are solvable. In addition, the fact that they are $p$-constrained and $p$-stable (except possibly if $p = 3$ and $SL(2, 3)$ is involved) is important. However, the main property that is used is the existence and conjugacy of Hall subgroups, without which most of the analysis would collapse. Other results, such as the fact that the Fitting subgroup of a solvable group contains its own centralizer, are occasionally used, but essentially very few results about solvable groups are needed.

Although our main theorem concerns involutions, I should remark that the effect of our hypotheses is to reduce almost the entire proof to questions about subgroups of odd order in our group $G$. This is meant in a much stronger sense than simply that we are interested in $O(C_G(x))$, $x$ an involution. Indeed, in the course of our arguments we shall be forced to consider many local subgroups $H$ which contain subgroups $X$ of $O(C_G(x))$ and it will be very important for us to know how $X$ is embedded in $H$. My point is that the critical configurations that must be considered will always turn out to lie in subgroups of $H$ of odd order and hence in solvable subgroups of $H$. That this is the case is by no means evident in advance.

### 2.5. Transitivity Theorems

One of the principal tools in the local analysis of the odd-order paper is the so-called Thompson transitivity theorem. In every general classification problem it seems that some transitivity theorem plays a key role, and this is the case in our main theorem as well. Transitivity theorems deal primarily with an abelian $p$-subgroup $A$ of a $G$ and the set of maximal $A$-invariant $q$-subgroups of $G$ for some prime $q \neq p$. In the Feit–Thompson terminology, $\mathcal{N}_G(A)$ denotes the set of $A$-invariant subgroups of $G$ disjoint from $A$ and $\mathcal{N}_G(A; q)$ denotes the subset of these that are $q$-groups. The maximal elements of $\mathcal{N}_G(A; q)$ we shall designate as $\mathcal{N}_G^*(A; q)$. (One writes simply $\mathcal{N}(A)$, $\mathcal{N}(A; q)$, and $\mathcal{N}^*(A; q)$ when there is no ambiguity.)

In the actual Thompson transitivity theorem, which we shall now state in a general form, $A$ is an element of $SCN_3(p)$.

Theorem. *Let $G$ be a group in which all $p$-local subgroups are $p$-constrained, $p$ a prime. If $A \in SCN_3(p)$, then $C_G(A)$ permutes the elements of $\mathcal{N}^*(A; q)$ transitively under conjugation for each prime $q \neq p$.*

The specific assumption of constraint enters into the proof of the following critical preliminary lemma.

LEMMA. *If $H$ is a $p$-local subgroup of $G$ containing $A$, then every element of* $\text{И}_H(A; q)$ *is contained in* $O_{p'}(H)$.

This is very important, for from the fact that $L = O_{p'}(H)$ is an $A$-invariant p$'$-group we know that $A$ leaves invariant some Sylow $q$-subgroup of $L$ and that any two of these are conjugate by an element of $C_L(A)$. Thus as a corollary of the lemma we see that $C_L(A)$ permutes the elements of $\text{И}^*_H(A; q)$ transitively under conjugation. Hence each $p$-local subgroup of $G$ containing $A$ satisfies the conclusion of the theorem. Roughly speaking, the theorem asserts that if this result holds locally, then it holds globally. It is only in the passage from the local to the global result that the assumption on the rank of $A$ being at least three enters.

There are two basic uses of this theorem in the local analysis. The first tells us about the Sylow $p$-subgroup of $N_G(Q)$ for $Q \in \text{И}^*(A; q)$.

COROLLARY. *$N_G(Q)$ contains a Sylow $p$-subgroup of $N_G(A)$ and hence of $G$.*

*Proof.* If $x \in N_G(A)$, then also $Q^x \in \text{И}^*(A; q)$. Hence by the theorem, $Q^x = Q^y$ for some $y$ in $C_G(A)$. But we know that $C_G(A) = A \times D$, where $D$ is a $p'$-group. Since $Q$ is $A$-invariant, we can clearly assume $y \in D$. Now $xy^{-1} \in N_G(Q)$ and so $x \in N_G(Q)D$. Since $x$ was arbitrary, we conclude that

$$N_G(A) = (N_G(Q) \cap N_G(A))D.$$

Since $D$ is a $p'$-group, it follows that $N_G(Q)$ contains a Sylow $p$-subgroup of $N_G(A)$. Furthermore, $A \lhd P$ for some Sylow $p$-subgroup $P$ of $G$ as $A \in SCN(p)$ and $P \subseteq N_G(A)$, so the corollary holds.

Some conjugate of $P$ in $N_G(A)$ normalizes $Q$, so $P$ normalizes an appropriate conjugate of $Q$. Hence the corollary can be rephrased to state that $P$ normalizes some element of $\text{И}^*(A; q)$. In particular, for example, if $\text{И}(P; q)$ is trivial, it follows that $\text{И}(A; q)$ is trivial. The effect of the corollary is that it reduces certain questions about the action of $P$ on $p'$-groups to that of the action of $A$ on $p'$-groups. The advantage of this is that $A$, being abelian, is much easier to work with than $P$ in the following sense: if $a \in A^\#$, $C_G(a)$ is a $p$-local subgroup of $G$ *containing* $A$. Hence we can confine our attention to $p$-local subgroups with this property. On the other hand, if $a$ is just some element of $P$, we don't know much about $C_P(a)$ in general and so without the transitivity theorem we have no means of regulating the situation.

As we have noted, any two elements of $\text{И}^*(A; q)$ are conjugate by an element of $D$. Furthermore, clearly $D \in \text{И}(A)$. Thus the transitivity theorem can be stated entirely in terms of elements of $\text{И}(A)$. As we shall see later, one is really after an affirmative answer to the following question: Do the elements of $\mathcal{N}(A)$ generate a $p'$-subgroup of $G$? Equivalently, do the elements of $\mathcal{N}(A)$ possess a *unique* maximal element? Obtaining a positive answer to this question is a critical step in every classification problem. In fact, in the case of our own main theorem, where $A$ is an abelian 2-subgroup of $G$ of rank at least 5, approximately two-thirds of the entire proof will be concerned with it.

I wish now to discuss an important modification of the Thompson transitivity theorem. As just mentioned, we wish to take $A$ as an abelian 2-subgroup of $G$ with $m(A) \geq 5$. By hypothesis, such a subgroup $A$ exists. However, it certainly does not follow that $A$ can be taken to be in $SCN(2)$—that is, $SCN_5(2)$ may be empty in $G$. Hence we may not be able to establish the critical lemma of the transitivity theorem for our subgroup $A$. (Note that we do have every 2-local subgroup of $G$ 2-constrained since this is a consequence of the assumption that the centralizer of every involution is 2-constrained.)

It is, however, possible to modify the concept of $\mathcal{N}(A)$ slightly so as to recover the critical lemma. Keep in mind that for each involution $a$ of $A$, we are only interested in $O(C_G(a))$. Suppose that $K \in \mathcal{N}(A)$. Then as $A$ is noncyclic,

$$K = \langle K \cap C_G(a) : a \text{ an involution of } A \rangle.$$

However, it certainly need *not* be the case that the group

$$K_0 = \langle K \cap O(C_G(a)) : a \text{ an involution of } A \rangle$$

is the same as $K$. Indeed, it may even happen that $K_0 = 1$. For example, consider the group $G = SL(n, q^m)$ ($q$ an odd prime and $n$ odd with $n \geq 3$). One checks directly that the subgroup

$$M = \left\{ \begin{pmatrix} & & & 0 \\ & & & 0 \\ & X & & \vdots \\ & & & 0 \\ 0 & 0 & \cdots & x \end{pmatrix} : x\det X = 1 \right\}$$

contains a Sylow 2-subgroup $S$ of $G$ and that $M$ leaves invariant the elementary abelian $q$-subgroup

$$K = \left\{ \begin{pmatrix} & & & 0 \\ & & & 0 \\ & I_{n-1} & & \vdots \\ & & & 0 \\ x_1 & x_2 & \cdots x_{n-1} & 1 \end{pmatrix} : x_i \in GF(q^m) \right\}.$$

In particular, if $A$ is any noncyclic abelian 2-subgroup of $S$, then $A$ normalizes $K$, so $K \in \mathcal{N}(A; q)$. On the other hand, one also checks that

$O(C_G(z))$ is a $q'$-group for any involution $z$ of $G$. Consequently $K \cap O(C_G(a)) = 1$ for any involution $a$ of $A$ and therefore $K_0 = 1$.

However, because our primary interest is in the group $K_0$ rather than $K$, it is natural to introduce the term $\mathsf{M}_0(A)$ to denote the set of those $K$ in $\mathsf{M}(A)$ such that

$$K = \langle K \cap O(C_G(a)) : a \text{ an involution of } A \rangle.$$

Likewise one defines $\mathsf{M}_0(A; q)$ and $\mathsf{M}_0^*(A; q)$.

If one works with the elements of $\mathsf{M}_0(A; q)$, one can establish an analogue of the critical lemma without $A$ being in $SCN(p)$. The pertinent argument will be given for the case $p = 2$ in the first section of Part 3. One can then obtain the following analogue of the transitivity theorem.

THEOREM. *Let $G$ be a group in which all $p$-local subgroups are $p$-constrained and let $A$ be an abelian $p$-subgroup of $G$ of rank at least* 3. *Then $O_{p'}(C_G(A))$ permutes the elements of $\mathsf{M}_0^*(A; q)$ transitively under conjugation for each prime $q \neq p$.*

In the case of our main theorem, this constitutes the first step in proving that the elements of $\mathsf{M}_0(A)$ generate a subgroup of $G$ of odd order—equivalently, that the subgroup

$$\langle O(C_G(a)) : a \text{ an involution of } A \rangle$$

has odd order.

## 2.6. Strongly Embedded Subgroups

We come now to one of the central tools in the entire development—groups which possess a strongly embedded subgroup. In fact, except in the case of groups of odd order, the aim of the local analysis and the purpose of investigating the maximal local subgroups is precisely to construct such a strongly embedded subgroup in $G$. Once this is accomplished, one is able to derive a contradiction and thereby rule out the existence of certain configurations. In our own case, such a strongly embedded subgroup will be constructed whenever $O(C_G(x)) \neq 1$ for some involution $x$ of $G$, and so our theorem will follow directly once this is accomplished.

To motivate the discussion, consider our own problem once again. We have, of course, hypothesized that $O(G) = 1$. But suppose for a moment instead that $O(G) \neq 1$ and that $\bar{G} = G/O(G)$ satisfies the conclusion of our theorem. Thus we assume that $O(C_{\bar{G}}(\bar{x})) = 1$ for every involution $\bar{x}$ of $\bar{G}$. Let $A$ be any noncyclic abelian 2-subgroup of $G$ and let us set

$$W_A = \langle O(C_G(a)) : a \text{ an involution of } A \rangle.$$

We claim that $W_A = O(G)$. Indeed, we know that

$$O(G) = \langle O(G) \cap C_G(a) : a \text{ an involution of } A \rangle$$

as $A$ is noncyclic. But obviously $O(G) \cap C_G(a) \subseteq \overline{O(C_G(a))}$. Hence $O(G) \subseteq W_A$. On the other hand, we also know that $\overline{C_G(a)} = C_{\bar{G}}(\bar{a})$, which implies that $\overline{O(C_G(a))} \subseteq O(C_{\bar{G}}(\bar{a}))$. However, the latter group is trivial by assumption, so $\overline{O(C_G(a))} = 1$ and consequently $O(C_G(a)) \subseteq O(G)$. Hence also $W_A \subseteq O(G)$ and so $W_A = O(G)$, as asserted.

Since A was arbitrary, we obtain as a consequence:

(1) $W_A$ is of odd order;
(2) $W_A = W_{A'}$ for any pair of noncyclic abelian 2-subgroups of $G$.

Now return to our hypothesis that $O(G) = 1$ and let us define $W_A$ exactly as above. Every involution $x$ of $G$ lies in some noncyclic abelian 2-subgroup of $G$. Hence, if each $W_A = 1$, then the conclusion of our main theorem will hold. Hence in proving the theorem, we can assume that some $W_A \neq 1$. Of course, there is now no reason why $W_A$ should have odd order—in fact, why $W_A$ should even be a proper subgroup of $G$. Clearly if we could show that $W_A$ did have odd order, it would represent an important step. It was just this problem that we were discussing in the preceding section.

Let us suppose we have somehow been able to show that $|W_A|$ is odd for some $A$ with $W_A \neq 1$. Heuristically the prototype situation we have considered above suggests that $W_A$ ought to be $O(G)$. But $O(G) = 1$ by assumption, so we cannot hope to establish this directly. However, it does suggest that we consider $N_G(W_A)$ and investigate its properties. Since $W_A \neq 1$ and $O(G) = 1$, $N_G(W_A)$ is, in any event, a proper subgroup of $G$.

The discussion so far depends upon the particular noncyclic abelian subgroup $A$ and hence $N_G(W_A)$ does not appear to be intrinsically related to $G$. However, suppose we were also able to demonstrate that $W_A = W_{A'}$ for every pair of noncyclic abelian 2-subgroups $A, A'$ of $G$ (or at least for every pair that lie in a fixed Sylow 2-subgroup of $G$). On setting $W = W_A$ and $M = N_G(W)$, it would then appear that $M$ is somehow intrinsically connected with $G$ and we would expect $M$ to have some special properties.

In fact, under the assumptions of our main theorem we shall later prove that $M$ satisfies the following conditions:

(1) $M$ is a proper subgroup of $G$ of even order;
(2) $C_G(x) \subseteq M$ for every involution $x$ of $M$;
(3) $M \cap M^g$ has odd order for every $g$ in $G-M$.

It is not surprising from the definition of $W_A$ and the fact that $W_A = W_{A'}$ is of odd order that we can establish these results.

The above conditions are precisely Bender's definition of a strongly embedded subgroup $M$ of a group $G$. In my book, using an equivalent definition, a number of elementary properties of groups with a strongly embedded subgroup are established. These results all appear in Thompson's work on $N$-groups. Although it was foreshadowed in the work of Feit and Suzuki, it was Thompson who first saw the full significance of strongly embedded subgroups for classification problems.

The key result here is due to Bender, who has classified all groups $G$ which contain a strongly embedded subgroup. The proof of this theorem is very difficult and divides into two parts. The first part represents a reduction to the case that the permutation representation of $G$ on the right cosets of $M$ is doubly transitive. Thus $G$ is doubly transitive, a subgroup fixing a letter has even order and a subgroup fixing two letters has odd order. Ultimately, Bender reduces this situation to the particular case in which $M$ is 2-closed. This special case of the problem had been previously treated by Suzuki as part of his work on groups in which the centralizer of every involution is 2-closed. The end result is the following.

THEOREM. *If $G$ is a group of 2-rank at least 2 with a strongly embedded subgroup $M$, then $G/O(G)$ possesses a normal subgroup of odd index isomorphic to $PSL(2, 2^n)$, $PSU(3, 2^n)$, or $Sz(2^n)$ $(n \geq 2)$.*

For most classification problems, it is the following corollary that is most useful.

COROLLARY. *If $G$ is a group with a strongly embedded subgroup $M$ and $O(G) = 1$ then $O(M) = 1$.*

In our own situation, the corollary yields a contradiction at once. Indeed, $M = N_G(W)$ and $W = W_A$ is nontrivial of odd order. Since $W$ is clearly normal in $M$, $W \subseteq O(M)$ and so $O(M) \neq 1$, contrary to the fact that $M$ is strongly embedded in $G$ and $O(G) = 1$.

## 2.7. CONCLUDING REMARKS

I have tried in this brief discussion to give you some feeling for the main tools that are used in carrying out local group-theoretic analysis. Obviously in specific classification problems—particularly those involving individual case analysis—more specialized results than those I have quoted are required. An advantage of our main theorem is that its proof depends only upon these general methods and needs no specialized results.

In carrying out this survey, I have tried at the same time to give an outline of the major steps needed to establish our main theorem. These are two in number.

I. For some noncyclic abelian 2-subgroup $A$ of $G$, show that

$$W_A = \langle O(C_G(a)) \colon a \text{ an involution of } A \rangle$$

is of odd order.

II. If $W_A \neq 1$, show that $M = N_G(W_A)$ is strongly embedded in $G$.

Part 3 will be taken up with problem I and Part 4 with problem II.

## 3. The Signalizer Functor Theorem

### 3.1. Signalizer Functors and Balanced Groups

We wish now to motivate two general concepts which are basic for the proof of our main theorem. We begin with the following key result.

PROPOSITION: *Let a and b be two commuting involutions of the group G and suppose that $C_G(b)$ is 2-constrained. Then*

$$O(C_G(a)) \cap C_G(b) \subseteq O(C_G(b)).$$

*Proof.* Set $C = C_G(a)$, $H = C_G(b)$, and $D = O(C) \cap H$. We must prove that $D \subseteq O(H)$. We use the following critical fact: $[D, H \cap C]$ has odd order. Indeed,

$$[D, H \cap C] \subseteq [O(C), C] \subseteq O(C),$$

the latter inclusion being a result of the normality of $O(C)$ in $C$. Thus $[D, H \cap C]$ is of odd order.

Now $a \in H$ as $a$ and $b$ commute. Let $T$ be an $\langle a \rangle$-invariant Sylow 2-subgroup of $O_{2',2}(H)$. $T$ exists inasmuch as $a \in S$ for some Sylow 2-subgroup $S$ of $H$, $a$ leaves $S \cap O_{2',2}(H)$ invariant, and $S \cap O_{2',2}(H)$ is a Sylow 2-subgroup of $O_{2',2}(H)$. Putting $R = C_T(a)$, we have that $[D, R]$ is of odd order by the preceding paragraph.

We put $\bar{H} = H/O(H)$. Since $O_{2',2}(H) = O(H)T$, $\bar{T} = O_2(\bar{H})$. To prove that $D \subseteq O(H)$, it will suffice to show that $\bar{D}$ centralizes $\bar{T}$, for then $\bar{D} \subseteq \bar{T}$ as $\bar{H}$ is 2-constrained. Since $|\bar{D}|$ is odd, this will force $\bar{D} = 1$, whence $D \subseteq O(H)$.

Since $|O(H)|$ is odd, $\overline{C_T(a)} = C_{\bar{T}}(\bar{a})$, so $\bar{R} = C_{\bar{T}}(\bar{a})$. Furthermore, $[\bar{D}, \bar{R}]$ is of odd order. But

$$[\bar{D}, \bar{R}] \subseteq [\bar{H}, \bar{T}] \subseteq \bar{T}$$

as $\bar{T} \lhd \bar{H}$. Thus $[\bar{D}, \bar{R}]$ is also a 2-group, whence $[\bar{D}, \bar{R}] = 1$ and $\bar{D}$ centralizes $\bar{R}$.

Finally $D$ centralizes $a$ and $|D|$ is odd, so $\bar{D}$ centralizes $\bar{a}$ and $|\bar{D}|$ is odd. But now we see that $\bar{D} \times \langle \bar{a} \rangle$ acts on the 2-group $\bar{T}$ with $\bar{D}$ centralizing $\bar{R} = C_{\bar{T}}(\bar{a})$. This is precisely the situation of one of Thompson's results described in Chapter 2 and we conclude that $\bar{D}$ centralizes $\bar{T}$.

Now let $A$ be an abelian 2-subgroup of the group $G$. For simplicity of notation we shall assume that $A$ is elementary abelian. Suppose that $C_G(a)$ is 2-constrained for every $a$ in $A^\#$ (i.e., for every involution of $A$). Then by the proposition,

$$O(C_G(a)) \cap C_G(b) \subseteq O(C_G(b))$$

for all $a, b$ in $A^\#$. (Note that $a, b$ commute as $A$ is abelian.)

In addition to this, $O(C_G(a))$ obviously has the following two properties:

$$O(C_G(a)) \lhd C_G(a)$$

and for any $g$ in $G$,

$$(O(C_G(a)))^g = O(C_G(a^g)).$$

These three conditions represent a particular case of a more general situation. Recall that in general classification problems one is not trying to prove that $O(C_G(a)) = 1$, but only that it is cyclic, and that there is some difficulty in deciding which subgroup of $O(C_G(a))$ to work with in order to derive a contradiction. The question therefore arises as to what properties a "good" subgroup of $C_G(a)$ must have. The answer is embodied in the concept of an $A$-signalizer functor $\theta$ on a group $G$, which you will see is a direct generalization of the conditions we have obtained above. Again for simplicity of exposition we limit ourselves to the elementary abelian case.

DEFINITION. Let $A$ be an elementary abelian 2-subgroup of the group $G$ of rank at least 3. Suppose that for each involution $a$ of $A$ there is associated a subgroup $\theta(C_G(a))$ of $O(C_G(a))$ with the following properties:

(a) $\theta(C_G(a)) \lhd C_G(a)$;
(b) $\theta(C_G(a)) \cap C_G(b) \subseteq \theta(C_G(b))$ for all involutions $b$ of $A$;
(c) if $B$ is a subgroup of index at most 4 in $A$, then

$$(\theta(C_G(a)))^g = \theta(C_G(a^g))$$

whenever $a \in B$ and $g \in N_G(B)$.

Under these conditions $\theta$ is called an *A-signalizer functor on G*.

In the case considered above, these conditions are clearly satisfied with $\theta(C_G(a)) = O(C_G(a))$. We express this by saying that $O$ is an $A$-signalizer functor on $G$. Obviously they are also satisfied if $\theta(C_G(a)) = 1$ for each $a$ in $A^{\#}$. In this case $\theta$ is called the *trivial* functor.

In these terms one can say that the study of the cores of the centralizers of involutions in simple groups $G$ comes down to showing that $G$ possesses no nontrivial $A$-signalizer functors for appropriate abelian 2-subgroups $A$ of $G$.

I shall refer to the main theorem of C.I. 1 as the *signalizer functor theorem*. When combined with the result of C.I. 2 and one of the results of C.I. 3, it can be stated in the following way.

SIGNALIZER FUNCTOR THEOREM. *Let $A$ be an elementary abelian 2-subgroup of $G$ of rank at least 5. If $G$ possesses the $A$-signalizer functor $\theta$, then the subgroup*

$$\langle \theta(C_G(a)) : a \in A^{\#} \rangle$$

*of $G$ is of odd order.*

We shall denote this subgroup as $W_A(\theta)$. Thus $W_A(O)$ is simply the subgroup we have previously designated as $W_A$.

The theorem thus gives us a sufficient condition for $W_A(\theta)$ to be of odd order when $A$ is elementary abelian of order at least 32. They hypotheses of the theorem—that is, the conditions for $\theta$ to be an $A$-signalizer functor—involve only conditions on the centralizers of involutions in $A$ and do not depend upon any other knowledge about $G$. On the other hand, in carrying out the second major step in the proof of our main theorem—that is, in establishing that $M = N_G(W_A(O))$ is strongly embedded in $G$ when $W_A(O) \neq 1$—we cannot expect to get by so easily. For strong embedding involves a condition on the centralizer of *every* involution of $M$, not only those of $A$. It is for this reason that it was necessary for us to assume that the centralizer of every involution of $G$ is 2-constrained.

Under this hypothesis on $G$, our proposition above shows that $G$ satisfies the following condition for every pair of commuting involutions $a$, $b$ of $G$

$$(*) \qquad O(C_G(a)) \cap C_G(b) = O(C_G(b)) \cap C_G(a).$$

Indeed (*) follows at once from the two inclusions

$$O(C_G(a)) \cap C_G(b) \subseteq O(C_G(b)) \quad \text{and} \quad O(C_G(b)) \cap C_G(a) \subseteq O(C_G(a)),$$

which hold as both $C_G(a)$ and $C_G(b)$ are 2-constrained.

This leads us to the following notion.

DEFINITION. A group $G$ is said to be *balanced* if condition (*) holds for every pair of commuting involutions $a$, $b$ of $G$.

There is obviously a close connection between balance and signalizer functors. Indeed, if $G$ is balanced, it is clear that $O$ is an $A$-signalizer functor on $G$ for every elementary abelian 2-subgroup $A$ of $G$ of rank at least 3.

In this terminology, $G$ is balanced if the centralizer of every involution of $G$ is 2-constrained. This is a critical property of $G$ for carrying out the second step in the proof or our main theorem. In the next chapter, we shall investigate balanced groups in more detail. For the remainder of this chapter, we shall concentrate on the proof of the signalizer functor theorem.

## 3.2. Outline of Proof

For the balance of the chapter, $A$ will denote an elementary abelian 2-subgroup of the group $G$ with $m(A) \geq 5$ and $\theta$ an $A$-signalizer functor on $G$; and it will be our objective to prove that

$$W_A(\theta) = \langle \theta(C_G(a)) : a \in A^{\#} \rangle$$

has odd order.

There is no particular advantage in restricting ourselves to the special case $\theta = O$. In fact, there is a definite notational disadvantage. Indeed, in the course of the proof it will be necessary for us to define the term $\theta(H)$ for subgroups $H$ of $G$ containing $A$ other than those of the form $C_G(a)$. The point is that when $\theta = O$, it is not necessarily true that $\theta(H)$ equals $O(H)$.

For example, set $H = C_G(A)$. Clearly $O(C_G(a)) \cap H \subseteq O(H)$ for each $a$ in $A^{\#}$. Our definition of $\theta(H)$ will be

$$\theta(H) = \langle \theta(C_G(a)) \cap H : a \in A^{\#} \rangle.$$

In the particular case $\theta = O$, this says

$$\theta(H) = \langle O(C_G(a)) \cap H : a \in A^{\#} \rangle.$$

Hence $\theta(H) \subseteq O(H)$. However, it is not necessarily the case that equality holds.

For this reason it is better to treat general signalizer functors from the outset.

Just as we defined $\mathcal{N}_0(A)$, $\mathcal{N}_0(A; p)$, and $\mathcal{N}_0^*(A; p)$ in the preceding chapter, we define more generally the terms $\mathcal{N}_\theta(A)$, $\mathcal{N}_\theta(A; p)$, and $\mathcal{N}_\theta^*(A; p)$. Thus $\mathcal{N}_\theta(A)$ is the set of $A$-invariant subgroups $K$ of $G$ of odd order such that

$$K = \langle K \cap \theta(C_G(a)) : a \in A^{\#} \rangle.$$

Clearly each element of $\mathcal{N}_\theta(A)$ is contained in $W_A(\theta)$. Hence $W_A(\theta)$ can equivalently be defined as

$$W_A(\theta) = \langle K : K \in \mathcal{N}_\theta(A) \rangle.$$

If one is going to be able to prove anything at all about the elements of $\mathcal{N}_\theta(A)$, one will certainly need to have available the following results.

Proposition. *The following conditions hold.*

(i) *If $K_i \in \mathcal{N}_\theta(A)$ $(1 \le i \le n)$, and $K = \langle K_i : 1 \le i \le n \rangle$ is of odd order, then $K \in \mathcal{N}_\theta(A)$.*

(ii) *If $K \in \mathcal{N}_\theta(A)$ and $x \in N_G(A)$, then $K^x \in \mathcal{N}_\theta(A)$.*

(iii) *If $K \in \mathcal{N}_\theta(A)$ and $L$ is an $A$-invariant subgroup of $K$, then $L \in \mathcal{N}_\theta(A)$.*

First, (i) is an immediate consequence of the definition of $\mathcal{N}_\theta(A)$. Moreover, (ii) follows directly from condition (c) in the definition of $A$-signalizer functor with $B = A$. Indeed, it implies that

$$(\theta(C_G(a)))^x = \theta(C_G(a^x))$$

for all $a$ in $A^{\#}$ and all $x$ in $N_G(A)$. Since $K \in \mathcal{N}_\theta(A)$, we have

$$K^x = \langle (K \cap \theta(C_G(a)))^x : a \in A^{\#} \rangle$$

and hence

$$K^x = \langle K^x \cap \theta(C_G(a^x)) : a \in A^{\#} \rangle.$$

But $a^x$ runs through $A^{\#}$ as $a$ does and so

$$K^x = \langle K^x \cap \theta(C_G(a)) : a \in A^{\#} \rangle.$$

Thus $K^x \in \mathcal{N}_\theta(A)$, as asserted.

Surprisingly enough, the proof of (iii) is not trivial. Essentially it requires knowing that

$$L \cap \langle K \cap \theta(C_G(a)) : a \in A^\# \rangle = \langle L \cap (K \cap \theta(C_G(a)) : a \in A^\# \rangle,$$

which is by no means obvious.

We actually establish (iii) as a corollary of the following general result (theorem 2.6 of C.I. 1), whose proof we shall discuss in section 4.

THEOREM. *If $K \in \mathcal{V}_\theta(A)$, then $K \cap \theta(C_G(a)) = C_K(a)$.*

Using this fact, we have

$$C_L(a) = L \cap C_K(a) = L \cap (K \cap \theta(C_G(a)) = L \cap \theta(C_G(a))$$

for each $a$ in $A^\#$. But $L = \langle C_L(a) : a \in A^\# \rangle$ as $A$ is noncyclic and we conclude that $L \in \mathcal{V}_\theta(A)$.

Note that without (iii) we could not even assert that $\theta(C_G(A))$, which has been defined above, is an element of $\mathcal{V}_\theta(A)$, for by definition

$$\theta(C_G(A)) = \langle C_G(A) \cap \theta(C_G(a)) : a \in A^\# \rangle.$$

The desired conclusion would follow from part (i) of the proposition (as $\theta(C_G(A))$ has odd order) provided we knew that each $C_G(A) \cap \theta(C_G(a)) \in \mathcal{V}_\theta(A)$. But to assert this requires (iii).

These basic properties of the elements of $\mathcal{V}_\theta(A)$ are used repeatedly.

The preceding theorem is important in another regard. Its proof represents the one place in the entire signalizer functor theorem in which the normality condition

$$\theta(C_G(a)) \lhd C_G(a)$$

is used. As a result, it is possible to replace condition (a) in the definition of $A$-signalizer functor by the condition that

$$K \cap \theta(C_G(a)) = C_K(a)$$

for each $a$ in $A^\#$ and each $K$ in $\mathcal{V}_\theta(A)$ without affecting the conclusion of the signalizer functor theorem. This observation will be used in connection with some extensions of the main theorem which we shall discuss in Part 5.

As we have strongly indicated in the preceding chapter, the proof of the signalizer functor theorem draws its major inspiration from the analysis of Chapter IV of the odd order paper and can properly be regarded as an abstract generalization of that analysis. We shall now briefly describe the procedure to be followed. First, one establishes yet another modification of the Thompson transitivity theorem (theorem 3.1 of C.I. 1).

THEOREM. *For each odd prime p, $\theta(C_G(A))$ acts transitively by conjugation on the elements of $\mathcal{V}_\theta^*(A; p)$.*

As you will see in a moment, it is important to have $\theta(C_G(A))$ here and not some subgroup of $C_G(A)$ properly containing $\theta(C_G(A))$. However, the theorem has an important corollary which we wish to establish before continuing our outline.

COROLLARY. *If $P \in \mathcal{N}_\theta^*(A; p)$, then $C_P(a)$ is a Sylow p-subgroup of $\theta(C_G(a))$ for each a in $A^\#$.*

*Proof.* Let $Q$ be an $A$-invariant Sylow $p$-subgroup of $\theta(C_G(a))$, so that by (iii) of the basic proposition $Q \in \mathcal{N}_\theta(A)$, whence, in fact, $Q \in \mathcal{N}_\theta(A; p)$. Choose $R$ in $\mathcal{N}_\theta^*(A; p)$ with $R \supseteq Q$. By the transitivity theorem, $R^x = P$ for some $x$ in $\theta(C_G(A))$. But $\theta(C_G(A)) \subseteq \theta(C_G(a))$ by definition of $\theta(C_G(A))$, so $Q^x$ is also a Sylow $p$-subgroup of $\theta(C_G(a))$. But $Q^x \subseteq C_P(a)$. Moreover, $C_P(a) = P \cap \theta(C_G(a))$ by the theorem quoted earlier, and consequently $Q^x = C_P(a)$. This establishes the corollary.

The transitivity theorem is a conjugacy theorem of the Sylow type. What we shall actually need to establish the signalizer functor theorem are theorems of the Hall type for the elements of $\mathcal{N}_\theta^*(A; p)$ as $p$ ranges over some set of primes. Let us denote by $\sigma$ the set of primes for which $\mathcal{N}_\theta(A; p)$ is nontrivial (equivalently, $\mathcal{N}_\theta^*(A; p)$ is nontrivial).

For any subset of $\tau$ of $\sigma$, we shall say that $G$ satisfies $E_\tau(A)$ provided there is an element $K$ in $\mathcal{N}_\theta(A)$ with the following properties:

(a) $K$ contains an element of $\mathcal{N}_\theta^*(A; p)$ for each prime $p$ in $\tau$;
(b) $K$ is a $\tau$-group—that is, $|K|$ is divisible only by primes in $\tau$.

We shall call $K$ an $S_\tau(A)$-*subgroup* of $G$. This is the direct analogue of a Hall $S_\tau$-subgroup of a group $G$.

We shall also say that $G$ satisfies $C_\tau(A)$ provided any two $S_\tau(A)$-subgroups of $G$ are conjugate by an element of $C_G(A)$ (it is not necessary to require conjugacy by $\theta(C_G(A))$ here).

The elements of $\mathcal{N}_\theta^*(A; p)$ are, of course, $S_p(A)$-subgroups of $G$ and the transitivity theorem tells us that $G$ satisfies $C_p(A)$ (for simplicity, one writes $E_p(A)$ and $E_{p,q}(A)$, etc. in place of $E_{\{p\}}(A)$ and $E_{\{p,q\}}(A)$).

The proof of the signalizer functor theorem amounts essentially to a proof of the following assertion:

$$G \text{ satisfies } E_\sigma(A).$$

Indeed, assume that this is true and let $K$ be an $S_\sigma(A)$-subgroup of $G$. Since $K \in \mathcal{N}_\theta(A)$, $K$ is of odd order. We shall argue that $K = W_A(\theta)$, which will establish that $W_A(\theta)$ is of odd order, as required.

We set $D = \theta(C_G(A))$. As noted above, $D \in \mathcal{N}_\theta(A)$. Hence certainly $D$ is a $\sigma$-group. Let $D_p$ be an $A$-invariant Sylow $p$-subgroup of $D$ for some prime $p$. Then we know that $D_p \in \mathcal{N}_\theta(A; p)$. We choose $P$ in $\mathcal{N}_\theta^*(A; p)$ with $D_p \subseteq P$ and we also let $Q$ be an element of $\mathcal{N}_\theta^*(A; p)$ with $Q \subseteq K$. $Q$ exists as $K$ is an $S_\sigma(A)$-subgroup of $G$ and $p \in \sigma$. By the transitivity theorem,

$P^x = Q$ for some $x$ in $D$. In particular,

$$D_p^x \subseteq Q \subseteq K.$$

But $D_p^x$ is a Sylow $p$-subgroup of $D$ as $x \in D$. Hence $K$ contains a Sylow $p$-subgroup of $D$ for each prime $p$ and so $D \subseteq K$.

Now let $L$ be an arbitrary element of $\mathcal{V}_\theta(A)$. Then just as with $D$ above, we have

$$L_p^x \subseteq Q \subseteq K$$

for some $x$ in $D$, where $L_p$ is an $A$-invariant Sylow $p$-subgroup of $L$. However, $x \in K$ as $D \subseteq K$ and so $L_p \subseteq K$. Hence $L \subseteq K$. Since

$$W_A(\theta) = \langle L : L \in \mathcal{V}_\theta(A) \rangle,$$

it follows that $W_A(\theta) \subseteq K$. The reverse inclusion is clear as $K \in \mathcal{V}_\theta(A)$. We conclude therefore that

$$K = W_A(\theta).$$

The preceding discussion thus shifts the focus of the signalizer functor theorem to the establishment of $E_\sigma(A)$. Clearly such a result will be proved inductively by establishing $E_\tau(A)$ for all subsets $\tau$ of $\sigma$. As in the corresponding problem in the odd order paper, the proof divides into two major parts:

I. Establish $E_{p,q}(A)$ and $C_{p,q}(A)$ for two distinct primes $p, q$ of $\sigma$.

II. Establish $E_\tau(A)$ and $C_\tau(A)$ under the assumption that $E_{p,q}(A)$ and $C_{p,q}(A)$ hold for every pair of primes $p, q$ in $\tau$.

Although we ultimately require only $E_\sigma(A)$ we need to know that $C_\tau(A)$ holds in order to establish $E_\tau(A)$.

The complete proof of I and II runs to approximately thirty journal pages. Hence it will not be possible to give all the details in these lectures. I shall try, however, to present some of the representative arguments in complete detail and to explain the key ideas of the proof.

### 3.3. A Reduction

The signalizer functor theorem is proved by induction on the order of $G$. This enables us to establish at the outset an important property of the $p$-local subgroups of $G$ containing $A$ ($p$ odd). This argument is the content of C.I. 2. We prove

Proposition. *If $H$ is a $p$-local subgroup of $G$, $p$ odd, which contains $A$, then*

$$W_A(H; \theta) = \langle \theta(C_G(a)) \cap H : a \in A^\# \rangle$$

*is of odd order.*

*Equivalently $W_A(H; \theta)$ can be described as the subgroup of $H$ generated by the elements of $\mathcal{V}_\theta(A)$ which lie in $H$.*

The proposition asserts essentially that the conclusion of the signalizer functor theorem holds for the $p$-local subgroups of $G$ ($p$ odd) containing $A$.

*Proof*

*Case 1.* $H = G$. In this case $W_A(H; \theta)$ is simply $W_A(\theta)$. Since $H = N_G(P)$ for some nontrivial $p$-subgroup $P$ ($p$ odd), we have $P \lhd G$. We set $\bar{G} = G/P$. For each $\bar{a}$ in $\bar{A}^{\#}$, we define

$$\bar{\theta}(C_{\bar{G}}(\bar{a})) = \overline{\theta(C_G(a))}.$$

It is not difficult to check that $\bar{\theta}$ defines an $\bar{A}$-signalizer functor on $\bar{G}$. Of course, one uses the fact that $\overline{C_G(a)} = C_{\bar{G}}(\bar{a})$ for any $a$ in $A^{\#}$ and that $\overline{N_G(B)} = N_{\bar{G}}(\bar{B})$ for any subgroups $B$ of $A$ in the verification. Since $\bar{A} \cong A$ and $|\bar{G}| < |G|$, we conclude by induction that the signalizer functor theorem holds for $\bar{G}$.

Hence

$$W_{\bar{A}}(\bar{\theta}) = \langle \bar{\theta}(C_{\bar{G}}(\bar{a})) : \bar{a} \in \bar{A}^{\#} \rangle$$

has odd order. However, it is clear from the definition of $\bar{\theta}$ that $W_{\bar{A}}(\bar{\theta})$ is precisely the image of $W_A(\theta)$ in $\bar{G}$. Thus $W_A(\theta) \subseteq X$, the complete inverse image of $W_{\bar{A}}(\bar{\theta})$ in $G$. Since $|P|$ and $|W_{\bar{A}}(\bar{\theta})|$ are each odd, so also is $|X|$ and we conclude that $W_A(\theta) = W_A(H; \theta)$ has odd order.

*Case 2.* $H \subset G$. This time we define $\phi$ by the condition

$$\phi(C_G(a)) = \theta(C_G(a)) \cap H$$

for each $a$ in $A^{\#}$ and verify that $\phi$ is an $A$-signalizer functor on $H$. In this case, we use the evident fact that $C_G(a) \cap H = C_H(a)$ and that $O(C_G(a)) \cap H \subseteq O(C_H(a))$. Since $|H| < |G|$, it follows again by induction that

$$W_A(\phi) = \langle \phi(C_H(a)) : a \in A^{\#} \rangle$$

has odd order. But clearly $W_A(\phi)$ is the same subgroup as $W_A(H; \theta)$ and so the proposition holds in this case as well.

At the time C.I. 1 was written, I was not aware of this simple proposition and so introduced an extra assumption called flatness to insure that the $p$-local ($p$ odd) subgroups of $G$ containing $A$ had this property. The above argument shows that such an assumption was superfluous.

For brevity we set $\theta(H) = W_A(H; \theta)$. We know now that $\theta(H)$ is a subgroup of $H$ of odd order and contains every element of $И_\theta(A)$ which lies in $H$. This is true for each $p$-local subgroup $H$ of $G$ containg $A$, $p$ odd.

### 3.4. Lemmas of Generation

At the base of the development are two rather technical, but general, lemmas (Lemmas 2.4 and 2.5 of C.I. 1) which assert in spirit that a certain group is generated by subgroups of it of a special type. The proofs, although not very difficult, are quite long. I prefer to state the results without proof and then to show you explicitly how they are applied. It is my feeling that lemmas of this type can be used in other group-theoretic situations.

LEMMA. *Let $B \times Z$ be the direct product of a noncyclic elementary abelian 2-group $B$ and a nontrivial elementary abelian $p$-group $Z$, $p$ an odd prime. Suppose $BZ$ acts on the $p'$-group $K$ of odd order in such a way that $K = [K, Z]$. Let $\mathscr{X}$ be the collection of $BZ$-invariant subgroups $X$ of $K$ such that*

(a) $$[X, Z] = X;$$

(b) $$|B : C_B(X)| \leq 4.$$

*Then*

$$C_K(Z) = \langle C_X(Z) : X \in \mathscr{X} \rangle.$$

The lemma tells us, in effect, that the centralizer of $Z$ on $K$ is generated by its centralizers on subgroups of $K$ which themselves centralize a large subgroup of $B$.

The conditions of the second lemma stem directly from the properties of our signalizer functor $\theta$.

LEMMA. *Let $B$ be an elementary abelian 2-group of rank at least 3 which acts on the group $K$ of odd order. Suppose that the following conditions hold.*

(a) *$K = \langle K(b) : b \in B^{\#} \rangle$, where $K(b)$ is a $B$-invariant normal subgroup of $C_K(b)$;*

(b) *if $b, b' \in B^{\#}$, then $C_{K(b)}(b') \subseteq K(b')$.*

*Then, if $F$ is any $B$-invariant subgroup of $K$, we have*

$$F = \langle F \cap K(b) : b \in B^{\#} \rangle.$$

With the aid of this last lemma we shall now establish the theorem stated without proof in section 2.

THEOREM. *Let $A$ be an elementary abelian 2-subgroup of the group $G$ of rank at least 3 and let $\theta$ be an $A$-signalizer functor on $G$. If $K \in \mathscr{V}_\theta(A)$, then*

$$C_K(a) = K \cap \theta(C_G(a))$$

*for all $a$ in $A^{\#}$.*

*Proof.* Let $B$ be a subgroup of $A$ of order 8 and set $K(a) = K \cap \theta(C_G(a))$ for $a$ in $A^{\#}$. By definition of $\mathscr{V}_\theta(A)$, we have $K = \langle K(a) : a \in A^{\#} \rangle$. Since $B$ is noncyclic, each $K(a) = \langle C_{K(a)}(b) : b \in B^{\#} \rangle$. But

$$C_{K(a)}(b) = (K \cap \theta(C_G(a))) \cap C_G(b) \subseteq K \cap \theta(C_G(b)) = K(b)$$

by the basic compatibility condition of signalizer functors. Hence $K = \langle K(b) : b \in B^{\#} \rangle$ and $C_{K(b)}(b') \subseteq K(b')$ for all $b$, $b'$ in $B^{\#}$. Furthermore, $K(b) = K \cap \theta(C_G(b)) \lhd K \cap C_G(b) = C_K(b)$ as $\theta(C_G(b)) \lhd C_G(b)$. Thus all the hypotheses of our second lemma are satisfied by $K$, $B$, and the subgroups $K(b)$.

We apply the lemma with $F = C_K(a)$, $a \in A^\#$, and conclude that

$$C_K(a) = \langle C_K(a) \cap K(b) : b \in B^\# \rangle.$$

However, $C_K(a) \cap K(b) = C_{K(b)}(a) \subseteq K(a) = K \cap \theta(C_G(a))$ for all $b$ in $B^\#$ and consequently $C_K(a) \subseteq K \cap \theta(C_G(a))$. The reverse inclusion being clear, the theorem follows.

As we have pointed out earlier, it is a consequence of this theorem that $A$-invariant subgroups of elements of $\mathsf{M}_\theta(A)$ are themselves elements of $\mathsf{M}_\theta(A)$.

To obtain a second consequence, we require a preliminary definition. In the course of the proof of our various $E$-theorems, it will be necessary to work with subgroups $B$ of index at most 4 in $A$ in much the same way as with $A$ itself. Accordingly we define $\mathsf{M}_\theta(B)$ for any noncyclic subgroup $B$ of $A$ as the set of $B$-invariant subgroups $K$ of $G$ of odd order such that

$$K = \langle K \cap \theta(C_G(b)) : b \in B^\# \rangle,$$

with analogous meanings given to $\mathsf{M}_\theta(B; p)$ and $\mathsf{M}_\theta^*(B; p)$.

As a corollary of the theorem, we have

Corollary. *If $B$ is a noncyclic subgroup of $A$ and $K \in \mathsf{M}_\theta(A)$, then $K \in \mathsf{M}_\theta(B)$.*

*Proof.* Since $B$ is noncyclic, $K = \langle C_K(b) : b \in B^\# \rangle$. By the theorem, $C_K(b) = K \cap \theta(C_G(b))$ as $K \in \mathsf{M}_\theta(A)$, so $K = \langle K \cap \theta(C_G(b)) : b \in B^\# \rangle$ and hence $K \in \mathsf{M}_\theta(B)$ by definition of $\mathsf{M}_\theta(B)$.

Condition (c) in the definition of $A$-signalizer functor has been stated as it is to enable us to work with subgroups $B$ of index at most 4 in $A$ in the same way as with $A$. We note, first of all, that because of our assumption on the rank of $A$, such a subgroup $B$ has rank at least 3. As a consequence, we can prove the following result.

Proposition. *If $B$ is a subgroup of index at most 4 in $A$, then the three parts of the proposition of section 2 hold for the elements of $\mathsf{M}_\theta(B)$.*

Now let $P \in \mathsf{M}_\theta^*(A; p)$ for some prime $p$ and let $P_0$ be an $A$-invariant subgroup of $P$. For suitable choices of $P_0$ and $B$, we shall be interested in $BP_0$-invariant $p'$-subgroups of $G$ of odd order. Accordingly we define $\mathsf{M}_\theta(BP_0)$ as the set of $P_0$-invariant elements of $\mathsf{M}_\theta(B)$ of order prime to $p$. Corresponding meanings are given to $\mathsf{M}_\theta(BP_0; q)$ and $\mathsf{M}_\theta^*(BP_0; q)$.

Our principal choice of $P_0$ is a minimal $A$-invariant subgroup of $Z(P)$, which we denote by $Z$. Then $Z$ is an elementary abelian $p$-group and $A$ acts irreducibly on $Z$. Hence if $A_0 = C_A(Z)$, then $A/A_0$ is cyclic, so $|A : A_0| \leq 2$. We choose $B$ to be any subgroup of index at most 2 in $A_0$. Thus $B$ has index at most 4 in $A$ and

$$BZ = B \times Z.$$

Because of the first lemma above, we shall be able to derive critical properties of elements of $\mathcal{N}_\theta(BZ)$ which satisfy the condition $[K, Z] = K$. Note first that $\theta(N_G(Z))$ is well defined as $N_G(Z)$ is a $p$-local subgroup of $G$ containing $A$ and $p$ is odd. We define $\theta(C_G(Z)) = C_G(Z) \cap \theta(N_G(Z))$. One checks easily that $\theta(C_G(Z))$ is the subgroup of $C_G(Z)$ generated by all elements of $\mathcal{N}_\theta(A)$ or $\mathcal{N}_\theta(B)$ which lie in $C_G(Z)$.

With $Z$ and $B$ as just defined, we shall prove

THEOREM. *If* $K \in \mathcal{N}_\theta(BZ)$ *and* $[K, Z] = K$, *then* $C_K(Z) \subseteq O_{p'}(\theta(C_G(Z))$.

Thus $C_K(Z)$ lies "as far down" in $C_G(Z)$ as possible. This result is very analogous in spirit to the key preliminary lemma of the Thompson transitivity theorem. As a consequence of it, one is able to establish a basic transitivity theorem for certain elements of $\mathcal{N}_\theta(BZ; q)$ for any prime $q$.

*Proof.* Put $C = C_G(Z)$. Let $\mathscr{X}$ be the collection of subgroups $X$ of $K$ which satisfy the conditions of the first lemma. Hence, by the lemma,

$$C_K(Z) = \langle C_X(Z) : X \in \mathscr{X} \rangle.$$

Thus to prove the theorem, it will suffice to show that each $C_X(Z) \subseteq O_{p'}(\theta(C_G(Z)))$.

The critical property of $X$ is that $|B : C_B(X)| \leq 4$. Since $m(B) \geq 3$, it follows that $C_B(X) \neq 1$. In addition, $[X, Z] = X$. We put $Y = C_X(Z)$ and we note that both $X$ and $Y$ are in $\mathcal{N}_\theta(B)$.

We first treat the case that $B$ has rank at least 4, in which case $C_B(X)$ is noncyclic. We let $T$ be a four-subgroup of $C_B(X)$ and let $u_i$ $(1 \leq i \leq 3)$ denote its involutions. We put $H_i = C_G(u_i)$ and $P_i = P \cap H_i$ $(1 \leq i \leq 3)$. Then $\langle A, X, Z, P_i \rangle \subseteq H_i$ as $T \subseteq B \subseteq A_0 = C_A(Z)$. Furthermore, $\langle X, Z, P_i \rangle \subseteq \theta(H_i)$ as each is in $\mathcal{N}_\theta(A)$. By an earlier result, we know now that $P_i$ is, in fact, a Sylow $p$-subgroup of $\theta(H_i)$. This is very important, for $Z \subseteq Z(P_i)$ as $Z \subseteq P_i \subseteq P$ and $Z \subseteq Z(P)$. Thus $Z$ is in the center of a Sylow $p$-subgroup of $\theta(H_i)$. But $\theta(H_i)$, being of odd order, is solvable and hence is $p$-constrained. A general consequence of this is that the centre of any Sylow $p$-subgroup always lies in $O_{p',p}(\theta(H_i))$. This is easily verified. In particular, it follows that $Z \subseteq O_{p',p}(\theta(H_i))$.

But $X \subseteq \theta(H_i)$ and therefore $[X, Z] \subseteq X \cap O_{p',p}(\theta(H_i))$. Since $[X, Z] = X$ is a $p'$-group, we conclude that $X \subseteq O_{p'}(\theta(H_i))$. Hence $[X, P_i]$ is a $p'$-group $(1 \leq i \leq 3)$. Since $Y = C_X(Z) \subseteq X$, it follows, in particular, that $[Y, P_i]$ is a $p'$-group $(1 \leq i \leq 3)$. It is this conclusion which we need to establish the theorem.

We turn now to $C$. We have $Y \subseteq C$ and, as $Z \subseteq Z(P)$, also $P \subseteq C$. Since $Y$ and $P$ are in $\mathcal{N}_\theta(B)$, it follows that $\langle Y, P \rangle \subseteq \theta(C)$. But $P \in \mathcal{N}_\theta^*(A; p)$ and so certainly $P$ is a Sylow $p$-subgroup of $\theta(C)$. Put $L = \theta(C)$, $R = P \cap O_{p',p}(L)$, and $\bar{L} = L/O_{p'}(L)$. Then $\bar{R} = O_p(\bar{L})$ and $C_{\bar{L}}(\bar{R}) \subseteq \bar{R}$

as $L$ is $p$-constrained. We shall argue that $\bar{Y}$ centralizes $\bar{R}$, which will force $\bar{Y} = 1$, whence $Y \subseteq O_{p'}(L) = O_{p'}(\theta(C_G(Z)))$, as required.

Now $P = \langle P_1, P_2, P_3 \rangle$ as $T$ is a four group. Likewise $R = \langle R_1, R_2, R_3 \rangle$, where $R_i = C_R(u_i) = P_i \cap R$ $(1 \leq i \leq 3)$. Thus $[Y, R_i]$ is a $p'$-group and hence so is $[\bar{Y}, \bar{R}_i]$. But $[\bar{Y}, \bar{R}_i] \subseteq [\bar{Y}, \bar{R}] \subseteq \bar{R}$ as $\bar{R} \lhd \bar{L}$. Thus $[\bar{Y}, \bar{R}_i]$ is also a $p$-group and so $[\bar{Y}, \bar{R}_i] = 1$. It follows that $\bar{Y}$ centralizes each $\bar{R}_i$ $(1 \leq i \leq 3)$ and therefore $\bar{Y}$ centralizes $\bar{R}$. The theorem is therefore proved in the case $m(B) \geq 4$.

The preceding argument appears in Theorem 4.2 of C.I. 1. In C.I. 3 we have shown that, with a slight modification, the argument can also be made to cover the case $m(B) = 3$. Note that in this case, $m(A) = 5$ and $m(A_0) = 4$. Since $A_0 Z = A_0 \times Z$, the preceding argument does hold for $A_0$. This is the critical observation we need.

We have shown above that $C_B(X) \neq 1$ in all cases. Choose $b \neq 1$ in $C_B(X)$ and set $H = C_G(b)$. Then, as with $H_i$ above, we obtain the inclusion $X \subseteq O_{p'}(\theta(H))$. Now $[X, Z] = X$ and so $X \subseteq K_0 = [O_{p'}(\theta(H)), Z]$. Moreover, $K_0 = [K_0, Z]$ by a general theorem. Hence we can take $K_0, A_0$ in the roles of $K, B$ in the theorem and conclude that $C_{K_0}(Z) \subseteq O_{p'}(\theta(C))$. But $Y = C_X(Z) \subseteq C_{K_0}(Z)$ and therefore $Y \subseteq O_{p'}(\theta(C))$ in this case as well.

A portion of the preceding proof can be used to help us establish the following result which appears in C.I. 3.

Lemma. *Let $P \in \mathcal{N}^*(A; p)$, let $T$ be a four subgroup of $A$, and let $Z$ be any subgroup of $C_{Z(P)}(T)$. Then if $K \in \mathcal{N}_\theta(A)$ and $Z \subseteq K$, we have $Z \subseteq O_{p',p}(K)$.*

*Proof.* With $H_i$ as above, we have seen that $Z \subseteq O_{p',p}(\theta(H_i))$ and that $P_i$ is a Sylow $p$-subgroup of $\theta(H_i)$ $(1 \leq i \leq 3)$.

Let $R$ be an $AZ$-invariant Sylow $p$-subgroup of $O_{p',p}(K)$ containing $Z$ and set $R_i = C_R(u_i)$. Also put $L_i = \theta(H_i)$ and $\bar{L}_i = L_i/O_{p'}(L_i)$ $(1 \leq i \leq 3)$. Then $\bar{Z} \subseteq O_p(\bar{L}_i) \subseteq \bar{P}_i$ and so $\bar{Z} \subseteq Z(O_p(\bar{L}_i))$. Furthermore, $R_i \subseteq \theta(H_i) = L_i$ and consequently $R_i$ normalizes $O_p(\bar{L}_i)$. Hence $[\bar{R}_i, \bar{Z}] \subseteq Z(O_p(\bar{L}_i))$ and therefore $[\bar{R}_i, \bar{Z}, \bar{Z}] = 1$. This yields the equality

$$[R_i, Z, Z] = 1 \qquad (i \leq i \leq 3),$$

the critical fact which we need.

Now consider $K$ and put $\bar{K} = K/O_{p'}(K)\Phi(R)$. Then we have $\bar{R} = \langle \bar{R}_1, \bar{R}_2, \bar{R}_3 \rangle = \bar{R}_1 \bar{R}_2 \bar{R}_3$ is elementary abelian. Since $[R_i, Z, Z] = 1$ for each $i$, it follows that

$$[\bar{R}, \bar{Z}, \bar{Z}] = 1.$$

But $\bar{K}$, being solvable of odd order, is both $p$-constrained and $p$-stable. The latter condition implies that $\bar{Z}$ centralizes $\bar{R}$, while the former, when combined with Burnside's $R/\Phi(R)$-theorem implies that $C_{\bar{K}}(\bar{R}) \subseteq \bar{R}$. Hence $\bar{Z} \subseteq \bar{R}$ and so $Z \subseteq O_{p'}(K)R = O_{p',p}(K)$.

The proof also provides a nice illustration of the use of $p$-stability.

## 3.5. A Relativized Transitivity Theorem

Let $P$, $Z$, $A_0$, $B$ be as in the preceding section, so that $B$ is elementary abelian of index at most 4 in $A$, $B \subseteq A_0 = C_A(Z)$, and $Z$ is a minimal $A$-invariant subgroup of $Z(P)$ with $P \in \mathscr{V}_\theta^*(A; p)$. We shall be interested in the elements $Q$ of $\mathscr{V}_\theta(BZ; q)$ and $\mathscr{V}_\theta(AZ; q)$, $q$ an odd prime distinct from $p$, with the property that

$$[Q, Z] = Q.$$

We designate the maximal such elements by $\mathscr{V}_\theta'(BZ; q)$ and $\mathscr{V}_\theta'(AZ; q)$ respectively. These can also be described in the following way. Let $Q \in \mathscr{V}_\theta'(BZ; q)$, say, and let $Q^* \in \mathscr{V}_\theta^*(BZ; q)$ with $Q^* \supseteq Q$. Then

$$Q_1 = [Q^*, Z] \supseteq [Q, Z] = Q.$$

But $[Q_1, Z] = Q_1$, so $Q_1 = Q$ by the maximality of $Q$. Thus $Q = [Q^*, Z]$ for some element $Q^*$ of $\mathscr{V}_\theta^*(BZ; q)$.

As we have remarked, one of the results of the preceding section enables us to prove in much the same way as the Thompson transitivity theorem the following fundamental relativized transitivity theorem.

Theorem. *$O_{p'}(\theta(C_G(AZ)))$ and $O_{p'}(\theta(C_G(BZ)))$ transitively permute the elements of $\mathscr{V}_\theta'(AZ; q)$ and $\mathscr{V}_\theta'(BZ; q)$ respectively under conjugation.*

Just as with the Thompson transitivity theorem one obtains as a consequence that for any $Q \in \mathscr{V}_\theta'(BZ; q)$

$$N_G(BZ) = (N_G(BZ) \cap N_G(Q))O_{p'}(\theta(C_G(Z))).$$

This in turn yields

Corollary. *$AC_P(B)$ normalizes some element of $\mathscr{V}_\theta'(BZ; q)$.*

The proof of the theorem is given in Theorem 4.3 of C.I. 1. I shall explain its significance. Note first that as $A$ normalizes some element of $\mathscr{V}_\theta'(BZ; q)$, it follows that every element of $\mathscr{V}_\theta'(AZ; q)$ is, in fact, an element of $\mathscr{V}_\theta'(BZ; q)$.

We would like, of course, to strengthen the preceding corollary to get $P$ itself to normalize some element of $\mathscr{V}_\theta'(AZ; q)$. This was really the basic consequence of the Thompson transitivity theorem. In that case, our group $P$ was contained in $N_G(A)$ ($A$ being normal in $P$). However, in the present situation, all we can manage to get into the picture is $P \cap N_G(BZ)$. Since $Z \subseteq Z(P)$ and $P$ is $B$-invariant, this is the same as $C_P(B)$. Unless we can get $P$ to normalize an element of $\mathscr{V}_\theta'(AZ; q)$, there will be very little chance of our being able to establish $E_{p,q}(A)$.

We come now to a central point in the argument. It will explain why we work with subgroups of index at most 2 in $A_0$ rather than $A_0$ itself. The reason is directly related to the Brauer–Wielandt formula which we discussed in Part 2. If $\mathscr{B}$ denotes the set of such subgroups of $A_0$, we know that

$|P|$ is completely determined by $|C_P(B)|$ as $B$ ranges over $\mathscr{B}$. It is this fact we need. Observe also that all the preceding results apply to *each* $B$ in $\mathscr{B}$.

Now choose $Q$ in $\mathsf{N}'_\theta(AZ; q)$, put $K = N_G(Q)$, and let $R$ be an $A$-invariant Sylow $q$-subgroup of $C_{\theta(K)}(Z)$. We shall prove for each $B$ in $\mathscr{B}$ that

$$|C_P(B)| \leq |C_R(B)|.$$

The Brauer–Wielandt theorem will then yield

$$|P| \leq |R|.$$

However, $P \in \mathsf{N}^*_\theta(A; p)$ and $R \in \mathsf{N}_\theta(A; p)$. Since every element of $\mathsf{N}^*_\theta(A; p)$ is conjugate to $P$, they all have the same order. Thus the only possibility is that $|P| = |R|$ and that $R \in \mathsf{N}^*_\theta(A; p)$. But both $P$ and $R$ lie in $C_G(Z)$, so they lie in $L = \theta(C_G(Z))$. Thus $P = R^y$ for some $y \in C_L(A)$. We have that $y$ centralizes $AZ$ and that $R$ normalizes $Q$. Hence $P$ normalizes $Q_1 = Q^y$. Moreover, also $Q_1 \in \mathsf{N}'_\theta(AZ; q)$ since

$$[Q_1, Z] = [Q^y, Z] = [Q^y, Z^y] = [Q, Z]^y = Q^y = Q_1$$

and $|Q_1| = |Q|$. Thus $P$ will normalize some element of $\mathsf{N}'_\theta(AZ; q)$, as desired.

To prove the assertion, take $B$ in $\mathscr{B}$. By the relativized transitivity theorem, $U = (C_P(B))^x$ normalizes $Q$ for some $x$ in $O_{p'}(\theta(C_G(BZ)))$. Thus $U \subseteq K = N_G(Q)$, $U \in \mathsf{N}_\theta(B)$, and $U$ centralizes $BZ$. Hence $U \subseteq \theta(K) \cap C_G(Z) = C_{\theta(K)}(Z) = K_0$. But $R$ is a Sylow $p$-subgroup of $K_0$ and is $B$-invariant. Hence $U^u \subseteq R$ for some $u$ in $C_{K_0}(B)$. Since $B$ centralizes $U^u$, it follows that

$$|C_R(B)| \geq |U^u| = |U| = |C_P(B)|.$$

We have therefore proved

THEOREM. *$P$ normalizes some element of $\mathsf{N}'_\theta(AZ; q)$.*

As a consequence, we obtain the following key result.

THEOREM. *If $P$ centralizes every element of $\mathsf{N}_\theta(AP; q)$, then $Z$ centralizes every element of $\mathsf{N}_\theta(AZ; q)$.*

*Proof.* $P$ normalizes some element $Q$ of $\mathsf{N}'_\theta(AZ; q)$, as we have just shown. Since $Q \in \mathsf{N}_\theta(A)$ and $Q$ is $P$-invariant, $Q \in \mathsf{N}_\theta(AP; q)$. Hence by our hypothesis, $Z$ centralizes $Q$. Since $Q = [Q, Z]$, we have $Q = 1$. But now the relativized transitivity theorem yields that 1 is the only element of $\mathsf{N}'_\theta(AZ; q)$. Hence if $U \in \mathsf{N}_\theta(AZ; q)$ and $U_0 = [U, Z]$, then $U_0 = [U_0, Z]$ and so $U_0 = 1$. Thus $Z$ centralizes $U$, as asserted.

In the proof of our $E$-theorems the set $\mathscr{B}$ is needed for the sole purpose of showing that $P$ normalizes some element of $\mathsf{N}'_\theta(BZ; q)$ (and in one other place to obtain an analogous conclusion).

Using the relativized transitivity theorem, one can establish the following additional result concerning the elements of $\mathsf{N}^*_\theta(AZ; q)$

THEOREM. *$C_G(AZ)$ permutes the elements of $\mathscr{N}_\theta^*(AZ; q)$ transitively under conjugation.*

For our $E$-theorems we do not require here that $O_{p'}(\theta(C_G(AZ)))$ acts transitively.

3.6. $E_{p,q}(A)$ AND $C_{p,q}(A)$

Section 19 of the odd order paper is devoted to a proof of $E_{p,q}$ ($G$ being the simple group of odd order under consideration) under the assumptions that $SCN_3(p)$ and $SCN_3(q)$ are non-empty and that a Sylow $p$-subgroup of $G$ centralizes every $q$-subgroup of $G$ which it normalizes, while a Sylow $q$-subgroup of $G$ centralizes every $p$-subgroup of $G$ which it normalizes. The proof of this theorem is quite difficult. On the other hand, the analogue of this result in our own case is very easy to prove as a consequence of our relativized transitivity theorem.

THEOREM. *Let $P \in \mathscr{N}_\theta^*(A; p)$ and $Q \in \mathscr{N}_\theta^*(A; q)$, $p \neq q$, and suppose that $P$ centralizes every element of $\mathscr{N}_\theta(AP; q)$, while $Q$ centralizes every element of $\mathscr{N}_\theta(AQ; p)$. Then $G$ satisfies $E_{p,q}(A)$.*

*Proof.* Let $Z_p$, $Z_q$ denote respectively a minimal $A$-invariant subgroup of $Z(P)$ and $Z(Q)$. Then the results of the preceding section apply to both $AZ_p$ and $AZ_q$; because of our hypotheses, we have that

$Z_p$ centralizes every element of $\mathscr{N}_\theta(AZ_p; q)$; and
$Z_q$ centralizes every element of $\mathscr{N}_\theta(AZ_q; p)$.

It is this fact which allows a simple proof.

There is actually a minor error in the published argument (Theorem 5.1 of C.I. 1). At one point we know that $P$ is permutable with a certain element of $\mathscr{N}_\theta(A; q)$ and we assert that without loss, this element can be chosen to lie in $Q$; but it is not clear that this normalization is possible under the given conditions. To correct for this, we use instead the general lemma from C.I. 3 established in Section 4.

We know that $A/C_A(Z_p)$ and $A/C_A(Z_q)$ are both cyclic, so there is an involution $a$ of $A$ which centralizes both $Z_p$ and $Z_q$. Putting $H = C_G(a)$, we have, as usual, $\langle Z_p, Z_q\rangle \subseteq \theta(H)$. Since $\theta(H)$ is solvable of odd order, it contains an $A$-invariant Hall $\{p, q\}$-subgroup $K$ and $\langle Z_p^x, Z_q^y\rangle \subseteq K$ for suitable $x$, $y$ in $C_{\theta(H)}(A)$. Since $P^x$, $Z_p^x$ and $Q^y$, $Z_q^y$ clearly have the same properties as $P$, $Z_p$ and $Q$, $Z_q$ respectively, we can assume that $\langle Z_p, Z_q\rangle \subseteq K$.

Now $K \in \mathscr{N}_\theta(A)$, so by the general lemma $Z_p \subseteq O_{p',p}(K) = O_{q,p}(K)$ and $Z_q \subseteq O_{q',q}(K) = O_{p,q}(K)$. But $O_q(K) \in \mathscr{N}_\theta(AZ_p; q)$ and so $Z_p$ centralizes $O_q(K)$. However, it is a routine matter to verify that $C_{O_{q,p}(K)}(O_q(K)) \subseteq O_p(K)$, whence $Z_p \subseteq O_p(K)$. Similarly $Z_q \subseteq O_q(K)$. But $O_p(K)$ centralizes $O_q(K)$ and so $Z_p$ centralizes $Z_q$.

Next put $N = N_G(Z_p)$, whence $\langle A, P, Z_q\rangle \subseteq N$, $\langle P, Z_q\rangle \subseteq \theta(N)$, and $P$ is a Sylow $p$-subgroup of $\theta(N)$ (as $P \in И_\theta^*(A; p)$). Let $L$ be an $A$-invariant Hall $\{p, q\}$-subgroup of $\theta(N)$ containing $P$. Then $Z_q^y \subseteq L$ for some $y$ in $C_{\theta(N)}(A)$, and, as above, we can assume without loss that $Z_q \subseteq L$. (However it is not clear that we can assume that $Q \cap L$ is a Sylow $q$-subgroup of $L$.) The same argument as in the preceding paragraph shows that $Z_q \subseteq O_q(L)$.

But by hypothesis $P$ centralizes $O_q(L)$ as $O_q(L) \in И_\theta(AP; q)$. In particular, $P$ centralizes $Z_q$. Finally putting $M = N_G(Z_q)$, we have $\langle P, Q, A\rangle \subseteq M$, $\langle P, Q\rangle \subseteq \theta(M)$, and $P, Q$ is a Sylow $p$-subgroup and Sylow $q$-subgroup of $\theta(M)$ whence, for some $u$ in $C_{\theta(M)}(A)$, $PQ^u$ is an $S_{p,q}(A)$ subgroup of $G$. Thus $G$ satisfies $E_{p,q}(A)$.

Clearly these conditions are satisfied if both $И_\theta(AP; q)$ and $И_\theta(AQ; p)$ are trivial. Hence the significance of the theorem is that in proving $E_{p,q}(A)$, it allows us to assume that either $И_\theta(AP; q)$ or $И_\theta(AQ; p)$ is nontrivial. For definiteness, let us assume that $И_\theta(AP; q)$ is nontrivial. Let $Q_1 \in И_\theta^*(AP; q)$, so that $Q_1 \neq 1$. Putting $H = N_G(Q_1)$, we have that $P$ is a Sylow $p$-subgroup of $\theta(H)$ and hence that $P$ is permutable with an $A$-invariant Sylow $q$-subgroup $Q_2$ of $\theta(H)$. Without loss we can assume that $Q_2 \subseteq Q$.

LEMMA. *If* $O_p(PQ_2) = 1$, *then* $Q_2 = Q$ *and* $G$ *satisfies* $E_{p,q}(A)$.

*Proof.* Under this assumption, as $PQ_2$ is both $q$-constrained and $q$-stable, Glauberman's $ZJ$-theorem applies and yields that $Z(J(Q_2)) \lhd PQ_2$. Hence, if $K = N_G(Z(J(Q_2)))$, we have $A \subseteq K$ and $PQ_2 \subseteq \theta(K)$, with $P$ a Sylow $p$-subgroup of $\theta(K)$.

Suppose that $Q_2 \subset Q$. Then $K \cap Q \supset Q_2$ as $Z(J(Q_2))$ is characteristic in $Q_2$. Also $K \cap Q \subseteq \theta(K)$. Hence, if $Q_3$ is an $A$-invariant Sylow $q$-subgroup of $\theta(K)$ which contains $Q_2$ and is permutable with $P$, then $Q_3 \supset Q_2$. (Such a $Q_3$ exists as $\theta(K)$ is $A$-invariant and solvable of odd order.) Since $Q_1$ is $P$-invariant, the $p$-constraint of $PQ_3$ is easily seen to force $Q_1$ to lie in $O_q(PQ_3)$. But $Q_1 \in И_\theta^*(AP; q)$ and this implies that $Q_1 = O_q(PQ_3)$. In particular, $Q_1 \lhd Q_3$ and so $Q_3 \subseteq H = N_G(Q_1)$, whence $Q_3 \subseteq \theta(H)$. But $Q_2 \subset Q_3$ and $Q_2$ is a Sylow $q$-subgroup of $\theta(H)$, a contradiction.

Thus we are reduced to treating the case that $P_1 = O_p(PQ_2) \neq 1$. This case is rather technical. In basic outline, it follows the proof of the corresponding result in the odd order paper. In fact, using essentially the same argument, one obtains the following conclusion.

If $X \in И_\theta(A)$ and $X$ contains $Q_2P_1$, then $Q_2$ is a Sylow $q$-subgroup of $X$.

In the odd order paper, the completion of the proof from this point is very delicate, whereas in our case it is quite simple. All we need to do is to produce an element $X$ of $И_\theta(A)$ which contains both $Q_2P_1$ and an element of $И_\theta^*(A; q)$. The preceding result will then force $Q_2 = Q$, whence $PQ_2 = PQ$ will be an $S_{p,q}(A)$-subgroup of $G$ and $G$ will satisfy $E_{p,q}(A)$.

To accomplish this, we need one preliminary result. Note that as $Q_2 \subseteq Q$, $Q_2 = N_Q(Q_1)$ and so certainly $Z(Q) \subseteq Q_2$. In particular, $Z_q \subseteq Q_2$. Thus $Z_q$ acts on $P_1$.

LEMMA. *If* $P_0 = [P_1, Z_q]$, *then either* $P_0 = 1$ *or* $P_0 \in \mathcal{U}'_\theta(AZ_q; p)$.

Using this lemma, we can easily construct the required $X$. Indeed, if $P_0 = 1$, then $Z_q$ centralizes $P_1$, whence $M = N_G(Z_q)$ contains both $Q$ and $P_1$, in which case $X = \theta(M)$ has the required properties. In the contrary case, we put $M = N_G(P_0)$. Note that $P_0 \lhd P_1$ as $P_0 = [P_1, Z_q]$, so $P_1 \subseteq M$. But $P_0$ is also $Q_2$-invariant as $Q_2$ normalizes both $P_1$ and $Z_q$. Thus $Q_2 \subseteq M$ and so $P_1 Q_2 \subseteq \theta(M)$. Furthermore, by one of the consequences of our main relativized transitivity theorem (with $p$ and $q$ interchanged), $Q$ normalizes some element of $\mathcal{U}'_\theta(AZ_q; p)$ and so $Q^x$ normalizes $P_0$ for some $x$ in $O_{p'}(\theta(C_G(AZ_q)))$ as $P_0 \in \mathcal{U}'_\theta(AZ_q; p)$. But then $Q^x \subseteq M$ and $Q^x \in \mathcal{U}_\theta(A)$, so $Q^x \subseteq \theta(M)$. Again $X = \theta(M)$ has the required properties.

As for $C_{p,q}(A)$, the proof is accomplished by much the same sort of argument, using the various relativized transitivity theorems of the preceding section. The key step is the following.

LEMMA. *Let* $R$ *be an* $S_{p,q}(A)$-*subgroup of* $G$ *containing* $P$. *Then*

(i) $O_q(R) \in \mathcal{U}^*_\theta(AP; q)$;

(ii) $[O_q(R), Z_p] \in \mathcal{U}'_\theta(AZ_p; q)$.

Hence, if $R$ and $R'$ are two $S_{p,q}(A)$-subgroups of $G$, we see that $O_q(R)$ and $O_q(R')$ are conjugate as are $[O_q(R), Z_p]$ and $[O_q(R'), Z_p]$ by our transitivity theorems. From this, one obtains the conjugacy of $R$ and $R'$. The details are given in section 5 of C.I. 1.

### 3.7. $E_\sigma(A)$ AND $C_\sigma(A)$

One now proceeds to establish $E_\tau(A)$ and $C_\tau(A)$ for all subsets $\tau$ of $\sigma$ by induction on the cardinality of $\tau$. In view of the results of the last section, one can assume that $\tau$ consists of at least three primes. In over-all outline the proof follows very closely the argument of the corresponding result in the odd order paper. Again, the arguments are similar in spirit to those we have given. Unfortunately the technicalities involved are considerable and it is not possible to present the details here. The complete proofs are given in Section 6 of C.I. 1.

I wish here to mention only one point, where our argument is forced to deviate from that in the odd order paper. Yet another relativized transitivity theorem is required to be used in conjunction with the Brauer–Wielandt formula as was done in Section 5. This argument occurs at the very end of the proof of $E_\tau(A)$ when determining the order of a Sylow $p$-subgroup of a certain subgroup $\theta(N_1)$ defined in the course of the proof.

In the given situation, one cannot work with elements of $\mathcal{N}_\theta(AZ; q)$ and $\mathcal{N}_\theta(BZ; q)$ as we did before, but instead we must consider elements of $\mathcal{N}_\theta(AZJ^*(P); q)$ and $\mathcal{N}_\theta(BZJ^*(P); q)$. Here $ZJ^*(P)$ is the subgroup of $P$, constructed by Glauberman, which enjoys the same properties as $Z(J(P))$, but which in addition contains its own centralizer in $P$. Thus one can establish the following lemma concerning $ZJ^*(P)$.

LEMMA. *If $K$ is an element of $\mathcal{N}_\theta(A)$ containing $P$, then*

(i) $O_{p'}(K)ZJ^*(P) \lhd K$. *In particular,* $ZJ^*(P) \subseteq O_{p',p}(K)$.

(ii) $C_K(ZJ^*(P)) \subseteq O_{p',p}(K)$.

Condition (ii) would not necessarily hold if we used $Z(J(P))$ in place of $ZJ^*(P)$.

The specific transitivity theorem we need is the following. (Here $B$ has the same meaning as in Section 5 relative to a minimal $A$-invariant subgroup $Z$ of $P$).

THEOREM. *Assume that $Z$ centralizes every element of $\mathcal{N}_\theta^*(AZ; q)$. Then $O_{p'}(\theta(C_G(AZJ^*(P))))$ and $O_{p'}(\theta(C_G(BZJ^*(P))))$ transitively permute the elements of $\mathcal{N}_\theta^*(AZJ^*(P); q)$ and $\mathcal{N}_\theta^*(BZJ^*(P); q)$ respectively under conjugation.*

Since $C_P(ZJ^*(P)) \subseteq ZJ^*(P)$, $Z(P) \subseteq ZJ^*(P)$ and, in particular, $Z \subseteq ZJ^*(P)$. Because of our hypothesis on the elements of $\mathcal{N}_\theta(AZ; q)$, any element $Q$ of $\mathcal{N}_\theta^*(AZJ^*(P); q)$ or $\mathcal{N}_\theta^*(BZJ^*(P); q)$ necessarily lies in $\theta(C_G(Z))$ and, because of the preceding lemma, it follows, in fact, that $Q \in O_{p'}(\theta(C_G(Z)))$. From this, one obtains the theorem at once.

Without our assumption on $Z$, the validity of the theorem would not be clear, for we do not have

$$BZJ^*(P) = B \times ZJ^*(P), \quad \text{as we had } BZ = B \times Z,$$

and so our generational lemma is not available. Fortunately, the theorem is needed only under the given hypotheses on $Z$.

I think that I have covered all the key points in the proof of the signalizer functor theorem in this outline; and I hope that what I have said will assist those of you who are interested in reading the complete proof in C.I. 1. The remarkable fact is that the assumptions of the theorem enable one to reduce the basic content of the proof to arguments used in the study of groups of odd order. Perhaps, then, this outline will also serve to give you some insights into this portion of the odd order paper.

In conclusion, we should like to point out that Bender has achieved a simplified proof of some of the main results of Chapter IV of the odd order paper. It would be of interest to know whether his arguments can be adapted to give a shorter proof of the signalizer functor theorem.†

## 4. Balanced Groups

### 4.1. The Balanced Theorem

All the results of this and the next part, aside from some concluding comments in the final two sections, have been obtained jointly with John H. Walter. As we have seen in the last chapter, if the centralizer of every involution of $G$ is 2-constrained, then $\theta = O$ is an $A$-signalizer functor on $G$ for every elementary abelian 2-subgroup $A$ with $m(A) \geq 3$. Moreover, if $m(A) \geq 5$, the signalizer functor theorem yields that

$$W_A = \langle O(C_G(a)) : a \in A^{\#} \rangle$$

has odd order. This, as we have described earlier, is the first of the two major steps in the proof of our main theorem. To complete its proof, it will now suffice to demonstrate the following result.

Theorem. *Let $G$ be a group of 2-rank at least 5 with $O(G) = 1$ in which the centralizer of every involution of $G$ is 2-constrained. Assume that $W_A$ has odd order for some noncyclic elementary abelian 2-subgroup $A$ of $G$. Then $O(C_G(x)) = 1$ for every involution $x$ of $G$.*

We have shown in Part 3 that such a group $G$ is, in fact, balanced; that is, if $a$ and $b$ are any two commuting involutions of $G$, then

$$O(C_G(a)) \cap C_G(b) = O(C_G(b)) \cap C_G(a).$$

Moreover, we have noted in Chapter I that a result of MacWilliams implies that $SCN_3(2)$ is non-empty. Hence the preceding theorem will be a special case of the following more general assertion, which we shall refer to as the *balanced theorem.*

The balanced theorem. *Let $G$ be a balanced group such that $O(G) = 1$, $SCN_3(2)$ is non-empty, and $W_A$ has odd order for some noncyclic elementary abelian 2-subgroup of $G$. Then $O(C_G(x)) = 1$ for every involution $x$ of $G$.*

Although it *may be* possible to establish the balanced theorem in this degree of generality, its proof involves some delicate minimal cases which have not yet been eliminated in all cases. However, in any specific application, one will most likely have enough additional information about the structure of the centralizers of involutions to avoid these difficult configura-

† Goldschmidt's proof does, in fact, utilize Bender's central idea.

tions. In particular, when the centralizers of involutions are 2-constrained, the proof is very simple and will be given in complete detail. We shall make some comments concerning extensions to more general cases in Section 4.

The assumption that $SCN_3(2)$ is non-empty is used primarily to establish a specific property of $G$—namely, that $G$ is *connected*, a concept we shall now define. In fact, in the 2-constrained case, it would suffice to assume that $G$ is connected.

DEFINITION. Let $S$ be a 2-group of 2-rank at least 2 and let $A$, $A'$ be two noncyclic elementary abelian subgroups of $S$. We say that $A$ and $A'$ are *connected* provided there exists a sequence of noncyclic elementary abelian 2-subgroups

$$A = A_1, A_2, \ldots, A_n = A'$$

of $S$ such that either

$$A_i \subseteq A_{i+1} \quad \text{or} \quad A_{i+1} \subseteq A_i$$

$(1 \leq i \leq n-1)$. We say that $S$ is *connected* if every pair of noncyclic elementary abelian subgroups of $S$ is connected. Furthermore, an arbitrary group $G$ is said to be *connected* if a Sylow 2-subgroup of it is connected.

We have the following

LEMMA. *If $SCN_3(2)$ is non-empty in $G$, then $G$ is connected.*

*Proof.* Let $S$ be a Sylow 2-subgroup of $G$. By assumption, $S$ contains an elementary abelian normal subgroup $B$ of order 8. It will suffice to argue that this implies the connectedness of $S$.

Let $A$, $A'$ be two noncyclic elementary abelian subgroups of $S$. We shall determine a connected sequence $A_i$ $(1 \leq i \leq 11)$ whose terminal members are $A$ and $A'$. Since $B \lhd S$, $Z = B \cap Z(S) \neq 1$. We set $A_1 = A$, $A_2 = AZ$, we let $A_3$ be a four subgroup of $A_2$ containing $Z$ or contained in $Z$, and we set $A_4 = A_3 C_B(A_3)$ and $A_5 = C_B(A_3)$. The critical point here is that $A_5$ is noncyclic. Indeed, if $A_3 \subseteq Z$, then $A_5 = B$ as $Z \subseteq B$. In the contrary case, $|Z| = 2$, $A_5 = \langle Z, a \rangle$ for some involution $a$, and $A_5 = C_B(a)$. But $a$ acts as a linear transformation of $B$ regarded as a 3-dimensional vector space over $GF(2)$. Since $|\langle a \rangle| = 2$, $a$ has at least two Jordan blocks and consequently $C_B(a) = A_5$ is noncyclic.

Next set $A_6 = B$, let $A_{11} = A'$, $A_{10} = A'Z$, let $A_9$ be a four subgroup of $A'Z$ containing $Z$ or contained in $Z$, let $A_8 = A_9 C_B(A_9)$, and $A_7 = C_B(A_8)$. Then $A_7$ is also noncyclic and we check directly that the sequence $A_i$ $(1 \leq i \leq 11)$ has the required properties.

A simple group $G$ in which $SCN_3(2)$ is empty may be either connected or non-connected. For example, if the Sylow 2-subgroups of $G$ are dihedral of order at least 8, quasi-dihedral, wreathed, or if $G \cong J_2$ or $J_3$, then $G$ is not connected. On the other hand, if $G \cong PSL(4, q)$, $PSU(4, q)$, or $G_2(q)$, $q$ odd, $G$ is connected; moreover, in these cases, $SCN_3(2)$ is empty for certain congruences of $q$ (mod 8).

## 4.2. Balanced and Connected Groups

In this section we shall establish a number of important, but elementary properties of an arbitrary balanced and connected group $G$. $W_A$ is given the usual meaning, but we make no assumptions concerning its order. The following result is basic.

Proposition. *If $A$ and $A'$ are noncyclic elementary abelian 2-subgroups of the same Sylow 2-subgroup of $G$, then*

$$W_A = W_{A'}.$$

*Proof.* Since $G$ is connected, there is a connected sequence of noncyclic elementary abelian 2-subgroups of $G$ with $A$ and $A'$ as terminal members. To prove the proposition, it will clearly suffice to show that $W_T = W_{T'}$, where $T$ and $T'$ are adjacent terms of that sequence. By definition $T' \subseteq T$ or $T \subseteq T'$, say $T' \subseteq T$. Then clearly $W_{T'} \subseteq W_T$.

We establish the opposite inclusion. Let $t \in T^{\#}$. Since $T$ is abelian, $T'$ normalizes $O(C_G(t))$ and, as $T'$ is noncyclic, we have

$$O(C_G(t)) = \langle O(C_G(t)) \cap C_G(t') : t' \in (T')^{\#} \rangle.$$

But $G$ is balanced and so for each $t'$ in $(T')^{\#}$,

$$O(C_G(t)) \cap C_G(t') \subseteq O(C_G(t')) \subseteq W_{T'}.$$

Thus $O(C_G(t)) \subseteq W_{T'}$ for each $t \in T^{\#}$ and therefore $W_T \subseteq W_{T'}$. The proposition follows.

To exploit the proposition, we need the following easy lemma.

Lemma. *For any four subgroup $T$ of $G$ and any $x$ in $G$, we have*

$$(W_T)^x = W_{T^x}.$$

*Proof.* Since $O(C_G(t^x)) = (O(C_G(t)))^x$ for each $t$ in $T^{\#}$, the lemma is an immediate consequence of the definition of $W_T$ and $W_{T^x}$.

We now prove.

Lemma. *If $T$ is a four subgroup of $G$ and $x$ is an element of $G$ such that $\langle T, T^x \rangle$ is a 2-group, then*

$$(W_T)^x = W_T.$$

*Proof.* By the lemma, $(W_T)^x = W_{T^x}$. But by assumption $T$ and $T^x$ lie in a Sylow 2-subgroup of $G$, so by the proposition $W_{T^x} = W_T$ and the lemma follows.

This enables us to obtain a key conclusion.

PROPOSITION. *If T is a four subgroup of G and R is a 2-subgroup of G containing T, then*

$$N_G(R) \subseteq N_G(W_T).$$

*Proof.* If $x \in N_G(R)$, then $T^x \subseteq R$ as $T \subseteq R$, so $\langle T, T^x \rangle$ is a 2-group. Hence $(W_T)^x = W_T$ by the preceding lemma, and so $x \in N_G(W_T)$. Since $x$ was arbitrary, the proposition is valid.

We can rephrase this last result in a convenient way. Let $W$ be the common subgroup $W_A$ as $A$ ranges over the noncyclic elementary abelian 2-subgroups of a Sylow 2-subgroup $S$ of $G$ and set $M = N_G(W)$. Since $S$ permutes the set of these $A$ among themselves under conjugation, we see that $S$ normalizes $W$. Thus, $S \subseteq M$.

PROPOSITION. *If R is a 2-subgroup of G which contains a four subgroup of M, then* $N_G(R) \subseteq M$.

*Proof.* Let $T$ be a four subgroup of $M$ contained in $R$. It will clearly suffice to prove that $N_G(R^m) \subseteq M$ for any $m$ in $M$. But $T^m \subseteq S$ for some $m$ in $M$ as $S$ is a Sylow 2-subgroup of $M$, so without loss we can assume that $T \subseteq S$. Then $W_T = W$. But by the preceding proposition, we have

$$N_G(R) \subseteq N_G(W_T) = N_G(W) = M.$$

In particular, the proposition tells us that $N_G(S) \subseteq M$ and that $N_G(T) \subseteq M$ for any four subgroup $T$ of $M$.

We can squeeze a bit more out of the proposition by introducing the following general concept.

DEFINITION. A group $X$ will be said to be *k-generated*, $k$ a positive integer, if $X$ is generated by its subgroups $N_X(R)$ as $R$ ranges over the subgroups of 2-rank at least $k$ of a fixed Sylow 2-subgroup of $X$.

The preceding proposition clearly has the following

COROLLARY. *If H is a 2-generated subgroup of G such that* $H \cap M$ *contains a four group, then* $H \subseteq M$.

Unfortunately the proposition and its corollary do not tell us in general that $C_G(x) \subseteq M$ for every involution $x$ of $M$, which is what we shall need to show that $M$ is strongly embedded in $G$

## 4.3. THE 2-CONSTRAINED CASE

The results of the preceding section enable us to give a very simple proof of the balanced theorem under the assumptions of our main theorem. However, most of the arguments are general; since we intend to discuss the non-constrained case in the next section, we shall assume that the centralizer of every involution is 2-constrained only in the one place that it is needed.

We assume now that $G$ is a balanced and connected group in which $O(G) = 1$ and is such that $|W_A|$ is odd for some noncyclic elementary abelian 2-subgroup $A$ of $G$. We let $S$ be a Sylow 2-subgroup of $G$ containing $A$ and fix this notation for the section.

PROPOSITION. *The balanced theorem holds if* $W_A = 1$.

*Proof.* Suppose that $W_A = 1$ and let $t$ be an arbitrary involution of $S$. Let $z$ be an involution of $Z(S)$. If $t \neq z$, $T = \langle t, z \rangle$ is a four group. If $t = z$, $T = \langle t, a \rangle$ is a four group for any involution $a \neq t$ of $A$. But by our basic proposition $W_A = W_T$ and so $W_T = 1$. Since $O(C_G(t)) \subseteq W_T$, it follows that $O(C_G(t)) = 1$. Since any involution $x$ of $G$ is conjugate to some involution $t$ of $S$, the balanced theorem follows.

Hence for the remainder of its proof, we may assume that $W_A \neq 1$. We put $W = W_A$ and $M = N_G(W)$ and fix this notation as well. We know that $W = W_{A'}$ for any noncyclic elementary abelian 2-subgroup $A'$ of $S$.

Using Bender's results on strong embedding we prove the

PROPOSITION. *There exists an involution $t$ of $M$ such that*

$$C_G(t) \nsubseteq M.$$

*Proof.* Suppose false. By the final result of the preceding section, $N_G(S) \subseteq M$. Furthermore, since $W \subseteq O(M)$ and $W \neq 1$, while $O(G) = 1$, we see that $M \subset G$. Hence, to prove that $M$ is strongly embedded in $G$, it remains to show that $|M \cap M^g|$ is odd for every $g$ in $G - M$.

Assume, then, that $M \cap M^g$ contains an involution $x$ for some $g$ in $G - M$. Without loss we can assume that $x \in S$. Since $M$ contains the centralizer in $G$ of each of its involutions, so does $M^g$ and hence $C_G(x) \subseteq M^g$. Thus $Z(S) \subseteq M^g$ and so we can suppose to begin with that $x \in Z(S)$, whence $S \subseteq M^g$. Therefore $S^{g^{-1}} \subseteq M$ and so $S^{g^{-1}} = S^m$ for some $m$ in $M$. But then $mg \in N_G(S) \subseteq M$ and hence $g \in M$, contrary to our choice of $g$. We conclude that $M$ is strongly embedded in $G$.

The corollary of Bender's classification theorem now yields that $O(M) = 1$ inasmuch as $O(G) = 1$, contrary to the fact that $W \subseteq O(M)$ and $W \neq 1$.

The following result, when combined with the preceding proposition, will complete the proof of the balanced theorem in the 2-constrained case and will therefore complete the proof of our main theorem.

PROPOSITION. *If $t$ is an involution of $M$ and $C_G(t)$ is 2-constrained, then* $C_G(t) \subseteq M$.

*Proof.* Without loss we can assume that $t \in S$. As noted above, $t \in T$ for some four subgroup $T$ of $S$. Put $H = C_G(t)$, let $R$ be a Sylow 2-subgroup of $H$ containing $T$, and put $Q = R \cap O_{2',2}(H)$. Then $Q$ is a Sylow 2-subgroup of $O_{2',2}(H)$ and $H = O(H)N_H(Q)$ by the Frattini argument. Furthermore, $O(H) = O(C_G(t)) \subseteq W_T = W$, so $O(H) \subseteq M$. Hence the desired conclusion $H \subseteq M$ will follow provided we show that $N_H(Q) \subseteq M$.

Since $T \subseteq R$ and $T \subseteq M$, we know that $N_G(R) \subseteq M$. In particular, $R$ and hence $Q$ is contained in $M$. If $Q$ contains a four subgroup, we will also have $N_G(Q) \subseteq M$. Hence we can suppose that $Q$ does not contain a four subgroup. Thus $Q$ is of 2-rank 1 and so $Q$ is either cyclic or generalized quaternion. Since $R \supseteq T$, we have $R \supset Q$ and hence $O(H)Q \subset H$. Putting $\bar{H} = H/O(H)$, we have $\bar{Q} = O_2(\bar{H})$ and $\bar{Q} \subset \bar{H}$. In particular, these conditions imply that $\bar{H}$ is not a 2-group. But by hypothesis, $H$ is 2-constrained and so $C_{\bar{H}}(\bar{Q}) \subseteq \bar{Q}$. We see then that some element of $\bar{H}$ of odd order must induce a nontrivial automorphism of $\bar{Q}$. However, if $\bar{Q}$ is cyclic or generalized quaternion of order at least 16, it is well known that its automorphism group is a 2-group. Hence $\bar{Q}$ must be quaternion of order 8 and $\bar{H}/\bar{Q}$ must be isomorphic to $S_3$. Since $\bar{H}$ contains the four group $\bar{T}$, the only possibility is that $\bar{H} \cong GL(2, 3)$, which implies that $\bar{R}$, and hence also $R$, is quasi-dihedral of order 16.

However, in this case $\langle t \rangle = Z(R)$. Since $R \subseteq M$, we can assume without loss that $R \subseteq S$. Since $Z(S)$ is characteristic in $S$, it follows that $N_S(R) \subseteq H$. Since $R$ is a Sylow 2-subgroup of $H$, this forces $S = R$ and so $S$ is quasi-dihedral. But this is impossible as a quasi-dihedral group is not connected, while $S$ is connected by hypothesis.

Since $C_G(t)$ contains a four subgroup of $M$ for every involution $t$ of $M$, the results of the preceding section also yield the following.

PROPOSITION. *If $t$ is an involution of $M$ such that $C_G(t)$ is 2-generated, then $C_G(t) \subseteq M$.*

Thus the balanced theorem holds if the centralizer of every involution of $G$ is 2-generated. As a consequence, our argument actually yields the following.

THEOREM. *Let $G$ be a balanced group of 2-rank at least 5 in which $O(G) = 1$. If the centralizer of every involution of $G$ is 2-generated, then $O(C_G(x)) = 1$ for every involution $x$ of $G$.*

We conclude this section with a further property of our group $G$.

PROPOSITION. $O_{2',2}(G) = 1$.

*Proof.* By assumption, $O(G) = 1$. Hence if the proposition is false, $R = O_2(G) \neq 1$. Suppose that $R$ contains a four subgroup $T$. Since $R \subseteq S$, $T \subseteq M$ and consequently $G = N_G(R) \subseteq M$, as $T \subseteq R$. However, this is impossible as $M$ is a proper subgroup of $G$.

Hence $R$ does not contain a four group and so $R$ is either cyclic or generalized quaternion. In particular, $R$ contains a unique involution $t$ and $\langle t\rangle = Z(R)$, so $\langle t\rangle \lhd G$. We conclude that $G = C_G(t)$. Since $O(G) = 1$, this in turn yields that $O(C_G(t)) = 1$. Now let $U$ be any four subgroup of $S$ containing $t$. Since $G$ is balanced, for any $u$ in $U^\#$ we have that

$$O(C_G(u)) = O(C_G(u)) \cap G = O(C_G(u)) \cap C_G(t) \subseteq O(C_G(t)) = 1.$$

Thus $W_U = 1$. But $W_U = W_A = W$ and $W \neq 1$.

## 4.4. The General Case

One cannot expect the centralizer of every involution of $G$ always to be 2-generated. Indeed, if $H = C_G(t)$, $t$ an involution of $G$, it may very well happen that $\bar{H} = H/O(H)$ possesses a perfect normal subgroup $\bar{L}$ with the following properties:

(a) $\bar{L}/Z(\bar{L}) \cong PSL(2, 2^n)$, $PSU(3, 2^n)$, or $Sz(2^n)$ for some $n \geq 2$;
(b) $C_{\bar{H}}(\bar{L})$ has a cyclic or generalized quaternion Sylow 2-subgroup;
(c) All involutions of $\bar{H}$ lie in $\bar{L}C_{\bar{H}}(\bar{L})$.

If this is the case and $S$ is a Sylow 2-subgroup of $H$, it is easy to see that as $R$ ranges over the subgroups of $S$ of 2-rank at least 2 the subgroup of $H$ generated by the groups $N_H(R)$ is precisely $O(H)N_H(L \cap S)$, where $L$ denotes the inverse image of $\bar{L}$ in $H$. It follows, therefore, that $H$ is not 2-generated.

Any group $H$ which satisfies the above three conditions will be called *exceptional.*

(It also turns out that $H$ is not 2-generated if $\bar{H} \cong \hat{A}_9$, the perfect central extension of $A_9$ by a group of order 2. However, groups in which the centralizer of an involution is of this form have been completely classified by Harada and myself).

Hence the argument of the preceding section does not apply if the centralizer of some involution of $G$ is exceptional in this sense. Remarkably enough, if the composition factors of a group $H$ are among the presently known simple groups with $H$ of 2-rank at least 3, it appears to be universally true that $H$ is either 2-generated or exceptional or else $G$ is of 2-rank 3 and $Z^*(G) \neq 1$ (in which case $O_{2',2}(G) \neq 1$). Thus in any specific application it would seem to be sufficient to establish the balanced theorem under the assumption that the centralizer of every involution of $G$ is either 2-generated or exceptional.

By the results of the preceding section, it suffices in this case to prove that the centralizer of no involution $t$ of $G$ is exceptional. However, we note that the automorphism group of $PSL\,(2, 2^{2m})$ of $PSU\,(3, 2^{2m})$, $m \geq 2$, contains an involution with an exceptional centralizer. To rule out these groups, it is necessary to assume, in addition, that $G$ has no normal subgroups of Index 2 (when $t$ is non-central).

At the present time this result has, in fact, been established in all but a very few special cases when $\bar{L}$ is a Suzuki group and $t$ is non-central. In particular, we have the following:

PROPOSITION. *The balanced theorem holds if the centralizer of every involution of G is either 2-generated or central and exceptional.*

This result will give the following further extension of our main theorem.

THEOREM. *Let G be a balanced group of 2-rank at least 5 in which $O(G) = 1$. If the centralizer of every involution of G is either 2-generated or central and exceptional, then $O(C_G(x)) = 1$ for every involution x of G.*

On the other hand, to treat the balanced theorem in complete generality, it is not enough to consider only the centralizers of involutions of $M$, but instead one must consider *all* nontrivial 2-subgroups $T$ of $M$ such that $N_G(T) \nsubseteq M$ and especially the subset $\mathscr{T}$ of those $T$ which are maximal with this property. It is then not difficult to establish the following.

PROPOSITION. *Let $T \in \mathscr{T}$, put $H = N_G(T)$, and let R be a Sylow 2-subgroup of H such that $R \cap M$ is a Sylow 2-subgroup of $H \cap M$. Then the following conditions hold:*

(i) *T is cyclic or generalized quaternion;*
(ii) *$R \subseteq M$;*
(iii) *If R has 2-rank at least 3, then $O(H) \subseteq M$;*
(iv) *$(H \cap M)/T$ is strongly embedded in $H/T$.*

Condition (iv) together with Bender's theorem will give the precise structure of $H/T$. If $H$ is nonsolvable, one is able to prove that $H$ is, in fact, exceptional, in which case the situation is very similar to that of the preceding proposition and can be treated in essentially the same way. However, when $A$ is solvable, there are certain minimal configurations whose elimination seems to be very difficult. These occur when $R$ has 2-rank exactly 2, in which case $O(H)$ need not be contained in $M$. Of course, since $H$ is solvable, $|T| > 2$; since otherwise $H = C_G(t)$, where $T = \langle t \rangle$, and as $H$ is 2-constrained, it would follow from our previous results that $H \subseteq M$ contrary to the definition of $\mathscr{T}$. To obtain a contradiction in this case, a detailed analysis of the structure of the normalizers of some of the proper subgroups of $T$ seems necessary; again using Bender's theorem.

## 4.5. NON 2-CONSTRAINED GROUPS

I would now like to describe a larger class of balanced groups than that in which the centralizer of every involution is 2-constrained. To do this, it will first be necessary to make some general remarks about non 2-constrained groups. Although theses remarks apply equally well to groups that are not $p$-constrained ($p$ an arbitrary prime), for simplicity we shall limit ourselves to the case $p = 2$. The general situation is treated in a paper written jointly with John Walter which will appear in the *Illinois Journal of Mathematics.*

So, let $H$ be a group in which $O(H) = 1$. If $H$ is 2-constrained, then $C_H(O_2(H)) \subseteq O_2(H)$. We have seen in numerous arguments how crucial this condition is. Hence, if $H$ is not 2-constrained, it is natural to ask whether there is some other, easily described, subgroup $K$ of $H$ containing $O_2(H)$ with the property that $C_H(K) \subseteq K$. It turns out that there is a *unique* characteristic subgroup $K$ which is minimal with these properties, and this has a very simple structure.

To describe it, consider first the special case where we also have $O_2(H) = 1$. Then $C_H(O_2(H)) = H$ and, assuming that $H$ is nontrivial, $H$ is certainly not 2-constrained. Since $O(H) = 1$, it follows that $O_p(H) = 1$ for every prime $p$, even or odd. But now consider a minimal normal subgroup $N$ of $H$. Clearly $N$ is characteristically simple and so, by an elementary result, is the direct product of isomorphic simple groups. However, $N$ is not an elementary abelian $p$-group for any prime $p$ inasmuch as $O_p(H) = 1$ for all $p$. Hence the components of $N$ must be non-abelian. Furthermore, if $N_1$ and $N_2$ are two distinct minimal normal subgroups of $H$, then $N_1 \cap N_2 = 1$. Since $[N_1, N_2] \subseteq N_1 \cap N_2$, it follows that $N_1$ centralizes $N_2$. We see then that if $K$ denotes the product of all minimal normal subgroups of $H$, then $K$ is the direct product of non-abelian simple groups. This in turn implies that $C_K(K) = 1$. But $C_H(K) \lhd H$. Hence if $C_H(K) \neq 1$, $C_H(K)$ would contain some minimal normal subgroup of $H$, contrary to the fact that $C_H(K) \cap K = C_K(K) = 1$. Thus $C_H(K) = 1$ and so $K$ satisfies our requirement that $C_H(K) \subseteq K$.

On the other hand, it is not difficult to show that if $L$ is any normal subgroup of $H$ such that $C_H(L) \subseteq L$, then $L \supseteq K$. We conclude therefore in this case that there is a unique characteristic subgroup $K$ with the desired property and that $K$ is the direct product of non-abelian simple groups.

To treat the general case, we need a preliminary notion.

DEFINITION. A group $L$ is called (*perfect*) *quasi-simple* if $[L, L] = L$ and $L/Z(L)$ is a non-abelian simple group (equivalently, $Z(L) \subseteq [L, L]$ and $L/Z(L)$ is a non-abelian simple group). Furthermore, a central product of quasi-simple groups is called a *semi-simple* group and its factors are called its *components*. For convenience, the trivial group is also said to be semi-simple.

It is easily seen that the components of a semi-simple group are uniquely determined by the group itself.

Now let $H$ be an arbitrary group in which $O(H) = 1$ and put $C = O_2(H)C_H(O_2(H))$ and $\bar{C} = C/O_2(H)$. It is not hard to prove, first, that $O(\bar{C}) = O_2(\bar{C}) = 1$, and second, that if $\bar{K}$ denotes the product of all the minimal normal subgroups of $\bar{C}$ and $K$ its inverse image in $H$, then

$$K \supseteq O_2(H) \quad \text{and} \quad C_H(K) \subseteq K;$$

moreover, $K$ is the unique characteristic subgroup of $H$ that is minimal with these properties.

Thus $K$ is the subgroup of $H$ we are seeking. Observe now that $K = O_2(H)C_K(O_2(H))$ and that $\bar{K} = K/O_2(K)$ is the direct product of non-abelian simple groups. It follows directly from these conditions that

$$K = LO_2(H),$$

where $L$ is semi-simple, $L$ centralizes $O_2(H)$, and $L$ is characteristic in $H$.

Definition. The subgroup $L$ of the above group $H$ is called the 2-*layer* of $H$ and its components are called the 2-*components* of $H$. In addition, we denote the subgroup $K$ by $O_2^*(H)$.

It is this subgroup $O_2^*(H)$ which plays the role in an arbitrary group $H$ (with $O(H) = 1$) that $O_2(H)$ does in 2-constrained groups. More generally, if $O(H) \neq 1$, we denote by $O_{2',2}^*(H)$ the inverse image of $O_2^*(H/O(H))$ in $H$. It is this subgroup $O_{2',2}^*(H)$ with which one works when studying non-2-constrained groups.

It should be evident from the structure of $O_{2',2}^*(H)$ that specific properties of the 2-components of $H/O(H)$ will enter into the local group-theoretic analysis in a general classification problem. It is for this reason that the 2-constrained case is so much easier and less technical to treat.

Note incidentally that our above group $H$ (with $O(H) = 1$) is 2-constrained if and only if $O_2^*(H) = O_2(H)$ and hence if and only if $L = 1$. In other words, $H$ is 2-constrained if and only if it has no 2-components.

Recall now that in our preliminary discussion in Part 1 about the centralizer $N$ of an involution in a simple group $G$, we stated that, as far as the core $O(N)$ of $N$ is concerned, the central aim of the local analysis is to demonstrate that in abstract simple groups

$$O(N) \text{ is cyclic} \quad \text{and} \quad |N : C_N(O(N))| \leq 2.$$

If, in a given group $N$, this last condition alone is satisfied, it is quite easy to prove that

$$O_{2',2}^*(N) = LO_{2',2}(N),$$

where again $L$ is semi-simple, $L$ centralizes $O_{2',2}(N)$, and $L$ is characteristic in $N$. As above, we call $L$ the 2-*layer* of $N$ and call its components the 2-*components* of $N$. Once again $L$ is nontrivial if and only if $N$ is not 2-constrained.

In these terms we are better able to see the implications of the local analysis of $O(N)$ on $N$ itself. Indeed, this naturally associates with $N$ a certain semi-simple normal subgroup $L$. Clearly, then, when $L$ is nontrivial, the ensuing stage of the analysis, which I shall not attempt to describe in these talks, must obviously involve a closer determination of the structure of $L$. To indicate the goal of this analysis, let us list the situation as it exists in the presently known simple groups. To the best of my knowledge, either $N$ is 2-constrained or the following conditions hold.

(a) $O_{2',2}(N)$ is cyclic or quaternion if $G \not\cong A_n$, and is an elementary abelian 2-group if $G \cong A_n$.
(b) The 2-layer of $N$ consists of 1 or 2 components (except in certain degenerate cases in which the number is 3 or 4).
(c) $N/O^*_{2',2}(N)$ is a 2-group of order at most 4.

### 4.6. Groups of Characteristic 2 Type

Using the concepts of the preceding section, we can now define the desired class of balanced groups.

First of all, let $H$ be a group in which $O(H) = 1$ and $O_2(H) \neq 1$. Let $L$ be the 2-layer of $H$, so that $O_2^*(H) = LO_2(H)$, $L$ semi-simple, centralizing $O_2(H)$, and characteristic in $H$. Then $Z(L)$ is normal in $H$ and, as $O(H) = 1$, $Z(L)$ is a 2-group and consequently $Z(L) \subseteq O_2(H)$. Furthermore, from the structure of a semi-simple group, $L/Z(L)$ is the direct product of non-abelian simple groups. Hence, if we set $\bar{H} = H/O_2(H)$, we conclude that the image $\bar{J}$ in $\bar{H}$ of each 2-component $J$ of $H$ is a non-abelian simple group.

Observe next that $N_{\bar{H}}(\bar{J})/C_{\bar{H}}(\bar{J}) = \bar{J}^*$ can be identified with a subgroup of the automorphism group of the simple group $\bar{J}$. With this notation, we now introduce the following class of groups.

Definition. We shall say that $H$ is of *characteristic* 2 *type* or briefly a *$\mathscr{C}$-group* if for each 2-component $J$ of $H$ and each involution $\bar{x}$ of the corresponding group $\bar{J}^*$, we have

$$O(C_{\bar{J}^*}(\bar{x})) = 1.$$

More generally, if $O(H) \neq 1$ and $O_2(H) \neq 1$, we call $H$ a *$\mathscr{C}$-group* if $H/O(H)$ is a $\mathscr{C}$-group.

We wish to stress the fact that the above restriction on the 2-components $J$ is not solely a property of the simple group $\bar{J}$, but depends as well upon the embedding of $\bar{J}$ in $\bar{H}$. For example, if $\bar{J} \cong PSL(2, 4)$ (and hence $\bar{J} \cong PSL(2,5) \cong A_5$), then $\bar{J}^* \cong \bar{J}$ or $\bar{J}^* \cong PGL(2, 5)$ ($\cong S_5$). In the first case, $C_{\bar{J}^*}(\bar{x})$ is a four group for each involution $\bar{x}$ of $\bar{J}^*$, and so $\bar{J}$ satisfies the required condition. On the other hand, if $\bar{J}^* \cong PGL(2, 5)$ and $\bar{x}$ is an involution not in the derived group of $\bar{J}^*$, then $C_{\bar{J}^*}(\bar{x}) = \langle \bar{x} \rangle \times \bar{F}$, where $\bar{F} \cong S_3$ and so $O(C_{\bar{J}^*}(\bar{x})) \neq 1$ in this case.

If the 2-components of $H$ are all central extensions of any of the presently known simple groups of Lie type over fields of characteristic 2, then for appropriate embeddings of the 2-layer of $H$, $H$ will be a $\mathscr{C}$-group. In particular, this will be the case if each of the associated groups $\bar{J}^*$ is such that $\bar{J}^*/\bar{J}$ is of odd order. However, with certain specific exceptions, if $\bar{J}$ is a group of Lie type of odd characteristic or an alternating group, $H$ will not be a $\mathscr{C}$-group, no matter how $\bar{J}$ is embedded in $\bar{H}$. It is for this reason that we have used the term "characteristic 2 type" for this class of groups.

We have deliberately demanded that $O_2(H) \neq 1$ in the above definition so that no confusion can occur as a result of the following.

DEFINITION. A simple group $G$ is said to be of *characteristic* 2 *type* or briefly a $\mathscr{C}$-group provided the centralizer of each of its involutions is a $\mathscr{C}$-group in the preceding sense.

Remarkably, each of the sporadic simple groups appears to be a $\mathscr{C}$-group. On the other hand, again with certain exceptions, the alternating groups and the groups of Lie type of odd characteristic are not $\mathscr{C}$-groups. The groups of Lie type of even characteristic are trivially $\mathscr{C}$-groups since in each of them the centralizer of every involution is 2-constrained.

The following theorem shows the importance of $\mathscr{C}$-groups for us.

THEOREM. *If $G$ is a simple $\mathscr{C}$-group, then $G$ is balanced.*

We should like to sketch the proof of this result, which is quite straightforward, for it illustrates how some of the properties of the 2-layer enter into the analysis.

Let $a$, $b$ be two commuting involutions of $G$, put $H = C_G(b)$ and $D = O(C_H(a))$. Then

$$O(C_G(a)) \cap H \subseteq O(C_H(a)) = D.$$

Hence, to prove that $G$ is balanced, it will suffice to show, for each such $a$ and $b$, that the corresponding subgroup $D$ satisfies $D \subseteq O(H)$. This is solely a local property of $H$ and so it will be enough to prove that the image of $D$ is trivial in $H/O(H)$. Thus we are reduced to establishing the following result.

If $H$ is a $\mathscr{C}$-group in which $O(H) = 1$ and $O_2(H) \neq 1$ and $a$ is an involution of $H$, then $O(C_H(a)) = 1$.

We again put $D = O(C_H(a))$. Our aim will be to prove that $D$ centralizes $O_2^*(H)$. Indeed, since $C_H(O_2^*(H)) \subseteq O_2^*(H)$, we have, in fact, that $C_H(O_2^*(H)) = Z(O_2(H))$ from the structure of $O_2^*(H)$. Thus it will follow that $D \subseteq Z(O_2(H))$ and, as $|D|$ is odd, this will force the desired conclusion $D = 1$.

It is immediate from the definition of $D$ that $D$ centralizes $C_{O_2(H)}(a)$ and therefore $D$ centralizes $O_2(H)$ by Thompson's lemma. Since $O_2^*(H) = LO_2(H)$, where $L$ is the 2-layer of $H$, it will therefore suffice to prove that $D$ centralizes $L$. Because $L$ is perfect, it follows easily from the three-subgroup lemma that we need only show that $\bar{D}$ centralizes $\bar{L}$ in $\bar{H} = H/O_2(H)$.

We can assume that $\bar{L} \neq 1$, so that $\bar{L}$ is the direct product of a positive number of non-abelian simple groups. We note that $\bar{a}$ induces by conjugation a permutation of these components. We can also assume that $\bar{a} \neq 1$. Indeed, otherwise $a \in O_2(H)$ and so $a$ centralizes $L$. But then $[D, L] \subseteq D \cap L$ and $[D, L] \lhd L$, whence $[D, L] = 1$ and $D$ centralizes $L$.

If $\bar{J}$ is a component of $\bar{L}$, there are two cases to consider: $\bar{J}^{\bar{a}} \neq \bar{J}$ and $\bar{J}^{\bar{a}} = \bar{J}$. In the first case, one shows first that $\bar{D}$ centralizes $C_{\bar{J}\bar{J}^{\bar{a}}}(\bar{a})$, the latter group being the "diagonal" of $\bar{J} \times \bar{J}^{\bar{a}}$. From this one concludes easily that $\bar{D}$ centralizes $\bar{J}$. Likewise $\bar{D}$ centralizes $\bar{J}$ if $\bar{a}$ centralizes $\bar{J}$.

In the remaining case, $\bar{D}$ leaves $\bar{J}$ invariant and if we put $\tilde{J}^* = N_{\bar{H}}(\bar{J})/C_{\bar{H}}(\bar{J})$ and give the obvious meanings to $\tilde{D}$ and $\tilde{a}$, we have that $\tilde{a}$ is an involution and that $\tilde{D} \subseteq O(C_{J^*}(\tilde{a}))$. But by hypothesis, $H$ is a $\mathscr{C}$-group and therefore $O(C_{J^*}(\tilde{a})) = 1$ by the definition of a $\mathscr{C}$-group. Thus $\tilde{D} = 1$ and so $\bar{D} \subseteq C_{\bar{H}}(\bar{J})$. Since $\bar{J}$ was arbitrary, we conclude in all cases that $\bar{D}$ centralizes $\bar{L}$, as required.

Clearly we want to apply the balanced theorem to $\mathscr{C}$-groups. The easier form of the theorem will suffice if one is prepared to add to the definition of $\mathscr{C}$-group the assumption that the centralizer $H = C_G(t)$ of every involution $t$ of $G$ is either 2-generated or exceptional. It is easy to see that in order to verify this condition for $H$, it is enough to show that, for each 2-component $\bar{J}$ of $\bar{H} = H/O(H)$, the group $\bar{J}\langle\bar{t}\rangle$ is either 2-generated or exceptional provided the 2-rank of $H$ is at least 4.

But for the groups of Lie type over a field of characteristic 2, it is very easy to establish this conclusion. Indeed, if $J$ is a perfect central extension of such a group, then $J$ is exceptional if it is of Lie rank 1 since the only simple groups of Lie type of characteristic 2 and rank 1 are the groups $PSL(2, 2^n)$, $PSU(3, 2^n)$, and $Sz(2^n)$ $(n \geq 2)$. On the other hand, if $J$ is of Lie rank $r \geq 2$ and we denote by $M_1, M_2, \ldots, M_r$ the maximal parabolic subgroups of $J$ containing a given Sylow 2-subgroup $S$ of $J$, one knows that

$$J = \langle M_i : 1 \leq i \leq r\rangle. \tag{6.1}$$

However, each $M_i = N_{M_i}(O_2(M_i))$ and the 2-rank of $O_2(M_i)$ is certainly at least 2 $(1 \leq i \leq r)$. We conclude therefore from (6.1) that $J$ is 2-generated.

The given conditions seem also to hold in the sporadic groups.

Thus using the appropriate definition of $\mathscr{C}$-group, one would end up with the following result once one established the corresponding form of the balanced theorem.

*If $G$ is a simple $\mathscr{C}$-group of 2-rank at least 5, then $O(C_G(x)) = 1$ for every involution $x$ of $G$.*

## 5. Generalizations and Concluding Remarks

### 5.1. $k$-Balanced Groups and the Existence of Signalizer Functors

It is natural to ask whether there exists a generalization of the concept of balanced group which would hold for any group $G$ in which the centralizer of every involution has 2-components of presently known type. The fact that a $\mathscr{C}$-group $G$ turned out to be balanced, depended upon the following assertion, applied to the groups $C_G(t)/O(C_G(t))$, $t$ an involution of $G$.

If $H$ is a $\mathscr{C}$-group with $O(H) = 1$ (and $O_2(H) \neq 1$), then $O(C_H(a)) = 1$ for every involution $a$ of $H$.

Clearly an affirmative answer to our question requires some extension of this result when the 2-components of $H$ are *arbitrary* groups of known type. Such an extension seems indeed to exist. In fact, if $T$ is any elementary abelian 2-subgroup of such a group $H$ of order 8, it appears always to be the case that

$$\bigcap_{t \in T^{\#}} O(C_H(t)) = 1, \tag{A}$$

provided no 2-component of $H$ is a central extension of $A_n$ with $n = 2^k m+3$, $m$ odd and $k \geq 3$. If such components exist and $k$ is the maximal such integer occurring, (A) will hold if we assume instead that $T$ has rank $k+1$. Furthermore, if no 2-component of $H$ is a central extension of $A_{4r+3}$ or of $PSL(2, p^m)$ with $p$ odd and $m$ not a power of 2, condition (A) seems to hold for every four subgroup $T$ of $H$.

In view of this discussion, we introduce the following terminology.

Definition. If $T$ is an elementary abelian 2-subgroup of $G$, we put

$$\Delta(T) = \bigcap_{t \in T^{\#}} O(C_G(t)).$$

It is also convenient to let $\mathscr{E}_k(G)$ denote the set of elementary abelian 2-subgroups of $G$ of rank $k$.

Definition. A group $G$ will be called *k-balanced* provided

$$\Delta(T) \cap C_G(a) \subseteq O(C_G(a))$$

for any element $T$ in $\mathscr{E}_k(G)$ and any involution $a$ of $G$ which centralizes $T$.

Thus a 1-balanced group is nothing else than a balanced group in the former sense. We can rephrase the condition in a more symmetric way that will be useful to us.

Lemma. *If $G$ is a $k$-balanced group and $T, T'$ are elements of $\mathscr{E}_k(G)$ which centralize each other, then*

$$\Delta(T) \cap C_G(T') = \Delta(T') \cap C_G(T).$$

*Proof.* By symmetry it will be enough to show that

$$\Delta(T') \cap C_G(T) \subseteq \Delta(T).$$

But since $G$ is $k$-balanced,

$$\Delta(T') \cap C_G(t) \subseteq O(C_G(t))$$

for each $t$ in $T^\#$, whence

$$\Delta(T') \cap \bigcap_{t \in T^\#} C_G(t) \subseteq \bigcap_{t \in T^\#} O(C_G(t)).$$

Since $C_G(T) = \bigcap_{t \in T^\#} C_G(t)$ and $\Delta(t) = \bigcap_{t \in T^\#} O(C_G(t))$, the desired conclusion follows.

The significance of $k$-balanced groups is due to the fact that one can construct effective signalizer functors on them. Indeed, one has the following.

THEOREM. *Let $G$ be a $k$-balanced group of 2-rank at least $k+3$ and let $A \in \mathscr{E}_{k+3}(G)$. If for each $a$ in $A^\#$, we put*

$$\theta(C_G(a)) = \langle C_G(a) \cap \Delta(T) : T \in \mathscr{E}_k(A) \rangle,$$

*then $\theta$ is an $A$-signalizer functor on $G$.*

The proof depends critically upon an extension of the second generational lemma discussed in Section 4 of Part 3, which is proved by an entirely analogous argument.

LEMMA. *Let $K$ be a group of odd order acted on by the elementary abelian 2-group $A$ of rank at least $k+2$, and assume that the following conditions hold:* s

(a) $K = \langle K(T) : T \in \mathscr{E}_k(A) \rangle$, *where each $K(T)$ is a normal subgroup of $C_K(T)$;*

(b) $C_{K(T)}(T') \subseteq K(T')$ *for each $T$, $T'$ in $\mathscr{E}_k(A)$.*

*Then, if $F$ is any $A$-invariant subgroup of $K$, we have*

$$F = \langle F \cap K(T) : T \in \mathscr{E}_k(A) \rangle.$$

Applying this lemma in the present situation, we let $\mathscr{N}_\theta(A)$ denote, a usual, the set of $A$-invariant subgroups $K$ of $G$ of odd order such that

$$K = \langle K \cap \theta(C_G(a)) : a \in A^\# \rangle.$$

In view of the definition of $\theta(C_G(a))$, we see that $\mathscr{N}_\theta(A)$ can be defined equivalently as the set of $A$-invariant subgroups $K$ of $G$ of odd order such that

$$K = \langle K \cap \Delta(T) : T \in \mathscr{E}_k(A) \rangle.$$

Using the lemma, we can prove

PROPOSITION. *If $K \in \mathscr{N}_\theta(A)$ and $F$ is an $A$-invariant subgroup of $K$, then $F \in \mathscr{N}_\theta(A)$.*

*Proof.* Put $K(T) = K \cap \Delta(T)$ for each $T$ in $\mathscr{E}_k(A)$. Then by definition of $\mathscr{N}_\theta(A)$,

$$K = \langle K(T) : T \in \mathscr{E}_k(A) \rangle.$$

Since $O(C_G(t)) \lhd C_G(t)$ for each $t$ in $T^\#$ and $C_G(T) = \bigcap_{t \in T^\#} C_G(t)$, while $\Delta(T) = \bigcap_{t \in T^\#} O(C_G(t))$, it follows at once that $\Delta(T) \lhd C_G(T)$, whence $K(T) = K \cap \Delta(T) \lhd K \cap C_G(T) = C_K(T)$. Furthermore $C_{\Delta(T)}(T') \subseteq \Delta(T')$ by the first lemma, whence $C_{K(T)}(T') = C_{K \cap \Delta(T)}(T') \subseteq K \cap \Delta(T') = K(T')$. Hence all the hypotheses of the generational lemma are satisfied and we conclude that

$$F = \langle F \cap K(T) : T \in \mathscr{E}_k(A) \rangle = \langle F \cap \Delta(T) : T \in \mathscr{E}_k(A) \rangle.$$

Thus $F \in \mathscr{N}_\theta(A)$, as asserted.

We can now easily demonstrate that $\theta$ is an $A$-signalizer functor on $G$. Recall that we have noted earlier that the condition that $\theta(C_G(a)) \lhd C_G(a)$ in the definition of $A$-signalizer functor can be replaced by the condition that

$$K \cap C_G(a) = K \cap \theta(C_G(a))$$

for each $K$ in $\mathscr{N}_\theta(A)$ without affecting the conclusion of the signalizer functor theorem. We can verify this latter condition immediately. Indeed, taking $F = C_K(a)$ in the preceding proposition, we have

$$C_K(a) = \langle C_K(a) \cap \Delta(T) : T \in \mathscr{E}_k(A) \rangle.$$

But $C_K(a) \cap \Delta(T) = K \cap C_G(a) \cap \Delta(T) \subseteq K \cap \theta(C_G(a))$ for each $T$ in $\mathscr{E}_k(A)$ and consequently

$$C_K(a) \subseteq K \cap \theta(C_G(a)).$$

Since the reverse inclusion clearly holds the desired equality follows.

Next let $a, b \in A^\#$ and apply the proposition with $K = \theta(C_G(a))$ and $F = \theta(C_G(a)) \cap C_G(b)$ to obtain

$$\theta(C_G(a)) \cap C_G(b) = \langle \theta(C_G(a)) \cap C_G(b) \cap \Delta(T) : T \in \mathscr{E}_k(A) \rangle \subseteq \theta(C_G(b)).$$

Thus to complete the proof, it remains only to verify that

$$(\theta(C_G(a)))^x = \theta(C_G(a^x)) \tag{1.1}$$

for $a$ in $B^\#$ and $x$ in $N_G(B)$, where $B$ is any subgroup of index at most 4 in $A$. To establish this, it will suffice to prove that for any $a$ in $A^\#$

$$\theta(C_G(a)) = \langle C_G(a) \cap \Delta(T) : T \in \mathscr{E}_k(B) \rangle. \tag{1.2}$$

Indeed, assume that this holds. It is immediate from the definition of $\Delta(T)$ that $(\Delta(T))^x = \Delta(T^x)$ and consequently

$$(\theta(C_G(a)))^x = \langle C_G(a^x) \cap \Delta(T^x) : T \in \mathscr{E}_k(B) \rangle.$$

However, $T^x$ runs over $\mathscr{E}_k(B)$ as $T$ does since $x \in N_G(B)$, whence

$$(\theta(C_G(a)))^x = \langle C_G(a^x) \cap \Delta(T) : T \in \mathscr{E}_k(B) \rangle. \tag{1.3}$$

Since $a^x \in A^\#$, (1.3) and (1.2) together yield (1.1).

Finally we shall verify (1.2). It is here that we shall use our assumption that $A$ has 2-rank at least $k+3$. It implies that $m(B) \geq k+1$. As a consequence we have

$$\theta(C_G(a)) = \langle \theta(C_G(a)) \cap C_G(T) : T \in \mathscr{E}_k(B) \rangle,$$

whence

$$\theta(C_G(a)) = \langle C_G(a) \cap \Delta(T') \cap C_G(T) : T' \in \mathscr{E}_k(A),\ T \in \mathscr{E}_k(B) \rangle.$$

However, $\Delta(T') \cap C_G(T) \subseteq \Delta(T)$ for each such $T$, $T'$ and (1.2) follows.

Combining the preceding theorem with the signalizer functor theorem itself, we obtain the following basic property of $k$-balanced groups.

THEOREM. *Let $G$ be a $k$-balanced group of 2-rank at least max $(5, k+3)$ and let $A$ be any elementary abelian 2-subgroup of $G$ of maximal rank. If for each $a$ in $A^{\#}$, we put*

$$\theta(C_G(a)) = \langle C_G(a) \cap \Delta(T) : T \in \mathscr{E}_k(A) \rangle$$

*the subgroups*

$$\langle \theta(C_G(a)) : a \in A^{\#} \rangle \quad \text{and} \quad \langle \Delta(T) : T \in \mathscr{E}_k(A) \rangle$$

*coincide and have odd order.*

## 5.2. $k$-CONNECTED GROUPS

Obviously we must now consider whether the balanced theorem can be extended to $k$-balanced groups. Clearly to accomplish this it will be necessary to generalize the basic results that we proved about balanced, connected groups.

We have the following natural extension of connectedness.

DEFINITION. Let $S$ be a 2-group of 2-rank at least $k \geq 2$ and let $A$, $A'$ be two elementary abelian subgroups of $S$ of rank at least $k$. We say that $A$ and $A'$ are *$k$-connected* provided there exists a sequence of elementary abelian 2-subgroups

$$A = A_1, A_2, \ldots, A_n = A'$$

of $S$ with each $A_i$ of rank at least $k$ and such that either

$$A_i \subseteq A_{i+1} \quad \text{or} \quad A_{i+1} \subseteq A_i \qquad (1 \leq i \leq n-1).$$

We say that $S$ is *$k$-connected* provided the following two conditions hold:

(a) every such pair $A$, $A'$ of $S$ is $k$-connected;

(b) any elementary abelian 2-subgroup of $S$ of rank less than $k$ is contained in one of rank $k$.

Moreover, we say that an arbitrary group $G$ is *$k$-connected* if a Sylow 2-subgroup of $G$ is $k$-connected.

When $k = 2$, condition (*b*) is automatically satisfied, so 2-connectedness is the same as connectedness in our old sense. Essentially the same argument which showed earlier that a group $G$ in which $SCN_3(2)$ is non-empty is necessarily connected yields the following sufficient condition for $k$-connectedness.

LEMMA. *If $SCN_m(G)$ is non-empty with $m = 2^k+1$, then $G$ is $k$-connected.*

Consider now an arbitrary $k$-balanced, $k+1$-connected group and for any elementary abelian 2-subgroup $A$ of $G$ of rank at least $k+1$, put

$$W_A = \langle \Delta(T) : T \in \mathscr{E}_k(A) \rangle.$$

Then we have the following basic result.

PROPOSITION. *If $A$, $A'$ are any two elementary abelian 2-subgroups of rank at least $k+1$ in the same Sylow 2-subgroup of $G$, then*

$$W_A = W_{A'}.$$

*Proof.* As in the case of connected groups, it will suffice to consider the special case $A' \subseteq A$ and prove that $W_A \subseteq W_{A'}$. By definition of $W_A$, we need only show that $\Delta(T) \subseteq W_{A'}$ for each $T$ in $\mathscr{E}_k(A)$.

However, as $A'$ has rank at least $k+1$, we have

$$\Delta(T) = \langle C_{\Delta(T)}(T') : T' \in \mathscr{E}_k(A') \rangle.$$

Since $T$ and $T'$ centralize each other and $G$ is $k$-balanced,

$$C_{\Delta(T)}(T') \subseteq \Delta(T') \subseteq W_{A'}$$

for each $T'$ in $\mathscr{E}_k(A')$ and the desired conclusion follows.

Next let $W$ be the common value of $W_A$ and put $M = N_G(W)$. Then exactly as in the case of connected groups, the proposition yields the following consequences.

PROPOSITION. *If $R$ is 2-subgroup of $G$ such that $R \cap M$ has 2-rank at least $k+1$, then $N_G(R) \subseteq M$.*

COROLLARY. *If $H$ is a $k+1$-generated subgroup of $G$ such that $H \cap M$ has 2-rank at least $k+1$, then $H \subseteq M$.*

## 5.3. POSSIBLE EXTENSIONS OF THE MAIN THEOREM

It remains only to consider the question of whether $M = N_G(W)$ is strongly embedded in $G$ when $W = W_A \neq 1$. By the final corollary of Section 2, this will indeed be the case if $H = C_G(t)$ is $k+1$-generated and $H \cap M$ has 2-rank at least $k+1$. We have, of course, discussed the case $k = 1$ in considerable detail in our analysis of the balanced theorem, so we may assume that $k \neq 1$. Thus $G$ is not 1-balanced, which means that the centralizer of some involution of $G$ has a 2-component which is not 1-balanced (equivalently the centralizer of some involution of $G$ is not a $\mathscr{C}$-group).

If one assumes under these conditions that the 2-components of both the centralizer of every involution of $G$ and of the subgroup $M$ are of known type, it seems very likely that one can, in fact, establish the following result:

The centralizer of every involution of $G$ is $k+1$-generated. If true, this would enable one to prove that $M$ is, in fact, strongly embedded in $G$. (B)

The present discussion makes it quite plausible that one will be able to establish the following assertion.

If $G$ is a simple group whose proper subgroups have 2-components of known type (excluding $A_n$ with $n \equiv 3 \pmod 8$) and if $SCN_9(2)$ is non-empty, then

$$\Delta(T) = \bigcap_{t \in T^{\#}} O(C_G(t)) = 1$$

for every involution $t$ of $G$.

Indeed, to turn this conjecture into a valid theorem, all that is required is the verification of the assertions labeled (A) and (B) above. If $A_n$ (with $n \equiv 3 \pmod 8$) occurs as a 2-component of the centralizer of some involution of $G$, one would then obtain a suitable modification of the preceding assertion.

John Walter and I are presently investigating these questions. However, whatever the outcome turns out to be in the most general case, one can certainly verify these conditions in specific situations. In particular, if $G$ has abelian Sylow 2-subgroups (in which case the 2-components of the proper subgroups of $G$ are isomorphic to $PSL(2, q)$, $q = 2^n$ or $q \equiv 3, 5 \pmod 8$, to Janko's first group, or are of Ree type of characteristic 3), conditions (A) and (B) are easily established. Moreover, the analysis goes through as long as the 2-rank of $G$ is at least 6.

The conclusion that $\Delta(T) = 1$ for every elementary abelian 2-subgroup of $G$ or order 8 is very strong. In fact, in the abelian case, it can be used to give a new simpler, proof of John Walter's fundamental classification theorem of groups with abelian Sylow 2-subgroups in the case that the 2-rank is at least 6. (We remark parenthetically that the abelian problem in the case of 2-rank 3, 4, or 5 can be treated in a similar way, using variations of the signalizer functor theorem to be described in the next section.)

Clearly the above conjecture, if substantiated in general, would represent a very powerful tool for studying the cases of the centralizers of involutions in simple groups whose proper subgroups have 2-components of presently known type. Hopefully, it would enable one to demonstrate ultimately that $O(C_G(x))$ is cyclic for any such group $G$ of suitably high 2-rank and any involution $x$ of $G$.

## 5.4. Groups of Low 2-rank

How important is the assumption that the 2-rank of $G$ be at least 5 in the signalizer functor theorem? We have already noted in Part 1 that the classification of simple groups of 2-rank less than 3 is almost complete†. Hence the question concerns primarily simple groups of 2-rank 3 or 4. The main theorem of C.I. 3 is, in fact, an extension of the signalizer functor theorem to groups of 2-rank 4. We are presently working on a similar extension to groups of 2-rank 3. In both instances, stronger assumptions on the given $A$-signalizer functor $\theta$ are required in our approach.

To describe these, let $G$ be a group, $A$ an elementary abelian 2-subgroup of $G$ with $m(A) \geq 3$, and let $\theta$ be an $A$-signalizer functor on $G$. In the proof of the signalizer functor theorem, we were able to establish by induction that, for a large class of subgroups $H$ of $G$ containing $A$, the elements of $\mathscr{N}_\theta(A)$ contained in $H$ generated a subgroup of $H$ of odd order. This result, which was essential, formed the basis of the concept of *flatness* introduced in C.I. 1. What we need are certain extensions of this concept.

Recall that in Part 3 we have defined the term $\mathscr{N}_\theta(B)$ for any noncyclic subgroup $B$ of $A$. If $H$ is a subgroup of $G$ containing $B$, let us use the term $\mathscr{N}_\theta(B; H)$ for the set of elements of $\mathscr{N}_\theta(B)$ contained in $H$, with corresponding meanings for $\mathscr{N}_\theta^*(B; H)$, $\mathscr{N}_\theta(B; p; H)$, and $\mathscr{N}_\theta^*(B; p; H)$.

Definition. If $H$ is a proper subgroup of $G$ such that $B = A \cap H$ is noncyclic, we shall say that $H$ is *very strongly flat* if the elements of $\mathscr{N}_\theta(B; H)$ generate a subgroup of $H$ of odd order. Moreover, we shall say that $H$ is *strongly flat* if either $H$ is very strongly flat or if the following conditions are satisfied:

(a) $B$ is a four group;

(b) any two elements of $\mathscr{N}_\theta^*(B; p; H)$, $p$ an odd prime, are conjugate in $N_H(B)$;

(c) if $K_i$, $1 \leq i \leq m$, are the distinct elements of $\mathscr{N}_\theta^*(B; H)$, then

  (i) $K_i \cap K_j / K_i \cap K_j \cap O(H)$ is cyclic for all $i \neq j$;

  (ii) $K_i \cap K_j = K_j \cap K_k$ for all $i \neq j \neq k \neq i$.

Although this last set of conditions appears to be very complicated, it can actually be verified quite easily if $H$ has dihedral or quasi-dihedral Sylow 2-subgroups and in a few other cases in which it is not possible to prove in general that $H$ is very strongly flat.

Definition. We shall say that $\theta$ is *strongly flat* or *very strongly flat* if every $p$-local subgroup $H$ of $G$, $p$ odd, such that $A \cap H$ is noncyclic is respectively strongly flat or very strongly flat.

The main result of C.I. 3 asserts:

† See footnote to p. 73.

THEOREM. *If $A$ is an elementary abelian 2-subgroup of the group $G$ of rank 4 and $G$ possesses the strongly flat $A$-signalizer functor $\theta$, then*

$$W_A = \langle O(C_G(A) : a \in A^{\#} \rangle$$

*is of odd order.*

In the case of groups of 2-rank 3, it seems very likely that the following result can be established.

Let $A$ be an elementary abelian 2-subgroup of the group $G$ of rank 3 such that $|N_G(A)/C_G(A)|$ is divisible by 7. If $G$ possesses the very strongly flat $A$-signalizer functor $\theta$, then

$$W_A = \langle O(C_G(A)) : a \in A^{\#} \rangle$$

is of odd order.

This assumption on $N_G(A)$ is actually quite natural. Indeed, it seems likely that one will be able to classify the possible 2-groups of 2-rank 3 which can appear as a Sylow 2-subgroup of a simple group $G$. If the conclusion turns out to be as anticipated, one will obtain as a consequence that in such a simple group $G$:

Either $G$ possesses an elementary abelian 2-subgroup $A$ of rank 3 such that $|N_G(A)/C_G(A)|$ is divisible by 7 or else a Sylow 2-subgroup of $G$ is isomorphic to one of the Mathieu group $M_{12}$ and $G$ has more than one class of involutions.

However, in the latter case, a theorem of Brauer and Fong implies that $G$ is, in fact, isomorphic to $M_{12}$. Hence it will be enough to study simple groups of 2-rank 3 in which the given condition on $N_G(A)$ is satisfied.

These analogues of the signalizer functor theorem should thus be sufficient for the analysis of the cores of the centralizers of involutions in simple groups of 2-rank 3 and 4. For example, Harada and I have used the first result to show that $J_2$ and $J_3$ themselves are the only finite simple groups $G$ having a Sylow 2-subgroup isomorphic to that of $J_2$ or $J_3$, these Sylow 2-subgroups being isomorphic of order $2^7$ and 2-rank 4. In this classification problem $G$ always turns out to be balanced. However, such a group $G$ is not connected and consequently Harada and I required a variation of the balanced theorem for non-connected groups. Such a result, which I shall not attempt to state explicitly, is given in our paper with John Walter.

## 5.5. CENTRALIZERS OF $p$-ELEMENTS ($p$ ODD)

Finally we may ask whether it is possible to carry out an entirely similar analysis for the centralizers of elements of *odd* prime order. Moreover, how valuable would such a generalization of our results be?

It turns out that there are two serious difficulties that must be overcome to carry out the entire development for arbitrary primes $p$. Of course, the definition of an $A$-signalizer functor $\theta$ on a group $G$, with $A$ an elementary abelian $p$-subgroup of $G$, will obviously be stated in terms of a subgroup $\theta(C_G(a))$ of $O_{p'}(C_G(a))$ for each $a$ in $A^{\#}$ and the elements of $\mathscr{V}_\theta(A)$ will be $p'$-groups. But when $p = 2$, a $p'$-group is of odd order and so is solvable. Furthermore, we have seen how crucial the solvability of the various subgroups $\theta(H)$ of $G$ was for the proof of the signalizer theorem. However, for no odd $p$ is it true in general that a $p'$-group is solvable. Here then, is the first problem.

Secondly, the proof of the balanced theorem depended upon the fact that $M = N_G(W_A)$ was strongly embedded in $G$, the key requirement being that $C_G(x) \subseteq M$ for each involution $x$ of $M$. Clearly, what we shall obtain in general for $M$ is that $C_G(x) \subseteq M$ for each *p-element* $x \neq 1$ of $M$, with similar modifications in the remaining conditions. For brevity, we may call such a subgroup $M$ *strongly p-embedded* in $G$. In our original case, we were able to invoke Bender's classification theorem at this point to complete the proof. However, no corresponding classification theorem exists for groups which possess a strongly $p$-embedded subgroup with $p$ odd, nor does it seem likely that such a result can be obtained. So here is our second major problem.

Fortunately, there appears to be a possible solution to each problem. As we have just asserted, the general $p'$-group is not solvable for odd $p$. However, we may ask whether there exist any values of $p$ for which the simple $p'$-groups are at least of some very restricted type. This indeed seems to be the case for the prime $p = 3$. The only presently known simple $3'$-groups are the Suzuki groups $Sz(2^n)$ and it seems very likely from some work of Thompson on $3'$-groups that these are, in fact, the only such simple groups. I shall return to this topic shortly. Assuming for the time being that this classification of $3'$-groups can be completed, we can see what will be required to establish the analogue of the signalizer functor theorem for elementary abelian 3-groups $A$: namely, it will be necessary to extend the various arguments of the proof from solvable groups to groups whose only nonsolvable composition factors are Suzuki groups. Since the structure of the Suzuki groups is, apart perhaps from the groups $PSL(2, 2^n)$, the simplest of all the families of known simple groups, such an extension seems entirely feasible.

We should mention a key additional property of $3'$-groups which would definitely be used in carrying out this contemplated extension of the signalizer functor theorem—namely, the fact that a solvable $3'$-group is $q$-stable for all odd $q$. This is not true in general for solvable $p'$-groups with $p \geq 5$. But $q$-stability is essential for the many applications of Glauberman's $ZJ$-theorem that one will have to make.

Thus, if we can also obtain an extension of the balanced theorem, we see that we may be able to say something about the centralizer of elements of order 3 in arbitrary simple groups and of elements of order 5 in $3'$-groups.

Before analyzing this problem, we prefer to consider the question of which classification problems will require a knowledge of the centralizers of elements of order 3. To answer this question, let us recall the approximate structure we hope to obtain for the centralizer $N$ of an involution $z$ in a simple group $G$ in the case that $N/O(N)$ possesses a 2-component of Lie type of odd characteristic—namely, that

$$O^*_{2',2}(N) = LO_{2',2}(N),$$

where $L$ is semi-simple with one or two components, $L$ centralizes $O_{2',2}(N)$, and $O_{2',2}(N)$ is cyclic or quanterion. Thus the structure of $N$ will be essentially of the same shape as that of one of the presently known simple groups and so we shall be reduced precisely to the type of problem described in our opening lecture. Hence, under these conditions one does not seem to need information about centralizers of elements of order 3 to complete the given classification problem. The same is probably true if $N/O(N)$ has a 2-component which is an alternating group. We are left then with the cases that $N$ has no 2-components (and so is 2-constrained) or has 2-components which are either of Lie type of characteristic 2 or sporadic groups—just the situation which relates to the case of balanced groups.

Let me give some indication of the role that the centralizers of elements of order 3 may play in the further analysis of such groups. Consider, say, the case that $N$ is 2-constrained. The best we can expect our initial analysis to yield is that $O(N) = 1$. This leaves us very far from a knowledge of the structure of $N$. Thompson's study of $N$-groups shows vividly that the bulk of the work in determining $N$ remains to be done. After all, we have no information at this stage concerning either the structure of $O_2(N)$ or the group $N/O_2(N)$ which acts on $O_2(N)$.

Suppose now that $N$ contains an abelian subgroup $T$ of type (3, 3), in which case $O_2(N) = \langle C_{O_2(N)}(t) : t \in T^\# \rangle$. Hence a knowledge of the structure of $C_G(t)$ for each $t$ in $T^\#$ will have strong implications for the structure of $O_2(N)$. On the other hand, if no such $T$ exists, then a Sylow 3-subgroup of $N$ is necessarily cyclic and this has strong implications for the structure of $N/O_2(N)$. In substance, knowledge of the 3-local structure of $G$ should, in the general situation, give information about the 2-local structure.

The preceding discussion indicates that an extension of the balanced theorem to the case $p = 3$ will be required only when $G$ is a balanced group —or at least closely resembles such a group. We also want an extension of the theorem to the case $p = 5$ when $G$ is a $3'$-group. However, we note in this case that such a group $G$ is necessarily a $\mathscr{C}$-group as the only possible 2-components of the centralizers of its involutions are Suzuki groups and consequently $G$ is balanced. For simplicity, we shall limit the discussion to the case $p = 3$, the other case being entirely similar.

Suppose then that we have proved that our simple group $G$, which we shall assume to be balanced, possesses a strongly 3-embedded subgroup $M$. How shall we derive a contradiction? That is the question that remains.

We know that $M$ contains $C_G(x)$ for any 3-element $x \neq 1$ of $G$ and that $|M \cap M^g|$ is a $3'$-group for any $g$ in $G - M$. These conditions directly imply that if $Q$ is any nontrivial 3-subgroup of $G$ such that $M \cap Q \neq 1$, then $N_G(Q) \subseteq M$.

We thus see that if we define *k-generated for the prime* 3 by analogy with our earlier definition of $k$-generation (that is, using a Sylow 3-subgroup in place of a Sylow 2-subgroup) then any 1-generated (for 3) subgroup $H$ of $G$ such that $H \cap M$ has 3-rank at least 1 will be contained in $M$.

Let us assume now that $G$ satisfies the following two conditions:

(a) every maximal 2-*local* subgroup of $G$ is 1-generated for the prime 3;
(b) $M$ itself is a 2-local subgroup of $G$.

Under these conditions, the preceding discussion will yield the following conclusion:

$G$ possesses only one conjugacy class of maximal 2-local subgroups—namely, the conjugates of $M$ in $G$.

However, I have shown in the paper that appears in I.H.E.S., by means of a very simple argument, that if a group $G$ has only one class of maximal 2-local subgroups, then each member of the class is strongly embedded in $G$. Thus, under the given assumptions, our strongly 3-embedded subgroup $M$ will, in fact, be strongly embedded in the usual sense and Bender's theorem will again be applicable.

How reasonable are assumptions (a) and (b)? It is precisely because of condition (b) that we wished to restrict ourselves to balanced groups. For we anticipate that in such groups either (b) will hold or else the maximal 2-local subgroups of $G$ will have a more closely determined structure. Similarly for condition (a). Indeed, if $H$ is an arbitrary group (with composition factors among the presently known simple groups), then one of the following seems to hold:

(i) $H$ is 1-generated for the prime 3;
(ii) a Sylow 3-subgroup of $H$ is cyclic; or
(iii) $H/O_{3'}(H)$ possesses a normal subgroup of $3'$-index isomorphic to either $PSL(2, 3^n)$, $PSU(3, 3^n)$, or to one of Ree type of characteristic 3.

The groups under (iii) are the direct analogues of the *exceptional* groups that we encountered in the case of involutions. The point is that either the maximal 2-locals of $G$ are 1-generated for the prime 3 or would seem to be of a very restricted structure.

The upshot of the discussion in the case of balanced groups or, more particularly, $\mathscr{C}$-groups is to suggest that when the 3-rank of $G$ is suitably large, either one may be able to pin down the structure of the centralizers of elements of order 3 or else force some maximal 2-local subgroup of $G$ to have a special shape. But, in the first case, one will also force the structure of the maximal 2-local subgroups of $G$, since a knowledge of the centralizers of elements of order 3 will have strong implication for the structure of any subgroup of $G$ that is 1-generated for the prime 3.

Obviously, the remarks of this final section are of a very tentative nature. They are meant more as a possible direction for future research on simple groups of characteristic 2 type than a measure of any present concrete accomplishments concerning their structure.

## Bibliography

The bulk of the material covered in these talks and related results can be found in the references below. Additional pertinent papers are listed in the first reference below, particularly that portion of it that is connected with the discussion of Chapters 16 and 17 of that book.


Gorenstein, D. (1968). "Finite groups", Harper and Row, New York.

Gorenstein, D. (1969a). On the centralizers of involutions in finite groups. *J. Algebra* **11,** 243.

Gorenstein, D. (1969b). The flatness of signalizer functors in finite groups. *J. Algebra* **13,** 509.

Gorenstein, D. (1969). On finite simple groups of characteristic 2 type. *Publ. Inst. des Hautes Etudes Scient.* **36,** 5.

Gorenstein, D. (1970). On the centralizers of involutions in finite groups, II. *J. Algebra* **14,** 350.

Gorenstein, D. and Harada, K. (1970). A characterization of Janko's two new simple groups. *J. Fac. Science, Univ. of Tokyo* **16,** 331.

Gorenstein, D. and Walter, J. H. The $\pi$-layer of a finite group. *Illinois J. Math.* To be published.

Gorenstein, D. and Walter, J. H. Centralizers of involutions in balanced groups. To be published.


## Notes Added in Proof

1. David Goldschmidt has recently made some fundamental improvements of and simplifications in the proof of the signalizer functor theorem. Defining an $A$-signalizer functor on $G$, $A$ an elementary abelian $p$-subgroup of $G$, $p$ a prime, by the *single* condition that $\theta(C_G(a))$ be an arbitrary $A$-invariant subgroup of $O_{p'}(C_G(a))$ such that

$$\theta(C_G(a)) \cap C_G(b)) \subseteq \theta(C_G(b)),$$

he has been able to establish the theorem in the following general cases:

(1) $p = 2$ and $m(A) \geq 3$;

(2) $p$ odd, $m(A) \geq 4$, and $\theta(C_G(a))$ is solvable for all $a$ in $A^{\#}$.

Goldschmidt's results in the case $p = 2$ give significant refinements of various results discussed in this paper. Typical is the improvement of our main theorem, already noted in Section 1.6. Likewise its corollary can be similarly sharpened and, when combined with Janko and Thompson's classification of simple groups with $SCN_3(2)$ empty in which the centralizer of every involution is solvable, we obtain a complete classification of all simple groups in which the centralizer of every involution is of 2-length 1.

Equally important is the effect of his result on the study of simple groups of 2-rank 3 and 4. The cumbersome notions of *strong flatness* and *very strong flatness* can now be entirely dispensed with. As a consequence, the use of the signalizer functor theorem in the study of such groups is made considerably easier.

2. The classification of simple groups with wreathed Sylow 2-subgroups has now been completed, thus completing the classification of simple groups of 2-rank less than 3.

CHAPTER III

# Chevalley Groups and Related Topics

C. W. CURTIS



## 1. Introduction

It would be a major undertaking to describe the sources of all the results which now find a place in the theory of the Chevalley groups. I shall have to be content with a few remarks, which show the point of view I am adopting in these lectures. The basic reference is Chevalley's paper (1955), in which he constructed the groups bearing his name, and proved their simplicity. In that paper, Chevalley acknowledged the earlier work on the classical groups by Dickson and Dieudonné, in which the structure of the groups was worked out by a case-by-case investigation. For each field $k$ and semi-simple Lie algebra $\mathfrak{g}$ over the complex field, Chevalley showed how, because of the remarkable integrality conditions he had discovered, a certain subgroup of the adjoint group (generated by automorphisms exp $(adX)$, for $X \in \mathfrak{g}$) determined an automorphism group $G$ of a Lie algebra over the field $k$. Chevalley's technique for investigating these groups included the use of certain homomorphisms of the group $SL_2(k)$ into $G$, so that $G$ could be viewed as some sort of amalgamation of copies of $SL_2(k)$. The different copies of $SL_2(k)$ in $G$ are permuted by the Weyl group of a root system attached to the semi-simple Lie algebra $\mathfrak{g}$. Later, Tits showed how the main features of the groups could be axiomatized, with the Weyl group of the root system assuming a central role, and the Lie algebra fading into the background. The twisted versions of the Chevalley groups, discovered by Hertzig, Steinberg, Suzuki, Ree and Tits all turned out to satisfy Tits' axioms. All known infinite families of finite simple groups, except for the alternating groups, are obtained as Chevalley groups or as twisted types of Chevalley groups.

Our point of view will be that Chevalley groups are built starting from finite (crystallographic) reflection groups and their root systems. From each such root system $\Delta$, we shall first construct a Lie algebra, by giving generators and relations determined by the numerical invariants of the root system. After transferring the coefficients to an arbitrary field, the Chevalley groups are defined as groups of automorphisms of the Lie algebras, or more generally, as automorphism groups of representation modules of the Lie algebras.

The objective is, then, to describe the structure, generators and relations, formulas for the orders of the finite groups obtained, automorphisms and their sets of fixed points, conjugacy classes, and representations (ordinary and modular), for all the Chevalley groups and twisted types. Much of this program, but not all, has been carried out by general methods, without requiring a case-by-case analysis using the classification of root systems. These lectures are an introduction to this work, and especially to the contributions of Chevalley, Steinberg, and Tits. Sections 4 to 8 of this chapter are based on the Yale lecture notes of Steinberg (1967), to which the reader is referred for complete proofs, more examples, and full discussions of the topics surveyed in this publication.

Perhaps the most serious omission is my failure to include an account of Tits' main contributions to the subject, especially his classification of finite groups with $(B, N)$-pairs. A detailed presentation of this topic is now available (Tits, 1971). Also missing are the connections with the theory of algebraic groups, which seem to be essential for further study of the conjugacy classes and representations of the finite Chevalley groups (Tits, 1962; Springer and Steinberg, 1967).

No attempt is made to give complete references or proofs for the results included. In particular, all proofs of the results in the introductory sections 3.1 and 3.2 are omitted, and can be found in Bourbaki (1968), Jacobson (1962) or Serre (1966). Enough detailed information about root systems and Lie algebras is stated to make it possible to give in Sections 4–7 proofs of some of the most important points. These include Chevalley's Lemma ((4.4) below) on the normalization of the structure constants of a semi-simple Lie algebra, Steinberg's derivation of the properties of the subgroups $B$, $\mathfrak{U}$, $H$, and $N$ of a Chevalley group, leading to the Bruhat decomposition and the $(B, N)$-axioms (Section 5), and Solomon's proof of Chevalley's result on the multiplicative formula for the index $[G : B]$ in a Chevalley group.

The reader confronting this subject for the first time may find it profitable to study the survey article by Carter (1965) along with this one. There he will find a sketch of the theory of semi-simple Lie algebras, which is replaced in these lectures by the construction (in Section 3.2) of the Lie algebras by generators and relations starting from root systems. Carter's article also contains a detailed discussion of the various twisted types of Chevalley groups. The point of the discussion of twisted types given below (in Section 8) is to show that it is possible to state general theorems (due to Steinberg) which show that all the twisted types admit $(B, N)$-pairs, so that their simplicity can be proved by Tit's argument (1964) or Bourbaki (1968), and their orders determined by an extension of Solomon's Theorem.

## 2. Notation

| | |
|---|---|
| **Z, Q, R, C** | integers, rationals, real and complex fields. |
| $V^*$ | dual space. |
| $V^E$ | extension of the base field by a field $E$. |
| $k^*$ | multiplicative group of a field $k$. |
| $\langle A, B, \ldots \rangle$ | group generated by $A$, $B$, etc. |
| $Z(G)$ | center of $G$. |
| $\lvert A \rvert$ | cardinal number of the finite set $A$. |
| $(G, G)$ | commutator group of $G$. |
| $[G : H]$ | index of a subgroup $H$ in $G$. |

## 3. Background on Root Systems and Lie Algebras

### 3.1. Finite Reflection Groups and Root Systems

In this section, $V$ denotes a finite dimensional real euclidean space, with positive definite scalar product $( . , .)$. An orthogonal transformation

$r : V \to V$ is called a *reflection* if $r \neq 1$, and $r$ leaves a hyperplane pointwise fixed. If $\alpha \neq 0$ is a vector orthogonal to the hyperplane, then

$$r(x) = x - \frac{2(x, \alpha)}{(\alpha, \alpha)} \alpha \quad (x \in V),$$

and we write $r = r_\alpha$. We have $r_\alpha(\alpha) = -\alpha$, and $r_\alpha(x) = x$ if and only if $(x, \alpha) = 0$.

DEFINITION (3.1). A finite subset $\Delta \subset V$ is called a *root system* provided that

(i) $\Delta$ is a set of generators of $V$;
(ii) $0 \notin \Delta$, and $\alpha$ and $c\alpha \in \Delta$ for some real number $c$ implies that $c = \pm 1$;
(iii) for each $\alpha \in \Delta$, $r_\alpha(\Delta) = \Delta$.
In case $\Delta$ satisfies also the condition
(iv) $2(\alpha, \beta)/(\alpha, \alpha) \in \mathbf{Z}$ for all $\alpha, \beta \in \Delta$,

we say that $\Delta$ is a root system satisfying the *crystallographic condition* (iv).

With each root system $\Delta$ is associated the group $W(\Delta)$ generated by all $r_\alpha$ $(\alpha \in \Delta)$. Then $W(\Delta)$ is a finite group generated by reflections, and the properties of root systems can be used to investigate the structure of such groups. Indeed, let $W$ be an arbitrary finite group generated by reflections on $V$. Let $V_W$ be the subspace of vectors left fixed by all elements of $W$; then $W$ acts faithfully on the vector space $V' = V/V_W$, and $W \cong W'$, where $W'$ is a finite group generated by reflections on $V'$. Let $\Delta = \{\pm\alpha\}$ be the set of all unit vectors orthogonal to the reflecting hyperplanes of all reflections belonging to $W'$; then $\Delta$ is a root system according to definition (3.1), and $W' = W(\Delta)$.

The group $W(\Delta)$ is called the *Weyl group* of the root system $\Delta$.

Let $\mathfrak{g}$ be a semi-simple Lie algebra over $\mathbf{C}$, and let $\mathfrak{h}$ be a Cartan subalgebra. Then the set of roots of $\mathfrak{g}$ with respect to $\mathfrak{h}$ is a root system satisfying the crystallographic condition, and conversely, we shall prove that from any such root system, it is possible to construct a semi-simple Lie algebra.

Every dihedral group of order $2m$ is a finite group generated by reflections. The corresponding root system satisfies the crystallographic condition if and only if $m = 1, 2, 3, 4$ or $6$.

Root systems are classified using the concept of a *base* of a root system.

DEFINITION (3.2). Let $\Delta$ be a root system in $V$. A subset $\Pi \subset \Delta$ is called a *base* of $\Delta$ provided that:

(a) $\Pi$ forms a basis of the vector space $V$;
(b) every root $\alpha \in \Delta$ can be expressed in the form

$$\alpha = \pm \Big( \sum_{\beta \in \Pi} c_\beta \beta \Big)$$

with coefficients $c_\beta \geq 0$.

PROPOSITION (3.3).

(a) *Every root system contains a base.*

(b) *If* $\Delta$ *is a root system and* $\Pi$ *is a base of* $\Delta$, *then* $\Delta = W_0 \cdot \Pi$, *where* $W_0$ *is the subgroup of* $W$ *generated by the reflections* $\{r_\alpha : \alpha \in \Pi\}$.

(c) *If* $\Pi$ *is a base of* $\Delta$, *then* $W(\Delta)$ *is generated by the reflections* $\{r_\alpha : \alpha \in \Pi\}$.

(d) *Any two bases of a root system* $\Delta$ *are conjugate by an element of the Weyl group (i.e.,* $W(\Delta)$ *acts transitively on the set of bases).*

(e) *Suppose that* $\Delta$ *satisfies the crystallographic condition, and let* $\Pi$ *be a base of* $\Delta$. *Then the coefficients* $c_\beta$ *in* (3.2) *are all integers.*

*Definitions* (3.4). We shall call dim $V$ the *rank* of $\Delta$. The roots in $\Pi$ will be called *fundamental roots* (or *simple roots*); the reflections $\{r_\alpha : \alpha \in \Pi\}$, the *fundamental reflections.* The set of fundamental reflections will be denoted by $R$. For an element $w \in W(\Delta)$, $l(w)$ will denote the minimal length of an expression of $w$ as a product of elements of $R$. Using (3.3), we define the *positive roots* in $\Delta$ (with respect to $\Pi$) to be the roots $\{+(\sum c_\beta \beta) : c_\beta \geq 0, \beta \in \Pi\}$, and the negative ones $\{-(\sum c_\beta \beta)\}$. Letting $\Delta^\pm$ denote the positive (resp. negative) roots, we define, for each $w \in W$, the sets

$$\Delta_w^\pm = \{\alpha \in \Delta^+ : w(\alpha) \in \Delta^\pm\},$$

and the numerical function $n(w) = |\Delta_w^-|$.

PROPOSITION (3.5). *Assume the notations of* (3.4).

(a) *Let* $\alpha \in \Pi$. *Then* $r_\alpha$ *permutes the positive roots* $\alpha' \neq \alpha$.

(b) *The numerical functions* $n(w)$ *and* $l(w)$ *coincide.*

(c) $|\Delta_w^-| > 0$ *if* $w \neq 1$.

(d) *There exists a unique element* $w_0 \in W$ *such that* $w_0(\Delta^+) = \Delta^-$. *The element* $w_0$ *is an involution* ($w_0^2 = 1$).

(e) *Suppose that* $w_1, w_2, \ldots, w_k \in R$ *(possibly with repetitions) and that* $l(w_2 \ldots w_k) = k-1$, $l(w_1 w_2 \ldots w_k) \leq k-1$. *Then for some* $m \leq k$, $w_1 w_2 \ldots w_{m-1} = w_2 \ldots w_m$.

Using statement (*e*), we can derive the structure of $W$ as an abstract group.

PROPOSITION (3.6). *The pair* $(W, R)$ *is a* Coxeter system, *i.e.,* $W$ *has the presentation*

$$\langle w_1, \ldots, w_n : (w_i w_j)^{n_{ij}} = 1, \quad \text{and} \quad n_{ii} = 1 \quad (1 \leq i \leq n)\rangle,$$

*where* $R = \{w_1, \ldots, w_n\}$.

Finally, we describe briefly the classification of root systems satisfying the crystallographic condition.

DEFINITION (3.7). Let $\Delta$ be a root system satisfying the crystallographic condition, and let $\Pi = \{\alpha_1, \ldots, \alpha_n\}$ be a base of $\Delta$. The *Cartan matrix* of $\Pi$

is the matrix $(A_{ij})$ where

$$A_{ij} = \frac{2(\alpha_i, \alpha_j)}{(\alpha_j, \alpha_j)}.$$

The integers $A_{ij}$ are called the *Cartan integers* of $\Pi$.

The matrix of Cartan integers satisfies the conditions

(i) $A_{ii} = 2$; $A_{ij} \le 0$ if $i \ne j$; and $A_{ij} = 0$ if and only if $A_{ji} = 0$;
(ii) $\det(A_{ij}) \ne 0$; and
(iii) if $\Pi = \{\alpha_1, \ldots, \alpha_n\}$ is the base, then the root system $\Delta$ is the set of vectors obtained from $\Pi$ by applying to $\Pi$ the elements of the group generated by the linear transformations $r_j : \alpha_i \to \alpha_i - A_{ij}\alpha_j$ $(1 \le j \le n)$.

A root system $\Delta$ with base $\Pi$ is *indecomposable* if it is impossible to split up $\Pi$ into two non-empty subsets which are orthogonal to each other. By Prop. (3.3) (d), this definition does not depend on the choice of $\Pi$.

Let $\alpha_i, \alpha_j \in \Pi$; then the angle $\theta_{ij}$ between them satisfies

$$\cos\theta_{ij} = \frac{(\alpha_i, \alpha_j)}{|\alpha_i||\alpha_j|}$$

where $|\alpha| = (\alpha, \alpha)^{\frac{1}{2}}$, so that

$$A_{ji} = 2\frac{|\alpha_j|}{|\alpha_i|}\cos\theta_{ij},$$

and $A_{ij}A_{ji} = 4(\cos\theta_{ij})^2$. It follows from the crystallographic condition that for $i \ne j$, $4(\cos\theta_{ij})^2 \in \{0, 1, 2, 3\}$, and that the indecomposable root systems (satisfying the crystallographic condition) of rank two are given as follows.

| *Type* | *Cartan matrix* | $\Delta^+$ | $\lvert\beta\rvert/\lvert\alpha\rvert$ | *Weyl group* |
|---|---|---|---|---|
| $A_2$ | $\begin{pmatrix} 2 & -1 \\ -1 & 2 \end{pmatrix}$ | $\alpha, \beta, \alpha+\beta$ | 1 | $r_\alpha^2 = r_\beta^2 = (r_\alpha r_\beta)^3 = 1$ |
| $B_2$ | $\begin{pmatrix} 2 & -1 \\ -2 & 2 \end{pmatrix}$ | $\alpha, \beta, \alpha+\beta, 2\alpha+\beta$ | $\sqrt{2}$ | $r_\alpha^2 = r_\beta^2 = (r_\alpha r_\beta)^4 = 1$ |
| $G_2$ | $\begin{pmatrix} 2 & -1 \\ -3 & 2 \end{pmatrix}$ | $\alpha, \beta, \alpha+\beta, 2\alpha+\beta,$ $3\alpha+\beta, 3\alpha+2\beta.$ | $\sqrt{3}$ | $r_\alpha^2 = r_\beta^2 = (r_\alpha r_\beta)^6 = 1$ |

The classification of arbitrary indecomposable root systems is achieved using the *Dynkin diagram* of $\Pi$, in which nodes of a graph are made to correspond to the elements of $\Pi$, and are connected by 0, 1, 2 or 3 lines according as $A_{ij}A_{ji} = 0, 1, 2$ or 3. In case $A_{ij}A_{ji} > 1$, it is also necessary to specify which are the long roots and which are short. The possible Dynkin diagrams of indecomposable systems are as follows.

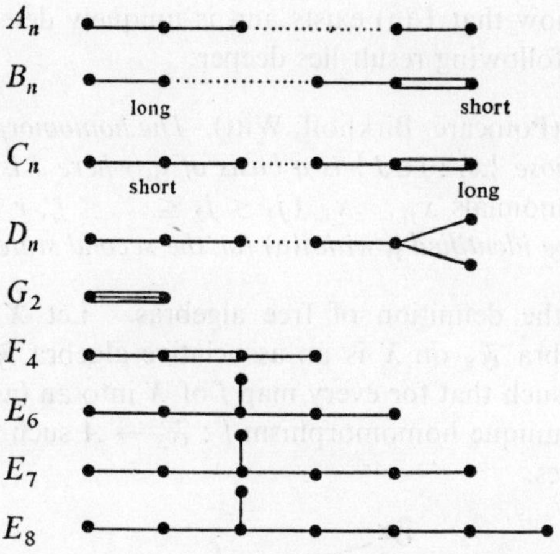


## 3.2. Construction of Certain Lie Algebras

Let g be a Lie algebra over a field $F$, with bracket operation $[xy]$. A *homomorphism* $\rho : \mathfrak{g} \to A$ of g into an associative algebra $A$ is a linear mapping $\rho$ such that

$$\rho([xy]) = \rho(x)\rho(y) - \rho(y)\rho(x) \quad (x, y \in \mathfrak{g}),$$

i.e., $\rho$ is a homomorphism of g into $A_L$, the Lie algebra obtained from $A$ by defining a bracket operation by $[ab] = ab - ba$. In case $A = \mathrm{Hom}_F(M, M)$ for some vector space $M$ over $F$, we call $\rho$ a *representation* of g, and $M$ becomes a *left* g-*module* if we define

$$x \cdot m = \rho(x) \cdot m \quad (x \in \mathfrak{g},\ m \in M).$$

Definition (3.8). The *universal enveloping algebra* $U(\mathfrak{g})$ of g is an associative algebra $U(\mathfrak{g})$ (with 1) together with a homomorphism $i : \mathfrak{g} \to U(\mathfrak{g})$ such that

(a) $\{1\} \cup i(\mathfrak{g})$ is a set of generators of $U(\mathfrak{g})$;

(b) for every homomorphism $f : \mathfrak{g} \to A$ of g into an associative algebra $A$, there exists a unique homomorphism of associative algebras

$\tilde{f} : U(\mathfrak{g}) \to A$ such that the diagram

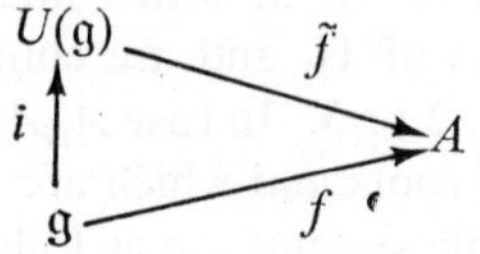


is commutative.

It is easy to show that $U(\mathfrak{g})$ exists and is uniquely determined up to isomorphism. The following result lies deeper.

THEOREM (3.9) (Poincaré, Birkhoff, Witt). *The homomorphism $i : \mathfrak{g} \to U(\mathfrak{g})$ is injective. Suppose $\{x_j : j \in J\}$ is a basis of $\mathfrak{g}$, where $J$ is some ordered set. The* standard monomials $x_{j_1} \ldots x_{j_r}$ $(j_1 \leq j_2 \leq \ldots \leq j_r,\ r \geq 0)$ *form a basis of $U(\mathfrak{g})$. (We have identified $\mathfrak{g}$ with $i(\mathfrak{g})$ for the second statement.)*

Let us recall the definition of free algebras. Let $X$ be a set; a free (associative) algebra $\mathfrak{F}_X$ on $X$ is an associative algebra $\mathfrak{F}_X$ together with a map $j : X \to \mathfrak{F}_X$ such that for every map $f$ of $X$ into an (associative) algebra $A$, there exists a unique homomorphism $\tilde{f} : \mathfrak{F}_X \to A$ such that the following diagram commutes.

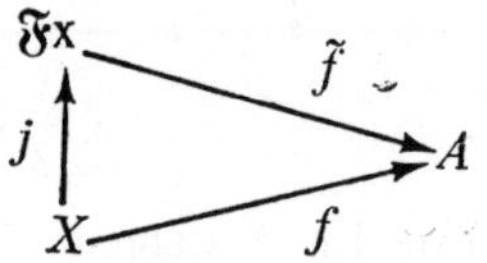


Similarly, a *free Lie algebra* over $F$ is defined.

It is easy to construct a free associative algebra over $X$. Form a vector space $M$ with basis $X$, and let $\mathfrak{F}_X$ be the tensor algebra over $M$. Thus,

$$\mathfrak{F}_X = F \cdot 1 \oplus M \oplus (M \otimes M) \oplus \ldots \oplus M^r \oplus \ldots,$$

where $M^r = M \otimes \ldots \otimes M$ ($r$ times), and multiplication is the obvious extension to $\mathfrak{F}_X$ of the map $(m, n) \to m \otimes n$ $(m \in M^k, n \in M^{k'})$. Then it can be verified that $\mathfrak{F}_X$ is a free algebra.

This construction, together with the Poincaré-Birkhoff-Witt theorem, enables us to construct a free Lie algebra on $X$. Namely, let $\mathfrak{F}_X$ be the free associative algebra on $X$, and let $\mathfrak{g}_X$ be the subalgebra of the Lie algebra $(\mathfrak{F}_X)_L$ generated by $X$.

PROPOSITION (3.10). *$\mathfrak{g}_X$ is a free Lie algebra on $X$. Moreover, $(\mathfrak{F}_X, i)$, where $i$ is the injection $\mathfrak{g}_X \to \mathfrak{F}_X$, is the universal enveloping algebra of $\mathfrak{g}_X$.*

Now let $(A_{ij})$ be the Cartan matrix of a root system $\Delta$ (satisfying the crystallographic condition) of rank $n$. Starting from the matrix $(A_{ij})$ and a field $F$ of characteristic 0, we shall first construct, after some preliminaries, a finite dimensional semi-simple Lie algebra over $F$, and ultimately the groups associated with $\Delta$.

We begin with the homomorphic image $\mathfrak{A}$ of the free Lie algebra over $F$ on $3n$ generators

$$\{X_1, \ldots, X_n, H_1, \ldots, H_n, Y_1, \ldots, Y_n\}$$

satisfying the relations

$$[H_iH_j] = 0 \quad (1 \leq i,\ \ j \leq n);$$
$$[X_iY_i] = H_i, \quad [X_iY_j] = 0 \quad (i \neq j);$$
$$[H_iX_j] = A_{ji}X_j;$$
$$[H_iY_j] = -A_{ji}Y_j.$$

PROPOSITION (3.11). *$\mathfrak{A} = \mathfrak{X} \oplus \mathfrak{H} \oplus \mathfrak{Y}$, where $\mathfrak{X}$ and $\mathfrak{Y}$ are the Lie algebras generated by the sets $\{X_i\}$ and $\{Y_i\}$ respectively, and $\mathfrak{H}$ is the vector space generated by the set $\{H_i\}$. Moreover,* $\dim \mathfrak{H} = n$.

For each $i$ $(1 \leq i \leq n)$, define a linear function $\alpha_i$ on $\mathfrak{H}$ by

$$\alpha_i(H_j) = A_{ij} \quad (1 \leq j \leq n).$$

Because $\det (A_{ij}) \neq 0$, $\{\alpha_i\}$ is a basis of the dual space $\mathfrak{H}^*$ of $\mathfrak{H}$, and it is possible to define linear transformations $\{r_i\}$ $(1 \leq i \leq n)$ on $\mathfrak{H}^*$ by

$$r_j(\alpha_i) = \alpha_i - A_{ij}\alpha_j.$$

Letting $V_0$ be the rational vector space generated by $\Pi = \{\alpha_1, \ldots, \alpha_n\}$, and $W$ the automorphism group of $V_0$ generated by $\{r_i\}$, it can be verified that $W \cdot \Pi$ can be identified with a root system $\Delta$ with base $\Pi$ in $V_0^{\mathbf{R}}$ whose Cartan matrix is $(A_{ij})$, and $W$ with the Weyl group $W(\Delta)$ of $\Delta$.

Recalling that in any Lie algebra, ad $x$ denotes the derivation $y \to [xy]$, we consider next the elements

$$\theta_{ij} = (\text{ad } X_i)^{-A_{ji}+1}(X_j) \quad (i \neq j),$$

in $\mathfrak{A}$. It can be proved that $[Y_k\theta_{ij}] = 0$ for each $k$ $(1 \leq k \leq n)$. Similarly, we define

$$\theta_{ij}^- = (\text{ad } Y_i)^{-A_{ji}+1}(Y_j) \quad (i \neq j).$$

Let $\mathfrak{U}$ be the ideal in $\mathfrak{X}$ generated by the $\{\theta_{ij}\}$, and $\mathfrak{B}$ the ideal in $\mathfrak{Y}$ generated by the $\{\theta_{ij}^-\}$. Then $\mathfrak{U} + \mathfrak{B}$ is an ideal in $\mathfrak{A}$, and we define the Lie algebra

$$\mathfrak{g} = \mathfrak{A}/(\mathfrak{U} + \mathfrak{B}).$$

Then $\mathfrak{g} = \mathfrak{n}^- \oplus \mathfrak{h} \oplus \mathfrak{n}^+$, where $\mathfrak{n}^-$, $\mathfrak{h}$, $\mathfrak{n}^+$ are the images in $\mathfrak{g}$ of $\mathfrak{Y}$, $\mathfrak{H}$, and $\mathfrak{X}$ respectively.

In order to investigate the structure of $\mathfrak{g}$, we consider the weights of the adjoint representation with respect to the subalgebra $\mathfrak{h}$; namely, for each $\lambda \in \mathfrak{h}^*$, we define

$$\mathfrak{g}_\lambda = \{x \in \mathfrak{g} : [hx] = \lambda(h)x, \quad h \in \mathfrak{h}\}.$$

Since $\mathfrak{h} \cong \mathfrak{H}$, the root system $\Delta$ and the group $W$ can be carried over to the dual space $\mathfrak{h}^*$ of $\mathfrak{h}$. Then it is proved in Serre (1966) that

(a) $\mathfrak{g}_\lambda \neq 0$ implies $\lambda = \pm \left\{ \sum_{\alpha_i \in \Pi} m_i \alpha_i \right\}$, with the $\{m_i\}$ non-negative integers

(b) $\dim \mathfrak{g}_\lambda = \dim \mathfrak{g}_{w\lambda}$ for all $\lambda$ and $w \in W$; and

(c) $\alpha \in \Delta$ implies that $\dim \mathfrak{g}_\alpha = 1$, and $\mathfrak{g}_\lambda = 0$ if $\lambda \notin \Delta \cup \{0\}$.

From these results, the following theorem follows easily, and provides the starting point for the construction of Chevalley groups.

THEOREM (3.12). (Harish-Chandra, Jacobson, Serre). *Let $\Delta$ be a root system satisfying the crystallographic condition, and let $(A_{ij})$ $(1 \le i, j \le n)$ be a Cartan matrix associated with some base $\Pi$ of $\Delta$. Let $\mathfrak{g}$ be the Lie algebra over an arbitrary field $F$ of characteristic zero, with generators*

$$\{X_1, \ldots, X_n, H_1, \ldots, H_n, Y_1, \ldots, Y_n\}$$

*and relations*

$$\begin{aligned} &[H_iH_j] = 0 \quad (1 \le i, j \le n) \\ &[X_iY_i] = H_i, \ [X_iY_j] = 0 \quad (i \neq j) \\ &[H_iX_j] = A_{ji}X_j, \ [H_iY_j] = -A_{ji}Y_j \\ &(\operatorname{ad} X_i)^{-A_{ji}+1}(X_j) = 0, \ (\operatorname{ad} Y_i)^{-A_{ji}+1}(Y_j) = 0 \quad (i \neq j). \end{aligned}$$

*Then $\mathfrak{g}$ is a finite dimensional Lie algebra, such that*

$$\mathfrak{g} = \mathfrak{h} + \sum_{\alpha \in \Delta} \mathfrak{g}_\alpha \qquad \text{(direct sum).}$$

*and $\{H_1, \ldots, H_n\}$ is a basis for $\mathfrak{h}$.*

*The subspaces $\mathfrak{g}_\alpha$ are one-dimensional. Let $\mathfrak{g}_\alpha = F \cdot x_\alpha$. Then the multiplication in $\mathfrak{g}$ is given by the formulas*

$$[\mathfrak{h}\ \mathfrak{h}] = 0,$$

$$[h\ x_\alpha] = \alpha(h)x_\alpha \quad (\alpha \in \Delta, h \in \mathfrak{h}),$$

$$[x_\alpha x_\beta] = \begin{cases} h_\alpha \neq 0 \text{ in } \mathfrak{h} & \text{if } \alpha+\beta = 0 \\ 0 & \text{if } \alpha+\beta \notin \Delta \text{ and } \alpha+\beta \neq 0 \\ n_{\alpha\beta}x_{\alpha+\beta} & \text{if } \alpha+\beta \in \Delta, \ n_{\alpha\beta} \in F. \end{cases}$$

It follows from what has been stated that $\mathfrak{g}$ is a split semi-simple Lie algebra over $F$, with Cartan subalgebra $\mathfrak{h}$ and root system $\Delta$ with respect to $\mathfrak{h}$. It can also be proved that every semi-simple Lie algebra over $\mathbf{C}$ is isomorphic to one constructed as above. We conclude with the following corollary of the theorem.

COROLLARY (3.13). *There exists an automorphism $\sigma$ of $\mathfrak{g}$ which is $-1$ on $\mathfrak{h}$, and is such that $\sigma(X_i) = -Y_i$, $\sigma(Y_i) = -X_i$ $(1 \le i \le n)$.*

It is sufficient to prove that $\sigma$, as defined on the generators, preserves the defining relations of $\mathfrak{g}$.

## 4. Integral Bases

In this chapter, let $F$ be an algebraically closed field of characteristic zero, and let $\mathfrak{g}$ be the Lie algebra over $F$ constructed in Section 3, Theorem (3.12). The purpose of this chapter is to construct certain integral bases for $\mathfrak{g}$, $U(\mathfrak{g})$, and for arbitrary $\mathfrak{g}$-modules $M$. For a finite dimensional vector space $V$ over $F$, we shall use the terminology "**Z**-*form of* $V$" for a finitely generated abelian group $V_0 \subset V$ such that $F \otimes_{\mathbf{Z}} V_0 \cong V$. In other words, $V_0$ is a free **Z**-module such that a **Z**-basis of $V_0$ is an $F$-basis of $V$. In case $V$ is an associative or Lie algebra over $F$, we require that a **Z**-form also be an associative or Lie algebra over **Z**.

### 4.1. THE CHEVALLEY BASIS OF $\mathfrak{g}$

We begin with some preliminary results, starting from Theorem (3.12) and its proof given in Serre (1966).

First of all, by $sl_2(F)$, (or simply $sl_2$), we mean the Lie algebra of 2 by 2 matrices of trace zero over $F$. Then $sl_2(F)$ is a simple 3-dimensional Lie algebra with basis $\{X, Y, H\}$ such that $[XY] = H$, $[HX] = 2X$, $[HY] = -2Y$.

We shall denote by $x_i$, $y_i$, $h_i$ the images in $\mathfrak{g}$ of $X_i$, $Y_i$, $H_i$.

LEMMA (4.1). *Let $\mathfrak{S}_\alpha$ $(\alpha \in \Delta)$ be the subalgebra of $\mathfrak{g}$ generated by $\mathfrak{g}_\alpha$ and $\mathfrak{g}_{-\alpha}$. Then $\mathfrak{S}_\alpha \cong sl_2$.*

*Proof.* The defining relations show that the lemma is true for $\alpha = \alpha_i \in \Pi$, for in that case $\mathfrak{g}_{\alpha_i} = F \cdot x_i$, $\mathfrak{g}_{-\alpha_i} = F \cdot y_i$, and there exists an epimorphism of $sl_2 \to \mathfrak{S}_{\alpha_i}$ such that $X \to x_i$, $Y \to y_i$, $H \to h_i$. Since $sl_2$ is simple, the map is an isomorphism.

For the rest of the proof, we use the fact, proved by Serre (1966) that for each $i$ $(1 \le i \le n)$, the automorphism of $\mathfrak{g}$,

$$\theta_i = \exp(\operatorname{ad} x_i) \exp -(\operatorname{ad} y_i) \exp(\operatorname{ad} x_i),$$

induces on $\mathfrak{h}$ the map

$$h \to h - \alpha_i(h) h_i.$$

It follows that

$$\theta_i(\mathfrak{g}_\alpha) = \mathfrak{g}_{r_i(\alpha)} \quad (\alpha \in \Delta),$$

since for all $h \in \mathfrak{h}$,

$$\begin{aligned}[h, \theta_i(x_\alpha)] &= \theta_i([h-\alpha_i(h)h_i, x_\alpha]) \\ &= \theta_i((\alpha(h)-\alpha(h_i)\alpha_i(h))\, x_\alpha) \\ &= r_i(\alpha)(h)\theta_i(x_\alpha).\end{aligned}$$

Therefore, $\mathfrak{S}_{r_i(\alpha)} = \theta_i(\mathfrak{S}_\alpha)$ for all $\alpha \in \Delta$, and from Prop. (3.3) the lemma follows.

DEFINITION. For each $\alpha$, let $H_\alpha$ be the unique element of $\mathfrak{h}$ such that

$$\alpha(H_\alpha) = 2, \qquad \lambda(H_\alpha) = 0 \quad \text{if } (\lambda, \alpha) = 0.$$

LEMMA (4.2).

(i) $r_\alpha(\lambda) = \lambda - \lambda(H_\alpha)\alpha$ *for all* $\lambda \in \mathfrak{h}^*$.

(ii) $\beta(H_\alpha) = 2(\alpha, \beta)/(\alpha, \alpha)$.

(iii) *The set* $\{H_\alpha\}_{\alpha \in \Delta}$ *is a crystallographic root system in* $\mathfrak{h}$. *Moreover,* $h_i = H_{\alpha_i}$ $(1 \le i \le n)$ *and* $\{h_1, \ldots, h_n\}$ *is a base of the root system* $\{H_\alpha\}$.

(iv) *For each* $\alpha$, $H_\alpha \in [\mathfrak{g}_\alpha \mathfrak{g}_{-\alpha}]$.

*Proof.* (i), (ii), (iii) are clear from the definitions. Statement (iv) is a corollary of the proof of Lemma (4.1).

LEMMA (4.3). *Let* $\alpha, \beta \in \Delta$, $\alpha \neq \pm\beta$. *Let p,q be the largest integers* $\ge 0$ *such that* $\beta + q\alpha \in \Delta$, $\beta - p\alpha \in \Delta$. *Then*

$$\beta(H_\alpha) = p - q.$$

For the proof, we let $\mathfrak{S}_\alpha = \langle x_\alpha, x_{-\alpha}, H_\alpha \rangle$ act on $\mathfrak{g}$ via the adjoint representation. Using the representation theory of $sl_2(F)$, together with the fact that $[x_\alpha \mathfrak{g}_\beta] \subset \mathfrak{g}_{\beta+\alpha}$, it can be shown that

$$V = \sum_{i=-\infty}^{\infty} \mathfrak{g}_{\beta + i\alpha}$$

is an $\mathfrak{S}_\alpha$-submodule of $\mathfrak{g}$, which is irreducible. Then, since $(\beta + q\alpha)(H_\alpha)$ and $(\beta - p\alpha)(H_\alpha)$ are the highest and lowest eigenvalues of $H_\alpha$ on this module, it follows that

$$(\beta + q\alpha)(H_\alpha) = -(\beta - p\alpha)(H_\alpha)$$

and the result follows (see Serre, 1966).

Now we are ready to obtain the Chevalley basis for $\mathfrak{g}$. We normalize the root elements $\{x_\alpha\}$ always so that $[x_\alpha, x_{-\alpha}] = H_\alpha$ $(\alpha \in \Delta)$ which is certainly possible by Lemma (4.2). The structure constants $\{n_{\alpha\beta}\}$ are not uniquely determined by this condition, but if we make another choice $\{x'_\alpha\}$, we have $x'_\alpha = u_\alpha x_\alpha$ $(\alpha \in \Delta)$ with

$$u_\alpha u_{-\alpha} = 1$$

for all $\alpha$. Then $[x'_\alpha x'_\beta] = n'_{\alpha\beta} x'_{\alpha+\beta}$ implies that

$$n'_{\alpha\beta} = \frac{u_\alpha u_\beta}{u_{\alpha+\beta}} n_{\alpha\beta}.$$

It follows that

$$n'_{\alpha\beta} n'_{-\alpha,\,-\beta} = n_{\alpha\beta} n_{-\alpha,\,-\beta}.$$

The next result is basic for the entire subsequent discussion.

LEMMA (4.4) (Chevalley). *Let $\alpha$, $\beta$, $\alpha+\beta \in \Delta$, and let $p$ be the largest integer $\geq 0$ such that $\beta - p\alpha \in \Delta$. Then*

$$n_{\alpha\beta} n_{-\alpha-\beta} = -(p+1)^2.$$

*Proof.* Applying the Jacobi identity to $x_\alpha$, $x_{-\alpha}$, $x_\beta$, we have

$$[[x_\alpha x_{-\alpha}]x_\beta] + [[x_{-\alpha} x_\beta]x_\alpha] + [[x_\beta x_\alpha]x_{-\alpha}] = 0$$

or

$$[H_\alpha x_\beta] + n_{-\alpha,\,\beta}[x_{\beta-\alpha} x_\alpha] + n_{\beta\alpha}[x_{\beta+\alpha} x_{-\alpha}] = 0,$$

and hence

$$\beta(H_\alpha) + n_{-\alpha,\,\beta} n_{\beta-\alpha,\,\alpha} + n_{\beta\alpha} n_{\beta+\alpha,\,-\alpha} = 0$$

Repeating the argument, we obtain

$$(\beta-\alpha)(H_\alpha) + n_{-\alpha,\,\beta-\alpha} n_{\beta-2\alpha,\,\alpha} + n_{\beta-\alpha,\,\alpha} n_{\beta,\,-\alpha} = 0,$$

and, since $n_{\beta,\,-\alpha} = -n_{-\alpha,\,\beta}$, upon adding the equations, we get

$$2\beta(H_\alpha) - \alpha(H_\alpha) + n_{\beta\alpha} n_{\beta+\alpha,\,-\alpha} + n_{-\alpha,\,\beta-\alpha} n_{\beta-2\alpha,\,\alpha} = 0$$

Continuing this process and adding the results for $\beta - k\alpha$ $(0 \leq k \leq p)$, we get

$$(p+1)\beta(H_\alpha) - (1+2+\ldots+p)\alpha(H_\alpha) + n_{\beta\alpha} n_{\beta+\alpha,\,-\alpha} + $$
$$+ n_{-\alpha,\,\beta-p\alpha} n_{\beta-(p+1)\alpha,\,\alpha} = 0.$$

Since $\beta(H_\alpha) = p - q$ (by (4.3)) and $n_{\beta-(p+1)\alpha,\,\alpha} = 0$, we have

$$n_{\alpha\beta} n_{\beta+\alpha,\,-\alpha} = -q(p+1). \tag{*}$$

Next apply the Jacobi identity to $x_{-\alpha}$, $x_{-\beta}$, $x_{\alpha+\beta}$ to obtain

$$-n_{-\alpha,\,-\beta} H_{\alpha+\beta} + n_{-\beta,\,\alpha+\beta} H_\alpha + n_{\alpha+\beta,\,-\alpha} H_\beta = 0.$$

Further information about the structure constants comes from the geometrical observation that

$$-\lambda(\alpha+\beta) H_{\alpha+\beta} + \lambda(\alpha) H_\alpha + \lambda(\beta) H_\beta = 0, \tag{**}$$

where $\lambda(\alpha) = (\alpha, \alpha)$ for $\alpha \in \Delta$. This statement is proved by first noting that $\lambda(\alpha)\beta(H_\alpha) = \lambda(\beta)\alpha(H_\beta)$ for all $\alpha$, $\beta \in \Delta$, and then checking that the expression in (**) is annihilated by all roots $\gamma \in \Delta$. Since $H_\alpha$ and $H_\beta$ are linearly independent, it follows that the two sets of coefficients of $H_{\alpha+\beta}$, $H_\alpha$ and $H_\beta$ are proportional, and hence that

$$n_{\alpha+\beta,\,-\alpha} \lambda(\alpha+\beta) = n_{-\alpha,\,-\beta} \lambda(\beta).$$

Combining this result with (*) yields the formula

$$n_{\alpha\beta}n_{-\alpha,-\beta} = -q(p+1)\lambda(\alpha+\beta)\lambda(\beta)^{-1}.$$

The lemma then follows from the result that

$$\frac{\lambda(\alpha+\beta)}{\lambda(\beta)} = \frac{p+1}{q},$$

which involves only the root system of rank 2 generated by $\alpha$ and $\beta$, and can be checked for the three possible such root systems.

LEMMA (4.5). *It is possible to choose basis elements $x_\alpha \in \mathfrak{g}_\alpha$ ($\alpha \in \Delta$) such that $[x_\alpha x_{-\alpha}] = H_\alpha$, and $n_{-\alpha,-\beta} = -n_{\alpha,\beta}$ for all $\alpha$ and $\beta$. For this choice of root elements $\{x_\alpha\}$, we have $n_{\alpha\beta} = \pm(p+1)$, with $p$ as in* (4.4).

*Proof.* By Corollary 3.13, there exists an automorphism $\sigma$ of $\mathfrak{g}$ such that $\sigma|\mathfrak{h} = -1$, and $\sigma(\mathfrak{g}_\alpha) = \mathfrak{g}_{-\alpha}$, for all $\alpha$. Let $\{x_\alpha\}$ be fixed choices of root elements for $\alpha \in \Delta^+$. Since $[x_\alpha, \sigma(x_\alpha)] \in F\cdot H_\alpha$, $[x_\alpha, \sigma(x_\alpha)] = t_\alpha H_\alpha$ for some $t_\alpha \in F^*$. Choose, for each $\alpha > 0$, a square root $u_\alpha$ of $-t_\alpha$, and define $X_\alpha = u_\alpha^{-1}x_\alpha$, $X_{-\alpha} = -\sigma(X_\alpha)$. Then $\sigma(X_{-\alpha}) = -X_\alpha$, and for $\alpha > 0$,

$$[X_\alpha X_{-\alpha}] = [u_\alpha^{-1}x_\alpha, -u_\alpha^{-1}\sigma(x_\alpha)] = H_\alpha.$$

Finally, since $\sigma$ is an automorphism,

$$n_{\alpha\beta}\sigma(X_{\alpha+\beta}) = \sigma([X_\alpha X_\beta]) = [-X_{-\alpha} - X_{-\beta}] = n_{-\alpha,-\beta}X_{-\alpha-\beta}$$

and the lemma is proved, since $\sigma(X_{\alpha+\beta}) = -X_{-\alpha-\beta}$.

THEOREM (4.6). *The elements $\{H_{\alpha_i} : \alpha_i \in \Pi\}$ together with elements $X_\alpha \in \mathfrak{g}_\alpha$ ($\alpha \in \Delta$) chosen to satisfy $[X_\alpha X_{-\alpha}] = H_\alpha$ and $[X_\alpha X_\beta] = \pm(p+1)X_{\alpha+\beta}$ (if $\alpha+\beta \in \Delta$) form a basis for a* **Z**-*form $\mathfrak{g}_{\mathbf{Z}}$ of $\mathfrak{g}$.*

*Proof.* It remains only to prove that the elements $H_\alpha$ are **Z**-linear combinations of the $H_{\alpha_i}$ ($1 \leq i \leq n$). This follows from the fact that the $\{H_\alpha : \alpha \in \Delta\}$ form a root system satisfying the crystallographic condition, with base $\{H_{\alpha_1}, \ldots, H_{\alpha_n}\}$ (by (4.2)).

From what has been shown, a knowledge of the signs of $\pm(p+1)$ appearing in Theorem (4.6) would suffice to construct $\mathfrak{g}$ from the root system. For a determination of the signs from a knowledge of $\Delta$, see Tits (1966).

## 4.2. A Z-FORM OF THE UNIVERSAL ENVELOPING ALGEBRA

The results in this section and the next are due to Kostant (1966) (see also Steinberg (1967). Let $\{X_\alpha : \alpha \in \Delta\}$ and $\{H_{\alpha_i}\}_{\alpha_i \in \Pi}$, be a *Chevalley basis* for $\mathfrak{g}$ (i.e., a basis satisfying the conditions of Theorem (4.6)). Let $\Delta^+ = \{\alpha_1, \ldots, \alpha_N\}$. For $Q = (q_1, \ldots, q_N)$, with the $q_i$ non-negative integers, put

$$e_{\pm Q} = \prod_{i=1}^{N}(X_{\pm\alpha_i}^{q_i}/(q_i)!).$$

(We are identifying $\mathfrak{g}$ with its image in $U(\mathfrak{g})$.) For $x \in \mathfrak{g}$ and $s \in \mathbf{Z}$, $s > 0$ put, in $U(\mathfrak{g})$,

$$\binom{x}{s} = \frac{x(x-1)\ldots(x-s+1)}{s!}$$

Let $n = \dim h$. For each $n$-tuple $P = (p_i)_{1 \le i \le n}$ ($p_i \in \mathbf{Z}$, $p_i \ge 0$) define the element of $U(\mathfrak{h})$

$$h_P = \prod_{i=1}^{n} \binom{H_{\alpha_i}}{p_i}.$$

Finally, let $\mathfrak{n}^{\pm}$ be the subalgebra of $\mathfrak{g}$,

$$\sum_{\alpha \in \Delta} \mathfrak{g}_{\pm\alpha}.$$

THEOREM (4.7). *The elements $\{e_{-Q} h_P e_S\}$, for all $Q$, $P$, $S$, form a basis of a $\mathbf{Z}$-form $U_{\mathbf{Z}}$ of $U(\mathfrak{g})$. The elements $\{e_{\pm Q}\}$ (resp $h_P$) form bases for $\mathbf{Z}$-forms $U(\mathfrak{n}^{\pm})_{\mathbf{Z}}$ of $U(\mathfrak{n}^{\pm})$ (resp $U(\mathfrak{h})_{\mathbf{Z}}$ of $U(\mathfrak{h})$). $U_{\mathbf{Z}}$ is generated by the elements $\{e_{\pm Q}\}$.*

### 4.3. Z-FORMS OF $\mathfrak{g}$-MODULES

Let $V$ be a $\mathfrak{g}$-module; then $V$ is a $U(\mathfrak{g})$-module by the universal property of $U(\mathfrak{g})$. Our objective is to show how to construct a $\mathbf{Z}$-form $M$ of $V$ with the property that $U_{\mathbf{Z}} \cdot M \subseteq M$. (Note that this is a much stronger requirement than $\mathfrak{g}_{\mathbf{Z}} \cdot M \subseteq M$). The existence of $\mathbf{Z}$-forms with this property is essential for the construction of Chevalley groups.

It is first necessary to recall some basic facts about finite dimensional $\mathfrak{g}$-modules. Let $V$ be such a module. An element $\mu \in \mathfrak{h}^*$ is called a *weight* of $V$ if

$$V_\mu = \{v \in V : h \cdot v = \mu(h)v \quad \text{for all } h \in \mathfrak{h}\} \neq 0.$$

If $\mu$ is a weight, then $\mu(H_\alpha)$ is an eigenvalue of $H_\alpha$ in a representation of $\mathfrak{S}_\alpha \cong sl_2$, and hence $\mu(H_\alpha) \in \mathbf{Z}$. Let $P(V)$ be the set of all weights of $V$. A weight $\mu$ is called *dominant* if $\mu(H_{\alpha_i}) \ge 0$ for $\alpha_i \in \Pi$.

PROPOSITION (4.8).

(a) *If $V$ is a $\mathfrak{g}$-module, then we have the direct decomposition*

$$V = \sum_{\mu \in P(V)} V_\mu$$

(b) *For $\alpha \in \Delta$, $X_\alpha V_\mu \subseteq V_{\mu+\alpha}$, for all $\mu \in P(V)$.*

(c) *$V$ is completely reducible, i.e., a direct sum of irreducible modules.*

(d) *Let $V$ be irreducible. Then there exists a unique weight $\Lambda$ such that $X_\alpha V_\Lambda = 0$ for all $\alpha > 0$. Then $\dim V_\Lambda = 1$, $V = U(\mathfrak{n}^-) \cdot V_\Lambda$, and $\Lambda$ is dominant. Every weight $\mu \in P(V)$ has the form*

$$\Lambda - \sum_{\alpha_i \in \Pi} m_i \alpha_i, \quad (m_i \in \mathbf{Z},\ m_i \ge 0).$$

(*The uniquely determined weight* $\Lambda$ *of* $V$ (*for a particular ordering of* $\Delta$) *is called the* highest weight *of* $V$).

(e) *Two irreducible* $\mathfrak{g}$*-modules are isomorphic if and only if they have the same highest weight.*

(f) *Given any linear function* $\Lambda \in \mathfrak{h}^*$ *such that* $\Lambda(H_{\alpha_i}) \geq 0$ $(1 \leq i \leq n)$ *there exists a finite dimensional irreducible* $\mathfrak{g}$*-module whose highest weight is* $\Lambda$.

For a proof, see Jacobson (1962) or Serre (1966).

DEFINITION. The **Z**-module generated by all weights of all $\mathfrak{g}$-modules is denoted by $\Gamma_{\text{univ}}$ (for universal); the **Z**-module generated by the weights of the adjoint representation, by $\Gamma_{\text{ad}}$; the **Z**-module generated by the weights of a representation $\rho$ (or a $\mathfrak{g}$-module $V$) is denoted by $\Gamma_\rho$ or $\Gamma_V$. We shall sometimes call $\Gamma_\rho$ a *lattice* of weights.

LEMMA (4.9).

(a) $\Gamma_{\text{univ}}$ *is a* **Z***-form of* $\mathfrak{h}^*$, *and has a basis consisting of the weights* $\lambda_j$; $\lambda_j(H_{\alpha_i}) = \delta_{ij}$.

(b) $\Gamma_{\text{univ}} \supset \Gamma_{\text{ad}}$, *and the Cartan matrix* $(A_{ij})$ *is a relation matrix*† *for the pair of* **Z***-modules* $(\Gamma_{\text{univ}}, \Gamma_{\text{ad}})$. *In particular,* $[\Gamma_{\text{univ}}; \Gamma_{\text{ad}}] = |\det(A_{ij})| < \infty$.

(c) *Let* $V$ *be a faithful* $\mathfrak{g}$*-module, then*

$$\Gamma_{\text{univ}} > \Gamma_V > \Gamma_{\text{ad}}.$$

*Proof.* (a) is obvious by (f) of Prop. (4.8).

For (b), evidently we have $\Gamma_{\text{univ}} \supset \Gamma_{\text{ad}}$, and if $\alpha_i \in \Pi$, $\alpha_i = \sum c_{ij}\lambda_j$, then $\alpha_i(H_{\alpha_j}) = c_{ij} = A_{ij}$.

For (c), since $V$ is faithful, $X_\alpha V \neq 0$ for each $\alpha \in \Delta$. Then for some $\mu \in P(V)$, $X_\alpha V_\mu \neq 0$, so that $\mu + \alpha \in P(V)$. Then $\alpha \in \Gamma_V$, and the lemma is proved.

DEFINITION (4.10). An *admissible* **Z***-form* of a $\mathfrak{g}$-module $V$ is a **Z**-form $M$ such that $U_{\mathbf{Z}} \cdot M \subseteq M$. (Since the elements $X_{\alpha_j}/j!$ generate $U_{\mathbf{Z}}$, it is sufficient to require that $(X_{\alpha_j}/j!)M \subset M$ for all $\alpha$ and $j$.

PROPOSITION (4.11). *Let* $V$ *be a finite dimensional* $\mathfrak{g}$*-module.*

(a) $V$ *has an admissible* **Z***-form* $M$. *If* $V$ *is irreducible, and* $v_0$ *is a vector in* $V_\Lambda$, ($\Lambda$ *the highest weight*), *then* $U_{\mathbf{Z}} \cdot v_0$ *is an admissible* **Z***-form of* $V$.

(b) *If* $M$ *is an admissible* **Z***-form of* $V$, *then*

$$M = \sum_{\mu \in P(V)} (V_\mu \cap M).$$

† A relation matrix of a pair of **Z**-modules with finite bases $(G, H)$ with $H \subseteq G$, is a matrix expressing a basis of $H$ in terms of a basis of $G$.

We shall sketch the proof of (a). By complete reducibility, we may assume that $V$ is irreducible with highest weight $\Lambda$. Let $v_0 \in V_\Lambda$. Since $\Lambda(H_\alpha) \in \mathbf{Z}$, we have, for all $\alpha$,

$$\binom{H_\alpha}{n} v_0 = \binom{\Lambda(H_\alpha)}{n} v_0 \in \mathbf{Z} v_0,$$

and it follows that $U_\mathbf{Z} v_0 = U(n^-)_\mathbf{Z} v_0$. From Prop. (4.8) (b), we have $X_\alpha^m V = 0$ for some sufficiently large $m$, and hence $U(n^-)_\mathbf{Z} v_0$ is finitely generated, and is an admissible $\mathbf{Z}$-form, provided we can show it is a $\mathbf{Z}$-form of $V$. We have to prove that if a set $\{v_i\} \subseteq U_\mathbf{Z} v_0$ is $\mathbf{Z}$-independent, then it is $F$-independent. For this it suffices to show that if

$$\sum_{1 \le i \le r} c_i v_i = 0 \quad (c_i \in F \backslash \{0\}), \qquad *$$

then $\sum m_i c_i = 0$ for some $m_i \in \mathbf{Z}$, not all zero. For in that case, we have, assuming $m_1 \neq 0$,

$$\sum_{i \ge 2} c_i (m_1 v_i - m_i v_1) = 0,$$

with $m_1 v_i - m_i v_1$ independent over $\mathbf{Z}$, and we are through by induction. To produce the $m_i$, we let $(v)_0$ be the coefficient of $v_0$ in the expression of $v$ in terms of $v_0$ and vectors belonging to other weights. From * we have

$$\sum c_i (x v_i)_0 = 0$$

for all $x \in U$. By irreducibility, $(x v_i)_0 \neq 0$ for some $x \in U_\mathbf{Z}$, and it suffices to show that $(x v_i)_0 \in \mathbf{Z}$ for $x \in U_\mathbf{Z}$. Since $v_i \in U_\mathbf{Z} v_0$, it suffices to show that $(x v_0)_0 \in \mathbf{Z}$ for $x \in U_\mathbf{Z}$, and taking $x$ in the form $e_{-Q} h_P e_{Q'}$, the result is obvious from what has been shown.

For the proof of (ii), see Steinberg (1967).

## 5. Basic Properties of Chevalley Groups

This section contains the definition of the Chevalley groups corresponding to $\mathbf{Z}$-forms of arbitrary $\mathfrak{g}$-modules. The groups originally defined by Chevalley (1955) correspond to a $\mathbf{Z}$-form of the module affording the adjoint representation of $\mathfrak{g}$. The point of the generalization given below is that from one construction we obtain not only the simple groups, coming from the adjoint representation, but all possible central extensions of the simple groups as well. In particular, the use of the universal central extension of a Chevalley group is practically essential for computation of the orders of the finite Chevalley groups, for the construction of automorphisms, for the study of centralizers of involutions, and for the representation theory.

### 5.1. The Chevalley Groups

Let $M$ be an admissible $\mathbf{Z}$-form of a faithful $\mathfrak{g}$-module $V$, as in Section 4.3. Let $k$ be an arbitrary field. Let $M_\mu = V_\mu \cap M$, $M^k = k \otimes_\mathbf{Z} M$,

$M_\mu^k = k \otimes_{\mathbf{Z}} M_\mu$, $(\mu \in P(V))$. Then $M^k$ is a vector space over $k$, and by Prop. (4.11), we have the direct decomposition

$$M^k = \sum_{\mu \in P(V)} M_\mu^k \tag{5.1}$$

Let $\alpha \in \Delta$, and let $\rho$ be the representation afforded by $M$. Let $\{m_1, \ldots, m_d\}$ be a $\mathbf{Z}$-basis of $M$. The endomorphism $\rho(X_\alpha)$ is nilpotent on $M$, by (4.8) (b). The important point, on which the existence of the groups depends, is that because $M$ is an admissible $\mathbf{Z}$-form, the matrix $e_\alpha(t)$ of

$$\exp t\rho(X_\alpha) = \sum_0^\infty \frac{t^i \rho(X_\alpha)^i}{i!}$$

with respect to the basis $\{m_1, \ldots, m_d\}$ belongs to $GL_d(Z[t])$, where $t$ is an indeterminate over $F$. Moreover,

$$e_\alpha(t+t') = e_\alpha(t) e_\alpha(t'), \tag{5.2}$$

where $t, t'$ are indeterminates.

Now let $u \in k$; then the matrix $e_\alpha(u)$ defines a $k$-linear transformation of $M^k$, which we denote by $x_\alpha(u)$. Moreover, from (5.2) we have for $u, u' \in k$,

$$x_\alpha(u+u') = x_\alpha(u) x_\alpha(u'), \tag{5.3}$$

and hence $\{x_\alpha(u) : u \in k\}$ is a subgroup of $GL(M^k)$ which we shall denote by $\mathfrak{X}_\alpha$.

Definition (5.4). The *Chevalley group* $G = G_{V,k}$ is the subgroup of $GL(M^k)$ generated by all subgroups $\mathfrak{X}_\alpha$, $\alpha \in \Delta$.

Apparently the Chevalley group depends on the choice of the admissible $\mathbf{Z}$-form $M$. It will be seen from the discussion to follow that in fact $G$ depends only on $k$ and the lattice of weights $\Gamma_V$ of the $\mathfrak{g}$-module $V$.

Our approach to the properties of Chevalley groups is taken from Steinberg (1967), and mainly uses a study of the relations satisfied by $\{x_\alpha(t) : \alpha \in \Delta, t \in \Delta\}$, together with calculations with exponentials, involving the Lie algebra $\mathfrak{g}_{\mathbf{Z}}$ acting on $M$, and the matrices $e_\alpha(t)$.

This procedure does not require the fact, proved by Chevalley in case $\Gamma_V = \Gamma_{\text{ad}}$, that for each $\alpha \in \Delta$, there exists a homomorphism $\mu_\alpha : SL_2(k) \to G$ such that for $t \in k$

$$\mu_\alpha \begin{pmatrix} 1 & t \\ 0 & 1 \end{pmatrix} = x_\alpha(t), \qquad \mu_\alpha \begin{pmatrix} 1 & 0 \\ t & 1 \end{pmatrix} = x_{-\alpha}(t).$$

The existence of these homomorphisms can be proved from the discussion below (see Steinberg, 1967).

Theorem (5.5). *Let $\alpha, \beta \in \Delta$, and suppose that $\alpha+\beta \in \Delta$. Then there exist integers $\{c_{ij}\}$ such that in the ring of formal series $U[[t,u]]$ (where $U = U(\mathfrak{g})$),* *we have the commutator formula*

$$(\exp tX_\alpha, \exp uX_\beta) = \prod_{i, j \geq 1} \exp (c_{ij} t^i u^j X_{i\alpha + j\beta}),$$

*the product being taken over some ordering of the roots of the form* $\{i\alpha + j\beta\}_{i, j \geq 1}$ *starting with* $\alpha + \beta$. *Moreover,* $c_{11} = n_{\alpha\beta}$, *and the* $\{c_{ij}\}$ *depend only on the ordering of the roots* $\{i\alpha + j\beta\}_{i, j \geq 1}$ *and not on t or u.*

We give a sketch of a proof of this result. We first note that in any associative algebra $A$, in which $\{\exp a : a \in A\}$ is defined and obeys the usual rules, for $a \in A$ and

$$d_a = L_a - R_a = \operatorname{ad}_A a,$$

we have

$$\exp d_a = L_{\exp a} R_{\exp(-a)},$$

i.e., for all $x \in A$,

$$\exp \operatorname{ad}_A a \cdot x = (\exp a) x \exp(-a). \tag{5.6}$$

The result is immediate since $L_x$ and $R_y$ commute, so that $\exp(L_a - R_a) = \exp L_a \exp(-R_a)$.

Now, letting $(A, B) = ABA^{-1}B^{-1}$, we define $f(t, u) = (\exp tX_\alpha, \exp uX_\beta) \prod \exp(-c_{ij} t^i u^j X_{i\alpha + j\beta})$, the product being taken in the reverse order from that in the statement of the Theorem, and with $\{c_{ij}\}$ arbitrary elements of $F$.

The idea of the proof is to differentiate, using the product rule, and obtain

$$\begin{aligned} t\frac{d}{dt} f(t, u) = {} & \\ & tX_\alpha f(t, u) + (\exp tX_\alpha)(\exp uX_\beta)(-tX_\alpha \exp(-tX_\alpha), (\exp(-uX_\beta)) \\ & + \sum_{k,l} (\exp tX_\alpha, \exp uX_\beta) \prod_{i\alpha + j\beta > k\alpha + l\beta} \exp(-c_{ij} t^i u^j X_{i\alpha + j\beta}) \\ & (-c_{kl} k t^k u^l X_{k\alpha + l\beta}) \exp(-c_{kl} t^k u^l X_{k\alpha + l\beta}) \prod_{i\alpha + j\beta < k\alpha + l\beta} \exp(-c_{ij} t^i u^j X_{i\alpha + j\beta}). \end{aligned}$$

From (5.6), we have formulas such as

$$(\exp X_\gamma) X_\delta = (\exp \operatorname{ad}_U X_\gamma)(X_\delta) \exp X_\gamma,$$

and

$$(\exp \operatorname{ad}_U X_\gamma)(X_\delta) = X_\delta + n_{\gamma\delta} X_{\gamma + \delta} + \ldots.$$

It follows from these formulas that the new terms arising in the differentiation can be brought to one side, yielding the equation

$$t\frac{d}{dt} f(t, u) = Af(t, u)$$

where $A$ is a polynomial expression involving the $\{c_{ij}\}$, $t^i$, $u^j$ and $X_{i\alpha + j\beta}$. It is then possible to choose the $\{c_{ij}\}$ inductively to make $A = 0$; it will follow that the $c_{ij} \in \mathbf{Z}$, $c_{11} = n_{\alpha\beta}$, and that

$$t \frac{d}{dt} f(t, u) = 0.$$

Then $f(t, u) = f(0, u) = 1$ as required. (For further details, see Steinberg, 1967.)

COROLLARY (5.7) (Chevalley, 1955). *Let G be a Chevalley group associated with a field k. Then for $\alpha, \beta \in \Delta$, for which $\alpha+\beta \in \Delta$, there exist integers $\{c_{ij}\}_{i,j \geq 1}$ (with $c_{11} = n_{\alpha\beta}$, depending only on some ordering of the roots $\{i\alpha+j\beta : i, j \geq 1\}$) such that*

$$(x_\alpha(t), x_\beta(u)) = \prod x_{i\alpha+j\beta}(c_{ij}t^i u^j),$$

*for all $t, u \in k$.*

For example, if $\mathfrak{g} = sl_n(F)$, the Lie algebra of matrices of trace zero, and $V$ is the module on which the transformations in $\mathfrak{g}$ are defined, then for a suitable choice of $\mathfrak{h}$, $\Delta$, etc., the elements of $\mathfrak{X}_\alpha$ are given by

$$x_{ij}(t) = 1 + tX_{ij} \quad (i \neq j)$$

where $X_{ij}$ is the matrix with a one in the $(i, j)$ position and zero's elsewhere. The relations in (5.7) are then of the form

$$(x_{ij}(t), x_{kl}(u)) = \begin{cases} 1 & \text{or} \\ x_{il}(\pm tu). \end{cases}$$

These relations had already been shown by Dickson to give a presentation of $SL_n(k)$, for $k$ a finite field.

PROPOSITION (5.8). *Let $\mathfrak{U}$ be the subgroup of G generated by all $\mathfrak{X}_\alpha$ ($\alpha \in \Delta^+$). For each $i \geq 1$, let $\mathfrak{U}_i$ be the subgroup of U generated by all $\mathfrak{X}_\alpha$ with $ht(\alpha) \geq i$.*†

(i) $(\mathfrak{U}, \mathfrak{U}_i) \subseteq \mathfrak{U}_{i+1}$ *for all i (and hence the group $\mathfrak{U}$ is nilpotent).*

(ii) $\mathfrak{U}$ *is a subgroup consisting of unipotent transformations on $M^k$.*

(iii) *Every element of $\mathfrak{U}$ can be expressed in the form*

$$x = \prod_{\alpha \in \Delta^+} x_\alpha(t_\alpha)$$

*for some ordering $\ll$ of $\Delta^+$ such that $\alpha \ll \beta$ implies $ht(\alpha) \leq ht(\beta)$, with coefficients $t_\alpha$ uniquely determined by x.*

(iv) *For each $w \in W$, let $\mathfrak{U}_w^+ = \langle \mathfrak{X}_\alpha : \alpha \in \Delta^+, w(\alpha) \in \Delta^+ \rangle$, $\mathfrak{U}_w^- = \langle \mathfrak{X}_\alpha : \alpha \in \Delta^+, w(\alpha) \in \Delta^- \rangle$. Then $\mathfrak{U} = \mathfrak{U}_w^+ \mathfrak{U}_w^-$, and $\mathfrak{U}_w^+ \cap \mathfrak{U}_w^- = \{1\}$.*

The proof of part (i) is immediate from Corollary (5.7). For part (ii), use the fact that $M^k = \Sigma M_\mu^k$ (by (5.1)) so that a $k$-basis for $M^k$ consisting of vectors from the subspaces $M_\mu^k$ can be chosen. Order the basis in such a way

† $ht(\alpha)$, the *height* of $\alpha$, is defined by

$$ht(\alpha) = \Sigma m_i, \quad \text{where } \alpha = \sum_{\alpha_i \in \Pi} m_i \alpha_i \quad (\alpha \in \Delta^+).$$

that $v \in M_\mu^k$ always precedes $v' \in M_{\mu'}^k$ in case $\mu' - \mu$ is a sum of positive roots. The matrices representing all $x \in \mathfrak{U}$ will be subdiagonal with respect to such a basis, and hence unipotent.

For (iii), the existence of such a factorization of $x$ follows from part (i) and Corollary (5.7). For the uniqueness, let $x = \Pi x_\alpha(t_\alpha)$, and consider a fixed $\alpha_0 \in \Delta^+$. Since $\rho$ is a faithful representation, $\rho(X_{\alpha_0}) \neq 0$, and there exists $v \in M_\mu^k$ for some $\mu$ with $\rho(X_{\alpha_0})v \neq 0$. Expanding the elements $x_\alpha(t) = 1 + t\rho(X_\alpha) + \ldots$, we see that

$$xv = v + t_{\alpha_0}\rho(X_{\alpha_0})v + y,$$

where $y$ is a sum of weight vectors belonging to weights different from $\mu$ and $\mu + \alpha_0$. Since vectors coming from different "weight spaces" $M_\mu^k$ are linearly independent, $t_{\alpha_0}$ is uniquely determined by $x$.

Finally, one proves $\mathfrak{U} = \mathfrak{U}_w^+ \mathfrak{U}_w^-$ in (iv) by proving inductively that $\mathfrak{U}_i = (\mathfrak{U}_w^+)_i (\mathfrak{U}_w^-)_i$, where $(\mathfrak{U}_w^+)_i = \mathfrak{U}_w^+ \cap \mathfrak{U}_i$ etc. The fact that $\mathfrak{U}_w^+ \cap \mathfrak{U}_w^- = \{1\}$ follows from the fact that the argument for the uniqueness of the coefficients $t_\alpha$ in $\Pi x_\alpha(t_\alpha)$ given above holds for an arbitrary ordering of the roots in $\Delta^+$.

COROLLARY (5.9). *If $k$ is a finite field of characteristic $p$, $\mathfrak{U}$ is a $p$-group.*

DEFINITION (5.10). For $\alpha \in \Delta$, $t \in k^*$ put

$$w_\alpha(t) = x_\alpha(t)x_{-\alpha}(-t^{-1})x_\alpha(t), \qquad h_\alpha(t) = w_\alpha(t)w_\alpha(1)^{-1}.$$

(Note that the form of $w_\alpha(t)$ is suggested by the $\theta_i$ in Lemma (4.1).)

PROPOSITION (5.11). *Let $\alpha, \beta \in \Delta$.*

(a) $w_\alpha(t)\rho(X_\beta)w_\alpha(t)^{-1} = ct^{-\beta(H_\alpha)}\rho(X_{r_\alpha(\beta)})$,† *with* $c = \pm 1$. *Moreover, $c = c(\alpha, \beta)$ is independent of $t$, $k$ and the representation, and $c(\alpha, \beta) = c(\alpha, -\beta)$.*

(b) *Let $v \in M_\mu^k$. Then there exists $v' \in M_{r_\alpha(\mu)}^k$, independent of $t$, such that*
$$w_\alpha(t)v = t^{-\mu(H_\alpha)}v'.$$

(c) *$h_\alpha(t)$ acts diagonally on $M_\mu^k$ as multiplication by $t^{\mu(H_\alpha)}$.*

*Proof.* We first prove that $w_\alpha(t)\rho(H)w_\alpha(t)^{-1} = \rho(H) - \alpha(H)\rho(H_\alpha)$ for all $H \in \mathfrak{h}^k$. If $\alpha(H) = 0$ then $\rho(H)$ commutes with $x_\alpha(t)$ and $x_{-\alpha}(t)$ and the result follows. It is therefore sufficient to prove the result for $H = H_\alpha$, and in that case one simply computes the whole thing out, using $H_\alpha = [x_\alpha x_{-\alpha}]$.

We next prove (b), first for $M$, with $t$ an indeterminate, and then passing to $M^k$. We have, for $v \in M_\mu$,

$$w_\alpha(t)v = \sum_{-\infty}^{\infty} t^i v_i \quad (v_i \in M_{\mu + i\alpha}).$$

† We continue to use the notation $\rho(X_\beta)$ for the endomorphism $\rho(X_\beta) \otimes 1$ of $M \otimes_{\mathbf{Z}} k$.

Using the computation above, we have

$$w_\alpha(t)v \in M_{r_\alpha(\mu)},$$

and hence the only non-zero term in the sum occurs when $i = -\mu(H_\alpha)$.

Applying part (b) to the adjoint representation of $\mathfrak{g}_{\mathbf{Z}}$ we obtain

$$w_\alpha(t)X_\beta w_\alpha(t)^{-1} = ct^{-\beta(H_\alpha)}X_{r_\alpha(\beta)},$$

where $c \in \mathbf{Z}$, and is independent of $t$. Since $w_\alpha(t)$ is an automorphism of $\mathfrak{g}_{\mathbf{Z}}$, upon setting $t = 1$, we get that $cX_{r_\alpha(\beta)}$ is primitive in $\mathfrak{g}_{\mathbf{Z}}$, and hence $c = \pm 1$. Applying $\rho$ we obtain part (a).

We next compute

$$\begin{aligned} H_{r_\alpha(\beta)} &= w_\alpha(1)H_\beta w_\alpha(1)^{-1} = [w_\alpha(1)X_\beta w_\alpha(1)^{-1}, w_\alpha(1)X_{-\beta}w_\alpha(1)^{-1}] \\ &= c(\alpha, \beta)c(\alpha, -\beta)H_{r_\alpha}(\beta), \end{aligned}$$

and obtain

$$c(\alpha, \beta)c(\alpha, -\beta) = 1.$$

Finally, part (c) follows from (b).

By exponentiating the expressions in Prop. (5.11) we obtain the following result.

PROPOSITION (5.12).

(a) $w_\alpha(1)h_\beta(t)w_\alpha(1)^{-1} = h_{r_\alpha(\beta)}(t)$.
(b) $w_\alpha(1)x_\beta(t)w_\alpha(1)^{-1} = x_{r_\alpha(\beta)}(ct)$, *$c$ as in Prop.* (5.11) (a).
(c) $h_\alpha(t)x_\beta(u)h_\alpha(t)^{-1} = x_\beta(t^{\beta(H_\alpha)}u)$.

PROPOSITION (5.13). *Let $H$ be the subgroup of $G$ generated by all $h_\alpha(t)$, and $B = \langle \mathfrak{U}, H \rangle$. Then*

(i) *$\mathfrak{U} \lhd B$, and $B = \mathfrak{U} \cdot H$.*
(ii) *$\mathfrak{U} \cap H = \{1\}$.*

*Proof.* (i) follows from the relations (5.12) (c), while (ii) follows since $\mathfrak{U}$ acts on $M^k$ as a group of unipotent transformations, and $H$ acts as a group of diagonal ones.

PROPOSITION (5.14). *Let $N = \langle w_\alpha(t) : \alpha \in \Delta, t \in k^* \rangle$ and $H = \langle h_\alpha(t) : \alpha \in \Delta, t \in k^* \rangle$. Then $H \lhd N$, and there exists an isomorphism $\phi : W \to N/H$ between the Weyl group of $\Delta$ and $N/H$ such that $\phi(r_\alpha) = w_\alpha(t)H$ for all $\alpha \in \Delta$. Moreover, $\phi(w) = nH$ implies $n\mathfrak{X}_\alpha n^{-1} = \mathfrak{X}_{w(\alpha)}$ ($w \in W$, $\alpha \in \Delta$).*

*Proof.* By (5.12) (a), $H \lhd N$. By the definitions of $h_\alpha(t)$, the coset $w_\alpha(t)H$ is independent of $t$, for all $\alpha$, and letting $\tilde{w}_\alpha$ denote the coset $\tilde{w}_\alpha(t)H$, it is easily checked using the relations already proved that

$$\tilde{w}_\alpha^2 = 1, \qquad \tilde{w}_\alpha \tilde{w}_\beta \tilde{w}_\alpha = \tilde{w}_{r_\alpha(\beta)} \qquad *$$

for all $\alpha$, $\beta \in \Delta$. It follows from Prop. (3.6) that the relations * are a set of defining relations for the Weyl group, and hence there exists a surjective homomorphism $\phi : W \to N/H$ such that $\phi(r_\alpha) = \tilde{w}_\alpha$ for all $\alpha$. Now let $w \in \ker \phi$. Writing $w = r_{\alpha_1}\ r_{\alpha_2} \ldots$, we have $w_{\alpha_1}(1)w_{\alpha_2}(1)\ldots = h \in H$. Conjugating by this element, we obtain $\mathfrak{X}_\alpha = \mathfrak{X}_{w\alpha}$ for all $\alpha \in \Delta$. The proof is completed by showing that if $\alpha,\beta$ are distinct in $\Delta$ then $\mathfrak{X}_\alpha \neq \mathfrak{X}_\beta$, and this is almost immediate from what has been shown. The last statement follows from Prop. (5.12) (b).

The next result is the basis for Tits' axiomatization of groups of Lie type, to be discussed in Section 6. It is useful to define $\mathfrak{U}^- = \langle \mathfrak{X}_\alpha : \alpha \in \Delta^- \rangle$, and to adopt the

CONVENTION. Let $\phi(w) = nH$ for $n \in N$; then we shall write $Bw$, $wB$, $BwB$ for $Bn$, $nB$ and $BnB$ respectively (this is justifiable since $H \subseteq B$).

PROPOSITION (5.15).

(i) *Let* $\alpha \in \Pi$; *then* $B \cup Br_\alpha B$ *is a group.*

(ii) *For* $\alpha \in \Pi$, $w \in W$,

$$r_\alpha Bw \subseteq BwB \cup Br_\alpha wB$$

(iii) $B \cap \mathfrak{U}^- = \{1\}$.

(iv) $B \cap N = H$.

*Proof.* Letting $G_\alpha = B \cup Br_\alpha B$, we have $G_\alpha = G_\alpha^{-1}$, and $G_\alpha$ is a group provided that $G_\alpha G_\alpha \subseteq G_\alpha$. For this, in turn, it is sufficient to prove that

$$r_\alpha B r_\alpha \subseteq B \cup Br_\alpha B,$$

(which is a special case of (ii)).

From the definition of $w_\alpha(t)$, it follows that $\mathfrak{X}_{-\alpha} \subseteq G_\alpha$. Then $B = \mathfrak{X}_\alpha \mathfrak{U}_{r_\alpha}^- H$ and

$$r_\alpha B r_\alpha = r_\alpha \mathfrak{X}_\alpha r_\alpha^{-1} r_\alpha \Big( \prod_{\beta \in \Delta^+ \setminus \{\alpha\}} \mathfrak{X}_\beta \Big) r_\alpha^{-1} H \subseteq G_\alpha B = G_\alpha$$

since $r_\alpha \mathfrak{X}_\beta r_\alpha^{-1} = \mathfrak{X}_{r_\alpha(\beta)} \subseteq \mathfrak{U}$ if $\beta > 0$, $\beta \neq \alpha$, by Prop. (3.5) (a).

In order to prove (ii), suppose first that $\alpha \in \Delta^+_{w^{-1}}$ (i.e., $w^{-1}(\alpha) > 0$). Then

$$r_\alpha Bw = r_\alpha \Big( \prod_{\beta \in \Delta^+ \setminus (\alpha)} \mathfrak{X}_\beta \Big) r_\alpha^{-1} r_\alpha w w^{-1} \mathfrak{X}_\alpha w H \subseteq Br_\alpha wB.$$

In case $w^{-1}(\alpha) < 0$, $w^{-1} r_\alpha(\alpha) > 0$, and

$$r_\alpha Bw = r_\alpha B r_\alpha^2 w \subseteq Br_\alpha w \cup Br_\alpha B r_\alpha w \subseteq Br_\alpha wB \cup BwB$$

by (i) and the preceding argument.

The proof of (iii) is immediate because, for the basis selected in the proof of part (ii) of Prop. (5.8), $B$ is represented by lower triangular matrices, and $\mathfrak{U}^-$ by unit upper triangular ones.

In order to prove (iv), let $w \in W$ ($w \neq 1$) and let $\phi(w) = nH$. Then for some $\alpha > 0$, $w(\alpha) < 0$, and if $w \in N \cap B$, we would have

$$n\mathfrak{X}_\alpha n^{-1} \subseteq B \cap \mathfrak{U}^-$$

which is impossible by (iii).

We conclude this section with the following important theorem.

THEOREM (5.16). (Bruhat decomposition of a Chevalley group). *We have*

$$G = \bigcup_{w \in W} BwB = \bigcup_{w \in W} Bw\mathfrak{U}_w^-$$

*and the mapping $w \to BwB$ is a bijection between $W$ and the double cosets of $B$ in $G$. If $\{n_w : w \in W\}$ be a transversal of $W$ in $N$, then expressions of elements of $G$ in the form*

$$g = uhn_w x,$$

*with $u \in \mathfrak{U}$, $h \in H$, $n_w \in wH$, $x \in \mathfrak{U}_w^-$, are unique.*

*Proof.* The existence of the decomposition $G = \bigcup BwB = \bigcup Bw\mathfrak{U}_w^-$ follows from Prop. (5.15) and Prop. (5.8) (iv). The proof that $BwB = Bw'B$ implies $w = w'$ is done by induction on $l(w)$, using as a start the fact that $B = Bw'B$ implies $w' = 1$ by part (iv) of Prop. (5.15) (the case $l(w) = 0$).

Finally, suppose that

$$unh_w x = u_1 h_1 n_{w_1} x_1,$$

with $u$, $u_1 \in \mathfrak{U}$, $h$, $h_1 \in H$, and $x \in \mathfrak{U}_w^-$, $x_1 \in \mathfrak{U}_{w_1}^-$. Then $n_w = n_{w_1}$ by the first part of the theorem. We then obtain

$$h_1^{-1} u_1^{-1} h u = n_{w_1} x_1 x^{-1} n_w^{-1} \in B \cap \mathfrak{U}^-.$$

Since $B \cap \mathfrak{U}^- = \{1\}$ by (iv) of Prop. (5.15), we have $x_1 = x$. The uniqueness of $u$ and $h$ follow since $\mathfrak{U} \cap H = \{1\}$.

## 5.2. THE STRUCTURE OF $H$

The definition of Chevalley groups corresponding to arbitrary $\mathfrak{g}$-modules now yields precise information about the role played by the subgroup $H$.

PROPOSITION (5.17). *Let $G$ be a Chevalley group, corresponding to the faithful $\mathfrak{g}$-module $V$, with lattice of weights $\Gamma_V$.*

(i) *For each $\alpha$, $h_\alpha(tt') = h_\alpha(t)h_\alpha(t')$, for $t$, $t' \in k^*$.*

(ii) *$H$ is an abelian group, generated by the elements $\{h_{\alpha_i}(t) : \alpha_i \in \Pi, t \in k^*\}$.*

(iii) $\prod_{i=1}^n h_{\alpha_i}(t_i) = 1$ *if and only i,*

$$\prod_{i=1}^n t_i^{\mu(H_{\alpha_i})} = 1$$

*for all $\mu \in \Gamma_V$.*

(iv) $Z(G) = \left\{ \prod_{i=1}^n h_i(t_i) : \prod_{i=1}^n t_i^{\beta(H_i)} = 1, \quad \beta \in \Gamma_{\text{ad}} \right\}.$

*Proof.* The first three results all follow from the fact that by Prop. (5.11) (c), $h_\alpha(t)$ acts on $M_\mu^k$ as multiplication by $t^{\mu(H_\alpha)}$. For example, (ii) follows from the formula

$$t^{\mu(H_\alpha)} = t^{\mu(\Sigma m_i H_{\alpha_i})} = t^{\Sigma m_i \mu(H_{\alpha_i})}$$

if $H_\alpha = \sum m_i H_{\alpha_i}$. For (iv), we use first the fact (left as an exercise) that $Z(G) \subseteq H$. We then observe that $\Pi h_i(t_i)$ commutes with $x_\beta(u)$ if and only if $\Pi\, t_i^{\beta(H_{\alpha_i})} = 1$ by Prop. (5.12) (c).

COROLLARY (5.18).

(i) *If* $\Gamma_V = \Gamma_{\text{ad}}$, *then* $Z(G) = 1$.

(ii) *If* $\Gamma_V = \Gamma_{\text{univ}}$, *then every* $h \in H$ *can be expressed in the form*

$$h = \prod_{i=1}^{n} h_{\alpha_i}(t_i)$$

*with uniquely determined* $t_i \in k^*$. *In particular, if* $k$ *is finite,* $|H| = (|k|-1)^n$ *(for* $\Gamma_V = \Gamma_{\text{univ}}$*).*

COROLLARY (5.19). *Let* $k$ *be a finite field of* $q$ *elements, of characteristic* $p$. *Then*

(i) $|G| = q^N |H| \left(\sum_{w \in W} q^{l(w)}\right)$.

*where* $N = |\Delta^+|$.

(ii) *In particular, if* $\Gamma_V = \Gamma_{\text{univ}}$, *then*

$$|G| = q^N (q-1)^n \left(\sum_{w \in W} q^{l(w)}\right).$$

(iii) $H$ *is an abelian* $p'$*-subgroup of* $G$, *and* $\mathfrak{U}$ *is a* $p$*-Sylow subgroup.*

These results all follow from the Bruhat decomposition (Theorem (5.16)) combined with Prop. (5.17). The additive formula for $[G:B]$ also requires the fact that $|\Delta_w^-| = l(w)$, by Prop. (3.5) (b).

THEOREM (5.20) (Steinberg, 1967). *Let* $G_V$ *and* $G_{V'}$ *be Chevalley groups constructed from the same Lie algebra* $\mathfrak{g}$ *and field* $k$. *Suppose* $\Gamma_{V'} \supseteq \Gamma_V$. *Then there exists a homomorphism* $\phi : G_{V'} \to G_V$ *with central kernel such that* $\phi(x'_\alpha(t)) = x_\alpha(t)$ *for all* $\alpha \in \Delta$ *and* $t \in k$ *(we are letting* $\{x'_\alpha(t)\}$ *and* $\{x_\alpha(t)\}$ *denote the generators of* $G_{V'}$ *and* $G_V$ *respectively). In particular, if* $\Gamma_{V'} = \Gamma_V$, *then* $G_{V'} \cong G_V$.

*Proof.* This result is an important dividend of the approach we have taken to the Chevalley groups. Briefly, the argument goes as follows. One first proves that the relations

$$\left.\begin{aligned} x_\alpha(t+t') &= x_\alpha(t)x_\alpha(t') \\ (x_\alpha(t),\, x_\beta(u)) &= \prod x_{i\alpha+j\beta}(c_{ij}t^iu^j) \\ w_\alpha(1)h_\beta(t)w_\alpha(1)^{-1} &= h_{r_\alpha(\beta)}(t) \\ w_\alpha(1)x_\beta(t)w_\alpha(1)^{-1} &= x_{r_\alpha(\beta)}(ct) \\ h_\alpha(t)x_\beta(u)h_\alpha(t)^{-1} &= x_\beta(t^{\beta(H_\alpha)}u) \end{aligned}\right\} \quad (5.21)$$

together with a set of relations for $H$, form a set of relations for $G_V$. This is shown by letting $G_V^*$ be the group with generators $\{x_\alpha(t) : \alpha \in \Delta,\ t \in k\}$ and relations (5.21), with $w_\alpha(t)$ and $h_\alpha(t)$ defined as usual. Then from the preceding discussion, using the homomorphism $\pi : G_V^* \to G_V$, one gets a Bruhat decomposition in $G_V^*$, with uniqueness of expression as in Theorem (5.16). It follows that $\ker \pi \subseteq H^*$, and since the relations for $H$ are included in the defining relations of $G_V^*$, $\pi|H^*$ is an isomorphism.

The next step is to observe that if $\Gamma_{V'} \supseteq \Gamma_V$, then the relations for $H$ contain the relations for $H'$, by Prop. (5.17). It follows that a homomorphism $\phi : G_{V'} \to G_V$ exists, such that $\phi(x_\alpha'(t)) = x_\alpha(t)$ for all $t$ and $\alpha$.

Now let $x \in \ker \phi$, and write $x = uhn_wu_1$ according to the Bruhat decomposition. Then

$$\phi(x) = \phi(u)\phi(h)\phi(n_w)\phi(u_1) = 1$$

implies that $w = 1$, and $x = h$. Applying $\phi$ to $hx_\alpha(t)h^{-1}$, we see that $h \in Z(G_{V'})$. This completes the proof.

The preceding result settles the question of the dependence of a Chevalley group on the module $V$, admissible $\mathbf{Z}$-form, etc., and shows that for a given field $k$ and root system $\Delta$, there are only finitely many Chevalley groups, up to isomorphism.

Definition. We denote by $G_U$ and $G_A$ Chevalley groups for which the lattices of weights are $\Gamma_{\text{univ}}$ or $\Gamma_{\text{ad}}$ respectively. $G_U$ is called a *universal Chevalley group*, and $G_A$ an *adjoint Chevalley group*.

Corollary (5.22). *Let $G_U$ be a universal Chevalley group, and $G_A$ an adjoint Chevalley group. Then*

$$G_U/Z(G_U) \cong G_A.$$

*Moreover,* $Z(G_U) \cong \operatorname{Hom}(\Gamma_{\text{univ}}/\Gamma_{\text{ad}}, k^*)$.

Some facts (from Steinberg, 1967) about the Chevalley groups corresponding to particular root systems are collected below for the case of a finite field of $q$ elements.

| *Type of* $\Delta$ | $\Gamma_{\text{univ}}/\Gamma_{\text{ad}}$ | $\lvert Z(G_U)\rvert$ | $G_A$ | $G_U$ |
|---|---|---|---|---|
| $A_n$ | $Z_{n+1}$ | $(n+1, q-1)$ | $PSL_{n+1}$ | $SL_{n+1}$ |
| $B_n$ | $Z_2$ | $(2, q-1)$ | $PSO_{2n+1} = SO_{2n+1}$ | $\text{Spin}_{2n+1}$ |
| $C_n$ | $Z_2$ | $(2, q-1)$ | $PSp_{2n}$ | $Sp_{2n}$ |
| $D_{2n+1}$ | $Z_4$ | $(4, q-1)$ | $PSO_{4n+2}$ | $\text{Spin}_{4n+2}$ |
| $D_{2n}$ | $Z_2 \times Z_2$ | $(4, q-1)$ | $PSO_{4n}$ | $\text{Spin}_{4n}$ |
| $E_6$ | $Z_3$ | $(3, q-1)$ | | |
| $E_7$ | $Z_2$ | $(2, q-1)$ | | |
| $E_8$ | 1 | 1 | | |
| $F_4$ | 1 | 1 | | |
| $G_2$ | 1 | 1 | | |

For the identifications of $G_A$ see Ree (1957) and Carter (1965). The orthogonal groups are associated with quadratic forms of maximal index.

### 5.3. Generators and Relations, Central Extensions

For construction of automorphisms and extensions, it is useful to have presentations of the Chevalley groups. The following results, due to Steinberg (1962), give complete solutions to the problems of finding presentations and central extensions for the finite Chevalley groups. Steinberg's work (1962, 1967) also handles the infinite Chevalley groups, but we shall limit the discussion below to the finite case.

Theorem (5.23). *Let $G$ be a universal Chevalley group, associated with an indecomposable root system $\Delta$ and a finite field $k$. Consider the group $G^*$ with generators $\{x_\alpha(t) : \alpha \in \Delta, t \in k\}$ and relations* (A) *and* (B) *in case $|\Pi| > 1$ and* (A) *and* (B′) *in case $|\Pi| = 1$:*

(A) $x_\alpha(t)x_\alpha(t') = x_\alpha(t+t')$ $(t, t' \in k, \alpha \in \Delta)$;

(B) $(x_\alpha(t), x_\beta(u)) = \sum_{i, j \geq 1} x_{i\alpha+j\beta}(t^i u^j c_{ij})$ $(\alpha, \beta \in \Delta, \alpha+\beta \neq 0)$;

(B′) $w_\alpha(t)x_\alpha(u)w_\alpha(t)^{-1} = x_{-\alpha}(-t^{-2}u)$ $(t \in k^*, \quad u \in k)$, *where* $w_\alpha(t) = x_\alpha(t)x_{-\alpha}(-t^{-1})x_\alpha(t)$.

*Then the natural epimorphism $\phi : G^* \to G$ is an isomorphism.*

Definition (5.24). A *central extension* of a group $G$ is a pair $(\tilde{G}, \psi)$, with $\tilde{G}$ a group and $\psi$ an epimorphism $\psi : \tilde{G} \to G$, such that $\ker \psi \subseteq Z(\tilde{G})$.

Definition (5.25). A *universal central extension* of a group $G$ is a central extension $(\tilde{G}, \psi)$ such that for any central extension $(G_1, \psi_1)$ of $G$, there exists a unique homomorphism $\phi : \tilde{G} \to G_1$, such that the following diagram is commutative.

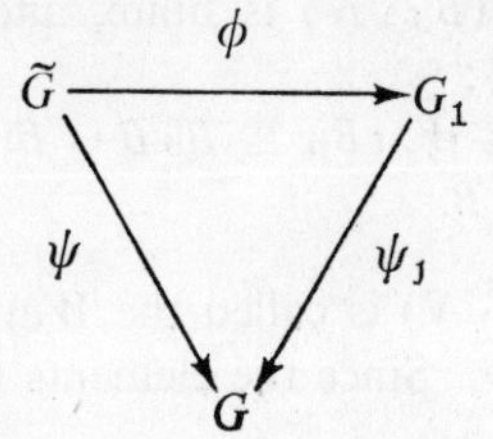

PROPOSITION (5.26). *$(\tilde{G}, \psi)$ is a universal central extension of $G$ if and only if, for every central extension $(G_1', \psi_1)$ of a group $G_1$, and homomorphism $\eta : G \to G_1'$, there exists a unique homomorphism $\tilde{\eta} : \tilde{G} \to \tilde{G}_1$ such that the diagram below commutes.*

$$\begin{array}{ccc} \tilde{G} & \xrightarrow{\tilde{\eta}} & G_1' \\ \downarrow{\scriptstyle \psi} & & \downarrow{\scriptstyle \psi_1} \\ G & \xrightarrow[\eta]{} & G_1 \end{array}$$

EXAMPLE. Suppose $G_1 = PGL(V)$, $G_1' = GL(V)$, where $V$ is a vector space. A *projective representation* $\eta$ of $G$ is a homomorphism $\eta : G \to PGL(V)$. If $(\tilde{G}, \psi)$ is a universal central extension of $G$, then every projective representation $\eta : G \to PGL(V)$ can be lifted to an ordinary representation $\tilde{\eta} : \tilde{G} \to GL(V)$.

THEOREM (5.27). *Let $\Delta$ be an indecomposable root system, and $k$ a finite field, such that $|k| > 4$, and if $|\Pi| = 1$, then $|k| \neq 9$. Let $G_U$ be the corresponding universal Chevalley group (with presentation given in Theorem (5.23)), and let $\psi : G_U \to G$ be the natural homomorphism. Then $(G_U, \psi)$ is a universal central extension of $G$.*

## 6. (*B*, *N*)-pairs and Simplicity

The first part of this section contains statements of the results about $(B, N)$-pairs, which will be required. All these theorems are due to Tits (1962, 1964), and now appear in Chapter 4 of Bourbaki (1968). The second section contains the applications of the theorems to Chevalley groups.

### 6.1. (*B*, *N*)-PAIRS

DEFINITION (6.1). A group $G$ is said to admit a $(B, N)$-*pair* (or *Tits system*) if there exist subgroups $B$ and $N$ of $G$ such that

(i) $G = \langle B \cup N \rangle$, $B \cap N \lhd N$;

(ii) the group $W = N/(B \cap N)$ is finite, and is generated by a set of involutions $R = \{r\}$;

(iii) for all $r \in R$ and $w \in W$, $rBw \subseteq BwB \cup BrwB$; and

(iv) for all $r \in R$, $rBr \neq B$.

The group $W = N/(B \cap N)$ is called the *Weyl group* of the $(B, N)$-pair. We shall write $H = B \cap N$. Since the elements $w \in W$ belong to $N/H$, and

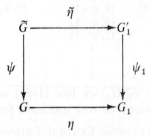

$H \subseteq B$, the sets $wH$, $wB$, $BwB$, etc., are unambiguously defined in $G$. It is unnecessary to assume that the group $W$ is finite; in these lectures, however, we are interested only in the applications to finite groups.

PROPOSITION (6.2). *Let $G$ be a group with a $(B, N)$-pair, and Weyl group $W$. Then*

$$G = \bigcup_{w \in W} BwB,$$

*and $BwB = Bw'B$ if and only if $w = w'$.*

The double coset decomposition of $G$ given in this theorem is called the *Bruhat decomposition* of $G$.

DEFINITION (6.3) A *parabolic subgroup* of a group with a $(B, N)$-pair is a subgroup containing $B$.

PROPOSITION (6.4). (Bourbaki, 1968, IV, Section 2, no. 6). *Let $G$ be a group with a $(B, N)$-pair. There is a bijection between the family of subsets $J \subseteq R$ and the parabolic subgroups of $G$. The parabolic subgroup $G_J$ corresponding to a set $J \subseteq R$ is the subgroup $BW_JB$, where $W_J$ is the subgroup of $W$ generated by $J$. Each parabolic subgroup is its own normalizer.*

COROLLARY (6.5). *$R$ is a minimal set of generators for $W$. An element $w \in W$ belongs to the set $R$ if and only if $B \cup BwB$ is a group.*

The following two results of Tits (1964) will be used to prove simplicity of the adjoint Chevalley groups (see also Bourbaki, 1968).

PROPOSITION (6.6). *Let $G$ be a group with a $(B, N)$-pair. Suppose $G_1$ is a subgroup of $G$ such that if $B_1 = B \cap G_1$, then $HB_1 = B$. Then*

(i) *$HG_1 = G_1H = G_J$ for some $J \subseteq R$;*
(ii) *if $N_1 = N \cap G_1$, then $\{B_1, N_1\}$ is a $(B, N)$-pair for $G_1$ whose Weyl group is $W_J$, and the set of distinguished generators is $J$;*
(iii) *if $G_1 \lhd G$, then $J = R$ and $HG_1 = G$.*

PROPOSITION (6.7). *Let $G$ be a group with a $(B, N)$-pair. If $G_1$ is any normal subgroup, then $G_1$ defines a partition of the set $R$ into two subsets, $R = R' \cup R''$ with $R' = \{r \in R : BrB \cap G_1\} \neq \emptyset$, and $R'' = \{r \in R : BrB \cap G_1 = \emptyset\}$. Moreover, we have*

(i) *$G_1B = G_{R'}$;*
(ii) *every element of $R'$ commutes with every element of $R''$;*
(iii) *$W = W_{R'} \times W_{R''}$ (direct product).*

Corollary (6.8). *Let $G$ be a group with a $(B, N)$-pair. Let $Z$ be the intersection of all conjugates of $B$, and $G_1$ the subgroup generated by all conjugates of $U$ for some subgroup $U$ of $B$. Suppose that*

(a) $U \lhd B$;
(b) $U \neq (U, U)$;
(c) $G_1 = (G_1, G_1)$;
(d) *$R$ is indecomposable (i.e., admits no non-trivial decomposition as in Prop. (6.7)).*

*Then every normal subgroup of $G$ either contains $G_1$ or is contained in $Z$*

*Proof.* Let $S \lhd G$, and suppose $S \nsubseteq Z$. Then $S \nsubseteq B$, and by Prop. (6.7), $SB = G$. Since $S \lhd G$, $SU = US$, and $SU$ contains all conjugates of $U$ by elements of $SB = G$. Then $G_1 \subseteq SU$, and $SU = SG_1$. Thus

$$U/(U \cap S) \cong SU/S \cong SG_1/S \cong G_1/(G_1 \cap S).$$

Since $U \neq (U, U)$ and $G_1 = (G_1, G_1)$, we conclude that $G_1 = G_1 \cap S$ and $S \supseteq G_1$ as required.

A little more work yields the following corollaries.

Corollary (6.9). *Let $G$, $U$, $G_1$, $Z$ etc., be as in Corollary (6.8). Suppose also that $B = UH$. Then $G_1/(G_1 \cap Z)$ is a simple group.*

Corollary (6.10). *Suppose $U$ has a conjugate $U^g$ such that $U^g \cap B = \{1\}$. Then $G_1/Z(G_1)$ is a simple group.*

## 6.2. The Chevalley–Dickson Theorem

Theorem (6.11). *Let $G$ be an adjoint Chevalley group, and assume the root system $\Delta$ is indecomposable. If $|k| = 2$, suppose that $\Delta$ is not of type $A_1$, $B_2$ or $G_2$. If $|k| = 3$, suppose that $\Delta$ is not of type $A_1$. Then $G$ is a simple group.*

*Proof.* By Prop. (5.15), it follows that $G$ has a $(B, N)$-pair, and the indecomposability of $\Delta$ implies that $(W, R)$ is indecomposable. By (5.8) (i) and (5.13), $\mathfrak{U}$ is a solvable normal subgroup of $B$ such that $B = \mathfrak{U}H$. Letting $w_0 \in W$ be such that $w_0(\Delta^+) = \Delta^-$ (from (3.5) (d)), we have, letting $\phi(w_0) = n_0 H$, $\mathfrak{U}^{n_0} \cap B = \mathfrak{U}^- \cap B = \{1\}$ by (5.14) and (5.15) (iii). The subgroup generated by the conjugates of $\mathfrak{U}$ is $G$ by (5.14) and the definition of $G$. Finally, $Z(G) = \{1\}$, by Corollary (5.18). By Corollary (6.10), it is sufficient to prove that $G = (G, G)$.

First assume $|k| \geq 4$, and choose $t \in k^*$ with $t^2 \neq 1$. Then by (5.21),

$$(h_\alpha(t), x_\alpha(u)) = x_\alpha((t^2 - 1)u),$$

and $\mathfrak{X}_\alpha \subseteq (G, G)$ for all $\alpha \in \Delta$. Thus $G = (G, G)$ in this case. In case $|k| < 4$, further calculations are needed (Steinberg, 1967).

## 7. The Orders of the Finite Chevalley Groups

Let $G$ be a Chevalley group associated with $\mathfrak{g}$ and $k$, for which the corresponding lattice is $\Gamma_{\text{univ}}$. By Corollary (5.19), we have, letting $q = |k|$,

$$|G| = q^N(q-1)^n \Big( \sum_{w \in W} q^{l(w)} \Big).$$

Chevalley (1955) proved a remarkable multiplicative formula for $\sum q^{l(w)}$, namely

$$\sum q^{l(w)} = \prod_{i=1}^{n} \frac{(1-q^{d_i})}{1-q}$$

where the $\{d_i\}$ are certain integers, connected with the topology of a compact Lie group associated with $\mathfrak{g}$. Solomon (1966) obtained an equally remarkable direct proof of this formula, entirely within the framework of finite groups generated by reflections. We shall essentially give Solomon's proof of this result, as simplified at one point by Steinberg (1967). Proofs of all the results stated in this chapter are given in Steinberg (1967).

The whole development, including the origin of the *exponents* $\{d_i\}$, is of importance not only because of the formula for $|G|$ in the case of a Chevalley group, but because the same method also gives multiplicative formulas for the orders of the twisted types of groups, and more generally, for subgroups of $G$ fixed under certain types of endomorphisms (Steinberg, 1968).

### 7.1. Invariants and Exponents of Finite Reflection Groups

We begin with the following result of Chevalley (1955).

THEOREM (7.1). *Let $W$ be a finite group generated by reflections on a euclidean space $V$, as in Section 3. Let $S$ be the algebra of polynomial functions on $V$. Then $W$ can be identified with a group of automorphisms of $S$, and we denote by $I(S)$ the algebra of invariants under $W$ in $S$. Then $I(S)$ is generated by $n = \dim V$ algebraically independent, homogeneous polynomials $I_1, \ldots, I_n$.*

DEFINITION (7.2). The degrees $\{d_i\}$ of the invariants $\{I_i\}$ are called the *exponents* of $W$. (The next result shows that they are uniquely determined by $W$.)

THEOREM (7.3). *The exponents $\{d_i\}$ of $W$ are uniquely determined and satisfy*

(i) $|W| = \prod_{i=1}^{n} d_i$,

(ii) $\sum_{i=1}^{n} (d_i - 1) = N$, *where $N$ is the number of positive roots in the root system associated with $W$.*

Assuming Theorem (7.1), we shall give a proof of Theorem (7.3).

LEMMA (7.4). (Shephard and Todd, (1954)). *Let $I_1, \ldots, I_n$ be algebraically independent generators of $I(S)$ of degrees $d_1, \ldots, d_n$ respectively. Then*

$$\prod_{i=1}^{n}(1-t^{d_i})^{-1} = \frac{1}{|W|}\sum_{w\in W}\det(1-wt)^{-1}$$

*is an identity in t.*

*Proof.* We can assume for this proof that the base field has been extended to **C**. Let $\varepsilon_1, \ldots, \varepsilon_n$ be the eigenvalues of $w$ and $x_1, \ldots, x_n$ the corresponding eigenfunctions of the semi-simple linear transformation $w$ acting on the polynomial functions of degree one. Then

$$\det(1-wt)^{-1} = \prod_{i=1}^{n}(1+\varepsilon_i t+\varepsilon_i^2 t^2+\ldots),$$

and the coefficient of $t^m$ is

$$\sum_{p_1+p_2+\ldots=m}\varepsilon_1^{p_1}\varepsilon_2^{p_2}\ldots$$

This expression is just the trace of $w$ acting on the homogeneous polynomials over $V$ of degree $m$, since the monomials $x_1^{p_1}x_2^{p_2}\ldots$ form a basis for this space. Upon averaging over $W$, the coefficient of $t^m$ gives the dimension of the space of invariant homogeneous polynomials of degree $m$. (To see this, let $\chi$ be the character of the representation of $W$ on the homogeneous polynomials of degree $m$. By the orthogonality relations, $|W|^{-1}\sum\chi(w)$ is the multiplicity of the trivial representation of $W$ in this representation.) This dimension is also equal to the number of monomials $I_1^{q_1}I_2^{q_2}\ldots$ of degree $m$, i.e., the number of solutions of $q_1d_1+q_2d_2+\ldots = m$. But this is the coefficient of $t^m$ in

$$\prod_{i=1}^{m}(1-t^{d_1})^{-1}.$$

This completes the proof of the lemma.

*Proof of Theorem* (7.3). We first establish (i) and (ii). We have

$$\det(1-wt) = \begin{cases}(1-t)^n & \text{if } w = 1,\\ (1-t)^{n-1}(1+t) & \text{if } w \text{ is a reflection,}\\ \text{a polynomial not divisible by } (t-1)^{n-1} & \text{otherwise.}\end{cases}$$

We also use the fact, left to the reader to verify, that the number of reflections in $W$ is the same as the number of positive roots $N$. Substituting this information in Lemma (7.4) we obtain

$$\prod_{i=1}^{n}(1+t+\ldots+t^{d_i-1})^{-1} = \frac{1}{|W|}\left(1+N\frac{(1-t)}{(1+t)}+(1-t)^2f(t)\right),$$

where $f(t)$ is differentiable at $t = 1$. Putting $t = 1$, we obtain

$$\prod(d_i)^{-1} = |W|^{-1},$$

which is (i). Next we have, upon differentiating and putting $t = 1$,

$$(\prod d_i^{-1})\sum -\frac{(d_i-1)}{2} = |W|^{-1}\left(-\frac{N}{2}\right),$$

so, using (i), we obtain (ii).

Finally, we prove the uniqueness of the exponents. Let $I'_1, \ldots, I'_n$ be another set of algebraically independent, homogeneous generators of $I$, and let $\{d'_i\}$ be their degrees. Then

$$I'_i = F_i(I_1, \ldots, I_n)$$

where $\{F_i\}$ are polynomials such that the Jacobian

$$\det\left(\frac{\partial F_i}{\partial I_j}\right) \neq 0.$$

By rearranging the $I'_i$ we can assume

$$\frac{\partial I'_i}{\partial I_i} \neq 0$$

and hence $d'_i \geq d_i$ for all $i$. Using (ii) we obtain $d_i = d'_i$, and the theorem is proved.

It can be shown that for the various crystallographic root systems the exponents are as follows.

| $W$ | $d_i$'s |
|---|---|
| $A_n$ | $2, 3, \ldots, n+1$ |
| $B_n$, $C_n$ | $2, 4, \ldots, 2n$ |
| $D_n$ | $2, 4, \ldots, 2n-2, n$ |
| $E_6$ | 2, 5, 6, 8, 9, 12 |
| $E_7$ | 2, 6, 8, 10, 12, 14, 18 |
| $E_8$ | 2, 8, 12, 14, 18, 20, 24, 30 |
| $F_4$ | 2, 6, 8, 12 |
| $G_2$ | 2, 6 |

The table is derived by means of the following result due to Coxeter (1951). Coxeter's proof used the classification of the root systems; proofs without using the classification have been given by Steinberg (1959) and Coleman (1958).

PROPOSITION (7.5). *Let $W$ and the $\{d_i\}$ be as above, and let $w = r_{\alpha_1} \ldots r_{\alpha_n}$, where $\{\alpha_1, \ldots, \alpha_n\} = \Pi$. Let $h$ be the order of $w$. Then*

(i) *$N = (nh)/2$;*

(ii) *$w$ contains $\omega = e^{2\pi i/h}$ as an eigenvalue but not 1;*

(iii) *if the eigenvalues of $w$ are $\{\omega^{m_i}\}$ $(1 \leq m_i \leq h-1)$ then $\{m_i+1\} = \{d_i\}$.*

## 7.2. Solomon's Theorem

We are now ready to prove Solomon's theorem, which will give the desired multiplicative formula for $|G_U|$.

THEOREM (7.6). *Let $W$ be a finite reflection group and let $\{d_i\}$ be the exponents of $W$. Let $t$ be an indeterminate over $\mathbf{Q}$ and put*

$$W(t) = \sum_{w \in W} t^{l(w)},$$

$$D(t) = \prod_{i=1}^{n} \frac{(1-t^{d_i})}{(1-t)}.$$

*Then $W(t) = D(t)$ in $\mathbf{Q}[t]$.*

The proof involves a series of lemmas. As in Section 3, we let $V$ be the underlying vector space. From ((Bourbaki, 1968) 37, Ex. 3) we have

LEMMA (7.7). *Let $\pi \subseteq \Pi$, $W_\pi = \langle r_\alpha, \alpha \in \pi \rangle$.*

(a) *$n(w)$, and $l(w)$ are the same for $w \in W_\pi$ regardless of whether we consider the root systems supported by $\pi$ or $\Pi$.*

(b) *Let $W'_\pi = \{w \in W : w(\alpha) \geq 0$ for all $\alpha \in \pi\}$. Then $W = W'_\pi W_\pi$ and every element of $W$ has a unique expression according to this factorization,*

$$w = w'w'' \quad (w' \in W'_\pi,\ w'' \in W_\pi).$$

(c) *In the above factorization*

$$l(w) = l(w') + l(w'').$$

Now let

$$W_\pi(t) = \sum_{w \in W_\pi} t^{l(w)}.$$

LEMMA (7.8).

$$\sum_{\pi \subseteq \Pi} (-1)^{|\pi|} \frac{W(t)}{W_\pi(t)} = t^N.$$

*Proof.* We have from (7.7)

$$\frac{W(t)}{W_\pi(t)} = \sum_{w \in W_\pi'} t^{l(w)}.$$

Therefore the contribution of a term to the left-hand side, associated with $w \in W$, is

$$\Bigg( \sum_{\pi \subseteq \Pi,\, w(\pi) > 0} (-1)^{|\pi|} \Bigg) t^{l(w)}.$$

If $\pi_w = \{\alpha \in \Pi : w(\alpha) > 0\}$ then if $\pi_w \neq \emptyset$ the coefficient is the binomial expression

$$(1 + (-1))^{|\pi_w|} = 0.$$

The coefficient is non-zero (and equal to 1) for the unique element $w_0$ for which $\pi_{w_0} = \emptyset$.

Let $C = \{v \in V;\ (v,\alpha) \geq 0$ for all $\alpha \in \Pi\}$, and for $\pi \in \Pi$, set $C_\pi = \{v \in V : (v, \alpha) = 0$ for $\alpha \in \pi$, $(v, \beta) > 0$ for $\beta \in \Pi - \pi\}$. $C$ (and its transforms by $W$) are called *chambers*; each one is the closure of a maximal convex subset of $V \backslash \cup \langle\alpha\rangle^\perp$, where $\langle\alpha\rangle^\perp = \{v \in V : (\alpha, v) = 0\}$.

PROPOSITION (7.9).

(i) *$C$ is a fundamental domain for $W$ in the sense that each $v \in V$ is congate (under $W$) to one and only one element in $C$.*

(ii) *If $v \in C$ and $w(v) = v$ for $w \in W$, then $w$ is a product of reflections, $\{r_\alpha, \alpha \in \Pi\}$ also fixing $v$.*

The proof is given, for example, in Steinberg (1967).

As a corollary to (7.9) we have

LEMMA (7.10). *The following subgroups of $W$ are equal.*

(a) $W_\pi$.

(b) *The stabilizer of $C_\pi$.*

(c) *The pointwise stabilizer of $C_\pi$.*

(d) *The stabilizer of any point in $C_\pi$.*

(One proves (a) $\subseteq$ (b) $\subseteq$ (c) $\subseteq$ (d) $\subseteq$ (a)).

LEMMA (7.11). *In the complex cut out of $V$ by a finite number of hyperplanes let $n_i$ be the number of i-cells.*† *Then*

$$\sum(-1)^i n_i = (-1)^{\dim V}.$$

*Proof.* The result is proved by induction. If an extra hyperplane $H$ is added, each original $i$-cell cut in two by $H$ has corresponding to it in $H$ an $(i-1)$-cell separating the two parts, so that $\sum(-1)^i n_i$ remains unchanged.

LEMMA (7.12) (Steinberg). *In the complex $K$ cut from $V$ by the reflecting hyperplanes, let $n_\pi(w)$ ($\pi \in \Pi$, $w \in W$) denote the number of cells $W$-congruent to $C_\pi$ and fixed by $w$. Then*

$$\sum_{\pi \subseteq \Pi} (-1)^{|\pi|} n_\pi(w) = \det w.$$

*Proof.* Each cell of $K$ is congruent under $W$ to exactly one $C_\pi$. By (7.10) every cell fixed by $w$ lies in $V^w = \{v \in V : wv = v\}$. Therefore, applying (7.11) and the fact that $\dim C_\pi = \dim V - |\pi|$, we have, on $V^w$,

$$\sum_{\pi \subseteq \Pi} (-1)^{n-|\pi|} n_\pi(w) = (-1)^{\dim V^w}$$

† An $i$-cell is an open set homeomorphic to $\mathbf{R}^i$.

where $n = \dim V$ and

$$\sum_{\pi \subseteq \Pi} (-1)^{|\pi|} n_\pi(w) = (-1)^{n-\dim V^w} = \det w$$

since

$$w \sim \begin{bmatrix} \overbrace{1 \; \cdot \; \cdot \; 1}^{V_w} & & & & \\ & -1 & & & \\ & & \ddots & & \\ & & & -1 & \\ & & & & \begin{pmatrix} \cos\theta - \sin\theta \\ \sin\theta - \cos\theta \end{pmatrix} \\ & & & & & \ddots \end{bmatrix}$$

Now let $\chi$ be a character on $W_1 < W$, and let $\chi^W$ denote the induced character (see Curtis and Reiner, 1962).

(7.13). *Let $\chi$ be a character on $W$ and $\chi_\pi = (\chi|W_\pi)^W$ (for $\pi \subseteq \Pi$). Then*

$$\sum_{\pi \subseteq \Pi} (-1)^\pi \chi_\pi(w) = \chi(w) \det w$$

*for all $w \in W$.*

*Proof.* Assume first that $\chi = 1$. Then

$$xwx^{-1} \in W_\pi \Leftrightarrow xwx^{-1} \text{ fixes } C_\pi \text{ (by (7.10))}$$
$$\Leftrightarrow w \text{ fixes } x^{-1} C_\pi.$$

Therefore, in this case,

$$\chi_\pi(w) = n_\pi(w)$$

and the result follows from Lemma (7.12). For the general case, we have, for any $\chi$

$$\chi \cdot 1_\pi = \chi_\pi, \quad \text{where } \chi_\pi = (\chi|_{W_\pi})^W)$$

by the identity (Curtis and Reiner, 1962)

$$\phi \cdot \eta^G = ((\phi|H) \cdot \eta)^G,$$

and the result follows.

COROLLARY (7.14). $\det(w) = \sum_{\pi \subseteq \Pi} (-1)^{|\pi|} 1_{W_\pi}^W(w)$ *for all* $w \in W$

LEMMA (7.15). *Let $M$ be a finite dimensional* $\mathbf{R}W$*-module, $I_\pi(M)$ the space* *of $W_\pi$-invariants, $\hat{I}(M)$ the space of W-skew invariants (i.e., $m \in \hat{I}(M) \Leftrightarrow wm = \det(w)m$, $(w \in W)$). Then*

$$\sum_{\pi \subseteq \Pi} (-1)^\pi \dim I_\pi(M) = \dim \hat{I}(M).$$

*Proof.* Let $\chi$ be the character of $M$. Then

$$\dim I_\pi(M) = (1_{W_\pi}, \chi | W_\pi) = (1_W, \chi_\pi)$$

by Frobenius reciprocity (Curtis and Reiner, 1962). By (7.13) it is sufficient to prove that

$$(1_W, \chi \cdot \det) = \dim \hat{I}(M).$$

But

$$(1_W, \chi \cdot \det) = (\det, \chi) = \dim \hat{I}(M).$$

This completes the proof.

(7.16). *Let $p = \prod_{\alpha \in \Delta^+} \alpha$. Then $p$ is skew and divides every other skew polynomial on $V$.*

*Proof.* For $\alpha \in \Pi$,

$$r_\alpha p = -p = (\det r_\alpha)p,$$

by (3.5) (a). It follows from (3.3) (c) that $p$ is skew. If $f$ is skew and $\alpha$ a root then

$$r_\alpha f = (\det r_\alpha) f = -f.$$

Then $\alpha(x) = 0 \Leftrightarrow r_\alpha x = x \Rightarrow r_\alpha f(x) = -f(x) = 0$, and $f$ vanishes on the nullspace of $\alpha$. Then $f \in \operatorname{rad}(\alpha) = (\alpha)$ and $\alpha | f$, by the nullstellensatz. Then $\Pi\alpha | f$ by unique factorization.

LEMMA (7.17). *Let $D(t) = \Pi(1-t^{d_i})/(1-t)$, and for $\pi \subseteq \Pi$ let $\{d_{\pi i}\}$ and $D_\pi(t)$ be defined for $W_\pi$. [Note that the number of degrees $\{d_{\pi i}\}$ is n even if $\pi \subseteq \Pi$—some are equal to one]. Then*

$$\sum_{\pi \subseteq \Pi} (-1)^{|\pi|} \frac{D(t)}{D_\pi(t)} = t^N.$$

*Proof.* We have to show

$$\sum_{\pi \subseteq \Pi} (-1)^{|\pi|} \prod (1-t^{d_{\pi i}})^{-1} = t^N \prod_i (1-t^{d_i})^{-1}.$$

From the proof of (7.4), letting $S_k$ be the space of homogeneous polynomials of degree $k$, we see that the coefficient of $t^k$ on the left is

$$\sum_{\pi \subseteq \Pi} (-1)^\pi \dim I_\pi(S_k).$$

Now, by (7.16), every skew polynomial $f$ in $S_k$ has the form

$$f = pf_1,$$

and $f_1 \in S_{k-N}$. Therefore

$$\dim \hat{I}(S_k) = \dim I(S_{k-N}).$$

It follows that the coefficient of $t^k$ on the right side is

$$\dim I(S_{k-N}) = \dim \hat{I}(S_k)$$

and the result follows from (7.15).

*Proof of Theorem* (7.6). Write (7.17) as

$$\frac{(t^N - (-1)^{|\Pi|})}{D(t)} = \sum_{\pi \subseteq \Pi} \frac{(-1)^{|\pi|}}{D_\pi(t)}$$

and (7.8) as

$$\frac{(t^N - (-1)^{|\Pi|})}{W(t)} = \sum_{\pi \subseteq \Pi} \frac{(-1)^{|\pi|}}{W_\pi(t)}$$

The theorem follows by induction on $|\pi|$.

## 8. Automorphisms and Twisted Types

In this chapter, we first indicate how the automorphisms of Chevalley groups are constructed, using the generators and relations for universal Chevalley groups given in Chapter 3. Steinberg's classification of the automorphisms is stated. Finally, a sketch is given of a uniform method for constructing and working out the structure of the twisted types of Chevalley groups. The details of the construction of the graph automorphisms, which are crucial in order to obtain the twisted types, are lengthy, and involve case by case discussions in the proofs, although the final results can be stated independently of the classification of types of root systems. An excellent survey of the twisted types, in greater detail than the discussion to follow, and from a different point of view, has been given by Carter (1965).

Steinberg (1968) investigated the structure of the subgroup $G_\sigma$ of elements in a semi-simple algebraic group $G$ fixed under an endomorphism $\sigma$ which fixes simultaneously a Borel subgroup of $G$ and a maximal torus in the Borel subgroup. As applications, the structure of centralizers of semi-simple elements in Chevalley groups, and the structure of the twisted types, are obtained. In particular, the centralizers of involutions, for Chevalley groups of odd characteristic, are determined in this way (cf. Ree, 1965; Iwahori, 1970).

The proofs of the theorems in this section are to be found in Steinberg (1967, 1968).

### 8.1. AUTOMORPHISMS

We begin with the construction of the automorphisms that come from automorphisms of the Dynkin diagram.

THEOREM (8.1).

(a) *Let $\Delta$ be an indecomposable root system, $\sigma$ an angle-preserving permutation of the fundamental roots, such that $\sigma \neq 1$. If all roots have the same length, then $\sigma$ extends to an angle-preserving permutation of $\Delta$. If not, then $\sigma$ interchanges long and short roots. In the unequal root lengths case, $\sigma$ extends to an angle-preserving permutation of the roots which interchanges long and short roots.*

(b) *The possibilities for $\sigma$ are:*

*one root length*

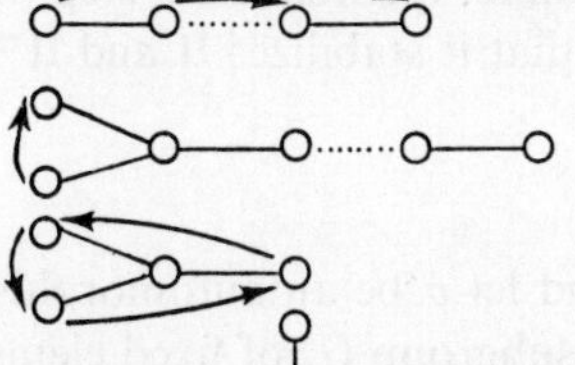

$A_n (n \geq 2)$ $(\sigma^2 = 1)$

$D_n (n \geq 4)$ $(\sigma^2 = 1)$

$D_4$ $(\sigma^3 = 1)$

$E_6$ $(\sigma^2 = 1)$

*two root lengths*

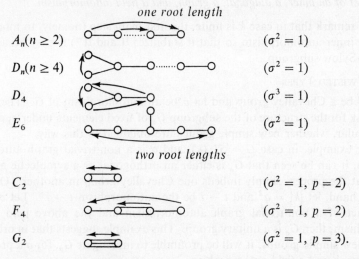

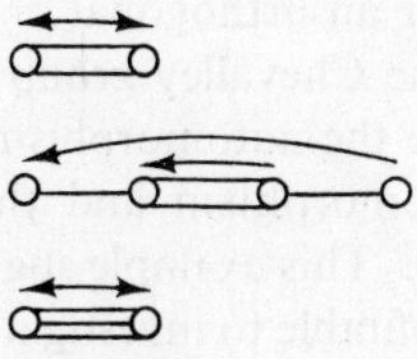

$C_2$ $(\sigma^2 = 1, p = 2)$

$F_4$ $(\sigma^2 = 1, p = 2)$

$G_2$ $(\sigma^2 = 1, p = 3)$.

(c) *Let $k$ be a field, and $G$ a universal Chevalley group associated with $\Delta$ and $k$. Let $\sigma$ be as above. If two root lengths occur, let $k$ be perfect of characteristic $p = (\alpha_0, \alpha_0)/(\beta_0, \beta_0)$, $\alpha_0$ long, $\beta_0$ short. Then there exists an automorphism $\phi$ of $G$ and signs $\varepsilon_\alpha$ (with $\varepsilon_\alpha = 1$ if $\alpha$ or $-\alpha$ is in $\Pi$) such that*

$$\phi(x_\alpha(t)) = \begin{cases} x_{\sigma\alpha}(\varepsilon_\alpha t) \text{ if } \alpha \text{ is long or all roots have one length.} \\ x_{\sigma\alpha}(\varepsilon_\alpha t^p) \text{ if } \alpha \text{ is short.} \end{cases}$$

Automorphisms of the Chevalley groups of this type, along with 1, are called *graph automorphisms.*

PROPOSITION (8.2). *Let $G$ be a Chevalley group, parametrized by a field $k$, with root system $\Delta$.*

(i) *For each $\alpha \in \Pi$, let $d_\alpha$ be an arbitrary element of $k^*$. Then $f: \alpha \to d_\alpha$ can be extended to a homomorphism $f$ of $\Gamma_{\text{ad}} \to k^*$. There exists a unique automorphism $\phi$ of $G$ such that*

$$\phi(x_\alpha(t)) = x_\alpha(f(\alpha)t) \quad (\alpha \in \Delta, \ t \in k).$$

(ii) *Let $\gamma$ be an automorphism of $k$. Then there exists a unique automorphism $\psi$ of $G$ such that*

$$\psi(x_\alpha(t)) = x_\alpha(t^\gamma) \quad (t \in k, \alpha \in \Delta).$$

Automorphisms of types (i) and (ii) are called *diagonal* and *field* automorphisms, respectively.

THEOREM (8.3). *Let $G$ be a Chevalley group, with an indecomposable root system $\Delta$ and a perfect field $k$. Let $\sigma$ be an automorphism of $G$. Then $\sigma$ is a product of an inner, a diagonal, a graph, and a field automorphism.*

We remark that in case $k$ is finite, the first main step (namely, to modify $\sigma$ by an inner automorphism so that it stabilizes $\mathfrak{U}$ and $\mathfrak{U}^-$) is easy because $\mathfrak{U}$ is a $p$-Sylow subgroup.

## 8.2. TWISTED TYPES

Let $G$ be a Chevalley group and let $\sigma$ be an automorphism of $G$. Then one can ask for the structure of the subgroup $G_\sigma$ of fixed elements under $\sigma$, and in particular, whether new simple groups are obtained in this way.

For example, in case $G = SL_n(k)$, and $\sigma$ is a non-trivial graph automorphism, it can be seen that $G_\sigma$ is either an orthogonal or a symplectic group, so that the process simply imbeds one Chevalley group in another. On the other hand, let $|k| = q^2$ and $t \to \bar{t}$ be the automorphism $t \to t^q$. Let $\sigma$ be a product of a non-trivial graph automorphism and the above field automorphism; then $G_\sigma$ is a unitary group. This example suggests that in order to find new simple groups, it will be profitable to investigate $G_\sigma$, for $\sigma$ a product of a graph and a field automorphism.

We now proceed to construct the twisted types of Chevalley groups. We first twist Coxeter groups and root systems, following Steinberg (1967, 1968).

Let $\Delta$ be a root system contained in a real euclidean vector space $V$. We do not assume that $\Delta$ satisfies the crystallographic condition. Let $\Pi$ be a base of $\Delta$, and $\Delta^+$, $\Delta^-$ the positive and negative roots determined by $\Pi$. Suppose $\sigma$ is an automorphism of $V$, and that $\sigma$ permutes the positive multiples of elements of $\Delta$, $\Delta^+$, and $\Pi$. Let $\rho$ be the corresponding permutation of the roots. Then $\sigma$ has finite order and normalizes $W$.

Let $V_\sigma$, $W_\sigma$ denote the fixed points in $V$ and $W$ respectively.

Let $\bar{\alpha}$ denote the average of the elements in the $\sigma$-orbit of $\alpha \in \Delta$; then

$$(\beta, \bar{\alpha}) = (\beta, \alpha) \quad (\beta \in V_\sigma).$$

Then, if we write

$$V = V_\sigma \oplus V_\sigma^\perp$$

and express

$$\alpha = \alpha' + \alpha''$$

according to this decomposition, we have $\alpha' = \bar{\alpha}$.

THEOREM (8.4) (Steinberg, 1967). *Let $\Delta, \Delta^+, \Pi, \sigma$ be as above.*

(i) *The restriction of $W_\sigma$ to $V_\sigma$ is faithful.*
(ii) *$W_\sigma|V_\sigma$ is a finite group generated by reflections.*
(iii) *If $\Delta_\sigma$ is the set of projections of $\Delta$ on $V_\sigma$ then $\Delta_\sigma$ is almost a root system for $W_\sigma$ (i.e., the reflections $\{r_{\bar{\alpha}} : \bar{\alpha} \in \Delta_\sigma\}$ generate $W_\sigma|V_\sigma$ and $r_{\bar{\alpha}}|\Delta_\sigma (= \Delta_\sigma)$. (But it may happen that $\Delta_\sigma$ violates the condition that $\bar{\alpha}$ and $c\bar{\alpha} \in \Delta_\sigma$ imply that $c = \pm 1$.)*
(iv) *If $\Pi_\sigma$ is the projection of $\Pi$ on $V_\sigma$, then $\Pi_\sigma$ is a base of $\Delta_\sigma$ (after multiples have been discarded, if necessary), i.e., $\Pi_\sigma$ is a linearly independent set, and the elements of $\Delta_\sigma$ are linear combinations of elements of $\Delta_\sigma$ with coefficients of the same sign.*

To discard multiples in a systematic way, if several projections $\bar{\alpha}_1, \bar{\alpha}_2, \ldots$ all have the same direction, we take the shortest one.

*Examples.* (The notation used to describe the twisted root systems will be used later for twisted groups.)
$^2A_{2n-1}$, $^2A_{2n}$. (i) $\sigma$ of order 2. For $W$ of type $A_{2n-1}$, we get $W_\sigma$ of type $C_n$. If $W$ is of type $A_{2n}$, $\alpha$ of order 2, then $W_\sigma$ is of type $BC_n$.
$^2D_n$. $\sigma$ of order 2, $W$ of type $D_n$, $W_\sigma$ is of type $B_{n-1}$.
$^3D_4$. $\sigma$ of order 3, $W$ of type $D_4$, $W_\sigma$ is of type $G_2$.
$^2E_6$. $\sigma$ of order 2, $W$ of type $E_6$, $W_\sigma$ is of type $F_4$.
$^2C_2$. $\sigma$ of order 2, $W$ of type $C_2$, $W_\sigma$ is of type $A_1$.
$^2G_2$. $\sigma$ of order 2, $W$ of type $C_2$, $W_\sigma$ is of type $A_1$.
$^2F_4$. $\sigma$ of order 2, $W$ of type $F_4$, $W_\sigma$ is of type $D_{16}$, (dihedral of order 16).
(Note that in the case of $^2F_4$, the root system of $W_\sigma$ does not satisfy the crystallographic condition.)

Now let $G$ be a universal Chevalley group over a finite field $k$ of characteristic $p$, and let $\sigma$ be an automorphism of $G$ which is a product of a graph automorphism and a field automorphism $\theta$ of $k$, such that if $\rho$ is the corresponding permutation of roots, then

(i) if $\rho$ preserves lengths then the order of $\theta$ = the order of $\rho$;
(ii) if $\rho$ does not preserve lengths then

$$\theta = \sqrt{\tau}, \text{ where } \tau(x) = x^{1/p} \quad (x \in k).$$

(For a discussion of why these cases are natural ones to consider, see Steinberg, 1968 and Carter, 1965.) We have $p = 2$ if $G$ is of type $C_2$ or $F_4$ and $p = 3$ if $G$ is of type $G_2$. The automorphism $\sigma$ satisfies

$$\sigma x_\alpha(t) = \begin{cases} x_{\rho\alpha}(\varepsilon_\alpha t^\theta) & |\alpha| \geq |\rho\alpha| \\ x_{\rho\alpha}(\varepsilon_\alpha t^{p\theta}) & |\alpha| < |\rho\alpha| \end{cases}$$

with $\varepsilon_\alpha = \pm 1$, and $\varepsilon_\alpha = 1$ if $\pm\alpha \in \Pi$.

Now $\sigma$ preserves $\mathfrak{U}$, $H$, $B$, $\mathfrak{U}^-$ and $N$ and hence defines an automorphism of $W$. Since $\rho$ preserves angles, it agrees up to constant multiples with an isometry of the space generated by the roots, and we can apply Theorem (8.4).

THEOREM (8.5). *Let $G$ be a universal Chevalley group, $k$ a finite field, and let $\sigma$ be defined as above. Let $W_\sigma$ be the subgroup of $W$ fixed under the automorphism induced by $\sigma$, and $G_\sigma$, $B_\sigma$, $\mathfrak{U}_\sigma$, etc. denote subgroups of fixed elements in $G$.*

(a) *For each $w \in W_\sigma$, $\mathfrak{U}_w^- = \mathfrak{U} \cap w^{-1}\mathfrak{U}w$* is fixed by $\sigma$.

(b) *For each $w \in W_\sigma$, there exists $n_w \in N_\sigma$, and actually $n_w \in \langle \mathfrak{U}_\sigma, \mathfrak{U}_\sigma^- \rangle \cap N_\sigma$, such that $n_w H = w$.*

(c) *Letting $\{n_w : w \in W_\sigma\}$ be chosen in* (b), *we have*

$$G = \bigcup_{w \in W_\sigma} B_\sigma n_w \mathfrak{U}_{w\sigma}^-,$$

*with the expressions $bn_w u$, $b \in B_\sigma$, $u \in \mathfrak{U}_{w\sigma}^-$, unique.*

(d) *$G_\sigma$ has a $(B, N)$-pair $\{B_\sigma, N_\sigma\}$, with Weyl group $W_\sigma = N_\sigma/H_\sigma$.*

(e) *$G_\sigma = \langle \mathfrak{U}_\sigma, \mathfrak{U}_\sigma^- \rangle$.*

(f) *$G_\sigma/Z(G_\sigma)$ is a simple group, except for the cases $^2A_2(4)$, $^2C_2(2)$, $^2G_2(3)$ and $^2F_4(2)$.*

(g) *$Z(G)_\sigma = Z(G_\sigma)$.*

Because of (d), the proof of (f) can be proved in exactly the same way as the simplicity of $G/Z(G)$ for an untwisted Chevalley group (see Section 6).

We shall conclude this section with a general theorem, due to Steinberg (1968), which gives a multiplicative formula for the orders of the groups $G_\sigma$ in (8.5).

Since $k$ is finite, of characteristic $p$, we can choose $a$ to be minimal such that $\theta(t) = t^{p^a}$, for $t \in k$. Then

$$\begin{aligned} |k| &= p^{2a} && \text{for types } {}^2A_n,\ {}^2D_n,\ {}^2E_6; \\ |k| &= p^{3a} && \text{for type } {}^3D_4; \\ |k| &= p^{2a+1} && \text{for types } {}^2C_2,\ {}^2F_4,\ {}^2G_2. \end{aligned}$$

Now define

$$q = \begin{cases} p^a & \text{for types } {}^2A_n,\ {}^2D_n,\ {}^2E_6,\ {}^2D_4. \\ p^{a+\frac{1}{2}} & \text{for types } {}^2C_2,\ {}^2D_4,\ {}^2G_2. \end{cases}$$

Next let $V$ denote the vector space generated by the roots, and $\sigma_0$ the automorphism of $V$ permuting the rays through the roots according as $\rho$ permutes the roots. Then $\sigma_0$ normalizes $W$, and acts on the ring $I(S)$ of polynomial invariants of $W$ (and on the subspaces of homogeneous elements in $I(S)$). Upon extending the base field if necessary, it is possible to choose the basic invariants $\{I_j\}_{1 \le j \le n}$ so that

$$\sigma_0(I_j) = \lambda_j I_j \quad (1 \le j \le n)$$

for some roots of unity $\{\lambda_j\}$. Finally, let

$$\{\varepsilon_{0j}\}_{1 \le j \le n}$$

be the eigenvalues of $\sigma_0$ on $V$, $N = |\Delta^+|$, and let $d_j = \deg I_j$ $(1 \le j \le n)$. The following theorem can be proved along much the same lines as Solomon's theorem (Section 7.2), although at every step the details are more subtle.

THEOREM (8.6). *Let G be a universal Chevalley group with indecomposable root system* $\Delta$, *and a finite field k of characteristic p. Let* $\{\sigma, q, \lambda_j, \varepsilon_{0j}\}$ *etc., be defined as above. Then*

(a) $\displaystyle\sum_{w \in W_\sigma} t^{|\Delta_w^-|} = \prod_{j=1}^{n} \frac{(1-\lambda_j t^{d_j})}{(1-\varepsilon_{0j} t)}$,

(*where t is an indeterminate*).

(b) $|\mathfrak{U}_\sigma| = q^N$, $|\mathfrak{U}_{w\sigma}^-| = q^{|\Delta_w^-|}$.

(c) $\displaystyle |G_\sigma| = q^N \prod_{j=1}^{n} (q^{d_j} - \lambda_j) = q^N \left( \prod_{j=1}^{n} (q - \varepsilon_{0j}) \right) \sum_{w \in W_\sigma} q^{|\Delta_w^-|}$.

For tables of the orders of the twisted groups, and identifications of ${}^2A_n$ and ${}^2D_n$ with unitary and orthogonal groups, see Steinberg (1959, 1967, 1968) and Carter (1965). The groups of type ${}^2C_2$ were discovered by Suzuki (1960), and the groups of types ${}^2G_2$ and ${}^2F_4$ by Ree (1961).

The isomorphisms between Chevalley groups (including the twisted types) are described in Steinberg (1967).

## 9. Representations of Chevalley Groups

Since a Chevalley group $G$ associated with a field $k$ is defined by a representation on a vector space over $k$, it is natural to begin the study of representations of $G$ by classifying the irreducible $\bar{k}G$-modules, where $\bar{k}$ is an algebraic closure of $k$. This was first done by Steinberg (1963), starting from the representations of $G$ coming from the Lie algebra, (see also Borel (1970) for a survey of these results.) We shall describe below a more direct, but perhaps less explicit, construction of the irreducible modules, due to Richen (1969) and Curtis (1965); the reader is referred to these works for details. Still unsolved is the problem of finding the dimensions and Brauer characters of the irreducible modules.

The theory of complex representations of the finite Chevalley group $G$ consists so far of scattered partial results, with no general survey of the characters or conjugacy classes available, except for Green's definitive work on $GL_n(k)$, (1955). Some general methods for extracting irreducible characters and representations from induced linear representations from various subgroups will be described briefly, with some examples.

### 9.1. Modular Representations of Finite Groups with Split $(B, N)$-pairs

In order to study the representations of the Chevalley groups and twisted types simultaneously, it seems useful to axiomatize the situation as follows.

Definition (9.1). Let $G$ be a finite group, and $p$ a prime number. The group $G$ is said to have a *split $(B, N)$-pair of characteristic $p$* and *rank $n$* provided that

(i) $G$ has a $(B, N)$-pair, such that

$$H = \bigcap_{m \in N} mBm^{-1},$$

with $n = |R|$, the number of distinguished generators of the Weyl group;

(ii) $B = UH$, with $U$ a normal $p$-subgroup of $B$, and $H$ an abelian group, of order prime to $p$.

We first remark that for a group with a $(B, N)$-pair, it is no loss of generality to assume that

$$H = \bigcap_{m \in N} mBm^{-1}$$

(Bourbaki, 1968). It has been shown previously in these lectures that all finite Chevalley groups and twisted types possess split $(B, N)$-pairs of characteristic equal to the characteristic of the field associated with the group. We remark that finite simple groups with split $(B, N)$-pairs of ranks 1† or 2 have not been classified (in contrast to the situation for rank $\geq 3$, where Tits has classified all finite simple groups with $(B, N)$-pairs). The known finite simple groups with split $(B, N)$-pairs of rank 1 or 2 are as follows

*rank* 1: $A_1(q)$ $(SL_2(q))$, ${}^2A_2(q^2)$ $(\cong U_3(q^2))$, ${}^2C_2$ (Suzuki groups), ${}^2G_2$ (Ree groups).

*rank* 2: $A_2(q)$, $B_2(q)$, $G_2(q)$, ${}^2A_3(q^2)$ $(\cong U_4(q^2))$, ${}^2A_4(q^2)^3$, $D_4(q)$, and ${}^2F_4$.

Now let $G$ be an arbitrary finite group with a split $(B, N)$-pair. Then $(W, R)$ is a Coxeter system (Bourbaki, 1968) and $W$ is isomorphic to the Weyl group of a uniquely‡ determined system of roots $\Delta$, such that the set $R$ corresponds to a set of fundamental reflections of the Weyl group $W(\Delta)$ with respect to some base $\Pi$. We shall identify $W$ with $W(\Delta)$ and $R$ with the set

† See, however, forthcoming papers by Hering, Kantor, O'Nan and Seitz.

‡ Except in the case $B-C$, since the Weyl groups of types $B_n$ and $C_n$ are isomorphic.

of fundamental reflections $\{r_\alpha : \alpha \in \Pi\}$. We let $w_0$ denote the unique element of $W$ such that $w_0(\Delta) = \Delta^-$. We next define "one parameter subgroups" in $G$.

DEFINITION (9.2). For $r = r_\alpha \in R$, let $U_\alpha = U \cap r_\alpha w_0 U w_0^{-1} r_\alpha^{-1}$.

*Exercise.* In case $G$ is a Chevalley group, show that $U_\alpha = \mathfrak{X}_\alpha$ for $\alpha \in \Pi$.

Note that $H$ normalizes the subgroups $\{U_\alpha : \alpha \in \Pi\}$, so that $W = N/H$ acts on the set of $N$-conjugates of the $\{U_\alpha : \alpha \in \Pi\}$.

PROPOSITION (9.3). *The subgroups $U_\alpha$ are nontrivial for all $\alpha \in \Pi$. Moreover, the permutation representation of $W$ on $\Delta$ is equivalent to the permutation representation of $W$ on the set of $N$-conjugates of $\{U_\alpha : \alpha \in \Pi\}$. More precisely, the map*

$$\phi : w(\alpha) \to n_w U_\alpha n_w^{-1} \quad (\textit{for } n_w H = w)$$

*is a bijection such that, on the set $\Delta$,*

$$\phi \circ w' = i(n_{w'}) \circ \phi$$

*for all $w' \in W$, where $i(n_{w'})$ denotes the inner automorphism determined by a representative $n_{w'} \in N$ of $w' \in W$.*

Thus we can speak unambiguously of subgroups $U_\alpha$, for all $\alpha \in \Delta$, and have the familiar rule

$$n_w U_\alpha n_w^{-1} = U_{w(\alpha)} \quad (w \in W, \alpha \in \Delta).$$

Most of the other structural properties of Chevalley groups derived in Section 3 carry over to groups with split $(B, N)$-pairs (Richen, 1969).

Our objective is to describe, for a group $G$ with a split $(B, N)$-pair of characteristic $p$, the irreducible $KG$-modules, where $K$ is an algebraically closed field of characteristic $p$.

Let $M$ be an arbitrary left $KG$-module. We first observe that $M$ contains a line (= one dimensional subspace) stabilized by $B$. To see this, let $M_1$ be an irreducible $KB$-submodule of $M$. Since $U \lhd B$, $M_1|U$ is a completely reducible $KU$-module, by Clifford's theorem. But $U$ is a $p$-group, and so acts trivially on $M_1$. Then $M_1$ is an irreducible $B/U \cong H$ module, and hence one-dimensional, since $H$ is abelian and $K$ is algebraically closed.

By analogy with the representation theory of semi-simple Lie algebras described in Section 4 (and of semi-simple algebraic groups), one might expect irreducible $KG$-modules to be parametrized in a bijective fashion by the one-dimensional representations of $B$ afforded by lines stabilized by $B$. Such a bijective correspondence is not available, however. For example, in case $G = SL_2(p)$, $G$ has irreducible modules of dimensions 1, 2,..., $p$, and both the modules of dimension 1 and $p$ contain unique lines stabilized by $B$ and affording the trivial representation of $B$.

The data needed to describe the modules is contained in the following definition.

*Notations.* For a subset $X \subseteq G$, let

$$\bar{X} = \sum_{x \in X} x.$$

Also, let $\{(w) : w \in W\}$ denote a fixed set of coset representatives of $W$ in $N$.

DEFINITION (9.4). An element $m \neq 0$ in a left $KG$-module $M$ is called a *weight element* of weight $(\chi, \mu_1, \ldots, \mu_n)$, where $n = |R|$, and $\chi : B \to K^*$ is a homomorphism, and the $\mu_i \in K$, provided that

$$bm = \chi(b)m \quad (b \in B),$$

*and*

$$\bar{U}_{\alpha_i} r_{\alpha_i} m = \mu_i m, \quad \text{for } \alpha_i \in \Pi \qquad (1 \le i \le n).$$

THEOREM (9.5) (Richen, 1969). *Let $G$ be a finite group with a split $(B, N)$-pair of rank $n$ and characteristic $p$, and let $K$ be an algebraically closed field of characteristic $p$.*

(a) *Every left KG-module contains a weight element.*
(b) *If $m \in M$ is a weight element, then $KGm = KU^- m$, where $U^- = (w_0)U(w_0)^{-1}$.*
(c) *Each irreducible KG-module contains a unique line fixed by B.*
(d) *If $M_1$ and $M_2$ are irreducible modules containing weight elements of the same weight, then $M_1 \cong M_2$. Conversely, each irreducible module determines a unique weight.*

DEFINITION (9.6). A finite group with a split $(B, N)$-pair is *restricted* in case a set of coset representatives $\{(w) : w \in W\}$, can be chosen such that $(r_\alpha) \in U_\alpha U_{-\alpha} U_\alpha$, for $\alpha \in \Pi$.

We note that the Chevalley groups and twisted types all possess restricted split $(B, N)$-pairs. We assume from now on that the coset representatives $\{(w) : w \in W\}$ satisfy the condition $(r_\alpha) \in U_\alpha U_{-\alpha} U_\alpha$, for $\alpha \in \Pi$. Also for $\alpha_i \in \Pi$ $(1 \le i \le n)$, let

$$H_i = H \cap \langle B, U_{-\alpha_i} \rangle \quad (1 \le i \le n).$$

THEOREM (9.7) (Richen, 1969). *Let $G$ be a finite group with a restricted split $(B, N)$-pair, and let $\{\chi, \mu_1, \ldots, \mu_n\}$ be a weight of an irreducible KG-module. Then $\mu_i = 0$ or $-1$. Moreover, $\chi | H_i \neq 1$ implies that $\mu_i = 0$. Conversely, for every set $(\chi, \mu_1, \ldots, \mu_n)$, satisfying* (a) *$\mu_i = 0$ or $-1$, and* (b) *$\mu_i = 0$ whenever $\chi | H_i \neq 1$, ($\chi$ being a linear representation of B), there exists an irreducible KG-module of weight $(\chi, \mu_1, \ldots, \mu_n)$. Finally, let $(\chi, \mu_1, \ldots, \mu_n)$ be a weight of an irreducible module and let*

$$J = \{r_{\alpha_j} \in R : \mu_j = 0 \ \ and \ \ \chi | H_j = 1\}.$$

*Then, letting*

$$H_\chi = \sum_{h \in H} \chi(h^{-1})h,$$

*the expression* $$m = \sum_{w \in W_J} U_{w_0 w} H_\chi(w^{-1})(w_0)\bar{U},$$

*(where* $U_w^- = U \cap w^{-1} w_0 U w_0^{-1} w$*) is a weight element in KG such that KGm is an irreducible module of weight* $(\chi, \mu_1 \ldots, \mu_n)$.

COROLLARY (9.8). *Let G be as in Theorem* (9.7).

(a) *The number of non-isomorphic irreducible KG-modules is* $\sum_{J \subseteq R} |H/H_J|$, *where* $H_J = \langle H_i : r_{\alpha_i} \in J \rangle$.

(b) *Suppose that* $H = \prod H_i$ *(direct product). Then the number of non-isomorphic irreducible modules is*

$$\prod_{i=1}^{n} (|H_i| + 1).$$

(c) *Suppose* $H = \langle H_i \rangle$. *For every irreducible module M,* dim $M \leq |U|$, *with equality if and only if the weight of M is* $(1_B, -1, \ldots, -1)$.

(d) *Let* $H = \langle H_i \rangle$ *as in* (c). *Then G has a unique p-block of defect* 0.

*Remarks.*

(a) Suppose $G$ is a universal Chevalley group over a field $k$ such that $|k| = q$ for some prime power $q = p^f$. Then the number of distinct irreducible $KG$-modules is, by Brauer's theorem, the number of $p$-regular conjugacy classes in $G$, and is equal to $q^n$ ($n = |R|$) (since $H = \prod H_i$ (direct product) in that case, and $H_i = \{h_{\alpha_i}(t) : t \in k^*\}$.)

(b) The fact that $G$ has at most one $p$-block of defect zero follows from (c), since $U$ is a $p$-Sylow subgroup of $G$ (Richen, 1969). The existence of at least one follows from the result of Curtis (1966) that there exists an irreducible complex character of $G$ whose degree is $[B : H] = |U|$.

We conclude with a reformulation of the isomorphism criterion for irreducible $KG$-modules.

PROPOSITION (9.9) (Curtis, 1970). *Let G satisfy the hypothesis of Theorem* (9.7). *Let M be an irreducible KG-module. Then M contains a unique line stabilized by B, so that if* $G_J$ *is the full stabilizer of the line, then* $G_J$ *is a parabolic subgroup, and M determines* $G_J$ *and the homomorphism* $\chi : G_J \to K^*$ *(afforded by the line) uniquely. Two irreducible KG-modules are isomorphic if and only if the parabolic subgroups and homomorphisms into* $K^*$ *associated with them, coincide.*

## 9.2. Representations and Characters of Chevalley Groups in the Complex Field

The characters of some individual Chevalley groups, and of all groups belonging to some infinite families, have been completely determined (Green, 1955 and Srinivasan, 1968). Rather than reporting on this work, we shall describe in this section a few general methods which apply to all Chevalley groups, some suggested by the above work, for splitting off irreducible components of induced characters from subgroups. We refer to Feit (1967) and Curtis and Reiner (1962) for notations and background from character theory.

(a) *Some irreducible components of* $1_B^G$. Let $G$ be a finite group with a $(B, N)$-pair, and let $B$ be the Borel subgroup in $G$, and $W$ the Weyl group. The Bruhat theorem asserts that there is a bijective correspondence $w \to BwB$ between $W$ and the family of $(B, B)$-double cosets. More generally, let $J, J'$ be subsets of the set of distinguished generators $R$ of $W$, and let $W_J = \langle J \rangle$, and $G_J$ the corresponding parabolic subgroup. An easy extension of the Bruhat theorem shows that there is a bijective correspondence between the $(W_J, W_{J'})$-double cosets in $W$ and the $(G_J, G_{J'})$ double cosets in $G$, given by

$$W_J w W_{J'} \to G_J w G_{J'} = B(W_J w W_{J'})B.$$

We shall combine this result with the following lemma.

Lemma (9.10). *Let $A$, $B$ be subgroups of a finite group $H$, and let $1_A$, $1_B$ be the principal characters of $A$ and $B$ respectively. Then $(1_A^H, 1_B^H)$ is equal to the number of $(A, B)$-double cosets in $H$.*

Proposition (9.11). *Let $G$ be a finite group with a $(B, N)$-pair, and let $W$ be the Weyl group of $G$. Suppose $\psi$ is an irreducible character of $W$ such that*

$$\psi = \sum_{J \subseteq R} n_J 1_{W_J}^W \quad (n_J \in \mathbf{Z}).$$

*Then*

$$\chi = \sum_{J \subseteq R} n_J 1_{G_J}^G$$

*is a generalized character of $G$ such that $\pm\chi$ is an irreducible character.*

The proof is immediate from what has been shown before, since by Lemma (9.10),

$$\begin{aligned} 1 = (\psi, \psi) &= \sum n_J n_{J'} (1_{W_J}^W, 1_{W_{J'}}^W) \\ &= \sum n_J n_{J'} (1_{G_J}^G, 1_{G_{J'}}^G) \\ &= (\chi, \chi). \end{aligned}$$

In particular, from Section 7.2, we have Solomon's result that the character $\varepsilon$ of $W$ given by $\varepsilon(r) = -1$ ($r \in R$) can be expressed in the form

$$\varepsilon = \sum_{J \subseteq R} (-1)^{|J|} 1_{W_J}^{W}.$$

One can then prove

PROPOSITION (9.12) (Curtis, 1966). *Let $G$, $W$ be as in* (9.11).

(a) *The character*

$$\chi = \sum_{J \subseteq R} (-1)^{|J|} 1_{G_J}^{G}$$

*is an irreducible character of $G$, of degree $[B : B \cap B^{w_0}]$.*

(b) *$\chi$ is the unique irreducible character of $G$ such that $(\chi, 1_B^G) > 0$ and $(\chi, 1_{G_J}^G) = 0$ for all $J \neq \phi$.*

This character is called the *Steinberg character* of $G$, and was discovered by Steinberg first for the groups $GL_n(k)$, and then for arbitrary Chevalley groups and twisted types (Steinberg, 1951, 1957). Its existence was used to prove Corollary (9.8) (d) on the existence of a block of defect zero for a finite group with a split $(B, N)$-pair.

Steinberg (1951) used the result of Frobenius that every irreducible character of the symmetric group $S_{n-1}$ (the Weyl group of $GL_n(k)$) has the form $\sum n_J \cdot 1_{W_J}^W$ ($n_J \in \mathbf{Z}$) (as in (9.11)) to prove that every irreducible character in $1_B^G$, for $G = GL_n(k)$, has the form $\sum n_J 1_{G_J}^G$. These characters played an important role in Green's determination of all characters of $GL_n(k)$. Unfortunately, this method yields only the Steinberg character and $1_G$ as constituents of $1_B^G$ for some Chevalley groups of other types (e.g., $B_2$), so that other methods are needed to find the components of $1_B^G$ in general.

(b) *Characters and centralizer rings.* We describe a few general theorems due to Curtis and Fossum (1968), which will be applied to the problem of decomposing $1_B^G$ (and other induced characters) in Chevalley groups.

Let $H$ be a subgroup of a finite group $G$, and let $\mathbf{C}G$ and $\mathbf{C}H$ denote their complex group algebras, with $\mathbf{C}H \subseteq \mathbf{C}G$. Let $\psi$ be an irreducible complex character of $H$, afforded by a minimal left ideal $\mathbf{C}He$, where $e$ is a primitive idempotent in $\mathbf{C}H$. Then $\mathbf{C}Ge$ affords the induced character $\psi^G$, and we have a natural isomorphism

$$e\mathbf{C}Ge \cong \mathrm{Hom}_{\mathbf{C}G}(\mathbf{C}Ge, \mathbf{C}Ge).$$

The subalgebra $e\mathbf{C}Ge$ is called the *centralizer ring* (or *Hecke algebra*) of the induced character $\psi^G$.

The next result relates characters of $G$ which are constituents of $\psi^G$ to characters of the centralizer ring.

PROPOSITION (9.13). *Let $\zeta$ be an irreducible character of $G$ such that $(\zeta, \psi^G) > 0$. Then the restriction $\zeta_E$ of $\zeta$ to the centralizer ring $E = e\mathbf{C}Ge$ is an* *irreducible character of E of degree* $(\zeta, \psi^G)$. *Conversely, each irreducible character of E is the restriction to E of a unique irreducible character of G which appears with positive multiplicity in* $\psi^G$.

The proposition states that there is a natural bijective correspondence $\zeta \to \zeta_E = \phi$ between irreducible constituents $\{\zeta\}$ of $\psi^G$ and characters $\{\phi\}$ of $E$, such that the degree of $\phi$ is equal to the multiplicity $(\zeta, \psi^G)$. In particular, in case $E$ is a commutative algebra, all the multiplicities $(\zeta, \psi^G) = 1$.

PROPOSITION (9.14) (Janusz, 1966). *Suppose that* $(\zeta, \psi^G) = 1$, *and let*

$$e(\zeta) = \zeta(1)|G|^{-1} \sum_{x \in G} \zeta(x^{-1})x.$$

*Then* $e(\zeta)e$ *is a primitive idempotent in* $\mathbf{C}G$ *such that* $\mathbf{C}Ge(\zeta)e$ *affords the character* $\zeta$.

A construction of the primitive idempotents yields all the values of the corresponding character, by the following result.

PROPOSITION (9.15) (Littlewood, 1940). *Let* $f = \sum \lambda(g)g$ *be a primitive idempotent in* $\mathbf{C}G$, *affording the irreducible character* $\zeta$. *Let* $\mathsf{C}$ *be a conjugacy class in G, and let* $g \in \mathsf{C}$. *Then*

$$\zeta(g^{-1}) = |C_G(g)| \sum_{g \in \mathsf{C}} \lambda(g).$$

Now we specialize to the case of a linear representation $\psi$ of $H$. Then the idempotent $e$ is given explicitly by $e = |H|^{-1}\sum\psi(h^{-1})h$. For $x \in G$, we write $H^x$ for $x^{-1}Hx$ and $x$ for $[H : H \cap H^x]$. $\Psi^x$ will denote the character defined by $\psi^x(h^x) = \psi(h)$ $(h \in H)$.

PROPOSITION (9.16). *Let*

$$G = \bigcup_{j \in I} Hx_jH$$

*(disjoint). Let* $J \subseteq I$ *be the set of indices* $\{j\}$ *such that* $\psi = \psi^{x_j}$ *on* $H \cap H^{x_j}$, *and let* $\beta_j = (\text{ind } x_j)ex_je$. *The elements* $\{\beta_j\}_{j \in J}$ *form a basis for* $e\mathbf{C}Ge$, *such that if* $\beta_i\beta_j = \sum a_{ijk}\beta_k$, *the* $a_{ijk}$ *are algebraic integers.*

COROLLARY (9.17). $\psi^G$ *is irreducible if and only if* $\psi^x \neq \psi$ *on* $H \cap H^x$ *for all* $x \notin H$.

The next result gives some orthogonality relations in the centralizer ring.

PROPOSITION (9.18). *Let* $\{\beta_j\}$ *be the basis of the centralizer ring* $e\mathbf{C}Ge$, *and let* $\{\hat{\beta}_j\}$ *be the "dual basis" consisting of the elements* $\hat{\beta}_j = (\text{ind } x_j)ex_j^{-1}e$ $(j \in J)$. *Let* $\zeta, \zeta'$ *be irreducible characters of G appearing in* $\psi^G$ *with positive multiplicity. Then*

$$\frac{\zeta(1)}{[G:H]}\sum_{j\in J}\frac{1}{\operatorname{ind} x_j}\zeta(\hat{\beta}_j)\zeta'(\beta_j)=\begin{cases}0 & (\zeta\neq\zeta')\\ (\zeta,\psi^G) & (\zeta=\zeta')\end{cases}.$$

From this result we obtain an explicit formula for $\zeta(1)$, computed entirely in terms of the centralizer ring.

COROLLARY (9.19). *Let* $\zeta$, $\psi$, *etc., be as in* (9.18). *Then*

$$\zeta(1)=\frac{[G:H](\zeta,\psi^G)}{\sum_{j\in J}\frac{1}{\operatorname{ind} x_j}\zeta(\hat{\beta}_j)\zeta(\beta_j)}$$

Finally, in case $(\zeta,\psi^G)=1$, we obtain, from (9.14) the following result.

COROLLARY (9.20). *Let* $(\zeta,\psi^G)=1$. *Then*

$$u=\frac{\zeta(1)}{[G:H]}\sum\frac{1}{\operatorname{ind} x_j}\zeta(\hat{\beta}_j)\beta_j$$

*is a primitive idempotent in* $\mathbf{C}G$ *such that* $\mathbf{C}Gu$ *affords* $\zeta$.

(c) *Applications to Chevalley groups.* A natural place to begin to apply the results of the last Section 9.2 (b) to Chevalley groups is with induced characters $\psi^G$, where $\psi$ is a linear character (a) on the unipotent subgroup $U$, (b) on the Borel subgroup $B$, and (c) on the parabolic subgroups $\{G_J\}$.

For example, if $\psi$ is a linear character on $\mathfrak{U}$ which is in "general position", in the sense that $\psi|\mathfrak{X}_{\alpha_i}\neq 1$ for all fundamental roots $\alpha_i\in\Pi$, then the centralizer ring of $\psi^G$ is commutative (Yokonuma, 1967)†. The problem of constructing the representations of this algebra has not as yet been completely solved. Yokonuma has also determined the structure of the centralizer ring $1_{\mathfrak{U}}^G$.

We shall discuss in somewhat more detail the structure of the centralizer ring of $1_B^G$, for an arbitrary finite group with a $(B, N)$-pair. In this case, the "standard basis" given in (9.16) for the centralizer ring $e\mathbf{C}Ge$, where

$$e=|B|^{-1}\sum_{b\in B}b,$$

consists of elements $\{\beta_w\}_{w\in W}$, such that

$$\beta_w=|B|^{-1}\sum_{x\in BwB}x\qquad(w\in W).$$

For the case of a permutation representation $1_B^G$, we shall use the notation $H_{\mathbf{C}}(G, B)$ for $e\mathbf{C}Ge$.

† This result was stated by Gelfand–Graev (1962), but proved only for $SL_n(k)$. The second assertion in this work is true for $SL_n(k)$ but not for all types of Chevalley groups (Steinberg, 1967).

PROPOSITION (9.21). (Iwahori, 1964 and Matsumoto, 1964). *Let $G$ be a finite group with a $(B, N)$-pair, and Coxeter system $(W, R)$. For $r \in R$, let $q_r = [B : B \cap B^r]$. Then the multiplication in $H_{\mathbf{C}}(G, B)$ is determined by the formulae*

$$\beta_r\beta_w = \beta_{rw} \quad (r \in R,\ w \in W,\ l(rw) > l(w)),$$
$$\beta_r\beta_w = q_r\beta_{rw} + (q_r - 1)\beta_w \quad (r \in R,\ w \in W,\ l(rw) < l(w)).$$

*Moreover, the algebra $H_{\mathbf{C}}(G, B)$ has a presentation with generators $\{\beta_r : r \in R\}$ and relations*

$$\beta_r{}^2 = q_r e + (q_r - 1)\beta_r \quad (r \in R),$$

*(where*

$$e = |B|^{-1} \sum_{b \in B} b$$

*is the identity element)*
*and*

$$\underbrace{\beta_r\beta_s\beta_r\cdots}_{m(r,\,s)} = \underbrace{\beta_s\beta_r\beta_s\cdots}_{m(r,\,s)}$$

*in case the order of the product $rs$ in $W$ is $m(r, s)$.*

COROLLARY (9.22). *The one dimensional representations $\phi$ of $H_{\mathbf{C}}(G,B)$ are determined as follows. $\phi(\beta_r)$ is either $-1$ or $q_r$. If there is only one root length in the root system of $W$, then there are exactly two one dimensional representations of $H_{\mathbf{C}}(G, B)$, while if there are two root lengths, there are four such representations.*

In particular, the homomorphism ind : $\beta_r \to q_r$ $(r \in R)$, corresponds to $1_G$, in the sense of Prop. (9.13). The homomorphism $\varepsilon : \beta_r \to -1$ for all $r \in R$, corresponds to the Steinberg character $\chi$. In fact this seems to be a good way to define the Steinberg character. The fact that the degree $\chi(1) = [B : H]$ follows from the formula (9.19), without making use of Solomon's formula $\varepsilon = \sum(-1)^{|J|} 1_{W_J}^W$.

EXAMPLE. Let $G$ be a Chevalley group, associated with a finite field $k$ of $q$ elements. We shall give a list of the degrees of the irreducible characters in $1_B^G$ corresponding to one dimensional representations of $H_{\mathbf{C}}(G, B)$. Besides the trivial character $1_G$, there exists, in all cases, the Steinberg character, whose degree is $q^N$, where $N$ is the number of positive roots. For the indecomposable root systems having two root lengths, there are two other irreducible characters $\zeta_1$, $\zeta_2$, whose degrees are as follows.

$$(B_n),\ (C_n) \qquad \zeta_1(1) = \frac{q(q^{n-1}+1)(q^n+1)}{2(q+1)}$$

$$\zeta_2(1) = \frac{q^{(n-1)^2}(q^{n-1}+1)(q^n+1)}{2(q+1)}$$

$$(F_4) \qquad \zeta_1(1) = \zeta_2(1) = \frac{q^4(q^3+1)^2(q^2+1)(q^4+1)(q^6+1)}{8(q+1)^2}$$

$$(G_2) \qquad \zeta_1(1) = \zeta_2(1) = \frac{q(q^4+q^2+1)}{3}$$

These characters are rational on the whole group, and their values can be computed in principle, but not yet explicitly, using (9.15). The complete story depends on a better understanding of the conjugacy classes. The values of the Steinberg character on all classes have been determined by Steinberg (1968), in case $G$ is a Chevalley group or a twisted type $G_\sigma$.

The scope of these methods is extended considerably by applying the same ideas to the Hecke algebras $H_{\mathbf{C}}(G, G_J)$ for arbitrary parabolic subgroups (Curtis, Iwahori and Kilmoyer, (1971)). A key point in this work is Tits' introduction of a generic ring associated with a Coxeter system (Bourbaki, 1968) and its use to prove the isomorphism of the Hecke algebras $H_{\mathbf{C}}(G, B)$ with the group algebra of the Weyl group $\mathbf{C}W$. In fact, it can be shown by the same method, that for all subsets $J \subseteq R$, $H_{\mathbf{C}}(G, G_J) \cong H_{\mathbf{C}}(W, W_J)$, (Iwahori, 1969 and Curtis, Iwahori and Kilmoyer, 1971).

## Bibliography


Borel, A. (1970). Properties and linear representations of Chevalley groups. Topics in finite and algebraic groups. Springer Lecture Note Series. **131.**

Bourbaki, N. (1968). "Groupes et algèbres de Lie". Act. Sci. Ind. 1337, Hermann, Paris.

Carter, R. (1965). Simple groups and simple Lie algebras. *J. London Math. Soc.* **40,** 193.

Carter, R. "Chevalley groups". To be published.

Chevalley, C. (1955). Sur certains groupes simples. *Tôhoku Math. J.* (2) **7,** 14.

Chevalley, C. (1955). Invariants of finite groups generated by reflections. *Amer. J. Math.* **77,** 778.

Coleman, A. J. (1958). The Betti numbers of the simple groups. *Can. J. Math.* **10,** 349.

Coxeter, H. S. M. (1951). The product of the generators of a finite group generated by reflections. *Duke Math. J.* **18,** 765.

Curtis, C. W. (1965). Irreducible representations of finite groups of Lie type. *Journal für Math.* **219,** 180.

Curtis, C. W. (1966). The Steinberg character of a finite group with a $(B, N)$-pair. *J. Algebra* **4,** 433.

Curtis, C.W. (1970). Modular representations of finite groups with split $(B, N)$-pairs Topics in finite and algebraic groups. Springer Lecture Note Series. **131.**

Curtis, C. W. and Fossum, T. (1968). On centralizer rings and characters of representations of finite groups. *Math. Zeit.* **107,** 402.

Curtis, C. W., Iwahori, N. and Kilmoyer, R. (1971). Hecke algebras and characters of parabolic type of finite groups with $(B, N)$-pairs. To appear.

Curtis, C. W. and Reiner, I. (1962). Representation Theory of Finite Groups and Associative Algebras. John Wiley and Sons, New York.

Feit, W. (1967). Characters of finite groups. Benjamin, New York.
Gelfand, I. M. and Graev, M. I. (1962). Construction of irreducible representations of simple algebraic groups over finite fields. *Doklady*, **147,** 529; *Soviet Math.* **3.**
Green, J. A. (1955). The characters of the finite general linear groups. *Trans. Amer. Math Soc.* **80,** 402.
Iwahori, N. (1964). On the structure of a Hecke ring of a Chevalley group over a finite field. *J. Fac. of Sci. Univ. Tokyo* **10,** 215.
Iwahori, N. (1969). On some properties of groups with $(B, N)$-pairs, pp. 203–212. Theory of Finite Groups, R. Brauer and C. H. Sah (eds). Benjamin, New York.
Iwahori, N. (1970). Centralizers of involutions in finite Chevalley groups. Topics in finite and algebraic groups. Springer Lecture Note Series. **131.**
Jacobson, N. (1962). Lie algebras. John Wiley and Sons, New York.
Janusz, G. (1966). Primitive idempotents in group algebras. *Proc. Amer. Math. Soc.* **17,** 520.
Kostant, B. (1966). Groups over **Z.** Algebraic groups and discontinuous subgroups. *Proc. Symp. pure math.* **9,** A. M. S. Providence, R. I.
Littlewood, D. E. (1940). Theory of group characters and matrix representations of groups, Oxford.
Matsumoto, H. (1964). Générateurs et relations des groupes de Weyl généralisé. *C. R. Acad. Sci. Paris* **258,** 3419.
Ree, R. (1957). On some simple groups defined by C. Chevalley. *Trans. Amer. Math. Soc.* **84,** 392.
Ree, R. (1961). A family of simple groups associated with the simple Lie algebra of type $(F_4)$. *Amer. J. Math.* **83,** 401.
Ree, R. (1961). A family of simple groups associated with the simple Lie algebra of type $(G_2)$. *Amer. J. Math.* **83,** 432.
Ree, R. (1965). Classification of involutions and centralizers of involutions in certain simple groups. *Proc. Int. Conf. Austral. Nat. Univ. Canberra*, 281.
Richen, F. (1969). Modular representations of split $(B, N)$-pairs. *Trans. Amer. Math. Soc.* **140,** 435.
Serre, J-P. (1966). Algèbres de Lie semi-simples complexes. Benjamin, New York.
Shepard, G. C. and Todd, J. A. (1954). Finite unitary reflection groups. *Can. J. Math.* **6,** 376.
Solomon, L. (1966). The orders of the finite Chevalley groups. *J. Algebra* **3,** 376.
Springer, T. A. and Steinberg, R. (1970). Conjugacy classes. Springer Lecture Notes Series. **131.**
Srinivasan, B. (1968). The characters of the finite symplectic group $Sp(4, q)$. *Trans. Amer. Math. Soc.* **131,** 488.
Steinberg, R. (1951). A geometric approach to the representations of the full linear group over a Galois field. *Trans. Amer. Math. Soc.* **71.** 274.
Steinberg, R. (1957). Prime power representations of finite linear groups II. *Can. J. Math.* **9,** 347.
Steinberg, R. (1959). Finite reflection groups. *Trans. Amer. Math. Soc.* **91,** 493.
Steinberg, R. (1959). Variations on a theme of Chevalley. *Pacific J. Math.* **91,** 875.
Steinberg, R. (1962). Générateurs, relations et revêtements de groupes algébriques. Colloque sur la théorie des groupes algébriques, Bruxelles.
Steinberg, R. (1963). Representations of algebraic groups. *Nagoya Math. J.* **22,** 33.
Steinberg, R. (1967). Lectures on Chevalley groups (mimeographed notes). Yale University.

Steinberg, R. (1968). Endomorphisms of algebraic groups. *Memoirs Amer. Math. Soc. No.* 80.

Suzuki, M. (1960). A new type of simple groups of finite order. *Proc. Math. Acad. Sci. U.S.A.* **46,** 868.

Tits, J. (1962). Théorème de Bruhat et sous-groupes paraboliques. *C. R. Acad. Sci. Paris* **254,** 2910.

Tits, J. (1962). Groupes simples et géometries associées. *Proc. Int. Cong. Math.*

Tits, J. (1964). Algebraic and abstract simple groups. *Ann. of Math.* (2) **80,** 313.

Tits, J. (1966). Sur les constantes de structure et le théorème d'existence des algèbres de Lie semi-simples. *Inst. Hautes Études Sci., Math. Publ.* **31,** 21.

Tits, J. Buildings of spherical types and finite (*B*, *N*)-pairs. Springer Lecture Note Series. To appear.

Yokonuma, T. (1967). Sur le commutant d'une représentation d'un groupe de Chevalley fini. *C. R. Acad. Sci. Paris*, **264,** 433.

Yokonuma, T. (1967). Sur la structure des anneaux de Hecke d'un groupe de Chevalley fini. *C. R. Acad. Sci. Paris*, **264,** 344.

CHAPTER IV

# Finite Complex Linear Groups of Small Degree

H. S. LEONARD, JR.



## 1. Introduction

Throughout we denote by $(G, X)$ a finite group $G$ together with a faithful representation $X$ over the complex field, and we denote the degree of $X$ by $n$. We wish to consider here some aspects of the present status of the general question as to what can be said about the structure of $G$ when the degree $n$ is specified or restricted.

DEFINITION. If $X$ is irreducible and if, for every normal subgroup $H$, all constituents of the restriction $X_H$ of $X$ to $H$ are equivalent, then $X$ is called *quasi-primitive*..

If $X$ is irreducible but not quasi-primitive then $G$ has a proper normal subgroup $H$ not contained in the center $Z$ of $G$ such that $X$ is induced by a representation of a proper subgroup $K$ of $G$ with $H \subseteq K$ (Curtis and Reiner, 1962), so that in particular $X$ is imprimitive. Thus the assumption of quasi-primitivity is not a serious restriction. Furthermore it has the obvious and useful consequence that if $H$ is a normal abelian subgroup of $G$ then $H \subseteq Z$.

All the groups $G$ for which $n < 8$ have been determined by the following authors under the assumptions that the values of $X$ have determinant one and, for $n = 4$, 5 and 7, that $X$ is primitive, and for $n = 6$ that $X$ is quasi-primitive. Results and references in the cases $n = 2$, 3 and 4 were given by Blichfeldt (1917) and van der Waerden (1948) and in all six cases by Feit (1971).

| | |
|---|---|
| $n = 2$ | F. Klein and others, |
| $n = 2$ or 3 | C. Jordan and others, |
| $n = 4$ | Blichfeldt and others, |
| $n = 5$ | Brauer (1967), |
| $n = 6$ | Lindsey, II (1969), |
| $n = 7$ | Wales (1969, 1970). |

Among the groups of degree 6 is the group $G$ with $G/Z$ the Hall–Janko group of order $2^7 \cdot 3^3 \cdot 5^2 \cdot 7$ and $|Z| = 2$ (Lindsey II, 1968, 1969; Conway, 1969). The only groups of order divisible by primes greater than 7 are those with $G/Z = PSL(2, 11)$ or $PSL(2, 13)$. None are divisible by 49. These last two facts are consequences of Theorems 2 and 4, respectively, stated below, and they illustrate the way in which the various theorems which we shall discuss apply to groups $(G, X)$ of specified degree $n$.

## 2. Structure Theorems

Jordan's Theorem implies that if $n$ is specified and $X$ is quasi-primitive then to within isomorphism there are only finitely many possible groups $G/Z$.

THEOREM 1. (Jordan, 1878). *There exists a mapping $f$ of the natural number system such that if $(G, X)$ has degree $n$ then $G$ has a normal abelian subgroup $H$ of index at most $f(n)$.*

Specific bounds were obtained by Blichfeldt, Frobenius and Schur, but the problem of obtaining a substantially stronger bound remains open. Results and references may be found in Blichfeldt (1917), Curtis and Reiner (1962) and Speiser (1937). An analog of Jordan's Theorem for fields of prime characteristic has been obtained by Brauer and Feit (1966).

THEOREM 2. *Given $(G, X)$ of degree $n$ and given a prime $p$, suppose that $n < (2p+1)/3$. Then the Sylow $p$-subgroup $P$ of $G$ is abelian and one of the following holds:*

(a) $P \trianglelefteq G$;

(b) $p > 7$, $n \geq (p-1)/2$, *and $G$ has a normal subgroup $H$ such that* $P \cap H \trianglelefteq H$ *and* $G/H \simeq PSL(2, p)$; *if $X$ is irreducible then* $H = Z$.

(c) $p \leq 7$ *and* $G/Z \simeq PSL(2, p)$, $S_4$, $A_5$, $A_6$, *or* $A_7$ *(symmetric and alternating groups).*

The most difficult case $|P| = p$ of Theorem 2 is due to Brauer (1942) and Tuan (1944), and depends on modular representation theory. Thompson (1967) has provided a new more direct development of the necessary modular theory. The result when $|P|$ is arbitrary and $n < (p-1)/2$ was obtained by Feit and Thompson (1967), and in the form stated above the theorem was proved by Winter (1964), both proofs being by reduction of the problem to the case $|P| = p$. Under restrictions on $|Z|$, Feit (1967a) has obtained two extensions of Theorem 2 when $X$ is irreducible. When $|P| = p$, Brauer (1966), Brauer and Hayden (1963) and Hayden (1963) have obtained results when $n < (7p+1)/8$, which by Theorem 4 below, imply results for arbitrary $|P|$. Isaacs (1966) and Winter (1968) report other conditions implying the existence of a normal $p$-subgroup.

THEOREM 3. *Let G be a finite group having a cyclic Sylow subgroup P for some prime p. Suppose that G has a faithful representation L over a field of characteristic p of degree d. If* $d < \frac{2}{3}(p-1)$ *or* $p \geq 13$ *and* $d < \frac{3}{4}(p-1)$, *then every composition factor of G is a p-group, a p′-group, or is isomorphic to* $PSL(2, p)$.

For $p < 13$ or $p \geq 13$ and $d < \frac{7}{10}p - \frac{1}{2}$, Theorem 3 is part of a Theorem of Feit (1966). As he pointed out, the theorem is sharp when $p < 13$, because for $G/Z$ we may have $A_5$, $A_6$, $A_7$ or the first Janko group. The refinement stated here for $p \geq 13$ is among results obtained by Blau (1971a). We can verify (see Brauer, 1967) that Theorem 3 is also true for faithful complex representations $X$. Then, using Theorem 4, one can verify that for this version it is not necessary to assume $P$ is cyclic. Alternatively, this version of Theorem 3 is implied by the results of Brauer and Hayden mentioned above. On the other hand, if Theorem 3 can be strengthened sufficiently, it will yield an extension of part of the work of Brauer and Hayden (1963). Using modular theory, Blau (1971b) recently sharpened an important inequality of theirs.

THEOREM 4. *Given* $(G, X)$ *of degree n, let* $\pi$ *be a set of primes p for which* $n \leq p-1$, *and if* $n = 4$ *assume* $5 \notin \pi$. *Suppose that G has a Sylow* $\pi$*-subgroup M. Then either*

(a) *G has a normal abelian subgroup of order* $|M|$ *or* $|M|/p$ *for some prime p, or*

(b) $n+1 \in \pi$, *each composition factor of G is either cyclic or isomorphic to* $PSL(2, n+1)$, *and X is not quasi-primitive.*

This result is due to Feit (1964) in the case that $n+1 \notin \pi$ and to Lindsey (1971) in the general case. (If $n+1 \notin \pi$ then a theorem of Blichfeldt (1917) asserts the existence of a Sylow $\pi$-subgroup $M$ of $G$ (Burnside, 1955; Feit, 1967b; Speiser, 1937). Lindsey's proof includes a new proof of Feit's Theorem. Lindsey (1971) has made substantial progress assuming only that $n < \frac{3}{2}(p-1)$.

We call a subset $S$ of a group $G$ a *trivial intersection set* (t.i. set) if the intersection of every pair of distinct conjugates of $S$ contains at most the identity element of $G$. Although Theorem 4 suggests that stronger results should be obtainable when $M$ is a t.i. set, little is known about this case. There is a result of Feit (1964, lemma 4.2; 1967b), and two other results, which we state now.

THEOREM 5 (Leonard, 1965). *Given* $(G, X)$ *of degree n, suppose that G has a subgroup* $C = M \times H$ *such that* $M \neq \{1\}$ *and*

(i) $C - H$ *is a t.i. set with normalizer* $N(C-H) = N(C)$,

(ii) *$N(C)/H$ is a Frobenius group with Frobenius kernel $C/H$, and*
(iii) *$H \trianglelefteq N(C)$ and $M \trianglelefteq N(C)$.*

*If $n < (2m-k-1)/2$ where $m = |M|$ and $k$ is the class number of $M$, then one of the following holds:*

(a) *$N(C) = M \times H$;*
(b) *$M$ is a non-abelian p-group with $(M : M') < 4(N(C) : C)^2$;*
(c) *$M \lhd G$; that is, $N(C) = G$;*
(d) *$M$ is an elementary abelian p-group, and no proper subgroup of $M$ is normal in $N(C)$.*

*Furthermore, if $n \leq \sqrt{(m)}-1$ then we must have case* (b) *or case* (c).

Hypotheses (i), (ii) and (iii) are satisfied when $M$ is a Sylow $p$-subgroup of prime order, or $G = SL(2, p^a)$ and $M$ is a Sylow $p$-subgroup, or $G$ is one of Suzuki's simple groups $Sz(q)$ and $M$ is a Sylow 2-subgroup. The groups $Sz(q)$ have representations of small degree (Suzuki, 1962), so case (b) cannot be omitted in the first part of the theorem when $p = 2$. It is not known whether case (b) can be omitted when $p$ is odd or $n < \sqrt{(m)}-1$. It is conjectured that case (d) can be omitted. To prove this when $M$ is a Sylow $p$-subgroup and $N(C)/C$ is cyclic it would suffice to show that relative to the t.i. set $C-H$ the number of non-exceptional characters of $G$ in blocks of full defect equals the number of irreducible characters of $N(C)/M$. Some progress in case (d) has been made (Leonard, 1971).

By using the Brauer correspondence between blocks of G and blocks of $N(P)$, it can be shown that

THEOREM 6 (Leonard, 1968). *Given $(G, X)$ of degree $n$, suppose that for some prime $p$ the Sylow p-subgroup $P$ of $G$ is a t.i. set and that $C(V) \subseteq N(P)$, where $V$ is the group of $p'$-elements in $C(P)$. If $n \leq \sqrt{(|P|+1)}$ then $P \lhd G$.*

It is conjectured that the assumption $C(V) \subseteq N(P)$ is superfluous here. The theorem shows that, in any counterexample of minimal order to the conjecture, we would have $C(P) \subseteq PZ(G)$. It is also conjectured that when $p \neq 2$ it should be sufficient to assume $n < (|P|-1)/2$.

## 3. Concluding Remarks

We have concentrated here on methods of handling large prime factors of $|G|$. Some techniques of Blichfeldt (1917) and Brauer (1967) are more effective in the treatment of small primes. For further references to the early work in the theory of linear groups, see Blichfeldt (1917) and van der Waerden (1948). For results concerning solvable complex linear groups, see Dixon (1967), Huppert (1957), Ito (1954) and Winter (1967, 1968, 1970, 1971).

A result of Lindsey (1969a), that the only projective linear groups which cannot be described in terms of linear groups of smaller degree are those containing an irreducible normal simple subgroup, has in particular a direct bearing on the theory of solvable linear groups. For the structure of groups all of whose irreducible complex representations have small degree see Isaacs (1967, 1968) and Isaacs and Passman (1964). In his forthcoming papers, Lindsey (1971) has obtained various new results on linear groups of small degree.

As regards the determination of the finite linear groups over fields of characteristic $p$, we mention that the subgroups of $SL(2, q)$ were found by Dickson (see Huppert, 1967), and those of $PSL(3, q)$ by Mitchell, Hartley and Bloom (1967) (see also the references made by van der Waerden (1948)).

## BIBLIOGRAPHY


Blau, H. (1971a). Under the degree of some finite linear groups. *Trans. Am. Maths. Soc.* To appear.

Blau, H. (1971b). An inequality for complex linear groups of small degree. *Proc. Am. Math. Soc.* To be published.

Blichfeldt, H. F. (1917). "Finite collineation groups". University of Chicago Press, Chicago.

Bloom, D. M. (1967). The subgroups of $PSL(3, q)$ for odd $q$. *Trans. Am. Math. Soc.* **127,** 150.

Brauer, R. (1942). On groups whose order contains a prime number to the first power, II. *Am. J. Math.* **64,** 421.

Brauer, R. (1966). Some results on finite groups whose order contains a prime to the first power. *Nagoya Math. J.* **27,** 381.

Brauer, R. (1967). Über endliche lineare Gruppen von Primzahlgrad. *Math. Annalen,* **169,** 73.

Brauer, R. and Feit, W. (1966). An analogue of Jordan's theorem in characteristic $p$. *Ann. of Math.* (2), **84,** 119.

Brauer, R. and Hayden, S. (1963). On finite linear groups of degree $n$ whose order contains primes $p > n+1$. Symposium on Group Theory, Department of Mathematics, Harvard University.

Brauer, R. and Leonard, Jr., H. S. (1962). On finite groups with an abelian Sylow group. *Canad. J. Math.* **14,** 436.

Burnside, W. (1955). "Theory of groups of finite order". Dover Publications, New York.

Conway, J. H. (1969). A group of order, 8,315,553,613,086,720,000. *Bull. London Math. Soc.* **1,** 79.

Curtis, C. W. and Reiner, I. (1962). "Representation theory of finite groups and associative algebras". Interscience Publishers, New York.

Dixon, J. D. (1967). The fitting subgroup of a linear solvable group. *J. Austral. Math. Soc.* **7,** 417.

Dixon, J. D. (1967). Normal $p$-subgroups of solvable linear groups. *J. Austral. Math. Soc.* **7,** 545.

Feit, W. (1964). Groups which have a faithful representation of degree less than $p-1$. *Trans. Amer. Math. Soc.* **112,** 287.

Feit, W. (1966). Groups with a cyclic Sylow subgroup. *Nagoya Math. J.* **27,** 571.
Feit, W. (1967a). On finite linear groups. *J. Algebra*, **5,** 378.
Feit, W. (1967b). "Characters of finite groups". W. A. Benjamin, Inc., New York.
Feit, W. (1971). The current situation in the theory of finite simple groups. To be published.
Feit, W. and Thompson, J. G. (1961). On groups which have a faithful representation of degree less than $(p-1)/2$. *Pacific J. Math.* **11,** 1257.
Hayden, S. (1963). "On finite linear groups whose order contains a prime larger than the degree", Ph.D. dissertation, Harvard University.
Huppert, B. (1957). Lineare auflösbare Gruppen. *Math. Z.* **67,** 479.
Huppert, B. (1967). "Endliche Gruppen I". Springer, Berlin.
Isaacs, I. M. (1966). Extensions of certain linear groups. *J. Algebra*, **4,** 3.
Isaacs, I. M. (1967). Finite groups with small character degrees and large prime divisors. *Pacific J. Math.* **23,** 273.
Isaacs, I. M. (1968). Groups with small character degrees and large prime divisors II. Proceedings of the Group Theory Conference, University of Michigan, July 8–19.
Isaacs, I. M. and Passman, D. S. (1964). Groups with representations of bounded degree. *Canad. J. Math.* **16,** 299.
Ito, N. (1954). On a theorem of H. F. Blichfeldt. *Nagoya Math. J.* **5,** 75.
Jordan, C. (1878). Mémoire sur les équations differentielles linéaires à intégral algébrique. *Jour. Reine Angew. Math.* **84,** 89.
Leonard, Jr., H. S. (1965). On finite groups which contain a Frobenius factor group. *Illinois J. Math.* **9,** 47.
Leonard, Jr., H. S. (1968). On finite groups whose *p*-Sylow subgroup is a T.I. set. *Proc. Am. Math. Soc.* **19,** 667.
Leonard, Jr., H. S. (1971). Finite linear groups having an abelian Sylow subgroup. *J. Algebra.* To be published.
Lindsey II, J. H. (1968). On a projective representation of the Hall–Janko group. *Bull. Amer. Math. Soc.* **74,** 1094.
Lindsey II, J. H. (1969a). "Finite linear groups in six variables." Ph. D. dissertation, Harvard University.
Lindsey II, J. H. (1969b). Linear groups of degree 6 and the Hall–Janko group. Theory of Groups, a Symposium, edited by R. Brauer and C.-H. Sah. New York, W. A. Benjamin, Inc.
Lindsey II, J. H. (1971). A generalization of Feit's theorem. *Trans. Am. Math. Soc.* To be published.
Speiser, A. (1937). "Die Theorie der Gruppen von endlicher Ordnung," Berlin, Springer.
Suzuki, M. (1962). On a class of doubly transitive groups. *Ann. of Math.* (2), **75,** 105.
Thompson, J. G. (1967). Vertices and sources. *J. Algebra*, **6,** 1.
Tuan, H. F. (1944). On groups whose orders contain a prime to the first power. *Ann. Math.* **45,** 110.
van der Waerden, B. L. (1948). "Gruppen von linearen Transformationen." New York, Chelsea.
Wales, D. (1969). Finite linear groups of prime degree. *Canad. J. Math.* **21,** 1025.
Wales, D. (1970). Finite linear groups of degree seven, I, II. *Canad. J. Math.* **21,** 1042; *Pacific J. Math.* **34,** 207.
Winter, D. L. (1964). Finite groups having a faithful representation of degree less than $(2p+1)/3$. *Am. J. Math.* **86,** 608.

Winter, D. L. (1967). Finite $p$-solvable linear groups with a cyclic Sylow $p$-subgroup. *Proc. Am. Math. Soc.* **18,** 341.

Winter, D. L. (1968). On finite linear groups. *Math. Z.* **106,** 245.

Winter, D. L. (1970a). Solvability of certain $p$-solvable linear groups of finite order. *Pacific J. Math.* **34,** 827.

Winter, D. L. (1970b). Finite linear groups containing an irreducible solvable normal subgroup. *Proc. Am. Math. Soc.* **25,** 716.

Winter, D. L. (1971). Finite solvable linear groups. *Illinois J. Math.* To be published.

CHAPTER V

# Finite Groups with a Large Cyclic Sylow Subgroup

M. HERZOG



## 1. Introduction

The main result I would like to talk about is the following:

THEOREM 1. *Let $G$ be a finite non-$p$-solvable group with a cyclic Sylow $p$-subgroup $P$ of order $p^a$. Suppose that $o(G) < p^{3a}$. Then $G$ is isomorphic to one of the following groups:*

(i) $a = 1$, $PSL(2, p)$, $p > 3$;
(ii) $a = 1$, $PSL(2, p-1)$, $p = 2^m+1 > 5$ *a Fermat prime;*
(iii) $a = 1$, $SL(2, p)$, $p > 3$;
(iv) $a = 1$, $PGL(2, p)$, $p > 3$;
(v) $a = 1$, $PSL(2, p)\times M$, $p > 3$, $o(M) = 2$;
(vi) $a = 2$, $PSL(2, 8)$, $p = 3$.

This type of a problem was first dealt with by Brauer and Reynolds (1958). They proved the following

THEOREM 2 (restated). *Let $G$ be a finite group of order $g$ divisible by a prime $p > g^{\frac{1}{3}}$. Suppose that $G$ cannot be mapped homomorphically on the group $N_G(P)/W$, where $P$ is a Sylow $p$-subgroup of $G$ and $C_G(P) = W \times P$. Then $G$ is isomorphic to one of the groups in* (i)–(v).

This result was generalized in my paper (Herzog, 1969), where the following is shown.

THEOREM 3. *Under the assumptions of Theorem* 1, *assuming also that the Sylow $p$-subgroups constitute a TI-set, the conclusion of Theorem* 1 *holds.*

The proof of Theorem 1 is published (Herzog, 1970), where the following reduction theorem is proved.

THEOREM 4. *Suppose that $G$ is a perfect group with a cyclic Sylow $p$-subgroup $P$ of order $p^a$, and suppose that $o(G) < p^{3a}$. Then the Sylow $p$-subgroups of $G$ constitute a TI-set.*

Theorem 1 follows quite easily from Theorems 3 and 4. It is the proof of Theorem 4 which I would like to discuss here.

In Section 2 some elementary properties of groups with a cyclic Sylow $p$-subgroup are collected. The relevant results on characters in the principal $p$-block are given in Section 3. Section 4 deals with the proof of Theorem 4. We conclude with some general remarks on groups with large cyclic Sylow subgroups in Section 5.

## 2. Elementary Properties of Groups with a Cyclic Sylow $p$-subgroup

The finite group $G$ will be called a *$Cp^a$-group* if its Sylow $p$-subgroup $P$ is cyclic of order $p^a > 1$.

LEMMA 1 (Herzog, 1968). *Let $G$ be a $Cp^a$-group and $1 \neq P_0 \subset P$. Then we have:*

(a) $N_G(P_0)/C_G(P_0)$ *acts frobeniusly on* $P_0$;
(b) $[N_G(P_0) : C_G(P_0)] | p-1$;
(c) $C_G(P_0) \cap N_G(P) = C_G(P)$;
(d) $[N_G(P_0) : C_G(P_0)] = [N_G(P) : C_G(P)]$; *and*
(e) *the $p^a-1$ non-principal ordinary irreducible characters of $P$ are divided under conjugation by elements of $N_G(P)$ into $(p^a-1)/q$ transitivity classes of $q$ characters each, where $q = [N_G(P) : C_G(P)]$.*

LEMMA 2 (Herzog, 1970). *Let $G$ be a $Cp^a$-group. Then,*

$$P \cap G' = \begin{cases} \{1\} & \text{if } N_G(P) = C_G(P) \\ P & \text{if } N_G(P) \neq C_G(P). \end{cases}$$

Finally, the equivalence of the assumptions in Theorems 1 and 2 (when $a = 1$ in Theorem 1) results from the following:

LEMMA 3 (Herzog, 1970). *Let $G$ be a $Cp^a$-group. Then $G$ is $p$-solvable if and only if it can be mapped homomorphically on $N_G(P)/W$, where $W$ is the normal complement of $P$ in $C_G(P)$.*

REMARK. As noticed by Professor G. Higman, Lemma 3 holds more generally for finite groups with an abelian Sylow $p$-subgroup.

### 3. The Characters in the Principal p-block

Dade (1966) described the characters in blocks with a cyclic defect group in great detail. Using Dade's results and concentrating on the case where the Sylow $p$-subgroup of $G$ is cyclic, the following description of the principal $p$-block of $G$ was obtained (Herzog, 1968). We state here only the relevant information.

PROPOSITION 1. *Let $G$ be a $Cp^a$-group, $q = [N_G(P) : C_G(P)]$, $B =$ the principal p-block of $G$ and $\Lambda =$ a set of representatives of the classes of non-trivial ordinary characters of $P$ which are conjugate by elements of $N_G(P)$. Then, the following hold.*

(a) *$B$ contains $q+(p^a-1)/q$ ordinary irreducible characters which can be divided into two families:*

*the exceptional characters: $\{X_\lambda : \lambda \in \Lambda\}$ and*
*the non-exceptional characters: $\{X_i : i = 1, \ldots, q\}$.*

(b) *Let $\sigma \in P^{\#}$, $\pi$ be a $p'$-element of $C_G(\sigma)$ and $R$ be any set of coset representatives of $C_G(P)$ in $N_G(P)$. Then,*

$$X_\lambda(\sigma\pi) = -\varepsilon_0 \sum_{\tau \in R} \lambda^\tau(\sigma) \quad \textit{for } \lambda \in \Lambda$$

*and*

$$X_i(\sigma\pi) = \varepsilon_i \qquad \textit{for } i = 1, \ldots, q,$$

*where*

$$\varepsilon_j = \pm 1 \qquad \textit{for } j = 0, 1, \ldots, q.$$

(c) $x_0 = X_\lambda(1) = bp^a - \varepsilon_0 q$ *for $\lambda \in \Lambda$ and*
$x_i = X_i(1) = b_i p^a + \varepsilon_i$ *for $i = 1, \ldots, q$.*

(d) $\sum_{j=0}^{q} \varepsilon_j x_j = 0.$

### 4. Proof of Theorem 4

Let $q = [N_G(P) : C_G(P)]$; then $q | p-1$ and since $G$ is perfect, $P$ has no normal complement in $G$. Consequently, $q > 1$ and $p > 2$.

In view of Proposition 1, the principal $p$-block $B$ of $G$ contains the following ordinary irreducible characters: $(p^a-1)/q$ exceptional characters of degree $x_0 = bp^a - \varepsilon_0 q$ and $q$ non-exceptional characters $R_0, R_1, \ldots, R_u, S_1, \ldots, S_v$ with the following properties:

| | $R_0 = 1_G$ | $R_i\ (i = 1, \ldots, u)$ | $S_i\ (i = 1, \ldots, v)$ |
|---|---|---|---|
| 1 | 1 | $r_i p^a + 1$ | $s_i p^a - 1$ |
| $\sigma \in P^{\#}$ | 1 | 1 | $-1$ |

and

$$u + v + 1 = q \tag{1}$$

Part (d) of Proposition 1 also yields:

$$0 = \left(\sum_{i=1}^{u} r_i - \sum_{1}^{v} s_i + \varepsilon_0 b\right) p^a + u + 1 + v - q$$

which in view of (1) simplifies to

$$\sum_{i=1}^{v} s_i = \sum_{i=1}^{u} r_i + \varepsilon_0 b. \tag{2}$$

Theorem 4 obviously holds if $a = 1$. Therefore from now on we will assume that $a > 1$.

The proof depends upon the arrangement of non-exceptional characters in $B$. It is easy to see that one of the following cases has to hold.

(i) There exist in $B$ two non-principal, non-exceptional characters of degrees $yp^a + \varepsilon_1$ and $zp^a + \varepsilon_2$ such that

$$yp^a + \varepsilon_1 \neq zp^a + \varepsilon_2 \quad \text{and} \quad yz > 1.$$

(ii) All the non-principal, non-exceptional characters in $B$ are of the same degree $yp^a + \varepsilon$.

(iii) We have $u \neq 0$, $v \neq 0$ and $r_i = s_j = 1$ $(i = 1, \ldots, u \quad j = 1, \ldots, v)$.

We will consider here Case (ii) only; the proofs in Cases (i) and (iii) are of a similar type.

In Case (ii) we must have either

$$v = 0, \qquad \varepsilon = 1, \qquad u = q-1$$

or

$$u = 0, \qquad \varepsilon = -1, \quad v = q-1.$$

In both cases it follows from (2) that

$$y(q-1) = b, \qquad \varepsilon_0 = -\varepsilon.$$

Thus,

$$x_0 = bp^a - \varepsilon_0 q = y(q-1)p^a + \varepsilon q = q(yp^a + \varepsilon) - yp^a.$$

Consequently, $(x_0, yp^a + \varepsilon) = 1$ and since $x_0$, $yp^a + \varepsilon$ and $p^a$ divide $o(G)$ we must have

$$(y(q-1)p^a + \varepsilon q)(yp^a + \varepsilon)p^a \mid o(G).$$

Suppose that at least one of the following cases holds: (i) $y > 1$; (ii) $q > 2$; or (iii) the above division is proper. Then,

$$p^{3a} > o(G) \geqslant (2p^a - 4)(p^a - 1)p^a$$

which yields $p^a < 6$ in contradiction to the assumption $a > 1$. Therefore none of cases (i)–(iii) holds and consequently

$$o(G) = (p^a + 2\varepsilon)(p^a + \varepsilon)p^a \quad \text{and} \quad q = 2.$$

Let $Y$ denote the non-exceptional character of degree $p^a + \varepsilon$ and let $X$ be an exceptional character of degree $x_0 = p^a + 2\varepsilon$. Let $r^e$ be the $r$-share of the prime number $r \neq p$ in $o(G)$. As $(p^a+\varepsilon,\ p^a+2\varepsilon) = 1$, either $r^e | p^a+\varepsilon$ or $r^e | p^a+2\varepsilon$. It follows then from the properties of blocks of defect 0 (Curtis and Reiner, 1962, Thm. 86.3) that if $\gamma \in G$ is not a $p$-element, then

$$\text{either } Y(\gamma) = 0 \quad \text{or} \quad X(\gamma) = 0.$$

Let $\sigma \in P^{\#}$, $\pi$ be a $p'$-element of $C_G(\sigma)$ and let $\gamma = \sigma\pi$. Then by Proposition 1

$$Y(\sigma\pi) = \varepsilon \quad \text{and} \quad X(\sigma\pi) = \varepsilon(\lambda^{\tau_1}(\sigma)+\lambda^{\tau_2}(\sigma)),$$

where $\lambda$ is an irreducible character of $P$ and $\{\tau_1, \tau_2\}$ are coset representatives of $C_G(P)$ in $N_G(P)$. Thus $Y(\sigma\pi) \neq 0$ and $X(\sigma\pi) \neq 0$, the second inequality following from the fact that $\lambda^{\tau_1}(\sigma)$ and $\lambda^{\tau_2}(\sigma)$ are $p^a$-th roots of 1 and $p \neq 2$. Hence $\sigma\pi$ is a $p$-element and consequently $\pi = 1$ and $C_G(\sigma) = P$ for all $\sigma \in P^{\#}$. It follows immediately that the Sylow $p$-subgroups of $G$ constitute a TI-set. The proof of Theorem 4 in Case (ii) is complete.

## 5. General Remarks

Let $G$ be a finite group with a Sylow $p$-subgroup $P$ of order $p^a$. As far as I know, it has not been shown that, in general, if $G$ is simple, then

$$o(G) > p^{2a}.$$

However, one can show quite easily that

(a) If $O_p(G) = 1$ and either $P$ is abelian or $a = 3$ then:

$$o(G) > p^{2a},$$

(b) if $O_p(G) = 1$ and $a \geqslant 4$ then:

$$o(G) > p^{a+3}.$$

Any improvement of these results would be of considerable interest.

### Bibliography


Brauer, R. and Reynolds, W. F. (1958). On a problem of E. Artin. *Ann. of Math.* **68,** 713.

Curtis, C. W. and Reiner, I. (1962). "Representation Theory of Finite Groups and Associative Algebras". Interscience Publishers, New York.

Dade, E. C. (1966). Blocks with cyclic defect groups. *Ann. of Math.* **84,** 20.

Herzog, M. (1968). On finite groups with cyclic Sylow subgroups for all odd primes. *Israel J. Math.* **6,** 206.

Herzog, M. (1969). On finite groups with independent cyclic Sylow subgroups. *Pacific J. Math.* **29,** 285.

Herzog, M. (1970). On a problem of E. Artin. *J. Algebra* **15,** 408.

CHAPTER VI

# Construction of Simple Groups from Character Tables

G. HIGMAN



## 1. Introduction

The general problem to be considered here is, given what purports to be the character table of a group, to construct, up to isomorphism, all groups of which it is the character table. In practice this arises in two contexts. In characterizing a known simple group, it may be convenient to know that no other group has the same character table; this is the *uniqueness* problem. In constructing a new simple group, we may be able to find a character table, and we then want to know that there is a group corresponding to it; this is the *existence* problem. The methods used in the two problems, however, are not essentially different. Rather than attempting to describe these methods in general, I propose to deal with two special cases, and to hope that the relevant generalities will emerge.

## 2. Alternating Groups

First, I shall sketch a proof of

*If a group has the same character table as the alternating group $A_n$, then it is isomorphic to $A_n$.*

This is a theorem of Oyama (1964), but the proof I shall give differs in detail from his. I shall assume that $n \geqslant 9$; special proofs can fairly easily be devised for the smaller values of $n$.

Suppose, then, that $G$ is a group having the same character table as $A_n$. That is, there are one to one correspondences between the characters of $G$ and the characters of $A_n$, and between the classes of $G$ and the classes of $A_n$, so that if characters $\chi$ and $\phi$ correspond, and the classes containing $x$ and $y$, then $\chi(x) = \phi(y)$. These correspondences are not unique, but we suppose them chosen once and for all.

The element $x$ of $G$ is a $p$-element if and only if $\chi(x) \equiv \chi(1) \pmod{\mathfrak{p}}$ for all characters $\chi$ of $G$, where $\mathfrak{p}$ is the radical of $(p)$ in the field generated by the characters. It follows that if the classes containing $x$ and $y$ correspond, $x$ is a $p$-element if and only if $y$ is. More generally, elements $x_1$, $x_2$ in $G$ have conjugate $p'$-parts if and only if $\chi(x_1) \equiv \chi(x_2) \bmod \mathfrak{p}$ for all characters $\chi$, whence (by induction on the number of primes involved) if the classes containing $x$ and $y$ correspond, the primes dividing the order of $x$ are the same as those dividing the order of $y$.

Let $a$, $b$, $c$ be elements of $G$ in the classes corresponding to the classes $31^{n-3}$, $2^2 1^{n-4}$, and $51^{n-5}$ respectively. By what we have said, $a$ is a 3-element, $b$ a 2-element and $c$ a 5-element. However, if $x$ is an element of $A_n$ in the class $2^2 1^{n-4}$, and $y$ is a 2-element of $A_n$, $y \neq 1$, $y$ not conjugate to $x$, then it is easy to see (using the assumption that $n \geqslant 9$) that $|C(y)| < |C(x)|$. Since orders of centralizers are determined by the character table, if $g$ is a 2-element of $G$, $g \neq 1$, $g$ not conjugate to $b$, then $|C(g)| < |C(b)|$. If $b$ is not of order 2, we obtain a contradiction by putting $g = b^2$. So $b$ is of order 2, and similarly $a$ is of order 3 and $c$ of order 5.

Next, we recall that the class multiplication constants are determined by the character table. If $\#(\dot{x}_1\dot{x}_2 = x_3)$ denotes the number of ways $x_3$ can be written as a product of a conjugate of $x_1$ and a conjugate of $x_2$, then

$$\#(\dot{x}_1\dot{x}_2 = x_3) = \frac{|G|}{|C(x_1)||C(x_2)|}\sum \frac{\chi(x_1)\chi(x_2)\chi(x_3^{-1})}{\chi(1)},$$

where the sum is over all irreducible characters $\chi$. It follows that if the classes containing $x_1$, $x_2$, $x_3$ correspond to the classes containing $y_1$, $y_2$, $y_3$ then

$$\#(\dot{x}_1\dot{x}_2 = x_3) = \#(\dot{y}_1\dot{y}_2 = y_3).$$

In particular if the element $x$ of $G$ can be written as a product of two conjugates of $b$, then an element $y$ of the corresponding class of $A_n$ is a product of two elements in the class $2^2 1^{n-4}$, and so has order 1, 2, 3, 4, 5 or 6. Thus $x$ is a 2-element, a 3-element, a 5-element, or a (2, 3)-element, and cannot have order 15. As a consequence,

*If two conjugates of b both invert a, their product cannot be of order* 5.

For if $b_1$, $b_2$ both invert $a$, $b_1b_2$ centralizes $a$, and so, if $b_1b_2$ is of order 5, $ab_1b_2$ is of order 15. Since $ab_1$ is conjugate to $b_1$, this is impossible.

By a *natural* $A_4$ *in* $A_n$ we mean the subgroup stabilizing $n-4$ points, and by a *natural* $A_5$, the subgroup stabilizing $n-5$ points. Now an $A_4$ contained in $A_n$ is natural if and only if its involutions belong to the class $2^21^{n-4}$ and its elements of order 3 to the class $31^{n-3}$. Hence we define a *natural* $A_4$ *in* $G$ as a subgroup of $G$ isomorphic to $A_4$, in which the involutions are conjugate to $b$, and the elements of order 3 to $a$. For a similar reason, a *natural* $A_5$ *in* $G$ is defined as a subgroup isomorphic to $A_5$ in which the involutions are conjugate to $b$, the elements of order 3 to $a$, and the elements of order 5 to $c$. If $a_1$ is a conjugate of $a$ and $b_1$ a conjugate of $b$ such that $a_1b_1 = a$, then

$$a_1^3 = b_1^2 = (a_1b_1)^3 = 1,$$

and, since the relations $x^3 = y^2 = (xy)^3 = 1$ define $A_4$, $a_1$ and $b_1$ generate an $A_4$, necessarily natural. Thus the number of natural $A_4$'s containing $a$ is

$$\tfrac{1}{3}\,\#\,(\dot{a}\dot{b} = a),$$

and is therefore the same as the number of natural $A_4$'s in $A_n$ containing a fixed element of the class $31^{n-3}$. Similarly, using the definition $x^5 = y^2 = (xy)^3 = 1$ of $A_5$, the number of natural $A_5$'s in $G$ containing $a$ is

$$\tfrac{1}{6}\,\#\,(\dot{c}\dot{b} = a)$$

and is the same as the corresponding number in $A_n$.

That is, there are just $n-3$ natural $A_4$'s and just $\tfrac{1}{2}(n-3)(n-4)$ natural $A_5$'s in $G$ containing $a$. Since a natural $A_5$ containing $a$ contains two natural $A_4$'s containing $a$ and is generated by them, any two natural $A_4$'s containing $a$ generate a natural $A_5$. If $A^{(1)}$ and $A^{(2)}$ are two such natural $A_4$'s, we can identify $\langle A^{(1)}, A^{(2)}\rangle$ with the alternating group on the symbols 1, 2, 3, 4, 5, identifying $a$ with (123), and the involutions in $A^{(1)}$ and $A^{(2)}$ with

$$\begin{array}{lcl} (23)\,(14) & \text{and} & (23)\,(15) \\ (31)\,(24) & & (31)\,(25) \\ (12)\,(34) & & (12)\,(35) \end{array}$$

respectively. Clearly we can set up a correspondence between the involutions in $A^{(1)}$ and those in $A^{(2)}$ so that the product of corresponding involutions has order 3 and the product of non-corresponding involutions has order 5. We want to show that if we take further natural $A_4$'s $A^{(3)}, A^{(4)}, \ldots$ containing $a$, the correspondences so obtained are coherent. That is, that if $b_1$ in $A^{(1)}$

corresponds to $b_2$ in $A^{(2)}$ and to $b_3$ in $A^{(3)}$, then $b_2$ and $b_3$ correspond. But if $b_1$ corresponds to $b_2$, we may take the images of $b_1$ and $b_2$ in the above isomorphism to be (23) (14) and (23) (15) respectively. Thus $b_1b_2b_1$ has image (23) (45), and so inverts $a$. Similarly $b_1b_3b_1$ inverts $a$. Hence the product $b_1b_2b_3b_1$ cannot have order 5. That is, $b_2b_3$ cannot have order 5, so that $b_2$ and $b_3$ correspond, as required. Now let $A^{(i)}$ $(i = 1, 2, \ldots, n-3)$ be the natural $A_4$'s containing $a$, and choose an involution $b_i$ in each $A^{(i)}$ so that $b_i$ and $b_j$ correspond for all $i \neq j$. Then $a, b_1, \ldots, b_{n-3}$ satisfy the relations

$$a^3 = b_i^2 = (ab_i)^3 = (b_ib_j)^3 = (ab_ib_jb_i)^2 = 1 \quad (1 \leqslant i < j \leqslant n-3).$$

We show next that if $U$ is the group generated by $x, y_1, \ldots, y_{n-3}$ subject to the defining relations

$$x^3 = y_i^2 = (xy_i)^3 = (y_iy_j)^3 = (xy_iy_jy_i)^2 = 1 \quad (1 \leqslant i < j \leqslant n-3),$$

then $U$ is isomorphic to $A_n$, under an isomorphism mapping $x$ on (123) and $y_i$ on (12) $(3\,3+i)$, $i = 1, \ldots, n-3$. This is clear for $n = 4, 5$ so we suppose that $n > 5$ and use induction on $n$. Let $V$ be subgroup $\langle x, y_1, \ldots, y_{n-4}\rangle$ of $U$, and $W$ the subgroup $\langle y_1xy_1, \ldots, y_{n-4}xy_{n-4}\rangle$. By the inductive hypothesis, $V$ is a homomorphic image of $A_{n-1}$, and $W$ is the image under the same homomorphism of the stabilizer in $A_{n-1}$ of the symbol 3. It follows that $V = W \cup WxW$. From the relations

$$y_ixy_i \cdot y_{n-3} = y_{n-3} \cdot y_ix^{-1}y_i$$

we see that $y_{n-3}$ normalizes $W$. It follows that

$$y_{n-3}Vy_{n-3} \subset W \cup Wy_{n-3}xy_{n-3}W \subset V \cup Vy_{n-3}V,$$

the last inclusion following from $(xy_{n-3})^3 = 1$. Thus $U = V \cup Vy_{n-3}V$. The fact that $y_{n-3}$ normalizes $W$ shows that $Vy_{n-3}V$ consists of at most $n-1$ cosets $Vg$. Thus $|U| \leqslant n|V| \leqslant \frac{1}{2}n!$. Since the specified permutations do indeed satisfy the relations, we are home. (As the proof shows, we can omit all but one of the relations $(y_iy_j)^3 = 1$.)

Now it follows that the subgroup of $G$ generated by $a, b_1, \ldots, b_{n-3}$ is a homomorphic image of $A_n$, and since it is certainly not trivial, is therefore isomorphic to $A_n$. Since $|G| = |A_n|$, this subgroup is the whole of $G$, proving the theorem.

## 3. Janko's First Group

The second special case that I shall consider is that of Janko's first group. That is, we shall show that there exists one and, up to isomorphism, only one group with the following character table:

| | 175, 560 | 120 | 30 | 6 | 30 | 30 | 10 | 10 | 15 | 15 | 7 | 11 | 19 | 19 | 19 |
|---|---|---|---|---|---|---|---|---|---|---|---|---|---|---|---|
| | 1 | 2 | 3 | 6 | $5_1$ | $5_2$ | $10_1$ | $10_2$ | $15_1$ | $15_2$ | 7 | 11 | $19_a$ | $19_b$ | $19_c$ |
| $\chi_0$ | 1 | 1 | 1 | 1 | 1 | 1 | 1 | 1 | 1 | 1 | 1 | 1 | 1 | 1 | 1 |
| $\chi_1$ | 56 | 0 | 2 | 0 | $-2\alpha$ | $-2\beta$ | 0 | 0 | $\alpha$ | $\beta$ | 0 | 1 | $-1$ | $-1$ | $-1$ |
| $\chi_2$ | 56 | 0 | 2 | 0 | $-2\beta$ | $-2\alpha$ | 0 | 0 | $\beta$ | $\alpha$ | 0 | 1 | $-1$ | $-1$ | $-1$ |
| $\chi_3$ | 76 | 4 | 1 | 1 | 1 | 1 | $-1$ | $-1$ | 1 | 1 | $-1$ | $-1$ | 0 | 0 | 0 |
| $\chi_4$ | 76 | $-4$ | 1 | $-1$ | 1 | 1 | 1 | 1 | 1 | 1 | $-1$ | $-1$ | 0 | 0 | 0 |
| $\chi_5$ | 77 | 5 | $-1$ | $-1$ | 2 | 2 | 0 | 0 | $-1$ | $-1$ | 0 | 0 | 1 | 1 | 1 |
| $\chi_6$ | 77 | $-3$ | 2 | 0 | $\alpha$ | $\beta$ | $\alpha$ | $\beta$ | $\alpha$ | $\beta$ | 0 | 0 | 1 | 1 | 1 |
| $\chi_7$ | 77 | $-3$ | 2 | 0 | $\beta$ | $\alpha$ | $\beta$ | $\alpha$ | $\beta$ | $\alpha$ | 0 | 0 | 1 | 1 | 1 |
| $\chi_8$ | 120 | 0 | 0 | 0 | 0 | 0 | 0 | 0 | 0 | 0 | 1 | $-1$ | $\lambda$ | $\mu$ | $\nu$ |
| $\chi_9$ | 120 | 0 | 0 | 0 | 0 | 0 | 0 | 0 | 0 | 0 | 1 | $-1$ | $\mu$ | $\nu$ | $\lambda$ |
| $\chi_{10}$ | 120 | 0 | 0 | 0 | 0 | 0 | 0 | 0 | 0 | 0 | 1 | $-1$ | $\nu$ | $\lambda$ | $\mu$ |
| $\chi_{11}$ | 133 | 5 | 1 | $-1$ | $-2$ | $-2$ | 0 | 0 | 1 | 1 | 0 | 1 | 0 | 0 | 0 |
| $\chi_{12}$ | 133 | $-3$ | $-2$ | 0 | $-\alpha$ | $-\beta$ | $\alpha$ | $\beta$ | $-\alpha$ | $-\beta$ | 0 | 1 | 0 | 0 | 0 |
| $\chi_{13}$ | 133 | $-3$ | $-2$ | 0 | $-\beta$ | $-\alpha$ | $\beta$ | $\alpha$ | $-\beta$ | $-\alpha$ | 0 | 1 | 0 | 0 | 0 |
| $\chi_{14}$ | 209 | 1 | $-1$ | 1 | $-1$ | $-1$ | 1 | 1 | $-1$ | $-1$ | $-1$ | 0 | 0 | 0 | 0 |

(Here $\alpha$, $\beta$ are $\frac{1}{2}(-1\pm\sqrt{5})$ so that $Q[\alpha]$ is the real subfield of the field of the 5-th roots of unity, and $\lambda$, $\mu$, $\nu$ are the roots of $x^3+x^2-6x-7 = 0$, so that $Q[\lambda]$ is the cubic subfield of the field of the 19-th roots of unity.) Besides the characters themselves, the table includes the orders of the centralizers of elements (which can be obtained from the characters) and names for the characters and classes. The name for a class consists of the period of the elements in the class, with a distinguishing suffix where necessary. It is immediate from the table that for each prime $p$ dividing the order of the group, there is just one class of cyclic $p$-subgroups. Thus all elements have square-free order. We let $G$ be a group with this character table.

### 3.1. Local Subgroups

If $x$ is an involution in $G$, $C(x)$, of order 120, cannot be soluble, else $G$ would contain an element of order 30. Thus $C(x)/\langle x\rangle \simeq A_5$, and the extension splits, since $G$ has no elements of order 4. Thus $C(x) = \langle x\rangle \times A$, where $A \simeq A_5$. A Sylow 2-subgroup $P$ of $G$ containing $x$ has order 8. Its seven involutions are conjugate in $N(P)$, and $|C(x) \cap N(P)| = 24$, so that $|N(P)| = 168$. Thus $N(P)$ is the extension of $P$ by a non-abelian group of order 21, acting faithfully. Because $P$ is normalized by an element of order 7, there is just one class of subgroups of $G$ of order 4. If two Sylow 2-subgroups of $G$ have non-trivial intersection, they lie in the centralizer of the same involution, whence their intersection is of order 2. Thus a subgroup $V$ of order 4 lies in a unique Sylow subgroup $P$, so that $N(V) \leqslant N(P)$, whence $|N(V)| = 24$. It follows that $G$ has a single class of subgroups isomorphic to $A_4$, namely, the class containg $O^2(N(V))$.

If $y$ is an element of order 3, $C(y)$, being of order 30, has a characteristic subgroup $\langle z \rangle$, say, of order 5. Then $N(\langle y \rangle) \leqslant N(\langle z \rangle)$. A similar argument proves the reverse inequality. Thus Sylow 3-normalizers and Sylow 5-normalizers coincide, and are direct products $D_6 \times D_{10}$.

Sylow 7-normalizers, 11-normalizers and 19-normalizers are evidently Frobenius groups of orders 42, 110 and 114 respectively.

### 3.2. Subgroups Isomorphic to $A_5$

As we have seen, if $x$ is an involution, $C(x) = \langle x \rangle \times A$, where $A \simeq A_5$. Evidently $N(A) = C(x)$. If $z$ is an element of order 5 in $A$, and $z$ belongs also to the conjugate $A^g$, then $\langle z \rangle$ is a Sylow subgroup both of $A$ and of $A^g$, whence we may suppose $g$ chosen from $N(\langle z \rangle)$. Since $|N(\langle z \rangle)| = 60$ but $|N(\langle z \rangle) \cap N(A)| = 20$, $z$ lies in just 3 conjugates of $A$. But if $B$ is a subgroup of $G$ isomorphic to $A_5$ but not conjugate to $A$, $N(B) = B$, and the same argument shows that $z$ belongs to 6 conjugates of $B$. Thus if $G$ has $r+1$ classes of subgroups isomorphic to $A_5$, $z$ lies in $3(2r+1)$ subgroups isomorphic to $A_5$, whence, if $x$ is an involution and $y$ is of order 3, $\#\ (\dot{x}\dot{y} = z) = 15(2r+1)$. From the character table $\#\ (\dot{x}\dot{y} = z) = 45$, so that $r = 1$. That is, apart from those contained in centralizers of involutions, $G$ has a single class of subgroups isomorphic to $A_5$.

### 3.3. Other Possible Subgroups

To find out whether $G$ may have other subgroups, we look for possible permutation characters. If the character $\phi = \Sigma a_i \chi_i$ of $G$ is afforded by the permutation representation of $G$ on the cosets of a subgroup $H$, then $\phi$ satisfies, among others, the following conditions:

(i) $a_0 = 1$;
(ii) $a_i \geqslant 0$, and $a_i = a_j$ if $\chi_i$ is conjugate to $\chi_j$;
(iii) $\phi(1) = |G|/|H|$ divides $|G|$;
(iv) $\phi(x)$ is a non-negative rational integer for all $x \in G$;
(v) $\phi(x^i) \geqslant \phi(x)$ for all integers $i$, and all $x \in G$.

It is not hard to list the characters of small degree with these properties. Those of degree at most 266 are:

(a) $\chi_0 + \chi_3$, degree 77, $(|H| = 2280)$;
(b) $\chi_0 + \chi_3 + \chi_5$, degree 154, $(|H| = 1140)$;
(c) $\chi_0 + \chi_3 + 2\chi_5$, degree 231, $(|H| = 760)$;
(d) $\chi_0 + \chi_1 + \chi_2 + \chi_3 + \chi_5$, degree 266, $(|H| = 660)$.

The first three of these cannot be permutation characters, because there is no group of the required order which could be $H$. Indeed, a Sylow 19-normalizer in $H$ would have to have order dividing 6.19, whence $H$ would have to have 20 Sylow 19-subgroups. As a permutation group on these 20 subgroups, $H$ would have to be a transitive extension of a Frobenius group of degree 19 and order 6.19, 3.19 or 2.19, and it is well known that no such thing exists. But it is easy to check that character (d) is consistent with the presence of a subgroup isomorphic to the simple group of order 660.

Thus $G$ has no subgroup of order greater than 660. $G$ cannot involve the simple group of order 168, which has non-abelian Sylow 2-subgroups, or the simple groups of orders 360 and 504, whose orders are divisible by 9. Moreover, since $G$ has abelian Sylow 2-subgroups of order 8, a subgroup having $A_5$ as a composition factor is either $A_5$ itself or is contained in a local subgroup. Since all these have been determined, the proper subgroups of $G$ are now known exactly, apart from the question of simple subgroups of order 660.

### 3.4. Commuting Involutions

If $x$ is an involution, $C(x) = \langle x \rangle \times A$, where $A$ is isomorphic to $A_5$. Thus there are two conjugate classes of pairs $(x, x')$ of distinct commuting involutions, in one of which $x'$ is in the derived group of $C(x)$, and in the other of which it is not. Now the pair $(x, x')$ is not conjugate to the pair $(x', x)$; for if it were, we could effect the conjugation by a 2-element $t$, and $\langle x, t \rangle$ would then be a non-abelian 2-subgroup of $G$, which is impossible. Thus, for any pair $(x, x')$ of distinct commuting involutions, exactly one of the two statements "$x$ is in the derived group of $C(x')$" and "$x'$ is in the derived group of $C(x)$" is true.

### 3.5. A Digression on Coset Enumeration

We shall conclude the proof that $G$ exists and is unique by producing generators for $G$, showing that they satisfy certain relations, and proving that these relations define a group of the correct order. This proof consists of a coset enumeration, carried out on a computer. I do not propose to describe the computer programme here, but I think something must be said to make clear precisely what the computation achieves.

Let $G$ be generated by $x_1, x_2, \ldots, x_r$ subject to the relations $u_1 = u_2 = \ldots = u_s = 1$ and let $H$ be the subgroup generated by $v_1, v_2, \ldots, v_t$. Here the $u_i$ and $v_j$ are words in the $x_i$ and their inverses, and for convenience we include among the $u_i$ the trivial relations $x_i x_i^{-1} = 1$, etc. For any positive integer $n$, we can form the set $\Omega_n$ of words in the $x_i$ and their inverses of length at most $n$. We then define $E_n$ to be the weakest equivalence relation on $\Omega_n$ satisfying the following three conditions. First, if $w_1, w_2$ are words of length less than $n$ and $w_1 \equiv w_2$, then $w_1 x_i \equiv w_2 x_i$, and $w_1 x_i^{-1} \equiv w_2 x_i^{-1}$, for $i = 1, \ldots, r$. Because of this condition, we can define, for $y = x_i$ or $x_i^{-1}$, a partial map of the set of equivalence classes into itself, by the rule that, if $[w]$ denotes the class containing $w$, $[w]y = [wy]$. Observe that $[w]y$ is defined if and only if $[w]$ contains a word of length less than $n$. By composition, any word in the $n_i$ and their inverses determines a partial map of the set of equivalence classes into itself (though it may well be nowhere defined). Then our second condition is that, for any class $[w]$ and for $i = 1,\ldots, s$, if $[w]u_s$ is defined, $[w]u_s = [w]$, and our third condition is that if 1 denotes the empty word, then if $[1]v_i$ is defined $[1]v_i = [1]$ $(i = 1,\ldots, t)$. Clearly, if $[w_1] = [w_2]$, the cosets $Hw_1$ and $Hw_2$ in $G$ coincide; and, more generally, if $[w_1]w$ is defined and equal to $[w_2]$, the cosets $Hw_1w$ and $Hw_2$ coincide.

For a fixed value of $n$, the equivalence $E_n$ can clearly be generated mechanically. Having generated it, we can examine the equivalence classes, to see if any of them contain only words of length $n$. If there are such classes, no conclusion can be drawn. If there are no such classes the situation is said to be *terminal.* In this case, all the partial maps mentioned above are total, so that we have permutations $x_i$ of the set of equivalence classes for each $i$. The second condition on the equivalence ensures that these permutations generate a permutation representation of $G$, and the third condition ensures that the stabilizer of $[1]$ contains $H$. The fact that if $[w_1]w = [w_2]$ then $Hw_1w = Hw_2$ ensures both that the stabilizer is no bigger than $H$, and that the representation is transitive. Thus in this case, the number of equivalence classes is the index of $H$ in $G$.

Logically, coset enumeration amounts to carrying out the above process for $n = 1, 2, 3,\ldots$ successively, until either one runs out of computing facilities, or one reaches a value of $n$ for which the situation is terminal. Thus it is a mechanical process which may or may not terminate, which calculates the index of $H$ in $G$ whenever it does terminate. Obviously it cannot terminate unless this index is finite. If the index is finite then in theory the process terminates. Whether it terminates in practice depends, of course, on the computing facilities available, and, even more critically, on the sophistication of the programme used to translate the logical structure into actual computation.

### 3.6. $G$ AS A HOMOMORPH OF $G^{3,7,19}$

We return now to the discussion of a group $G$ with the character table given above. If $x, y, u$ are elements of $G$ of orders 2, 3, 7 respectively, then, from the character table $\#\,(\dot{x}\dot{y} = u) = 49$. The pairs $(x^a, y^b)$ with $x^a y^b = u$ are permuted regularly by conjugation by $u$, so that there are 7 conjugacy classes of pairs $(x, y)$ of elements of $G$ such that $x^2 = y^3 = (xy)^7 = 1$, (but not $x = y = 1$). But if $(x, y)$ is such a pair so is $(x^{-1}, y^{-1})$, and the map $(x, y) \to (x^{-1}, y^{-1})$ induces an involution on this set of conjugacy classes. Thus we can choose $(x, y)$ conjugate to $(x^{-1}, y^{-1})$. That is, $G$ contains elements $x, y, t$ satisfying

$$x^2 = y^3 = (xy)^7 = t^2 = (xt)^2 = (yt)^2 = 1$$

and, of course, also

$$(xyt)^m = 1,$$

where $m = 1, 2, 3, 5, 6, 7, 10, 11, 15$ or $19$. That is, in the notation of Coxeter (1939), the subgroup $\langle x, y, t\rangle$ is a homomorphic image of $G^{3,7,m}$ for one of these values of $m$. But Coxeter (*loc. cit.*) has determined the structure of $G^{3,7,m}$ for $m \leqslant 15$; in particular it collapses (has order 1 or 2) for $m \leqslant 7$, and for $m = 10$ or $11$, and is $PSL(2, 29)$ for $m = 15$. Since $\langle x, y, t\rangle$ contains elements of order 7 but not of order 29 none of these is possible, so that $m = 19$. Thus $\langle x, y, t\rangle$ contains elements of order 19 as well as of order 7, and from our survey of the proper subgroups of $G$, it follows that $\langle x, y, t\rangle$ is the whole of $G$. That is, $G$ is a homomorphic image of $G^{3,7,19}$. This homomorphism is not an isomorphism, because $PSL(2, 113)$ is also a homomorphic image of $G^{3,7,19}$. Indeed, it seems likely that $G^{3,7,19}$ is infinite, though I know no proof of this.

Thus to define $G$ we have to produce further relations. To obtain these note that $t$ and $xt$ are commuting involutions. Moreover $\langle yt, t\rangle$ is dihedral of order 6 and $\langle yt, xt\rangle$ dihedral of order 14, so that we can find $g, h$ so that $t^g = yt$ and $(xt)^h = yt$. Then both $(xt)^g$ and $t^h$ centralize $yt$, and, by the argument in (iv) above, one of them does and the other does not lie in the derived group of $C(yt)$. It follows that $(xt)^g t^h yt$ lies in the derived group of $C(yt)$, and so has order 1, 2, 3 or 5.

It is convenient at this point to introduce different generators for $G$. If we put $a = y^{-1}x^{-1}$, $b = xyt$, one verifies from the relations that $a^2b^2 = y$, so that $a$ and $b$ generate $G$. One verifies that the relations we have already are equivalent to

$$a^7 = b^{19} = (ab)^2 = (a^2b^2)^3 = (a^3b^2)^2 = (a^2b^3)^2 = 1 \qquad (*)$$

and that the new relation asserts that $a^4b^7$ is of order 1, 2, 3, or 5. Obviously $a^4b^7 = 1$ with relations (*) implies $a = b = 1$ and so is impossible. Also $(a^4b^7)^2 = 1$ with (*) gives $(b^2a^3)^2(a^4b^7)^2 = 1$ which (using $a^7 = 1$) reduces to $a^{-3}b^9a^3 = b^{-9}$, which, since $a$ is of odd order, implies $b = 1$ and hence $a = 1$. For the other two cases, $(a^4b^7)^3 = 1$ and $(a^4b^7)^5 = 1$, an enumeration of the cosets of $\langle b, a^2b^2a\rangle$ was carried out. (From (*), this subgroup is dihedral of order 38, or, of course a homomorphic image of this.) In the case of $(a^4b^7)^3 = 1$, the enumeration gave the answer 1, which shows that this relation together with (*) imples $a = b = 1$. However, in the case of

$$(a^4b^7)^5 = 1 \qquad (**)$$

the enumeration gave the answer 4620. Thus (*) and (**) do not imply

$b = 1$, whence $\langle b, a^2b^2a\rangle$ is indeed of order 38, and $\langle a, b\rangle$, subject to these relations, is of order 175,560.

(In fact, the coset enumerations were carried out with rather more complicated but equivalent sets of relations. In this form, it was not obvious that the case $n = 2$ could be eliminated, and the enumeration was performed for $n = 2$ also. It indicated collapse in this case, as it should have done. The enumerations were carried out by John McKay in the Atlas Laboratory at Chilton.)

### 3.7. Conclusions

What we have established is that if there exists a group with the character table from which we started out, it is generated by elements $a$, $b$ satisfying (*) and (**). Moreover these relations define a group of the correct order. It is immediate from this that up to isomorphism there is at most one group with the character table. To prove existence we have strictly to show that the group we have constructed has the right character table. However, we shall omit this; the group is of known smallish order, and coincides with its derived group (from the relations). The methods for constructing the character table in such a case are standard.

## Bibliography

Coxeter, H. S. M. (1939). The abstract groups $G^{m,n,p}$. *Trans. Amer. Math. Soc.* **45,** 73.

Oyama, T. (1964). On the groups with the same table of characters as the alternating groups. *Osaka J. Math.* **1,** 91.

CHAPTER VII

# Three Lectures on Exceptional Groups†

J. H. CONWAY



## 1. Introduction

The general theme is as follows. The first lecture describes certain exceptional properties of the groups $L_2(p)$, and a description of the Mathieu group $M_{12}$ and some of its subgroups, followed by a digression on the Janko group of order 175560. With the exception of the Janko group material, all the structure described appears within the group $M_{24}$, which is the subject of the second lecture, where it is constructed and its subgroups described in some detail. The information on $M_{24}$ is then found useful in the third lecture, on the group $\cdot 0$ and its subgroups. A postscript describes all the exceptional simple groups known to date.

† These notes do not correspond exactly to the lectures as given. They differ mainly in having more tabulated material, and a more detailed exposition of the subgroups of the groups described.

## 2. First Lecture

### 2.1. Some Exceptional Behaviour of the Groups $L_n(q)$

The general linear group $GL_n(q)$ is the group of all linear automorphisms of an $n$-dimensional vector space over the field $F_q$ of $q$ elements, $q$ being any prime power. The special linear group is the normal subgroup consisting of the automorphisms of determinant 1. The centre of either of these groups consists of operations of the form $x \to kx$ ($k \in F_q$), and we obtain the corresponding projective groups $PGL_n(q)$ and $PSL_n(q)$ by factoring out these centres. $PSL_n(q)$ is a simple group ($n \geq 2$) except in the two cases $n = 2$ and $q = 2$ or 3. It was called by Dickson the Linear Fractional group $LF(n, q)$, but we shall use Artin's abbreviation $L_n(q)$.

When $n = 2$, we take a base $y, z$ for the space, so that the operations of the group $SL_n(q)$ have the form $y \to ay+bz$; $z \to cy+dz$ ($ad-bc = 1$). The projective line $PL(q)$ consists of the $q+1$ values of the formal ratio $x = y/z$ (which are conveniently thought of as the $q$ field elements together with the formal ratio $\infty$) and, on the projective line, $L_2(q)$ becomes the group of all operations of the form

$$x \to \frac{ax+b}{cx+d}$$

with $ad-bc = 1$, or equivalently, with $ad-bc$ any non-zero square in $F_q$. We use the following names for subsets of $PL(q)$:

$$\Omega = PL(q), \qquad \Omega' = F_q = \Omega\backslash\{\infty\}, \qquad Q = \{x^2 : x \in F_q\},$$
$$N = \Omega\backslash Q, \qquad Q' = Q\backslash\{0\}, \qquad N' = N\backslash\{\infty\}.$$

In fact $L_2(q)$ is generated by the three operations

$$\alpha : x \to x+1 \qquad \beta : x \to kx \qquad \gamma : x \to -x^{-1}$$

provided that $Q'$ is the set of powers of the field element $k$; for $\beta^b\alpha^a$ takes $x$ to $k^bx+a$, while $\beta^b\alpha^a\gamma\alpha^c$ takes $x$ to $c-(k^bx+a)^{-1}$, and plainly every operation of $L_2(q)$ can be expressed in one of these forms.

The necessary set of defining relations varies slightly with the structure of $q$: when $q$ is a prime congruent to 3 modulo 4 we have

$$L_2(q) = \langle \alpha, \beta, \gamma : \alpha^q = \beta^{\frac{1}{2}(q-1)} = \gamma^2 = \alpha^\beta \cdot \alpha^{-k} = (\beta\gamma)^2 = (\alpha\gamma)^3 = 1 \rangle$$

since it is easy to see that these relations enable us to put every function of $\alpha, \beta, \gamma$ into one of the two forms above. ($\alpha^\beta$ denotes $\beta^{-1}\alpha\beta$.) In the cases $q = 3, 5, 7, 11$, we shall take $k = 1, 4, 2, 3$; for $p = 5$, the relation $(\alpha\beta\gamma)^5 = 1$ completes the above set.

It was proved by Galois (in a letter to Chevalier written on the eve of his fatal duel) that $L_2(p)$ cannot have a non-trivial permutation representation on fewer than $p+1$ symbols if $p > 11$. However, for $p = 3, 5, 7, 11$ there exist transitive representations on exactly $p$ symbols, and by studying these we shall illuminate some of the "unexpected" isomorphisms

$$L_2(3) \cong A_4, \quad L_2(4) \cong L_2(5) \cong A_5, \quad L_2(7) \cong L_3(2),$$
$$L_2(9) \cong A_6, \quad L_4(2) \cong A_8.$$

These isomorphisms are visible inside the Mathieu groups, and in the second lecture we shall find the pieces fitting together. This study is also interesting as an example of a common but puzzling phenomenon: the four cases have much in common but in each particular case there is something peculiar to that case, so that there is no completely general pattern.

It happens that the $p$ objects permuted by $L_2(p)$ ($p = 3, 5, 7, 11$) can in each case be taken as $p$ involutory permutations of the set $\Omega$. For $p = 3, 5, 7, 11$, respectively, define $\pi$ as

$$(\infty\, 0)(12), \qquad (\infty\, 0)(14)(23), \qquad (\infty\, 0)(13)(26)(45),$$
$$(\infty\, 0)(12)(36)(97)(5X)(48) \qquad (X \text{ denotes } 10)$$

and define $\pi_i$ as $\pi^{\alpha^i}$. (Mnemonic: $\pi$ interchanges $\infty$ with 0, and takes $x$ to $nx$ or $x/n$ according as $x \in Q'$ or $x \in N'$, where $n = 5, 4, 3, 2$ in the respective cases.) It is convenient to define $\pi_\infty$ as the identical permutation of $\Omega$.

THEOREM 1. *The group $L_2(p)$ ($p = 3, 5, 7, 11$) leaves invariant the set $\Pi$ consisting of the $p$ involutions $\pi_i$ ($i \in \Omega'$).*

*Proof.* $\Pi$ is obviously invariant under $\alpha$ and the mnemonic shows it to be invariant under $\beta$ also. So we need only check invariance under $\gamma$. Here the miraculous enters—we find that $(\pi_i)^\gamma = \pi_{i\delta}$, where $\delta$ is

$$(\infty)(0)(1)(2), \qquad (\infty)(0)(12)(34), \qquad (\infty)(0)(12)(36)(4)(5),$$
$$(\infty)(0)(19)(26)(3)(45)(78)(X)$$

in the four cases.

We now discuss the cases separately.

## 2.2. THE CASE $p = 3$

Here the $\pi_i$ are the elements of a Klein 4-group which is contained in $L_2(3)$, and being invariant, is a normal subgroup of $L_2(3)$ which is therefore not simple. In this case the representation on $p$ letters is not faithful, since we observe that $\delta$ is the identity. Since $L_2(3)$ is of order 12 and contains only even permutations of $\Omega$, we have the isomorphism $L_2(3) \cong A_4$.

## 2.3. THE CASE $p = 5$

The group $L_2(5)$ induces only even permutations of $\Pi$, and being of order 60, can only be the alternating group on $\Pi$, and we have the isomorphism $L_2(5) \cong A_5$. Transforming $\Pi$ by the operations of the $S_6$ on $\Omega$, we obtain just 6 such sets of 5 involutions. Each permutation of this $S_6$ defines a permutation of these six sets, and so $S_6$ can be regarded as a permutation group on two distinct systems of 6 objects. The $S_5$ fixing $\Pi$ is generated by the permutations $(\pi_i)^{\pi_j}$ and contains the original $L_2(5)$, and so does not fix any element of $\Omega$. It follows that the two permutation representations of $S_6$ on 6 objects are essentially distinct, in the sense that they are related by an *outer* automorphism of $S_6$. It is known that only for $n = 6$ does $S_n$ possess an outer automorphism.

The details are as follows. Let $\Pi_i = \Pi^{\pi_i}$ $(i \in \Omega)$. The sets $\Pi_i$ are the 6 sets of 5 involutions, and the permutation $\pi_i$ of $\Omega$ induces the permutation $(\Pi_\infty, \Pi_i)$ of these 6 sets, and symmetrically the permutation $(\infty, i)$ induces the permutation taking $\Pi_x$ to $\Pi_{x\pi_i}$. So there is an outer automorphism $\theta$ interchanging $(\infty, i)$ with $\pi_i$, and hence all the 15 involutions of shape $(ab)$ $(c)$ $(d)$ $(e)$ $(f)$ with those of shape $(uv)$ $(wx)$ $(yz)$. The former class correspond to Sylvester's *duads*, the latter to his *synthemes*, and the 6 sets $\Pi_i$ are the *synthematic totals*. We have $\theta^2 = 1$, and $(i, j)^\theta = \pi_{i,j}$, where $\pi_{i,j} = \pi_i^{\pi_j} = \pi_j^{\pi_i}$ except that $\pi_{\infty,i} = \pi_{i,\infty} = \pi_i$.

2.4. The Case $p = 7$

In this case the $\Pi_i$ are the elements of an elementary abelian group $E$ of order 8. $L_2(7)$ acts as a subgroup of the automorphism group $L_3(2)$ of $E$ ($L_3(2)$ since we can regard $E$ as a 3-dimensional vector space over $F_2$), and since $|L_3(2)| = |L_2(7)|$ we have the isomorphism $L_2(7) \cong L_3(2)$.

The group $E$ generates together with $L_2(7)$ a subgroup $F$ of $A_8$ of order $8.168$, and so of index 15 in $A_8$. Since $L_2(7)$ is generated by $\alpha$, $\beta$, $\gamma$ and is transitive on $\Pi$, the group $F$ is generated by $\alpha$, $\beta$, $\gamma$, $\pi_0$. But (a miracle!) we observe that $\gamma\delta = \pi_0$, so that $F$ is equally generated by $\alpha$, $\beta$, $\delta$, $\pi_0$. The group $F$ therefore contains two subgroups $L_3(2)$ complementary to $E$, namely the "original" $L_2(7) = \langle\alpha, \beta, \gamma\rangle$, and the "exceptional" $L_2(7) = L_3(2) = \langle\alpha, \beta, \delta\rangle$. These are not conjugate in $F$, for one of them is transitive on $\Omega$, and the other fixes $\infty$. It is exceptional for the holomorph of an elementary abelian group to exhibit this behaviour.

We can describe this situation in another way by saying that $F$ has an outer automorphism $\theta$ which fixes $E$ pointwise and also fixes the quotient group $F/E$ pointwise, without of course fixing $F$. In fact we have $\theta^2 = 1$, $\alpha^\theta = \alpha$, $\beta^\theta = \beta$, $\gamma^\theta = \delta$, $\delta^\theta = \gamma$, and $\pi_i^\theta = \pi_i$. It is not hard to see that every subgroup of $F$ isomorphic to $L_2(7)$ is conjugate to either $\langle\alpha, \beta, \gamma\rangle$ or $\langle\alpha, \beta, \delta\rangle$, so that $\theta$ is essentially the only outer automorphism of $F$. Another consequence of this situation is that $F$ has two essentially distinct faithful representations on 8 letters (compare the behaviour of $S_6$).

Since $F$ has index 15 in $A_8$ we obtain 15 sets of 7 involutions like $\Pi$ by transforming $\Pi$ by the elements of $A_8$. These 15 sets we call the *even* sets—if we transform instead by the elements of $S_8 \backslash A_8$ we obtain 15 further sets, the *odd* sets. Each of the 105 regular involutions of $A_8$ (i.e., those of shape $(ab)(cd)(ef)(gh)$) belongs to just one even set and just one odd set. There is therefore a natural graph with vertices the 30 sets and edges the 105 regular involutions of $A_8$, each involution joining the two sets containing it. Each vertex of either set is joined to just 7 vertices of the other set.

We can derive a similar graph from an elementary abelian group of order 16 by taking as even and odd vertices its 15 involutions and 15 subgroups of order 8, joining each subgroup to the 7 involutions it contains. To show that these two graphs are isomorphic we must turn our 15 even sets $\Pi_i$ into an elementary abelian group of order 16. We do this by defining the product $\Pi_i \Pi_j$ of two distinct even sets as the unique third set joined in the graph to all the vertices joined to both $\Pi_i$ and $\Pi_j$. There is a simple combinatorial proof that this does indeed define a group, and since each element of $A_8$ acts non-trivially on this group, $A_8$ is a subgroup of its automorphism group $L_4(2)$. Since $|L_4(2)| = \frac{1}{2}.8!$ we have the isomorphism $L_4(2) \cong A_8$. We do not go into further details here, since we shall give another proof of this isomorphism in the second lecture.

In this sort of work it is useful to have a simple way of specifying at a glance the structure of the groups that appear. We shall say that $G$ is a group $A.B$ (or $AB$, when no confusion can arise) when we mean that $G$ has a normal subgroup $A$ whose quotient is isomorphic to $B$. In this notation, the cyclic group of order $n$ is written simply as $n$, and the elementary abelian group of order $p^n$ just as $p^n$. Thus the group $F$ we have just discussed is a group of type $2^3 L_3(2)$.

## 2.5. The Case $p = 11$

In the cases $p = 3, 5, 7$ the group $\langle \alpha, \beta, \gamma, \delta \rangle$ was $A_4$, $S_6$, $2^3 L_3(2)$ respectively, and in the cases $p = 5, 7$ this group had an outer automorphism $\theta$ fixing $\alpha$ and $\beta$ and interchanging $\gamma$ with $\delta$. We assert that for $p = 11$ the group $\langle \alpha, \beta, \gamma, \delta \rangle$ is the Mathieu group on 12 letters, and that $\langle \alpha, \beta, \gamma \rangle$ and $\langle \alpha, \beta, \delta \rangle$ are two subgroups of type $L_2(11)$ not interchanged by any automorphism of $M_{12}$.

We shall show here that $\langle \alpha, \beta, \gamma, \delta \rangle$ is a proper subgroup of $A_{12}$, with a 6-dimensional projective representation over $F_3$. For the sake of consistency with our later notation, we transform $\alpha, \beta, \gamma, \delta$ by the permutation $\varepsilon$ which replaces $x$ by $-x$. This has the same effect as renaming our original permutations $\pi_i$ as $\pi_{-i}$.

Now take a 12-dimensional space $\mathscr{X}$ over $F_3$, with basis vectors $x_i$ $(i \in \Omega)$, and for $S \subseteq \Omega$, let $x_S$ denote $\Sigma x_i$ $(i \in S)$, with a similar notation in other cases. We consider the space $\mathscr{W}$ spanned by vectors $w_i$ $(i \in \Omega)$, where $w_\infty = x_\Omega$, and $w_i = x_{N-i} - x_{Q-i}$ $(i \in \Omega')$, where for instance $N - i$ means $\{n - i : n \in N\}$.

The *ternary Golay code* $\mathscr{C}$ is the set of all 12-tuples $(c_\infty, c_0, \ldots, c_X)$ with $\Sigma c_i x_i \in \mathscr{W}$.

THEOREM 2. *$\mathscr{C}$ is 6-dimensional, and $\Sigma c_i w_i = 0$ if and only if $(c_i) \in \mathscr{C}$.*

*Proof.* $w_\infty, w_1, w_3, w_4, w_5, w_9$ are linearly independent in $\mathscr{W}$, so $\mathscr{C}$ is at least 6-dimensional. But $w_N = w_Q = 0$, so that $w_\Omega = 0$ and $w_{N-i} = w_{Q-i} = 0$ for $i \in \Omega'$. Thus $(c_i) \in \mathscr{C}$ implies that $\Sigma c_i w_i = 0$, and the $w_i$ satisfy at least 6 linearly independent relations, from which the statements follow.

We define linear maps $A, B, C, D$ on $\mathscr{X}$ by:

$$A : x_i \to x_{i+1}, \qquad B : x_i \to x_{3i}, \qquad C : x_i \to \pm x_{-1/i}, \qquad D : x_i \to x_{i\Delta},$$

where $\Delta = \delta^\varepsilon = (\infty)\,(0)\,(1)\,(2X)\,(34)\,(59)\,(67)\,(8)$, and the sign $\pm$ is $+$ for $i \in Q$, $-$ for $i \in N$.

THEOREM 3. *$A, B, C, D$ preserve $\mathscr{W}$.*

*Proof.* We check the effects of these on the $w_i$, finding miraculously that

$$A : w_i \to w_{i-1}, \qquad B : w_i \to w_{3i}, \qquad C : w_i \to \mp w_{-1/i}, \qquad D : w_i \to w_{i\Delta},$$

where the sign $\mp$ is the opposite of the previous sign $\pm$.

It follows at once that $\alpha, \beta, \gamma, \Delta$ (and so $\alpha, \beta, \gamma, \delta$) generate a proper subgroup $M_{12}$ of $A_{12}$. The following portmanteau theorem, which we shall not prove, gives us considerable information about $M_{12}$. Most of it is easily proved by mimicking the methods of the next lecture for $M_{24}$.

THEOREM 4. *$M_{12}$ is a quintuply transitive group of order* 12.11.10.9.8. *The group $\langle A, B, C, D\rangle$ is a non-splitting extension $2M_{12}$, consisting precisely of those automorphisms of $\mathscr{X}$ which preserve $\mathscr{W}$. It has an outer automorphism $\theta$ satisfying*

$$\theta^2 = 1, \quad A^\theta = A^{-1}, \quad B^\theta = B, \quad C^\theta = C^{-1} = -C, \quad D^\theta = D$$

*whose adjunction completes it to a group $2M_{12}2$.*

The symmetry corresponding to $\theta$ is illuminated by defining $\mathscr{V} = \mathscr{X}/\mathscr{W}$, and vectors $v_i$ as the canonical images of the $x_i$ in $\mathscr{V}$ The group $2M_{12}2$ is then a group of linear automorphisms of the space $\mathscr{V} \oplus \mathscr{W}$, and as such permutes the 48 vectors $\pm v_i$, $\pm w_i$, the automorphism $\theta$ interchanging each $v_i$ with the corresponding $w_i$. The quotient group $M_{12}2$ permutes the 24 subgroups $V_i = \{0, v_i, -v_i\}$ and $W_i = \{0, w_i, -w_i\}$ of $\mathscr{V} \oplus \mathscr{W}$. We describe some of the subgroups of $2M_{12}2$ in these terms.

Fixing $V_\infty$ we have a subgroup $2M_{11}$ with orbits of sizes 2, 22, 24 on the 48 vectors, namely $\pm v_\infty$, $\pm v_i$ $(i \in \Omega')$, and $\pm w_i$ $(i \in \Omega)$. This must be a direct product $C_2 \times M_{11}$, since it has a subgroup $M_{11}$ of index 2 fixing $v_\infty$ and $-v_\infty$ separately, with orbits 1, 1, 22, 12, 12, namely $v_\infty$, $-v_\infty$, $\pm v_i$ $(i \in \Omega')$, and $w_i$ $(i \in \Omega)$, $-w_i$ $(i \in \Omega)$. Note that the permutation representation of $2M_{11}$ on 22 objects here is *not* the direct product of permutation representations of $C_2$ on 2 objects and $M_{11}$ on 11 objects, since the subgroup $M_{11}$ is still transitive on all 22 objects. There is a second conjugacy class of subgroups $M_{11}$ in $M_{12}$, obtained by interchanging the roles of the $v_i$ and $w_i$.

There is a subgroup $SL_2(11) = \langle A, B, C\rangle$ with two orbits of size 24, and a subgroup $L_2(11)$ with orbits of sizes 1, 1, 11, 11, 1, 1, 11, 11, namely $v_\infty$, $v_i$ $(i \in \Omega')$, $w_\infty$, $w_i$ $(i \in \Omega')$ and their negatives. The group $L_2(11)$ is completed to a direct product $2 \times L_2(11)$ by adjoining $-1$. In the quotient group $M_{12}2$ these both become $L_2(11)$'s, but one is maximal in $M_{12}$, and the other is not.

The subgroup $2 \times M_{10}$ fixing two subgroups $V_\infty$, $V_0$ has orbits of sizes 2, 2, 20, 24. It has a subgroup $M_{10}$ fixing $v_\infty$ and $-v_\infty$ separately, and this has a further subgroup $M'_{10}$ fixing also $v_0$, $-v_0$ separately. $M'_{10}$ has orbits of sizes 1, 1, 1, 1, 20, 6, 6, 6, 6, the fixed points being obvious, and the orbits of size 6 being $w_i$ $(i \in N)$, $w_i$ $(i \in Q)$ and their negatives. Coincidence of orders implies the isomorphism $M'_{10} \cong A_6$, but we can also see $M'_{10} \cong L_2(9)$ by translating the 10 subgroups $V_1, V_2 \ldots, V_X$ into the 10 points

$$0,\ 1,\ i,\ -i,\ 1-i,\ -1-i,\ i+1,\ \infty,\ i-1,\ -1$$

of the projective line $PL(9)$, when the permutations of $M'_{10}$ become linear fractional transformations. We therefore have $A_6 \cong M'_{10} \cong PSL_2(9)$. The three groups $S_6$, $M_{10}$, $PGL_2(9)$ are all distinct, being the three subgroups of index 2 in $\mathrm{Aut}(A_6)$ ($\mathrm{Aut}(A_6)/A_6$ is a 4-group). $S_6$ is completed to $\mathrm{Aut}(A_6)$ by our outer automorphism $\theta$, interchanging duads and synthemes, $M_{10}$ by an outer automorphism interchanging $V_\infty$ and $V_0$, and $PGL_2(9)$ by its field automorphism, interchanging $i$ with $-i$. The groups $S_6$, $M_{10}$, $PGL_2(9)$ may be distinguished as abstract groups by the numbers of classes of elements of orders 3 and 5. $S_6$ has two of order 3, one of order 5, $M_{10}$ has one of each, and $PGL_2(9)$ has one of order 3 and two of order 5.

### 2.6. A Presentation for $M_{12}$

Each of the triples $\alpha, \beta, \gamma$ and $\alpha, \beta, \delta$ satisfies the defining relations for $L_2(11)$. It is remarkable that the conjunction of the relations so obtained with the miraculous relation $(\gamma\delta)^2 = \beta^2$, or equivalently $(\delta\gamma\beta)^2 = 1$, gives a presentation for $M_{12}$, namely

$$\begin{aligned}\langle \alpha, \beta, \gamma, \delta : \alpha^{11} = \beta^5 = \gamma^2 = \delta^2 = \alpha^\beta \cdot \alpha^{-3} = (\alpha\gamma)^3 = (\alpha\delta)^3 \\ = (\beta\gamma)^2 = (\beta\delta)^2 = (\delta\gamma\beta)^2 = 1\rangle.\end{aligned}$$

This can be supplemented by the relations $\theta^2 = 1$, $\alpha^\theta = \alpha^{-1}$, $\beta^\theta = \beta$, $\gamma^\theta = \gamma^{-1}$, $\delta^\theta = \delta$ to yield a presentation for $M_{12}2$. We can get a presentation for $2M_{12}2$ by making products involving $\gamma$ into $-1$ instead of 1, where $-1$ is an involutory central element.

If in the presentation for $M_{12}$ we eliminate $\beta$ by means of the miraculous relation, we get the considerably simpler presentation

$$\langle \alpha, \gamma, \delta : \alpha^{11} = \gamma^2 = \delta^2 = (\alpha\gamma)^3 = (\alpha\delta)^3 = (\gamma\delta)^{10} = \alpha^{\gamma\delta\gamma\delta} \cdot \alpha^2 = 1 \rangle,$$

which we can further transform into

$$\langle \alpha, \gamma, \eta : \alpha^{11} = \gamma^2 = (\gamma\eta)^2 = (\alpha\gamma)^3 = (\eta\gamma\alpha)^3 = \eta^{10} = \alpha^{\eta^2} \cdot \alpha^2 = 1 \rangle$$

by replacing $\gamma\delta$ by $\eta$, eliminating $\delta$. We get yet another by eliminating $\gamma$ instead. (These are quite essentially distinct, though in appearance very similar, since we know that the subgroups $\langle \alpha, \beta, \gamma \rangle$ and $\langle \alpha, \beta, \delta \rangle$ are thoroughly distinct—indeed one is maximal and the other is not.)

## 2.7. Janko's Group of Order 175560

The way in which the exceptional representations of the small groups $L_2(p)$ can arise in other situations is well illustrated by considering the permutation representation of Janko's group $J$ of order 175560 on 266 letters. The stabilizer of a point is an $L_2(11)$, with orbits of sizes 1, 11, 110, 132, 12. Knowing this much, it is easy to construct $J$ and so verify its existence. The centralizer $\langle i \rangle \times A_5$ of an involution $i$ of $J$ is such that the $A_5$ is contained in an $L_2(11)$. We take a particular subgroup $L_2(11) = L$. Associated with $L$ is a set $S$ of 11 involutions centralizing $A_5$'s in $L$, and permuted like our 11 permutations $\pi_i$ of the projective line $PL(11)$—we let $\pi_i$ be the permutation of $PL(11)$ corresponding to the involution $i$ of $S$. The 12 cosets of $L$ which form an orbit under $L$ are permuted like the points $x$ of $PL(11)$—we let $L_x$ be the coset corresponding to $x \in PL(11)$. Any element of $L$ can be expressed as a product $f(\alpha, \beta, \gamma)$ of the generating permutations of $L$—the permutation it induces on $S$ is the corresponding product $f(\alpha, \beta, \delta)$.

Now the effect of an involution $i$ of $S$ on the 266 cosets must be capable of being described in a manner invariant under $L$. This limits the possibilities, and in fact forces the following unique situation. The 266 cosets are $L$, $Li$, $Lij$, $L_xj$, $L_x$, for $x \in PL(11)$, $i, j \in S$, $i \neq j$. Postmultiplication by the typical element of $L$ permutes these by permuting $x$ by $f(\alpha, \beta, \gamma)$ and $i$ and $j$ by $f(\alpha, \beta, \delta)$; so we need only describe the effect of postmultiplying by a typical $k \in S$.

Since $k^2$ is the identity and $Liji = Lij$, we need only consider expressions $Lijk$ and $L_xjk$ in which $i, j, k$ are distinct. Now the permutation $\pi_i \pi_j \pi_k$ has two fixed points $u$ and $v$ in $PL(11)$ if it has any, and $u$ and $v$ may be distinguished by the condition that the permutation $\pi_h$ which interchanges $u$ and $v$ also interchanges $u\pi_i$ and $v\pi_k$, but not $u\pi_k$ and $v\pi_i$. We define $u = [i, j, k]$, and then $[k, j, i]$ will be $v$. We then have

$$\begin{aligned} Lijk &= Lji \quad \text{if } \pi_i\pi_j\pi_k \text{ has no fixed point,} \\ Lijk &= L_xh \quad \text{if } [i, j, k] = x = [h, i, j], \\ L_xjk &= L_yh \quad \text{if } x = [h, k, j] \text{ and } y = [j, k, h], \\ L_xjk &= Lhi \quad \text{if } x\pi_j = [i, j, k] = [h, i, j]. \end{aligned}$$

These equations are quite easy to work with when we remember that the equation $[i, j, k] = u$ inverts to determine $\pi_i, \pi_j, \pi_k$ as interchanging the pairs $(u, u\pi_k\pi_j)$, $(u\pi_i, u\pi_k)$, and $(u, u\pi_i\pi_j)$ respectively. (Any pair of distinct points of $PL(11)$ is interchanged by just one of the permutations $\pi_i$.)

That $J$ is a proper subgroup of $A_{266}$ with the right order is also easy to verify. If we join each of the 266 points to the orbit of size 11 in its stabilizer we get a graph in which the joins are $(L, Lh)$, $(Li, Lhi)$, $(Lij, Lhij)$, $(L_x, L_y h)$, $(L_x j, L_y hj)$, where in each case $h$ varies arbitrarily and $y = x\pi_h$. It is not hard to verify that the operations defined above preserve this graph. If instead we join each point to the corresponding set of 12 points, we get the joins (in which $y$ and $h$ are arbitrary)

$$(L, L_y), \quad (Li, L_y i), \quad (Lij, L_y ij), \quad (L_x, L), \quad (L_x, L_x h),$$
$$(L_x j, Lj), \quad (L_x j, L_x hj).$$

## 3. Second Lecture

### 3.1. The Mathieu Group $M_{24}$

We define $M_{24}$ as the group obtained by adjoining the permutation $\delta : x \to x^3/9$ $(x \in Q)$ or $x \to 9x^3$ $(x \in N)$ to the group $L_2(23)$ acting on the projective line $\Omega = PL(23)$. We list the generators in full:

$$\alpha = (\infty)(0\ 1\ 2\ 3\ 4\ 5\ 6\ 7\ 8\ 9\ 10\ 11\ 12\ 13\ 14\ 15\ 16\ 17\ 18\ 19\ 20\ 21\ 22)$$
$$\beta = (\infty)(15\ 7\ 14\ 5\ 10\ 20\ 17\ 11\ 22\ 21\ 19)(0)(3\ 6\ 12\ 1\ 2\ 4\ 8\ 16\ 9\ 18\ 13)$$
$$\gamma = (\infty\ 0)(15\ 3)(7\ 13)(14\ 18)(5\ 9)(10\ 16)(20\ 8)(17\ 4)(11\ 2)(22\ 1)(21\ 12)(19\ 6)$$
$$\delta = (\infty)(14\ 17\ 11\ 19\ 22)(15)(20\ 10\ 7\ 5\ 21)(0)(18\ 4\ 2\ 6\ 1)(3)(8\ 16\ 13\ 9\ 12).$$

It is sometimes convenient to replace the pair $\gamma, \delta$ by the product $\gamma\delta^2$, which generates the same group, since $\gamma$ and $\delta$ plainly commute. We have $\gamma = (\gamma\delta^2)^5$, $\delta = (\gamma\delta^2)^{-2}$, and

$$\gamma\delta^2 = (\infty\ 0)(15\ 3)(14\ 2\ 22\ 4\ 19\ 18\ 11\ 1\ 17\ 6)(20\ 13\ 21\ 16\ 5\ 8\ 7\ 12\ 10\ 9),$$

or, algebraically, $\gamma\delta^2$ interchanges $\infty$ with 0, and otherwise takes $x$ to $-(\frac{1}{2}x)^2$ $(x \in Q)$, or $(2x)^2$ $(x \in N)$. The operations $\gamma, \delta$ plainly normalize the group $\langle\beta\rangle$, and in fact we have $\beta^\gamma = \beta^{-1}$, $\beta^\delta = \beta^3$, and $\beta^{\gamma\delta^2} = \beta^2$.

THEOREM 1. *$M_{24}$ is quintuply transitive on $\Omega$.*

*Proof.* By randomly multiplying the generators, we find permutations of cycle-shapes $1\,23$, $1^2 11^2$, $1^3 7^3$, $2^{12}$, $1^8 2^8$, and $4^6$ (for instance $\alpha$, $\beta$, $\delta\alpha^2$, $\delta$, $\gamma$, $(\alpha\delta)^3$, and $(\alpha^{13}\gamma\delta^2)^3$). The shapes $1\text{-}23$ and $2^{12}$ show that $M_{24}$ is transitive, and then that the stabilizer of any point must have a permutation of shape $1\,23$, and so be transitive on the remaining 23 points. In a similar way we see that the stabilizer of 2 points is transitive on the remaining 22 (using shapes $1^2 11^2$ and $1^3 7^3$), and the stabilizer of 3 points transitive on the remaining 21 (using $1^3 7^3$ and $1^4 5^4$), so that $M_{24}$ is quadruply transitive. Now the subgroup fixing a set of 4 points as a whole has permutations of shapes $1^4 5^4$ and $4^6$, so is transitive on the remaining 20 points, so that $M_{24}$ acts transitively on the set of all 5-element subsets of $\Omega$. But the subgroup fixing any 5-element set has permutations of shapes $1^4 5^4$ and $1^8 2^8$ which induce permutations of shapes 5 and $1^3 2$ on the 5-elements, and the quintuple transitivity follows from this, since any two permutations of shapes 5 and $1^3 2$ generate the full symmetric group on 5 letters.

We define $M_{24-k}$ as the pointwise stabilizer of a $k$-element subset of $\Omega$ in $M_{24}$ ($k \leq 5$).

To show that $M_{24}$ is a proper subgroup of the alternating group $A_{24}$, we turn the set $P(\Omega)$ of all subsets of $\Omega$ into a 24-dimensional vector space over $F_2$ (by defining the sum $A+B$ of two sets as their symmetric difference $(A \backslash B) \cup (B \backslash A)$) and show that $M_{24}$ preserves a 12-dimensional subspace. The *binary Golay code* $\mathscr{C}$ is the space spanned by the 24 sets $N_i$ ($i \in \Omega$), where $N_\infty = \Omega$, and otherwise $N_i = N - i = \{n - i : n \in N\}$. If $S \subseteq \Omega$, we write $N_S$ for $\Sigma N_i$ ($i \in S$).

THEOREM 2. *$\mathscr{C}$ is at most 12-dimensional.*

*Proof.* We can verify directly that $N_\Omega = N_N = 0$, whence $N_{N_i} = 0$ for each $i \in \Omega$, and $N_C = 0$ for each $C \in \mathscr{C}$. If $\mathscr{C}$ is $k$-dimensional, we therefore have at least $k$ independent linear relations between its generating sets $N_i$, so that $k \leq 24 - k$. (The relations above are consequences of well-known number-theoretical facts but, in any case, it is obvious *a priori* that each of $N_\Omega$, $N_N$ must have the form $aN + bQ + c\{\infty\} + d\{0\}$, so there is not much to check.)

Although we do not need it yet, and could derive it easily from any of several facts later on, we can use a similar calculation to find the exact dimension of $\mathscr{C}$, and a test for membership of $\mathscr{C}$.

THEOREM 3. *$\mathscr{C}$ is exactly 12-dimensional, and we have $C \in \mathscr{C}$ if and only if $N_C = 0$.*

*Proof.* We find $N_{\{-2,0,2,3\}} = \{0, 1, 2, 3, 4, 7, 10, 12\}$, a $\mathscr{C}$-set (i.e., member of $\mathscr{C}$) with least element 0. Adding $i$ ($i \leq 10$) to the elements of this we get $\mathscr{C}$-sets with least element $i$ ($0 \leq i \leq 10$). These obviously span an 11-dimensional space, and we get the extra dimension by adding any $\mathscr{C}$-set containing $\infty$.

THEOREM 4. *$M_{24}$ preserves $\mathscr{C}$.*

*Proof.* We have obviously $N_i\alpha = N_{i-1}$, $N_i\beta = N_{2i}$, and we verify the equations

$$N_0\gamma\delta^2 = N_\infty, \qquad N_\infty\gamma\delta^2 = N_0, \qquad N_1\gamma\delta^2 = N_{\{-1,0,1,3\}},$$
$$N_{-1}\gamma\delta^2 = N_{\{3,12,21\}}$$

from which we can deduce $N_i\gamma\delta^2 \in \mathscr{C}$ for all $i$, using $N_{2i}\gamma\delta^2 = N_i\beta\gamma\delta^2 = N_i\gamma\delta^2\beta^2$. (There is really only one non-trivial relation to be verified, since $\mathscr{C}$ is spanned by the $N_i$ $(i \in Q')$ together with $N_\infty = \Omega$.)

THEOREM 5. *There exist 8-element $\mathscr{C}$-sets, called (special) octads, and each non-empty $\mathscr{C}$-set is the symmetric difference of a strictly smaller $\mathscr{C}$-set with an octad. Each 5-element set is contained in just one octad.*

*Proof.* We have already found one octad, {0, 1, 2, 3, 4, 7, 10, 12}, and so each 5-element set is contained in at least one octad, by Theorem 1. Any $\mathscr{C}$-set with at least 5 points is therefore the symmetric difference of an octad containing those 5 points and a strictly smaller $\mathscr{C}$-set. If some non-empty $\mathscr{C}$-set has fewer than 5 elements, every set of the same cardinal would be a $\mathscr{C}$-set, by Theorem 1, and by taking symmetric differences we should obtain every two-element set, and so every set of even cardinal, as a $\mathscr{C}$-set, which cannot be. It follows that the octads are the smallest non-empty $\mathscr{C}$-sets, for from any smaller non-empty one we could obtain a still smaller one, and so on. No 5-element set can be contained in two distinct octads, for their symmetric difference would be a $\mathscr{C}$-set with at most 6 members.

The last sentence of Theorem 5 asserts that the octads form what is known as a *Steiner system* $S(5, 8, 24)$. We can deduce from it that there are just

$$\binom{24}{5}\Big/\binom{8}{5} = 759 \text{ octads,}$$

permuted transitively by $M_{24}$. Witt proved that the system $S(5, 8, 24)$—likewise the systems $S(4, 7, 23)$, $S(3, 6, 22)$, $S(5, 6, 12)$, $S(4, 5, 11)$ which are involved with the other Mathieu groups—is essentially unique, and defined $M_{24}$ as its automorphism group.

### 3.2. The Stabilizer of an Octad

$M_{24}$ contains permutations of shapes 13515 and $1^2 2\ 4\ 8^2$, for example, $\delta\alpha^{11}$ and $\delta\alpha^5$. The octad containing the 5-cycle in the first case (being fixed) can only be the union of the 5- and 3-cycles, and in the second case the octad containing the 4-cycle and a 1-cycle must be the union of the 4-, 2-, and 1-cycles. The fourth power of the second operation is therefore a permutation of shape $1^8 2^8$ whose fixed points form an octad. We can therefore suppose that $M_{24}$ contains the permutation $\lambda = (abcde)\ (fgh)\ (i)\ (jkl\ldots x)$ and a permutation $\mu$ of shape $1^8 2^8$ interchanging $i$ and $j$ and fixing $a, b, c, d, e, f, g, h$. The permutations $\mu^\lambda$, $\mu^{\lambda^2}, \ldots$ have the same fixed points but now interchange $i$ with $k$, $i$ with $l$, etc., and so the pointwise stabilizer of $\{a, b, c, d, e, f, g, h\}$ has at least 15 involutions and order at least 16. The pointwise stabilizer of

$\{a, b, c, d, e\}$ has at least three times this order, since it contains also the permutation $\lambda^5$, of order 3. It follows that $M_{24}$ has order at least $24.23.22.21.20.16.3 = 244823040$.

THEOREM 6. *The subgroup of $M_{24}$ fixing an octad setwise is a group $2^4A_8$, an extension of an elementary abelian group of order* 16 (*fixing the octad pointwise*) *by the alternating group on* 8 *letters, which is isomorphic to $L_4(2)$. $M_{24}$ has order exactly* 244823040, *and contains all permutations of $\Omega$ which preserve the Golay code $\mathscr{C}$.*

*Proof.* We consider the subgroup $H$ of index 16 which, in addition to fixing the octad $\{a, b, c, d, e, f, g, h\}$ as a whole, fixes the point $i$. Since $\{a, b, c, d, e, f, g, h, i\}$ contains just one non-empty $\mathscr{C}$-set, the $\mathscr{C}$-sets disjoint from it form a space of codimension 8, and so dimension 4, and it is easy to see that every non-trivial element of $H$ acts non-trivially on this space, so that $H$ is a subgroup of its automorphism group $L_2(4)$, of order (16–1) (16–2) (16–4) (16–8) = 20160. The group of permutations of $a, b, c, d, e, f, g, h$ induced by $H$ is transitive and contains the 3-cycle $(hgf)$ (induced by $\lambda^5$), and so must be the alternating group $A_8$ of order $\frac{1}{2}.8! = 20160$; the coincidence of orders establishes the isomorphism $H \cong A_8 \cong L_4(2)$. Since $20160 = 244823040/759.16$, the order of $M_{24}$ is exactly 244823040, and the subgroup fixing $a, b, c, d, e, f, g, h$ individually has order exactly 16, being transitive on the remaining 16 letters. It consists of the identity and the 15 involutions we found earlier, and so is an elementary abelian group. (It is the dual of the above 4-dimensional space of $\mathscr{C}$-sets, and is also permuted by the group $H$.) The upper bound by this argument for the order of $M_{24}$ applies also to the group of all permutations of $\Omega$ preserving $\mathscr{C}$, and establishes the identity of the two groups.

THEOREM 7. *Let $\{a_1, a_2, \ldots, a_8\}$ be an octad. Then the number of octads intersecting $\{a_1, \ldots, a_i\}$ in $\{a_1, \ldots, a_j\}$ (exactly) is the $(j+1)$th entry in the $(i+1)$th line of Table* 1.

TABLE 1. *How many octads?*

| | | | | | | | | |
|---|---|---|---|---|---|---|---|---|
| 759 | | | | | | | | |
| 506 | 253 | | | | | | | |
| 330 | 176 | 77 | | | | | | |
| 210 | 120 | 56 | 21 | | | | | |
| 130 | 80 | 40 | 16 | 5 | | | | |
| 78 | 52 | 28 | 12 | 4 | 1 | | | |
| 46 | 32 | 20 | 8 | 4 | 0 | 1 | | |
| 30 | 16 | 16 | 4 | 4 | 0 | 0 | 1 | |
| 30 | 0 | 16 | 0 | 4 | 0 | 0 | 0 | 1 |

TABLE 2. *How many dodecads?*

| | | | | | | | | |
|---|---|---|---|---|---|---|---|---|
| 2576 | | | | | | | | |
| 1288 | 1288 | | | | | | | |
| 616 | 672 | 616 | | | | | | |
| 280 | 336 | 336 | 280 | | | | | |
| 120 | 160 | 176 | 160 | 120 | | | | |
| 48 | 72 | 88 | 88 | 72 | 48 | | | |
| 16 | 32 | 40 | 48 | 40 | 32 | 16 | | |
| 0 | 16 | 16 | 24 | 24 | 16 | 16 | 0 | |
| 0 | 0 | 16 | 0 | 24 | 0 | 16 | 0 | 0 |

*Proof.* Since $M_{24}$ is transitive on those sets of any given cardinal which are contained in octads, the number of octads containing $\{a_1, \ldots, a_i\}$ is exactly

$$759 \cdot \binom{8}{i} \bigg/ \binom{24}{i} \quad (i \leq 5), \quad 1 \quad (i \geq 5),$$

and we have the rightmost entries. The others are deduced from these by repeated use of the rule that the sum of two adjacent entries in any line is the entry just above them in the previous line. Table 2 gives the number of umbral dodecads meeting $\{a_1, \ldots, a_i\}$ in $\{a_1, \ldots, a_j\}$ (see below).

### 3.3. The Structure of $\mathscr{C}$

THEOREM 8. *$\mathscr{C}$ has* $4096 = 1+759+2576+759+1$ *sets of cardinals* 0, 8, 12, 16, 24 *respectively.*

*Proof.* Since by Theorem 7 two distinct octads intersect in 0 or 2 or 4 points their symmetric difference has cardinal 16 or 12 or 8. The theorem follows using Theorem 5 and the fact that $\mathscr{C}$-sets come in complementary pairs.

### 3.4. The Structure of $P(\Omega)/\mathscr{C}$

THEOREM 9. *Every subset of $\Omega$ is congruent* modulo $\mathscr{C}$ either *to a unique set of cardinal at most* 3 or *to each of* 6 *distinct sets of cardinal* 4.

*Proof.* If $S$ has 5 or more members we obtain a congruent set with fewer members by taking the symmetric difference with an octad containing 5 members of $S$. If $S$ has 4 members we obtain 5 more 4-element sets congruent to $S$ by taking the symmetric difference of $S$ with the 5 octads which contain it (Theorem 7). If two distinct sets of 4 or fewer elements are congruent modulo $\mathscr{C}$, their symmetric difference is an octad, and so they must be disjoint and have 4 elements each, so the theorem is best possible.

Thus $4096 = 1 + 24 + \binom{24}{2} + \binom{24}{3} + \frac{1}{6}\binom{24}{4}$.

### 3.5. The Maximal Subgroups of $M_{24}$

The complete list of maximal subgroups of $M_{24}$ is now known. J. A. Todd (1966) observed eight distinct types in his paper on $M_{24}$, and Chang Choi (to be published) systematically enumerated the subgroups, at first confirming the completeness of Todd's list. But calculations of McKay and Livingstone led to the discovery of a new maximal subgroup of order 168 which was omitted from Chang Choi's list as the result of a trivial arithmetical error. With but two exceptions the groups are easily described in terms of the Golay code $\mathscr{C}$, and we describe them in some detail.

A 12-element $\mathscr{C}$-set is called an (*umbral*) *dodecad*, and a complementary pair of umbral dodecads is a *duum*. A triplet of mutually disjoint octads is a *trio*, and a system of 6 tetrads with the property that the union of any two is an octad is a *sextet*. (Mnemonics: du*um* = 2 *um*brals, tri*o* = 3 *o*ctads, sex*tet* = 6 *tet*rads.) In general, an *n-ad* denotes an *n*-element subset of $\Omega$, except that the terms *octad* and *dodecad* usually presuppose the corresponding adjectives *special* and *umbral*.

The maximal subgroups of $M_{24}$ can now be described as the stabilizers of the concepts *monad, dyad, triad, octad, sextet, trio, duum*, along with two further groups $L_2(23)$, $L_2(7)$. The $L_2(23)$ is the group we started from, and the $L_2(7)$ is the new maximal subgroup. We give theorems asserting the exact amount of transitivity of each of these groups on various configurations in $\Omega$. We say that a group is $a+b$ transitive on two sets $A$ and $B$ if it contains elements taking any $a$ elements of $A$, and simultaneously any $b$ elements of $B$, to preassigned positions.

THEOREM 10. *The group $2^4A_8$ fixing an octad is* $6+1$, $3+2$, *and* $1+3$ *transitive on the octad and its complement.*

*Proof.* The stabilizer of an octad *and* one point in its complement is the group $H = A_8$, sextuply transitive on the octad, and, in its role as $L_4(2)$, doubly transitive on the remaining 15 points of the complement, which can be regarded as the involutions of a group $2^4$ acted upon by $L_4(2)$. The stabilizer of two points of the complement is therefore a group $2^3L_3(2)$, which can only be the group we met in the first lecture, triply transitive on the octad. The stabilizer of one point in the octad together with two of the complement is $L_3(2)$, and is easily seen to be transitive on the remaining 14 points of the complement.

THEOREM 11. *$M_{24}$ is transitive on dodecads. The group fixing a dodecad setwise is the Mathieu group $M_{12}$, which is* $5+0$, $3+1$, $1+3$, $0+5$ *transitive on the dodecad and its complement. This group $M_{12}$ has index 2 in a group $M_{12}2$, the stabilizer of a duum.*

*Proof.* The group stabilizing the sets $\{a, b\}$ and $\{c, d, e, f, g, h\}$ is plainly a subgroup $2^4S_6$ inside our group $2^4A_8$, and it is easy to see that its subgroup $2^4$ is transitive on the 16 octads which by Theorem 7 intersect $\{a, b, c, d, e, f, g, h\}$ in $\{a, b\}$. If one of them is $\{a, b, i, j, k, l, m, n\}$, it follows that the stabilizer of the two sets $\{c, d, e, f, g, h\}$ and $\{i, j, k, l, m, n\}$ is a group $S_6$, being, by symmetry, sextuply transitive on each of the two sets. (The permutation representations on these sets cannot be permutation identical, for then—the notation being suitably chosen—there would be an element of order 3 fixing each of $a, b, c, d, e, i, j, k$, and our knowledge of the pointwise stabilizer of $\{a, b, c, d, e\}$ contradicts this.)

Since every dodecad is expressible in this way as the symmetric difference of two octads with two points in common (by Theorem 5), it follows that $M_{24}$ is transitive on dodecads. Again, any 5 points of a dodecad belong to one of a pair of octads whose symmetric difference is the dodecad, and we deduce that the stabilizer of a dodecad is quintuply transitive on the dodecad (and, by symmetry, on the complementary dodecad). (The permutation representations are not permutation identical, for the subgroup $S_6$ has two orbits of size 6 in one dodecad, but an orbit of size 2 in the other.)

When we have identified this group with $M_{12}$ we can say that the stabilizer of a point in one of the dodecads is a group $M_{11}$, triply transitive on the other. In fact, identification with our previous $M_{12}$ is unnecessary, since we could easily *define* $M_{12}$ as the stabilizer of a dodecad in $M_{24}$. But it is quite easy —the groups $V_\infty, V_0, \ldots, V_X, W_\infty, W_0, \ldots, W_X$ of the first lecture become the points

$\infty$, 15, 7, 14, 5, 10, 20, 17, 11, 22, 21, 19, 0, 3, 13, 18, 9, 16, 8, 4, 2, 1, 12, 6

of $\Omega$, and we easily check that the permutations $\alpha, \beta, \gamma, \Delta$ of that lecture do indeed correspond to permutations of $M_{24}$ as defined here. ($\alpha$ and $\beta$ become $\beta$ and $\delta$ rather plainly, so we need only check the effects of $\gamma, \Delta$ on the $N_i$, using the test of Theorem 3 for $\mathscr{C}$-sets. Only two of the calculations are really required.) The outer automorphism $\theta$ of the first lecture translates into the operation $\gamma$ of this, interchanging the two dodecads of the duum, here taken as $N$ and $Q$.

THEOREM 12. *$M_{24}$ is transitive on monads, dyads, and triads, whose stabilizers are respectively groups $M_{23}$, $M_{22}2$, and $M_{21}S_3$. These groups are $a+b$ transitive on the two appropriate sets if $a+b \leq 5$ and $a \leq 1, 2, 3$ respectively.*

*Proof*. These statements follow at once from the quintuple transitivity of $M_{24}$.

If we give the points 2, 3, . . ., 22 the coordinates

100, 010, 001, $ab0$, 111, $b11$, $1a1$, $1b1$, $0ba$, $11b$, 101, $ba0$, $b1a$, $a0b$, 110, $a1b$, 011, $0ab$, $a11$, $b0a$

then the 21 octads containing $\infty$, 0, 1 yield the 21 lines of the projective plane over $F_4 = \{0, 1, a, b\}$. Using standard techniques we can then identify $M_{21}$ with $L_3(4)$ and deduce the simplicity of $M_{21}, M_{22}, M_{23}, M_{24}$.

THEOREM 13. *$M_{24}$ is transitive on sextets. The stabilizer of a sextet is a group $2^6.3.S_6$, and the subgroup $2^6.3$ fixing the tetrads individually is $2+1+1+0+0+0$ and $3+1+0+0+0+0$ transitive on the tetrads (in any order).*

*Proof.* The transitivity follows at once from the quadruple transitivity of $M_{24}$ and the fact that a sextet is determined by any one of its tetrads (as, say, the family of all tetrads congruent to the given one modulo $\mathscr{C}$). Now in the stabilizer of the sextet containing $\{a, b, c, d\}$ and $\{e, f, g, h\}$, consider the subgroup fixing the four points *a*, *b*, *c*, *i*. This is contained within the group $H = A_8$ considered earlier, and so we can see what it is—a group $S_3$ permuting *f*, *g*, *h* in the obvious way. It permutes the three tetrads disjoint from $\{a, b, e, i\}$ in the same way, and so the group fixing the tetrads individually as well as the 4 points *a*, *b*, *e*, *i* has order 1. Since this group has index at most $4.3.4.4 = 192$ in the group of all permutations fixing the tetrads individually, the latter has order at most 192. But the number of sextets is

$$\frac{1}{6}\binom{24}{4} = 1771,$$

and so the stabilizer of a sextet has order $244823040/1771 = 192.6!$. We conclude that there are 192 permutations fixing the tetrads individually, and that all 6! permutations of the tetrads are induced by permutations fixing the sextet. The transitivity $2+1+1+0+0+0$ is also established. We leave the transitivity $3+1+0+0+0+0$ to the reader, and also the exact discussion of the group of order 192—Todd remarks that as well as the identity it contains 45 permutations of shape $1^8 2^8$, 18 of shape $2^{12}$, and 128 of shape $1^6 3^6$.

THEOREM 14. *$M_{24}$ is transitive on trios. The stabilizer of a trio is a group $2^6(S_3 \times L_3(2))$, and the subgroup $2^6 L_3(2)$ fixing the octads separately is $2+1+1$ and $3+1+0$ transitive on the three octads (in any order).*

*Proof.* The element $(abcde)(fgh)(i)(jkl\ldots x)$ of shape 1 3 5 15 permutes the 30 octads disjoint from $\{a, b, c, d, e, f, g, h\}$ (Theorem 7) in two orbits of size 15, consisting of those containing, and those not containing, the point *i*. Each octad therefore belongs to 15 trios, and $M_{24}$ is transitive on the trios, of which there must be

$$759.\frac{15}{3} = 3795.$$

Since the tetrads of a sextet can be grouped to form 15 distinct trios, and since $1771.15 = 3795.7$, each trio can be refined into 7 distinct sextets. Now each sextet corresponds naturally to an element of $P(\Omega)/\mathscr{C}$ (Theorem 9), and the 7 sextets refining a trio must therefore form, together with the empty set, a 3-dimensional vector space over $F_2$ (a subspace of $P(\Omega)/\mathscr{C}$). The stabilizer of a trio has therefore two interesting normal subgroups—one fixing the octads individually, and another fixing each of the 7 sextets refining the trio. The quotient by the first is $S_3$, since all permutations of the three octads are induced, and the quotient by the second is $L_3(2)$ or a subgroup thereof, since the permutations of the 7 sextets must preserve the vector space structure. The index of their intersection therefore divides 6.168, and so the intersection, fixing both the octads and the sextets individually, has order of form $64n$ (since $64.6.168.3795 = 244823040$). But the intersection is a subgroup of the group described in the previous theorem, and since an element $1^6 3^6$ cannot fix 7 sextets, it has order at most 64, this giving the structure of the group.

We make only the following remarks about the transitivity. If we fix the three octads individually, and then also a point of one of them, we obtain a subgroup $2^3 L_3(2)$ of index 8 in $2^6 L_3(2)$, which is represented on 7 points of the first octad (with kernel $2^3$) and in two distinct ways on the 8 points of the other two octads (faithfully in each case, related by the outer automorphism of the first lecture). Fixing a point in the second octad in addition, we come down to a group $L_3(2)$, represented on 7 points in the first and second octads, on 8 points in the third, and doubly transitively in each case.

THEOREM 15. *$M_{24}$ contains subgroups of type $L_2(23)$, doubly transitive on $\Omega$, and transitive on the set of* 759 *octads.*

*Proof.* Such is the subgroup $\langle \alpha, \beta, \gamma \rangle$. The double transitivity on $\Omega$ is obvious, and the transitivity on octads is proved as follows. An easy, if rather tedious, calculation, shows that at most 8 linear fractional transformations preserve some given octad. This octad therefore has at least $|L_2(23)|/8 = 759$ images under $L_2(23)$.

THEOREM 16. *$M_{24}$ contains subgroups $L_2(7)$, transitive on $\Omega$ but with imprimitivity sets of size* 3, *not contained in any of the above subgroups of $M_{24}$.*

*Proof.* Consider the permutation

$\Pi = (12\ 13\ 14)(21\ 7\ 8)(17\ 1\ 20)(2\ 19\ 15)(6\ 3\ 11)(\infty\ 5\ 10)(16\ 0\ 9)(4\ 22\ 8)$,

*not* in $M_{24}$. Direct calculation reveals that it commutes with the permutation $\gamma$ and also with the permutation

$$(12)(21\ 17\ 2\ 6\ \infty\ 16\ 4)(13)(7\ 1\ 19\ 3\ 5\ 0\ 22)(14)(18\ 20\ 15\ 11\ 10\ 9\ 8)$$

of $M_{24}$, which generate a subgroup of type $L_2(7)$ whose imprimitivity blocks are the eight 3-cycles of $\Pi$. It is also easy to show that this $L_2(7)$ preserves no monad, dyad, triad, octad, sextet, trio, or duum, and consideration of orders shows that it cannot be contained in the group $L_2(23)$.

Thus the new maximal subgroup of $M_{24}$ is easily described as the centralizer *in* $M_{24}$ of the permutation $\Pi$, *not* in $M_{24}$. So it is the *octern stabilizer*, the cycles of $\Pi$ being ordered triads, or *terns.*

In Chang Choi's analysis, subgroups of $M_{24}$ are divided into three classes —intransitive, transitive but imprimitive, and primitive—and his arguments for the second and third classes are very short. Here we present a quick way of dealing with the intransitive groups.

THEOREM 17. *Any intransitive subgroup of $M_{24}$ fixes a monad, dyad, triad, octad, dodecad, or sextet, and so is contained in one of the groups of Todd's list.*

*Proof.* Let $S$ be a non-empty proper subset of $\Omega$ invariant under the group. If $S \in \mathscr{C}$, then $S$ or its complement is an octad or dodecad fixed by the group. Otherwise, $S$ is congruent modulo $\mathscr{C}$ to a unique monad, dyad, or triad, or to the six tetrads of a sextet, fixed by the group, by Theorem 9.

Similar ideas enable us to deal with certain imprimitive groups, as we shall see in a moment. They also enable us to give a complete classification of the subsets of $\Omega$ under the action of $M_{24}$.

## 3.6. THE STRUCTURE OF $P(\Omega)$

THEOREM 18. *The subsets of $\Omega$ fall into 49 orbits under $M_{24}$, related as in Fig.* 1. *Each node corresponds to one orbit, and the nodes are joined by lines indicating the number of ways a set of one type can be converted to one of another by the addition or removal of a single point. Thus in an umbral heptad ($U_7$), there is one point whose removal leaves a special hexad ($S_6$), removal of any of the six others leaving an umbral hexad ($U_6$) instead. In the complement of an umbral heptad there are two points whose addition gives a transverse octad ($T_8$),and* 15 *yielding umbral octads ($U_8$).*

*Proof.* We discuss 8-element sets as an example. Each 8-element set $S$ is congruent modulo $\mathscr{C}$ to one of

(i) the empty set,
(ii) a unique 2-element set $T$,
(iii) each of the tetrads $T_0, \ldots, T_5$ of a sextet.

In case (i), $S$ is a *special octad* $S_8$. $M_{24}$ is transitive on these.

In case (ii), $S$ is obtained from the special octad $S+T$ by adding one point and subtracting another. Since the stabilizer of a special octad is $1+1$ transitive on it and its complement, $M_{24}$ is transitive on sets of this type, which we call *transverse octads* $T_8$.

In case (iii) $S+T_i$ is a special octad if $T_i$ contains two points of $S$ and two of its complement, and an umbral dodecad if $T_i$ is disjoint from $S$. Counting points of $S$ shows that there are four tetrads of the first kind and two of the second, so that $S$ can be obtained (in two ways) by removing four points from an umbral dodecad. Since the stabilizer of an umbral dodecad is quadruply transitive on the dodecad, $M_{24}$ is transitive on sets like $S$ (which we call *umbral octads* $U_8$).

In general, a set of cardinal $n < 12$ is called *special* ($S_n$) if it contains or is contained in a special octad, otherwise *umbral* ($U_n$) if it is contained in an umbral dodecad, and *transverse* ($T_n$) if not. A non-umbral dodecad is *extraspecial* ($S_{12}^+$) if it contains three special octads, *special* ($S_{12}$) if it contains just one, *penumbral* ($U_{12}^-$) if it contains all but one of the points of an umbral

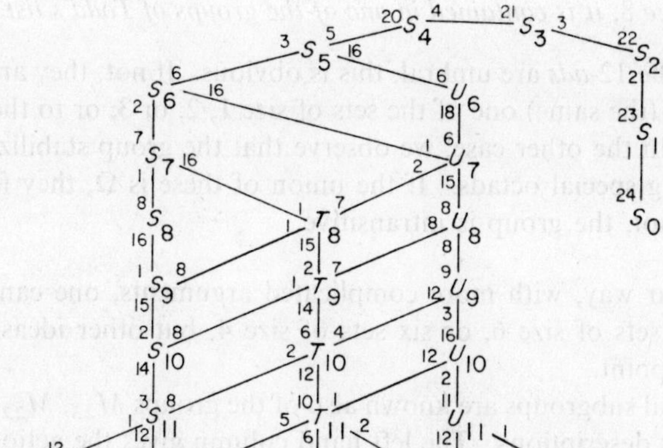

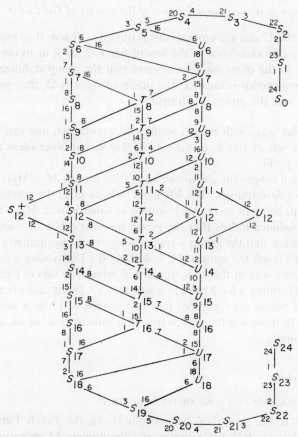

FIG. 1. The action of $M_{24}$ on $P(\Omega)$ (Theorem 18).

dodecad, and *transverse* ($T_{12}$) in all other cases. Sets of more than 12 points are described by the same adjectives as their complements.

Figure 1 enables us to read off many interesting relationships at a glance. Thus on removing any point from a $U_8$ we find a $U_7$, in which there is a unique further point whose removal leaves an $S_6$. The 8 points of a $U_8$ therefore split into 4 sets of 2 in a natural manner, so that the stabilizer of a $U_8$ acts imprimitively on the $U_8$. We can also see from the diagram that the stabilizer of any 8-element set also stabilizes a special octad. For $S_8$, this is the $S_8$ itself; for $T_8$, the unique $S_8$ sharing 7 points with the $T_8$; and for $U_8$, the union of the two tetrads whose addition separately would convert the $U_8$ to a $U_{12}$.

THEOREM 19. *If a subgroup of $M_{24}$ has two imprimitivity sets of size* 12, *or three of size* 8, *it is contained in one of the groups of Todd's list.*

*Proof.* If the 12-*ads* are umbral, this is obvious. If not, they are congruent modulo $\mathscr{C}$ to (the same) one of the sets of size 1, 2, or 3, or to the six tetrads of a sextet. In the other case, we observe that the group stabilizes the three corresponding special octads. If the union of these is $\Omega$, they form a fixed trio, and if not, the group is intransitive.

In a similar way, with more complicated arguments, one can handle the case of four sets of size 6, or six sets of size 4, but other ideas are needed beyond this point.

All maximal subgroups are known also of the groups $M_{23}$, $M_{22}$, $M_{12}$, $M_{11}$. Table 3 gives descriptions. The left hand column gives the action on the set $\Omega$, the groups being in every case regarded as subgroups of $M_{24}$. The symbol [8, 16] for instance means that the group has two orbits of sizes 8 and 16, while $[4^6]$ means that the group is transitive, with six imprimitivity sets of size 4. (Thus $[4^6]$ is *not* the same as [4, 4, 4, 4, 4, 4].) The symbol $6 \times 2$ denotes a set of 12 points with at the same time six imprimitivity sets of size 2 and two of size 6, so forming a $6 \times 2$ table in a natural way. In the case of subgroups of $M_{12}$, the 24 points are separated into two groups of 12 by a semicolon. It is very easy to recognize the groups from this information, so we refrain from further comment.

## 4. Third Lecture

### 4.1. THE GROUP ·0 AND SOME OF ITS SUBGROUPS

This group is the group of automorphisms of the Leech Lattice in 24-dimensional space $\mathbf{R}^{24}$. It contains as subquotients 12 exceptional simple groups, and it seems fair to say that these groups are easiest studied inside ·0.

### 4.2. THE GEOMETRY OF THE LEECH LATTICE

Let $\mathbf{R}^{24}$ be spanned by the orthonormal basis $v_i$ ($i \in \Omega = PL(23)$), and define $v_S$ as $\Sigma v_i$ ($i \in S$) whenever $S \subseteq \Omega$. We use $\mathscr{C}$ for the binary Golay code, $\mathscr{C}_8$ for the set of special octads, and $\mathscr{C}_{12}$ for the set of umbral dodecads, and finally $\Omega_n$ for the set of all $n$-ads. Let $\Lambda_0$ be the lattice spanned by the vectors $2v_C$ ($C \in \mathscr{C}_8$).

TABLE 3. *Maximal subgroups of the Mathieu groups*

| $M_{24}$ | | $M_{23}$ | | $M_{22}$ | | $M_{12}$ | | $M_{11}$ | |
|---|---|---|---|---|---|---|---|---|---|
| [24] | $L_2(23)$ | [23] | 23.11 | [1, 21] | $M_{21}$ | [12; 12] | $L_2(11)$ | [1, 11; 1, 11] | $L_2(11)$ |
| [1, 23] | $M_{23}$ | [1, 22] | $M_{22}$ | [2, 4⁵] | $2^4S_5$ | [1, 11; 12] | $M_{11}$ | [1, 1, 10; 6²] | $M_{10}$ |
| [2, 22] | $M_{22}.2$ | [2, 21] | $M_{21}2$ | [6, 16] | $2^4A_6$ | [12; 1, 11] | | [1, 2, 9; 3⁴] | $M_92$ |
| [3, 21] | $M_{21}.S_3$ | [3, 4⁵] | $2^4(3\times S_5)$ | [7, 15] | $A_7$ | [2, 10; 6²] | $M_{10}2$ | [1, 3, 8; 4, 8] | $M_8S_3$ |
| [4⁶] | $2^6.3S_6$ | [7, 16] | $2^4A_7$ | [8, 14] | $2^3L_3(2)$ | [6²; 2, 10] | | [1, 5, 6; 2, 10] | $S_5$ |
| [8³] | $2^6(L_3(2)\times S_3)$ | [8, 15] | $A_8$ | [11, 11] | $L_2(11)$ | [3, 9; 3⁴] | $M_9S_3$ | | |
| [12²] | $M_{12}.2$ | [11, 12] | $M_{11}$ | [10, 6²] | $M_{10}$ | [3⁴; 3, 9] | | | |
| [3⁸] | $L_2(7)$ | | | | | [4, 8; 4, 8] | $M_8S_4$ | | |
| [8, 16] | $2^4.A_8$ | | | | | [6×2; 6×2] | $2\times S_5$ | | |
| | | | | | | [4×3; 4×3] | $A_4\times S_3$ | | |
| | | | | | | [4³; 4³] | $2^2.2^3.S_3$ | | |

THEOREM 1. *$\Lambda_0$ contains all vectors $4v_T$ ($T \in \Omega_4$), and $4v_i - 4v_j$ ($i, j \in \Omega$). A vector belongs to $\Lambda_0$ if and only if its coordinate-sum is a multiple of* 16 *and the coordinates not divisible by* 4 *fall in the places of a $\mathscr{C}$-set, the coordinates being all even.*

*Proof.* If $T$, $U$, $V$ are three tetrads of a sextet, $\Lambda_0$ contains

$$2v_{T+U} + 2v_{T+V} - 2v_{U+V} = 4v_T.$$

We illustrate this by the addition sum

$$\begin{array}{rrr} 2222 & 2222 & 0000 \\ +2222 & 0000 & 2222, \\ -0000 & 2222 & 2222 \\ \hline =4444 & 0000 & 0000 \end{array}$$

and the sum

$$\begin{array}{rrrrr} 4 & 4 & 4 & 4 & 0 \\ -0 & 4 & 4 & 4 & 4 \\ \hline =4 & 0 & 0 & 0 & -4 \end{array}$$

proves the second statement similarly. The final statement then follows since $\mathscr{C}_8$ spans $\mathscr{C}$ and the set of vectors $4v_T$ and $4v_i - 4v_j$ spans the lattice of all points with coordinate-sum a multiple of 16 and each coordinate a multiple of 4.

The *Leech Lattice* $\Lambda$ is the lattice spanned by $v_\Omega - 4v_\infty$ with the vectors of $\Lambda_0$. The addition sum

$$\begin{array}{rrrr} -3 & 1 & 1 & 1\ldots \\ 4 & -4 & & \\ \hline 1 & -3 & 1 & 1\ldots \end{array}$$

shows that $\Lambda$ contains each vector $v_\Omega - 4v_i$.

THEOREM 2. *The vector* $(x_\infty, x_0, \ldots, x_{22})$ *is in $\Lambda$ if and only if*

(i) *the coordinates $x_i$ are all congruent modulo* 2, *to $m$, say;*
(ii) *the set of $i$ for which $x_i$ takes any given value modulo* 4 *is a $\mathscr{C}$-set;*
(iii) *the coordinate-sum is congruent to* $4m$ *modulo* 8.

*For $x, y \in \Lambda$ the scalar product $x.y$ is a multiple of* 8, *and $x.x$ a multiple of* 16.

*Proof.* (i), (ii), (iii) and the statement about scalar products are satisfied by the generating vectors of $\Lambda$, and so for all vectors of $\Lambda$, by linearity. If $x$ satisfies (i), (ii), (iii), we can subtract a suitable multiple of $v_\Omega - 4v_\infty$ to make all coordinates even and their sum a multiple of 16, and then $x \in \Lambda$ by Theorem 1.

We can use these conditions to enumerate the vectors in $\Lambda$ of any given short length. We use $\Lambda_n$ for the set of $x \in \Lambda$ with $x.x = 16n$, and we find, for instance, that $\Lambda_1$ is empty, while $\Lambda_2$ contains 196560 vectors, namely $2^7.759$ vectors of shape $((\pm 2)^8 0^{16})$ (the non-zero coordinates having positive product and being in the places of an octad), $2^{12}.24$ vectors of shape $(\mp 3(\pm 1)^{23})$ (the upper sign taken on a $\mathscr{C}$-set), and all the $2^2.\binom{24}{2}$ possible vectors of the shape $((\pm 4)^2 0^{22})$, in an obvious notation. The corresponding decompositions of $\Lambda_3$ and $\Lambda_4$ are also included in Table 4, these sets being orbits under the group $N$ below, and the signs of the coordinates being suppressed.

### 4.3. The Group $\cdot 0$, and its Subgroup $N$

We define $\cdot 0$ (pronounced "dotto") as the group of all euclidean congruences of $\mathbf{R}^{24}$ which fix the origin and preserve the lattice $\Lambda$ as a whole. Since each vector $8v_i \in \Lambda$, the elements of $\cdot 0$ have rational orthogonal matrices in which all the denominators divide 8. If $\pi$ is a permutation of $\Omega$, we extend $\pi$ to a congruence of $\mathbf{R}^{24}$ by defining $v_i\pi = v_{i\pi}$, and if $S \subseteq \Omega$, we define a congruence $\varepsilon_S$ by $v_i\varepsilon_S = v_i$ $(i \notin S)$ or $-v_i$ $(i \in S)$.

Theorem 3. *The following conditions on an element $\lambda$ of $\cdot 0$ are equivalent:*

(i) $v_i\lambda = \pm v_j$ *for some* $i, j \in \Omega$, *and some sign* $\pm$;
(ii) $\lambda = \pi\varepsilon_C$ *for some* $\pi \in M_{24}$ *and some* $C \in \mathscr{C}$.

*These operations form a subgroup* $N = 2^{12}M_{24}$.

*Proof.* Plainly (ii) implies (i). We show (i) implies (ii). Now (i) implies that the $i$th row of the matrix of $\lambda$ contains a single non-zero entry $\pm 1$ in its $j$th place. Since $\lambda$ is orthogonal, the $j$th place of every other row must be zero. Now the image of $4v_i + 4v_k$ under $\lambda$ is 4 times the sum of the $i$th and $k$th rows, and so has a coordinate $\pm 4$. Since this vector must be in $\Lambda_2$, it has shape $((\pm 4)^2 0^{22})$, and the $k$th row has also a single non-zero coordinate $\pm 1$. It follows that $\lambda = \pi\varepsilon_S$ for some permutation $\pi$ and some set $S$. But if $C \in \mathscr{C}$, the non-zero coordinates of $2v_C\lambda$ are in the places of $C\pi$, so $\pi$ preserves $\mathscr{C}$, and is in $M_{24}$. Again, the coordinates congruent to 3 modulo 4 in $(v_\Omega - 4v_\infty)\lambda$ are in the places of $S$, so that $S \in \mathscr{C}$, and $\lambda$ satisfies (ii). The final statement is obvious.

Suspecting that $N$ is a proper subgroup of $\cdot 0$ we seek an additional operation. If $\cdot 0$ is transitive on $\Lambda_4$, it will contain an element $\lambda$ taking $8v_\infty$ to some vector $4v_T$, and a row in the matrix of $\lambda$ will then have $\frac{1}{2}$ in every position of $T$ and zeros elsewhere. But 4 times the sum or difference of this row and any other must be a vector of $\Lambda_2$, and this limits the possibilities—if this vector of $\Lambda_2$ has shape $((\pm 2)^8 0^{16})$, the other row has four entries $\pm\frac{1}{2}$ in the places of some tetrad in the same sextet as $T$, and if the shape is $((\pm 4)^2 0^{22})$, we get entries $\pm\frac{1}{2}$ in the places of $T$. This suggests that we try a

TABLE 4. *The vectors of* $\Lambda_2, \Lambda_3, \Lambda_4$

| | | | | | | | | | |
|---|---|---|---|---|---|---|---|---|---|
| *Class* | $\Lambda_2^2$ | $\Lambda_2^3$ | $\Lambda_2^4$ | $\Lambda_3^2$ | $\Lambda_3^3$ | $\Lambda_3^4$ | $\Lambda_3^5$ | | |
| *Shape* | $(2^8 0^{16})$ | $(3\ 1^{23})$ | $(4^2 0^{22})$ | $(2^{12} 0^{12})$ | $(3^3 1^{21})$ | $(4\ 2^8 0^{15})$ | $(5\ 1^{23})$ | | |
| *No.* | $2^7.759$ | $2^{12}.24$ | $2^2\binom{24}{2}$ | $2^{11}.2576$ | $2^{12}\binom{24}{3}$ | $2^8.759.16$ | $2^{12}.24$ | | |
| *Class* | $\Lambda_4^{2+}$ | $\Lambda_4^{2-}$ | $\Lambda_4^3$ | $\Lambda_4^4$ | $\Lambda_4^{4+}$ | $\Lambda_4^{4-}$ | $\Lambda_4^5$ | $\Lambda_4^6$ | $\Lambda_4^8$ |
| *Shape* | $(2^{16} 0^8)$ | $(2^{16} 0^8)$ | $(3^5 1^{19})$ | $(4^4 0^{20})$ | $(4^2 2^8 0^{14})$ | $(4\ 2^{12} 0^{11})$ | $(5\ 3^2 1^{21})$ | $(6\ 2^7 0^{16})$ | $(8\ 0^{23})$ |
| *No.* | $2^{11}.759$ | $2^{11}.759.15$ | $2^{12}\binom{24}{5}$ | $2^4\binom{24}{4}$ | $2^9.759.\binom{16}{2}$ | $2^{12}.2576.12$ | $2^{12}\binom{24}{3}.3$ | $2^7.759.8$ | $2^1.24$ |

matrix which is the direct sum of six $4 \times 4$ matrices of $\pm\frac{1}{2}$s in the places of a sextet. The signs must be chosen carefully—if $\Xi$ is a sextet we let $\zeta = \zeta_\Xi$ be the map taking $v_i$ to $v_i - \frac{1}{2}v_T$ ($i \in T \in \Xi$), and then define $\xi = \xi_T$ as $\eta\varepsilon_T$, $\Xi$ being the sextet of $T$.

Now since two octads intersect in 0, 2, 4, or 8 points, any octad is either the union of two tetrads of $\Xi$ or contains two points from each of four tetrads, or has three points from one tetrad and one from each of the others. Supposing the coordinates properly ordered, we apply $\eta$:

$$
\begin{array}{rcccccc}
x = & 2222 & 2222 & 0000 & 0000 & 0000 & 0000 \\
x\eta = & \bar{2}\bar{2}\bar{2}\bar{2} & \bar{2}\bar{2}\bar{2}\bar{2} & 0000 & 0000 & 0000 & 0000 \\
x = & 2220 & 2000 & 2000 & 2000 & 2000 & 2000 \\
x\eta = & \bar{1}\bar{1}\bar{1}\bar{3} & 1\bar{1}\bar{1}\bar{1} & 1\bar{1}\bar{1}\bar{1} & 1\bar{1}\bar{1}\bar{1} & 1\bar{1}\bar{1}\bar{1} & 1\bar{1}\bar{1}\bar{1} \\
x = & 2200 & 2200 & 2200 & 2200 & 0000 & 0000 \\
x\eta = & 00\bar{2}\bar{2} & 00\bar{2}\bar{2} & 00\bar{2}\bar{2} & 00\bar{2}\bar{2} & 0000 & 0000 \\
x = & \bar{3}111 & 1111 & 1111 & 1111 & 1111 & 1111 \\
x\eta = & \bar{3}111 & \bar{1}\bar{1}\bar{1}\bar{1} & \bar{1}\bar{1}\bar{1}\bar{1} & \bar{1}\bar{1}\bar{1}\bar{1} & \bar{1}\bar{1}\bar{1}\bar{1} & \bar{1}\bar{1}\bar{1}\bar{1}
\end{array}
$$

($\bar{n}$ denotes $-n$). Since the vectors $x\eta$ are not all in $\Lambda$, $\eta$ is *not* in $\cdot 0$ (cf. Tits, 1970), but on changing the sign of any tetrad we get a vector of $\Lambda$, and so $\xi = \eta\varepsilon_T$ *is* in $\cdot 0$.

In my paper (Conway, 1969) I give a proof that $N$ is a maximal subgroup of $\cdot 0$ which at the same time computes the order of $\cdot 0$. The following method involves explicit calculations with the element $\xi_T$, but is perhaps as elegant.

THEOREM 4. *$\cdot 0$ is transitive on each of the three sets $\Lambda_2$, $\Lambda_3$, $\Lambda_4$, and $N$ has index $|\Lambda_4|/48$ in $\cdot 0$, which is generated by $N$ together with any element $\xi_T$.*

*Proof.* We have already shown that $\xi_T$ takes certain elements of $\Lambda_2^2$ to elements of $\Lambda_2^3$, and in a similar way we see that if $i$ and $j$ are in distinct tetrads of $\Xi$, $\xi_T$ takes $4v_i + 4v_j$ to a member of $\Lambda_2^2$, so that the subgroup $\langle N, \xi_T \rangle$ of $\cdot 0$ is transitive on $\Lambda_2$. Similar calculations (they are all easy!) establish the transitivity on $\Lambda_3$ and $\Lambda_4$. Now the particular vector $8v_\infty$ has just 48 images under $N$ (the vectors $\pm 8v_i$), and every operation of $\cdot 0$ taking $8v_\infty$ to one of these is in $N$. From these remarks the rest of the theorem follows. (If $\lambda \in \cdot 0$ we can find $\mu \in \langle N, \xi_T \rangle$ with $8v_\infty \mu = 8v_\infty \lambda$, so $\lambda \in N\mu$.)

The permutation representation of $\cdot 0$ on $\Lambda_4$ is imprimitive, since the vectors of $\Lambda_4$ come in naturally defined *coordinate-frames*, each consisting of 24 mutually orthogonal pairs of opposite vectors. We can distinguish these frames geometrically—the vectors of the frame containing $x$ are all those vectors of $\Lambda_4$ which are congruent to $x$ modulo $2\Lambda$. To prove this, it suffices by transitivity to consider the particular case $x = 8v_\infty$.

THEOREM 5. *Every vector of $\Lambda$ is congruent modulo $2\Lambda$ to one of:*

(i) *the zero vector;*
(ii) *each vector of a unique pair $x, -x$ $(x \in \Lambda_2)$;*
(iii) *each vector of a unique pair $x, -x$ $(x \in \Lambda_3)$;*
(iv) *each of the 48 vectors of a coordinate-frame.*

*Proof.* Let $x, y$ be two vectors of $\Lambda_0 \cup \Lambda_2 \cup \Lambda_3 \cup \Lambda_4$ congruent modulo $2\Lambda$, with $y \neq \pm x$. Then since $x \pm y \in 2\Lambda$ we have $(x\pm y).(x\pm y) \geq 128$, whence $x.y = 0$ and $x.x = y.y = 64$, since we know that $x.x$ and $y.y$ are both at most 64. It follows that $x$ and $y$ are members of $\Lambda_4$ both in the same coordinate-frame. We therefore have at least $|\Lambda_0| + \frac{1}{2}|\Lambda_2| + \frac{1}{2}|\Lambda_3| + \frac{1}{48}|\Lambda_4|$ distinct classes of $\Lambda/2\Lambda$, and since this number comes to exactly $2^{24}$ we have accounted for every class.

Let us say that a vector $x \in \Lambda_n$ has *type n*, and type $n_{ab}$ if also $x$ is the sum of two vectors of types $a$ and $b$.

THEOREM 6. *Every vector x of type n has some type $n_{ab}$ in which $a+b = \frac{1}{2}(n+k)$, where $k = 0, 2, 3$, or 4 (corresponding to the cases of Theorem 5), and these possibilities are exclusive. $\cdot 0$ is transitive on vectors of each of the types*

$$2, 3, 4, 5, 6_{22}, 6_{32}, 7, 8_{22}, 8_{32}, 8_{42}, 9_{33}, 9_{42}, 10_{33}, 10_{42}, 11_{43}, 11_{52}$$

*(which include all vectors of type $n < 12$).*

*Proof.* For the first part, we suppose $x$ congruent to $y \in \Lambda_0 \cup \Lambda_2 \cup \Lambda_3 \cup \Lambda_4$ modulo $2\Lambda$, and let $\frac{1}{2}(x+y)$ and $\frac{1}{2}(x-y)$ have types $a$ and $b$. For the second part, we give only a few typical samples.

*Type* $6_{32}$. Each such vector is $x-y$, where $x \in \Lambda_3$, $y \in \Lambda_2$, $x.y = -8$, and so $x+y \in \Lambda_4$. (This follows from the first part by a sign-change.) To within transformation by some element of $\cdot 0$, we can take $x+y$ as $(8, 0, 0,\ldots)$, from which we get $x = (5, 1, 1,\ldots)\varepsilon_C$, $y = (3, \bar{1}, \bar{1},\ldots)\varepsilon_C$ for some $C \in \mathscr{C}$, $\infty \notin C$ (this is the only way to express $(8, 0, 0,\ldots)$ as $x+y$, $x \in \Lambda_2$, $y \in \Lambda_3$). To within a further transformation we can suppose $C$ empty, so that $x-y$ is equivalent under $\cdot 0$ to $(2, 2, 2,\ldots)$.

*Type* 5. In this case the typical vector is $x-y$, where $x, y \in \Lambda_2$, $x.y = -8$, and so $x+y \in \Lambda_3$, and can be taken as $(5, 1, 1,\ldots)$. From this, without loss of generality, we get $x = (3, (-1)^7, 1^{16})$, $y = (2^8, 0^{16})$ or vice versa, or $x = (4^2, 0^{22})$, $y = (1, -3, 1^{22})$ or vice versa, so that $x-y$ is one of $\pm(1, (-3)^7, 1^{16})$ or $\pm(-3, -7, 1^{22})$, the coordinates $-3$ in the first case being in a special heptad $S_7$. Applying a suitable $\xi_T$ we see that

$$1\overline{333}\quad \overline{3333}\quad 1111\quad 1111\quad 1111\quad 1111$$

and

$$\bar{3}111\quad 7111\quad 1111\quad 1111\quad 1111\quad 1111$$

become

$$\overline{5111}\quad 3333\quad \overline{1111}\quad \overline{1111}\quad \overline{1111}\quad \overline{1111}$$

and

$$3\overline{111}\quad \bar{5}333\quad \overline{1111}\quad \overline{1111}\quad \overline{1111}\quad \overline{1111},$$

which are plainly equivalent under $M_{24}$.

These proofs get easier as we go further, since we then have more transitivities already known. We can avoid explicit calculations with $\xi_T$ by using instead the counting methods of Conway (1969).

## 4.4. SUBGROUPS OF $\cdot 0$

Many subgroups of $\cdot 0$ are easiest considered in relation to the infinite group $\cdot\infty$ of all euclidean congruences of $\Lambda$, including translations. Any finite subgroup $G$ of $\cdot\infty$ fixes a point (not necessarily a lattice point), and so there is a translation $t$ of $\mathbf{R}^{24}$ (not necessarily in $\cdot\infty$) such that $G^t \subseteq \cdot 0$. The stabilizer of two points whose difference is a vector of type $n$ is written $\cdot n$, the stabilizer of the vertices of a triangle whose sides have types $a, b, c$ is written $\cdot abc$, and so on—see Figure.

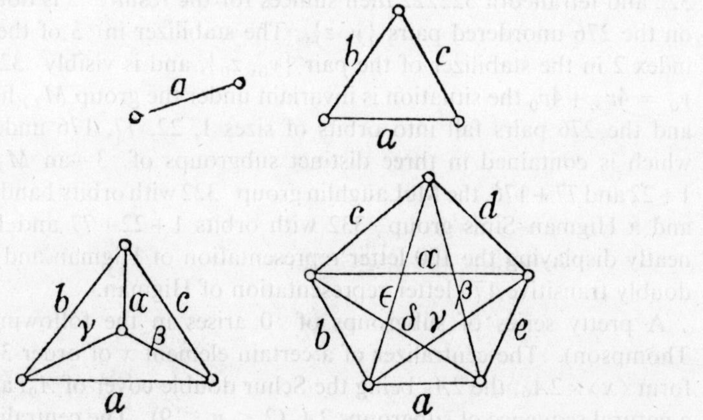


FIG. 2.

These groups are identified in Table 5. Many of the identifications are easy. Thus, $\cdot 6_{32}$ can be taken as the stabilizer in $\cdot 0$ of the vector $(2, 2, 2, \ldots)$. The argument of Theorem 6 shows that this is expressible in just 24 ways as a sum $x+y$ ($x \in \Lambda_3, y \in \Lambda_2$), and we can see all of them: $x = (5, 1^{23})$, $y = (-3, 1^{23})$ and their images under coordinate permutations. Since the vectors $x-y$ are the $8v_i$, we have $\cdot 6_{32} \subseteq N$, and so $\cdot 6_{32} = M_{24}$. At the same time, we get $\cdot 632 = M_{23}$, stabilizing the same two points together with $(5, 1^{23})$. The groups $\cdot 632$ and $\cdot 432$ are identical, since a parallelogram with sides $\sqrt{2}$ and $\sqrt{3}$ and one diagonal $\sqrt{6}$ has $\sqrt{4}$ for the other diagonal. Since $\cdot 432 = \cdot 632 = M_{23}$, the group $M_{23}$ must be contained in each of the groups $\cdot 2$, $\cdot 3$, $\cdot 4$, $\cdot 6_{32}$—we can see this also by taking suitable coordinates for the vectors involved, namely $(-3, 1^{23})$, $(5, 1^{23})$, $(8, 0^{23})$, and $(2^{24})$.

In a similar way we see that $M_{22}$ is contained in each of the groups $\cdot 222$, $\cdot 322$, $\cdot 332$. Since the former is the group $PSU_6(2)$ it follows that $M_{22}$ has a 6-dimensional projective representation over $F_4$ (and from this it follows fairly easily that the multiplicator of $M_{22}$ has order divisible by 3). The other two groups are the newly discovered simple groups of McLaughlin and of Higman and Sims, whose embedding in $\cdot 0$ was briefly discussed (Conway, 1969) (and in the lectures as given). Here we shall show instead that $\cdot 3$ has a doubly transitive permutation representation on 276 letters, the stabilizer of a point being a group $M^c.2$. To this end we consider the stabilizer of the vector $x = (5, 1^{23})$. There are 276 unordered pairs $\{y, z\}$ with $x = y+z$ $y, z \in \Lambda_2$, namely 23 with $y$ (say) of shape $(4^2, 0^{22})$ and 253 with $y$ of shape $(2^8, 0^{16})$, the coordinate $y_\infty$ being non-zero in each case.

Letting $\{y_0, z_0\}$ be one of these pairs, we have for each other pair $\{y, z\}$ that either $z-y_0$ or $y-y_0$ is of type 2. Assuming transitivity of $\cdot 0$ on triangles 322 and tetrahedra 322222 then suffices for the result: $\cdot 3$ is doubly transitive on the 276 unordered pairs $\{y, z\}$. The stabilizer in $\cdot 3$ of the point $y_0$ has index 2 in the stabilizer of the pair $\{y_0, z_0\}$, and is visibly $\cdot 322$. If we take $y_0 = 4v_\infty + 4v_0$ the situation is invariant under the group $M_{22}$ fixing $\infty$ and 0, and the 276 pairs fall into orbits of sizes 1, 22, 77, 176 under this group, which is contained in three distinct subgroups of $\cdot 3$—an $M_{23}$ with orbits $1+22$ and $77+176$, the McLaughlin group $\cdot 322$ with orbits 1 and $22+77+176$, and a Higman–Sims group $\cdot 332$ with orbits $1+22+77$ and 176, the latter neatly displaying the 100 letter representation of Higman and Sims and the doubly transitive 176 letter representation of Higman.

A pretty series of subgroups of $\cdot 0$ arises in the following way (J. G. Thompson). The centralizer of a certain element $x$ of order 3 in $\cdot 0$ has the form $\langle x \rangle \times 2A_9$, the $2A_9$ being the Schur double cover of $A_9$, and containing a natural sequence of subgroups $2A_n$ $(2 \leq n \leq 9)$. The centralizers $B_n$ of $2A_n$ are, for $n = 2, 3, \ldots, 9$, groups $\cdot 0$, $6S$, $2G_2(4)$, $2HJ$, $2U_3(3)$, $2L_3(2)$, $2A_4$, $C_6$, $HJ$ denoting the Hall–Janko simple group and $S$ the isolated simple group of Suzuki. It follows that $HJ$ has a multiplicator of order divisible by 2, and $S$ a multiplicator of order divisible by 6, and also that $HJ$ has a 6-dimensional projective representation over the field of $\sqrt{-3}$ and $\sqrt{5}$, while $S$ has a

TABLE 5.

| Name | Order | Structure | Name | Order | Structure |
|---|---|---|---|---|---|
| ·0 | $2^{22}\,3^{9}\,5^{4}\,7^{2}\,11.13.23$ | New perfect | ·222 | $2^{15}\,3^{6}\,5.7.11$ | $PSU_6(2)$ |
| ·1 | $2^{21}\,3^{9}\,5^{4}\,7^{2}\,11.13.23$ | New simple | ·322 | $2^{7}\,3^{6}\,5^{3}\,7.11$ | $M^c$ |
| ·2 | $2^{18}\,3^{6}\,5^{3}\,7.11.23$ | New simple | ·332 | $2^{9}\,3^{2}\,5^{3}\,7.11$ | $HS$ |
| ·3 | $2^{10}\,3^{7}\,5^{3}\,7.11.23$ | New simple | ·333 | $2^{4}\,3^{7}\,5.11$ | $3^5.M_{11}$ |
| ·4 | $2^{18}\,3^{2}\,5.7.11.23$ | $2^{11}\,M_{23}$ | ·422 | $2^{17}\,3^{2}\,5.7.11$ | $2^{10}.M_{22}$ |
| ·5 | $2^{8}\,3^{6}\,5^{3}\,7.11$ | $M^c.2$ | ·432 | $2^{7}\,3^{2}\,5.7.11.23$ | $M_{23}$ |
| $\cdot 6_{22}$ | $2^{16}\,3^{6}\,5.7.11$ | $\mathrm{PSU}_6(2).2$ | ·433 | $2^{10}\,3^{2}\,5.7$ | $2^4.A_8$ |
| $\cdot 6_{32}$ | $2^{10}\,3^{3}\,5.7.11.23$ | $M_{24}$ | ·442 | $2^{12}\,3^{2}\,5.7$ | $2^{1+8}A_7$ |
| ·7 | $2^{9}\,3^{2}\,5^{3}\,7.11$ | $HS$ | ·443 | $2^{7}\,3^{2}\,5.7$ | $M_{21}.2$ |
| $\cdot 8_{22}$ | $2^{18}\,3^{6}\,5^{3}\,7.11.23$ | ·2 | ·522 | $2^{7}\,3^{6}\,5^{3}\,7.11$ | $M^c$ |
| $\cdot 8_{32}$ | $2^{7}\,3^{6}\,5^{3}\,7.11$ | $M^c$ | ·532 | $2^{8}\,3^{6}\,5.7$ | $PSU_4(3).2$ |
| $\cdot 8_{42}$ | $2^{15}\,2^{2}\,5.7$ | $2^5.2^4.A_8$ | ·533 | $2^{4}\,3^{2}\,5^{3}\,7$ | $PSU_3(5)$ |
| $\cdot 9_{33}$ | $2^{5}\,3^{7}\,5.11$ | $3^5.M_{11}.2$ | ·542 | $2^{7}\,3^{2}\,5.7.11$ | $M_{22}$ |
| $\cdot 9_{42}$ | $2^{7}\,3^{2}\,5.7.11.23$ | $M_{23}$ | ·633 | $2^{6}\,3^{3}\,5.11$ | $M_{12}$ |
| $\cdot 10_{33}$ | $2^{10}\,3^{2}\,5^{3}\,7.11$ | $HS.2$ | *2 = !2 | $2^{19}\,3^{6}\,5^{3}\,7.11.23$ | $(\cdot 2)\times 2$ |
| $\cdot 10_{42}$ | $2^{17}\,3^{2}\,5.7.11$ | $2^{10}.M_{22}$ | *3 = !3 | $2^{11}\,3^{7}\,5^{3}\,7.11.23$ | $(\cdot 3)\times 2$ |
| $\cdot 11_{43}$ | $2^{10}\,3^{2}\,5.7$ | $2^4.A_8$ | *4 | $2^{19}\,3^{2}\,5.7.11.23$ | $(\cdot 4)\times 2$ |
| $\cdot 11_{52}$ | $2^{8}\,3^{6}\,5.7$ | $\mathrm{PSU}_4(3).2$ | !4 | $2^{22}\,3^{3}\,5.7.11.23$ | $2^{12}.M_{24}$ |
| | | | !333 | $2^{7}\,3^{9}\,5.11$ | $3^6.2.M_{12}$ |
| | | | !442 | $2^{15}\,3^{4}\,5.7$ | $2^{1+8}A_9$ |

*HS*, $M^c$, $p^n$ denote respectively the Higman–Sims group, the McLaughlin group, and the elementary group of order $p^n$. $A.B$ denotes an extension of the group $A$ by the group $B$. The notation is otherwise standard.

12-dimensional projective representation over the field of $\sqrt{-3}$. The latter can be obtained as follows. Take an element $\omega$ of order 3 with no fixed point, and so satisfying (as a matrix) the equation $\omega^2+\omega+1 = 0$. In the ring of $24\times 24$ matrices $\omega$ therefore generates a copy of the complex numbers in which it is identified with $e^{2\pi i/3}$. If we define, for $x \in \Lambda$, $x(a+be^{2\pi i/3}) = ax+b(x\omega)$, the Leech Lattice becomes the *complex Leech Lattice* $\Lambda_C$, a 12-dimensional lattice (module) over the ring $\mathbf{Z}(e^{2\pi i/3})$, whose automorphism group is the group $6S$. The complex Leech Lattice has a natural coordinate system which displays a remarkable analogy between it and the real Leech Lattice, with 2 everywhere becoming $\theta = \sqrt{-3} = \omega-\omega^2$.

The details are as follows. Define vectors $x_i$, $y_i$, $z_i$ ($i \in PL(11)$) by

$$x_i+y_i+z_i = 0, \qquad y_i = z_i\varepsilon_Q, \qquad z_\infty = v_\Omega+Lv_\infty,$$
$$z_i = (v_\infty+4v_{15})\varepsilon_{\{\infty,15,1,2,3,9,12,13\}}\alpha^{-i} \quad (i \in F_{11}).$$

Let $\omega$ be the operation of $\cdot 0$ taking $x_i \to y_i \to z_i \to x_i$ for each $i$. In terms of the coordinates $x_i$, and the *ternary* Golay code $\mathscr{C}$, $\Lambda_C$ is now spanned by the vectors $\theta v_C$ ($C \in \mathscr{C}$), $3v_i-3v_j$, and $v_\Omega+3v_i$ ($\Omega = PL(11)$). These are obtained from $2v_C$, $4v_i-4v_j$, $v_\Omega-4v_i$ in the real case by replacing 2 by $\theta$ and 4 by $\theta^2 = -3$. It is interesting that the order of the binary Golay code is $2^{12}$, and that of the ternary Golay code $\theta^{12} = 3^6$, and that $6S$ is generated by its monomial part together with a matrix of $3\times 3$ blocks whose entries have the form $\pm\omega^i/\theta$. (Compare the real case, when we have $4\times 4$ blocks with entries $\pm\frac{1}{2}$.)

## 4.5. Involutions in $\cdot 0$

Each involution in $\cdot 0$ is conjugate to an involution $\varepsilon_C$ ($C \in \mathscr{C}$). If $C$ is an octad, the centralizer preserves the corresponding 8-dimensional subspace, which intersects in a copy of the 8-dimensional lattice $E_8$. The subgroup of the centralizer which fixes every point of this space is therefore in $N$ (since the space contains vectors $v_i$), and is easily seen to be an extraspecial group $2^{1+8}$ of order $2^9$. The quotient by this extraspecial group is a subgroup of the Weyl group of $E_8$, and in fact it is the derived group $W(E_8)'$, so that the whole centralizer is a group $2^{1+8}(W(E_8))'$. If $C$ is a 16-*ad*, we get the same centralizer.

If $C$ is a dodecad, the fixed space can contain no vector of shape $((\pm 2)^8 0^{16})$ or $((\mp 3)\,(\pm 1)^{23})$, and so contains in $\Lambda_2$ only the $2^2.66$ vectors of shape $((\pm 4)^2 0^{22})$ whose non-zero coordinates are in the fixed space. We call two such vectors *skew* unless they are equal, opposite, or orthogonal. Then the only vectors of the set which are skew to all the vectors skew to $x = 4v_i+4v_j$ say, are just the four vectors $\pm 4v_i \pm 4v_j$ having the same non-zero coordinates as $x$. It follows (since $(4v_i+4v_j)+(4v_i-4v_j) = 8v_i$) that the centralizer of $\varepsilon_C$

is in $N$, and so is the group $2^{12}M_{12}$ centralizing $\varepsilon_C$ in $N$. (The centralizer of this type of involution $\varepsilon_C$ in $\cdot 3$ is $\langle\varepsilon_C\rangle \times M_{12}$, as is seen by taking the fixed vector of $\cdot 3$ to be $(2^{12}, 0^{12})$—compare the situation in Janko's group $J$.)

The only remaining involution in $\cdot 0$ is $\varepsilon_\Omega = -1$, which is central in the whole group.

## 4.6. A Connection Between $\cdot 0$ and Fischer's Group $F_{24}$

$\cdot 0$ can be regarded as a permutation group on the 196560 vectors of $\Lambda_2$. It has a subgroup $N = 2^{12}M_{24}$ permuting these in three orbits of sizes $2.759$, $2^{12}.24$, and $2^2\binom{24}{2}$. The elementary part $2^{12}$ of $N$ is permuted by $M_{24}$ like the binary Golay code $\mathscr{C}$. The extension $2^{12}M_{24}$ is splitting.

The group $F_{24}$ of B. Fischer is a permutation group on 306936 objects (the involutions in a certain conjugacy class). It has a subgroup $N^* = 2^{12}M_{24}$ permuting these in three orbits of sizes $2^5.759$, $2^0.24$, and $2^{10}.\binom{24}{2}$. The elementary part $2^{12}$ of $N^*$ is permuted by $M_{24}$ like the quotient $P(\Omega)/\mathscr{C}$. The extension $2^{12}M_{24}$ is non-splitting

These facts suggest that there is a relation between the two groups, and this is borne out by the fact that the subgroups of $M_{24}$ corresponding to the three orbits are identical, while the elementary subgroups are dual. The appropriate name for this relationship—whatever it is—is twinning, as we see from the following parable. "Once upon a time there was an egg (the elementary group $P(\Omega)$ of order $2^{24}$), which after fertilization (by the action of $M_{24}$) split into two (the group $\mathscr{C}$ and its quotient $P(\Omega)/\mathscr{C}$), which grew into two healthy twins ($\cdot 0$ and $F_{24}$)."

That the parable is meaningful is suggested by the existence of similar structure relating $\cdot 0$ to itself via its subgroup $6S$, with elements of order 3 like $\omega$ taking the place of Fischer's involutions. The elementary $2^{12}$ in the subgroup $N^*$ of $F_{24}$ has 24 of Fischer's 3-transpositions (the special involutions of his group), permuted in the natural way by $M_{24}$, the centralizer of any one being a group $2F_{23}$. In a similar way $\cdot 0$ has a group $3^6.2M_{12}$ in which the $3^6$ has 12 special subgroups of order 3 (isomorphic to the groups $V_i$ of the first lecture) permuted naturally by $M_{12}$, the centralizer of any one being the group $6S$.

**Acknowledgements**

I should like to thank Professor D. Livingstone and his colleagues for some remarks which I have made use of in the second lecture, and in particular, for the material of Table 3, and Professor J. G. Thompson for his sustained interest in the subject of the third lecture.

## Bibliography


Chang Choi. The maximal subgroups of $M_{24}$. To be published.

Conway, J. H. (1969). A group of order 8, 315, 553, 613, 086, 72000. *Bull. L.M.S.* **1**, 79.

Tits, J. (1970). Groupes Finis Simples Sporadiques. Seminaire Bourbaki No. 375.

Todd, J. A. (1966). A representation of the Mathieu Group $M_{24}$ as a collineation group. *Annali di Mat.*, Serie IV–Vol. LXXI.


## Postscript on the Known Exceptional Simple Groups—*Added in proof*

We discuss under five headings the 18 or 19 known non-abelian simple groups not occurring among the infinite families of Chevalley and twisted Chevalley groups and alternating groups. With each group we give its order, preceded and followed by lower bounds (almost all known to be exact) for its Schur multiplicator and outer automorphism group. Asterisks indicate groups belonging to one of the infinite families but included here by analogy.

*The Mathieu Groups*

| | | | |
|---|---|---|---|
| $M_{24}$ | $1(2^{10}3^{3}5.7.11.23)1$ | $M_{23}$ | $1(2^{7}3^{2}5.7.11.23)1$ |
| $M_{22}$ | $6(2^{7}3^{2}5.7.11)2$ | $*M_{21}$ | $4\times 4\times 3(2^{6}3^{2}5.7)S_3.2$ |
| $M_{12}$ | $2(2^{6}3^{3}5.11)2$ | $M_{11}$ | $1(2^{4}3^{2}5.11)1$ |
| $*A_6$ | $6(2^{3}3^{2}5)2\times 2$ | $*A_7$ | $6(2^{3}3^{2}5.7)2$ |

We have already discussed the isomorphisms $M_{21} \cong L_3(4)$ and $M_{10'} \cong A_6 \cong L_2(9)$, but the unusual multiplicators show that the usual definitions of these two groups are unrevealing. For alternating groups other than $A_6$, $A_7$ our symbol reads $2(n!/2)2$. The groups under our second and third headings are in some sense the Mathieu groups "writ large".

*The Fischer Groups*

| | | | |
|---|---|---|---|
| $F_{24'}$ | $1(2^{21}3^{16}5^{2}7^{3}11.13.17.23.29)2$ | $F_{23}$ | $1(2^{18}3^{13}5^{2}7.11.13.17.23)1$ |
| $F_{22}$ | $6(2^{17}3^{9}5^{2}7.11.13)2$ | $*F_{21}$ | $2\times 2\times 3(2^{15}3^{6}5.7.11)S_3$ |

Here $F_n$ has a conjugacy class of involutions any one of which has centralizer a group $2F_{n-1}$ and any two of which have product of order at most 3. A maximal commuting set of such involutions has exactly $n$ of them, generating a group $2^{n-12}$ whose normalizer is a group $2^{n-12}M_n$. We have the isomorphism $F_{21} \cong U_6(2)$.

*Lattice Stabilizers in* $\cdot 0$

| | | | |
|---|---|---|---|
| $\cdot 1 \cong C_1$ | $2(2^{21}3^{9}5^{4}7^{2}11.13.23)1$ | $\cdot 2 \cong C_2$ | $1(2^{18}3^{6}5^{3}7.11.23)1$ |
| $\cdot 3 \cong C_3$ | $1(2^{10}3^{7}5^{3}7.11.23)1$ | $*\cdot 222$ | $2\times 2\times 3(2^{15}3^{6}5.7.11)S_3$ |
| $\cdot 322 \cong M^c$ | $3(2^{7}3^{6}5^{3}7.11)2$ | $\cdot 322 \cong HS$ | $2(2^{9}3^{2}5^{3}7.11)2$ |

The first group contains $M_{24}$, the next two contain $M_{23}$, and the remainder contain $M_{22}$. We have the isomorphism $\cdot 222 \cong U_6(2)$.

*The Suzuki Chain*

| | | | |
|---|---|---|---|
| $Sz$ | $6(2^{13}3^7 5^2 7.11.13)2$ | $*G_2(4)$ | $2(2^{12}3^3 5^2 7.13)2$ |
| $J_2$ | $2(2^7 3^3 5^2 7)2$ | $*U_3(3)$ | $1(2^5 3^3 7)2$ |

These are the central quotients of the groups $B_n$ described near the end of the third lecture. We have the isomorphism $U_3(3) \cong G_2(2)'$.

*The Remaining Groups*

| | | | |
|---|---|---|---|
| $J_1$ | $1(2^3 3.5.7.11.19)1$ | $J_3$ | $3(2^7 3^5 5.17.19)2$ |
| $HTH$ | $1(2^{10}3^3 5^2 7^3 17)2$ | $Ly?$ | $1(2^8 3^7 5^6 7.11.31.37.67)1$ |

These were all discovered via their centralizers of involutions. $J_2$ (the Hall–Janko group) and $J_3$ (the Higman–Janko–McKay group) have involutions with the same centralizer, as do $HTH$ (the Held–Thompson–Higman group), $M_{24}$, and $L_2(5)$ . The centralizers of suitable elements of order 3 are, in the Held group, $3A_7$, and in Lyons' group (not yet proved to exist), $3M^c$, exhibiting the 3-parts of the multiplicators of these groups. $J_1$ is a subgroup of $G_2(11)$, and $G_2(5)$ is apparently a subgroup of Lyons' group.

It is perhaps worthy of note that the group $U_4(3)$ has an unusual multiplicator $(3\times 3\times 4)$. $3^2 U_4(3)$ is the centralizer of two commuting elements of order 3 in $\cdot 0$, and $U_4(3)$ is closely connected with a number of the groups above.

CHAPTER VIII

# Character Theory Pertaining to Finite Simple Groups

E. C. DADE



## 1. Introduction

The original idea behind these lectures was to prove directly, starting from "first principals", some beautiful, but deep, result about finite simple groups, whose proof would illustrate the "practical" uses of character theory in group theory. The advantages of this idea were evident. Since the number (twelve) of lectures was quite limited, it would be necessary to concentrate on those parts of representation theory which are really used to prove the theorem —essentially those parts dealing with the values of the ordinary irreducible characters and their relations with the structure of the group and its subgroups—and to use the minimum of ring-theoretic machinery. This would avoid the usual trap in which one spends so much time developing this machinery that none is left over for the groups. The main disadvantage was also evident—namely, this machinery is used currently in the literature. So a student attending this course would perhaps learn the proof of this one theorem, plus many auxiliary results, but would by no means be prepared to read published proofs.

A minor disadvantage was less evident at the time, but became serious during the lectures themselves. The result chosen as a goal—Glauberman's theorem about weakly closed involutions in 2-Sylow subgroups (Theorem 15.1 below)—has a very elegant proof, given the character–theoretic tools which were developed in the course. But it depends essentially upon the theorem of Brauer and Suzuki that groups with quaternion 2-Sylow subgroups are not simple (Theorem 14.11 below), whose proof is rather messy no matter how you do it, and which requires more machinery—in particular, Brauer's characterization of characters (Theorem 7.1 below)—than is needed for Glauberman's proof. In fact, due to the pressure of time, it was impossible to prove either of the last two theorems during the lectures. So part of the written account below—Sections 7 and 14, and the last part of Section 12, starting at (12.16)—were not actually given verbally.

The rest of this account represents more or less closely the lectures as given, although there are a large number of minor modifications and a few major ones. These will be evident to those who were present during the course, and of no interest to anyone else. So there is no need to list them here.

As mentioned above, the reader really needs more preparation than this before tackling articles in the literature. For example, he should read the chapter on modular representation (Chapter XII) in the book by Curtis and Reiner (1962), or the excellent lecture notes of Feit (1969) on the subject. Even in the theory of ordinary characters, he should be familiar with much more than there was time to mention here. A reading of Chapter V of the book by Huppert (1967) would be very profitable in this regard.

## 2. Characters

Let $V$ be a finite-dimensional vector space over a (commutative) field $F$. We denote by $GL(V)$ the group of all non-singular linear transformations of $V$. A (*linear*) *representation* of a finite group $G$ on $V$ is a homomorphism $R$ of $G$ into $GL(V)$. The *character* $\chi_R$ of such a representation $R$ is the function from $G$ to $F$ sending each $\sigma \in G$ into the trace $\operatorname{tr}(R(\sigma))$ of the corresponding linear transformation $R(\sigma)$ of $V$:

$$\chi_R(\sigma) = \operatorname{tr}(R(\sigma)), \quad \textit{for all } \sigma \in G. \tag{2.1}$$

We are interested in the characters as invariants of the group $G$, and in their relations with the algebraic structure of $G$. For example, we have

PROPOSITION 2.2. *$\chi_R(\sigma^\tau) = \chi_R(\sigma)$, for all $\sigma, \tau \in G$.*

*Proof.* Of course, $\sigma^\tau$ is the conjugate $\tau^{-1}\sigma\tau$ of $\sigma$ by $\tau$. Since $R$ is a homomorphism, we have $R(\sigma^\tau) = R(\tau)^{-1}R(\sigma)R(\tau)$. So the linear transformations $R(\sigma^\tau)$ and $R(\sigma)$ are similar. Therefore they have the same trace, which is the proposition.

The (finite) dimension of $V$ is called the *degree* $\deg(R)$ of the representation $R$. Since the homomorphism $R$ sends the identity $1 = 1_G$ of $G$ into the identity linear transformation $1_{V\to V}$ of $V$, we have

$$\chi_R(1_G) = \deg(R)\cdot 1_F = \overbrace{1_F + \ldots + 1_F}^{\deg(R)}. \tag{2.3}$$

In particular, if $F$ has characteristic zero, and if we identify the ordinary integers with their images in $F$, then $\chi_R(1) = \deg(R)$ is always a non-negative integer.

If $\deg(R) = \dim(V) = 1$, then $\chi_R$ is called a *linear character* of $G$. In this case the trace function is an isomorphism of $GL(V)$ onto the multiplicative group $F^\times$ of $F$. It follows that

*The linear characters of G are just the homomorphisms from G into* $F^\times$. (2.4)

Of course, the linear characters are all equal to $1_F$ on the derived group $G'$ of $G$ (since $F$ is commutative). So there aren't very many of them. In fact, the only linear character present for all finite groups $G$ is the *trivial character* $1 = 1_{G\to F} : \sigma \to 1_F$, for all $\sigma \in G$.

We should note that any function $R$ satisfying the following conditions is a representation of the group $G$ on the finite-dimensional vector space $V$:

$$R(\sigma) \in \mathrm{Hom}_F(V, V), \quad \textit{for all } \sigma \in G, \tag{2.5a}$$

$$R(\sigma)R(\tau) = R(\sigma\tau), \quad \textit{for all } \sigma, \tau \in G, \tag{2.5b}$$

$$R(1_G) = 1_{V\to V}. \tag{2.5c}$$

Indeed, (1.5b, c) imply that $R(\sigma^{-1})$ is a two-sided inverse to $R(\sigma)$, for al $\sigma \in G$. So $R$ is in fact a homomorphism of $G$ into $GL(V)$.

Any permutation representation of $G$ determines a linear representation, and hence a character, of $G$. Let $G$ act as permutations of a finite set $S$, with any $\sigma \in G$ taking each $s \in S$ into $s\sigma \in S$. Form the finite-dimensional vector space $FS$ having $S$ as a basis (the elements of $FS$ are just the formal linear combinations $\sum_{s\in S} f_s s$ with unique coefficients $f_s \in F$, and the operations are "coefficientwise"). Then any $\sigma \in G$ determines a unique linear transformation $R_S(\sigma)$ of $FS$ satisfying

$$sR_S(\sigma) = s\sigma, \quad \textit{for all } s \in S. \tag{2.6}$$

One verifies immediately that $R_S$ satisfies (2.5) and hence is a linear representation of $G$ on $FS$. Using the matrix of the linear transformation $R_S(\sigma)$ with respect to the basis $S$ of $FS$, we easily compute the character of this representation:

$$\chi_{R_S}(\sigma) = \left|\{s \in S : s\sigma = s\}\right| \cdot 1_F, \quad \textit{for all } \sigma \in G. \tag{2.7}$$

An important special case is the *regular representation* in which $S = G$ and $s\sigma$, for $s \in S = G$ and $\sigma \in G$, is the product in the group $G$. We then denote $R_S$ by Reg. In this case $s\sigma = s$ if and only if $\sigma = 1_G$. So the *regular character* $\chi_{\text{Reg}}$ has the values:

$$\chi_{\text{Reg}}(\sigma) = \begin{cases} |G| \cdot 1_F, & \text{if } \sigma = 1_G, \\ 0, & \text{if } \sigma \neq 1_G. \end{cases} \tag{2.8}$$

Any natural method for making new vector spaces from old ones can be used to construct new linear representations or characters from old ones. For example, let $R_1, R_2, \ldots, R_n$ be a finite number of representations of $G$ on finite-dimensional vector spaces $V_1, V_2, \ldots, V_n$, respectively, over $F$. The direct sum $V_1 \oplus \ldots \oplus V_n$ is again a finite-dimensional vector space over $F$, on which we have the representation $R_1 \oplus \ldots \oplus R_n$ of $G$ defined by

$$[R_1 \oplus \ldots \oplus R_n](\sigma) = R_1(\sigma) \oplus \ldots \oplus R_n(\sigma):$$
$$v_1 \oplus \ldots \oplus v_n \to v_1 R_1(\sigma) \oplus v_2 R_2(\sigma) \oplus \ldots \oplus v_n R_n(\sigma),$$
$$\textit{for all } \sigma \in G,\ v_1 \in V_1, \ldots, v_n \in V_n. \tag{2.9}$$

An elementary calculation gives

$$\chi_{R_1 \oplus \ldots \oplus R_n}(\sigma) = \chi_{R_1}(\sigma) + \ldots + \chi_{R_n}(\sigma), \quad \textit{for all } \sigma \in G. \tag{2.10}$$

In particular, *the sum of two characters of G is again a character of G.*

Suppose that $Q$, $R$ are representations of finite groups $H$, $G$ on finite-dimensional vector spaces $U$, $V$, all respectively, over $F$. Then the tensor product $U \otimes V$ (over $F$) is again a finite-dimensional vector space on which we have the representation $Q \otimes R$ of the direct product group $H \times G$ determined by

$$[Q \otimes R](\sigma \times \tau) = Q(\sigma) \otimes R(\tau) : u \otimes v \to uQ(\sigma) \otimes vR(\tau),$$
$$\textit{for all } \sigma \in H,\ \tau \in G,\ u \in U,\ v \in V. \tag{2.11}$$

The character $\chi_{Q \otimes R}$ can easily be computed:

$$\chi_{Q \otimes R}(\sigma \times \tau) = \chi_Q(\sigma)\chi_R(\tau), \quad \textit{for all } \sigma \in H,\ \tau \in G. \tag{2.12}$$

If $H = G$ in the preceding case, then the diagonal map $\sigma \to \sigma \times \sigma$ is a natural monomorphism of $G$ into $G \times G$ whose composition with $Q \otimes R$ gives the *inner Kronecker product* $Q * R$ of the two representations $Q$ and $R$ of $G$:

$$[Q * R](\sigma) = Q(\sigma) \otimes R(\sigma) : u \otimes v \to uQ(\sigma) \otimes vR(\sigma),$$
$$\textit{for all } \sigma \in G,\ u \in U,\ v \in V. \tag{2.13}$$

Of course $Q * R$ is a representation of $G$ on $U \otimes V$ whose character is given by

$$\chi_{Q*R}(\sigma) = \chi_Q(\sigma)\chi_R(\sigma), \quad \text{for all } \sigma \in G. \tag{2.14}$$

In particular, *the product of two characters of G is again a character of G.*

Returning to the case of representations $Q$, $R$ of two groups $H$, $G$ on $U$, $V$, respectively, we can define a representation $\text{Hom}\,(R^{-1}, Q)$ of $G \times H$ on the finite-dimensional vector space $\text{Hom}_F(V, U)$ by:

$$[\text{Hom}\,(R^{-1}, Q)](\sigma \times \tau) = \text{Hom}\,(R(\sigma^{-1}), Q(\tau)) : T \to R(\sigma^{-1}) T Q(\tau),$$
$$\textit{for all } \sigma \in G,\ \tau \in H,\ T \in \text{Hom}_F(V, U). \tag{2.15}$$

Its character is given by:

$$\chi_{\text{Hom}\,(R^{-1},\,Q)}(\sigma \times \tau) = \chi_R(\sigma^{-1})\chi_Q(\tau), \quad \textit{for all } \sigma \in G,\ \tau \in H. \tag{2.16}$$

As a special case, let $H = \langle 1 \rangle$ be the trivial group, $U$ be $F$, and $Q(1) = 1_{F \to F}$. Then $\text{Hom}_F(V, F)$ is the dual vector space $V^*$ to $V$ and $[\text{Hom}\,(R^{-1}, Q)]\,(\sigma \times 1)$, for $\sigma \in G$, is the dual linear transformation $R(\sigma^{-1})^*$ to the linear transformation $R(\sigma^{-1})$ of $V$. Hence the *dual representation* $R^{-*}$ to $R$, the representation on $V^*$ defined by

$$T[R^{-*}(\sigma)] = R(\sigma^{-1})T, \quad \textit{for all } \sigma \in G,\ T \in V^*, \tag{2.17}$$

has the character

$$\chi_{R^{-*}}(\sigma) = \chi_R(\sigma^{-1}), \quad \textit{for all } \sigma \in G. \tag{2.18}$$

### 3. Group Algebras

The deeper properties of the characters of a finite group $G$ come from a study of the group algebra $FG$ of $G$ over the field $F$, and of modules over this algebra. Those parts of the theory of algebras needed to carry out this study (essentially the theory of Wedderburn) are quite well known. A very concise account of them can be found in Chapter V of the book by Huppert (1967). So we shall just state the necessary definitions and results here, and refer the reader to that book for the proofs.

By an *algebra A over our field F*, we understand a finite-dimensional vector space $A$ over $F$ together with an $F$-bilinear associative product $(a, a') \to aa'$ from $A \times A$ to $A$. We assume, unless otherwise noted, that $A$ has a two-sided identity $1 = 1_A$ for this product. Then algebra multiplication and vector space addition make $A$ into a ring with identity, while scalar multiplication gives us a natural, identity-preserving homomorphism $f \to f1_A$ of the field $F$ into the center of the ring $A$.

Let $a_1, \ldots, a_n$ be a basis of $A$ (as a finite-dimensional vector space over $F$). Then there are unique *multiplication coefficients* $f_{ijk} \in F$, for $i, j, k = 1, \ldots, n$, such that:

$$a_i a_j = \sum_{k=1}^{n} f_{ijk} a_k, \quad \textit{for all } i, j = 1, \ldots, n. \tag{3.1}$$

Since the product in $A$ is $F$-bilinear, it is determined by the $n^2$ products $a_i a_j$ $(i, j = 1, \ldots, n)$. These products themselves are computed via (3.1) from the coefficients $f_{ijk}$. Hence the multiplication coefficients determine the algebra $A$ to within isomorphism.

The *group algebra FG* of a finite group $G$ over $F$ has the elements of $G$ as a basis. The corresponding multiplication coefficients are determined by the rule that the algebra product of two basis elements $\sigma, \tau \in G$ be their product $\sigma\tau$ in the group $G$. Using $F$-bilinearity, we see that the product of two arbitrary elements of $FG$ is given by:

$$\left(\sum_{\sigma\in G} f_\sigma \sigma\right)\left(\sum_{\tau\in G} g_\tau \tau\right) = \sum_{\rho\in G}\left(\sum_{\sigma,\tau\in G,\sigma\tau=\rho} f_\sigma g_\tau\right)\rho, \tag{3.2}$$

for any coefficients $f_\sigma$, $g_\tau \in F$. The associativity in $G$ easily implies that of multiplication in $FG$, and the identity $1_G$ of $G$ is also the identity for $FG$. So $FG$ is an algebra over $F$ in the sense in which we use the term.

By a *module* $M$ over an $F$-algebra $A$ we mean a right module over the ring $A$, which is unitary, in the sense that $m1_A = m$, for all $m\in M$, and finite-dimensional as a vector space over $F$. Of course, the vector space structure of $M$ comes from the $A$-module structure via the natural homomorphism of $F$ into the center of $A$, so that scalar multiplication is defined by:

$$fm = m(f1_A), \quad \textit{for all } f\in F,\ m\in M. \tag{3.3}$$

We note that the module product $(m, a) \to ma$ is an $F$-bilinear map of $M\times A$ into $M$, and that any $A$-homomorphism of modules is also an $F$-linear map.

If $M$ is an $FG$-module, then each $\sigma\in G$ defines a linear transformation $R(\sigma)$ of the finite-dimensional vector space $M$ by

$$mR(\sigma) = m\sigma, \quad \textit{for all } m\in M. \tag{3.4}$$

The module identities $m(\sigma\tau) = (m\sigma)\tau$, for all $\sigma, \tau\in G$, $m\in M$, and $m1_G = m$, for all $m\in M$, imply that $R$ satisfies (2.5), i.e. that $R$ is a representation of $G$ on $M$.

Conversely, let $R$ be a representation of $G$ on a finite-dimensional vector space $M$ over $F$. Then there is a unique $FG$-module structure for $M$ which is consistent with both (3.4) and the vector space structure of $M$. Indeed, the module product must be given by

$$m\left(\sum_{\sigma\in G} f_\sigma\sigma\right) = \sum_{\sigma\in G} f_\sigma mR(\sigma), \quad \textit{for all } m\in M \textit{ and any } f_\sigma\textit{'s} \in F. \tag{3.5}$$

Thus there is a complete identity between $FG$-modules and representations of $G$ on finite-dimensional vector spaces over $F$.

A module $M$ over an algebra $A$ is *irreducible* if $M \neq \{0\}$, and if $M$ and $\{0\}$ are the only $A$-submodules of $M$. The (Jacobson) *radical* $J(A)$ of the algebra $A$ is the two-sided ideal of $A$ defined by

$$J(A) = \{a\in A : Ma = 0, \quad \textit{for all irreducible } A\textit{-modules } M\}. \tag{3.6}$$

The algebra $A$ is called semi-simple if $J(A) = 0$. In the case of group algebras $FG$ we have

PROPOSITION 3.7. *The group algebra $FG$ of a finite group $G$ over a field $F$ is semi-simple if and only if the characteristic of $F$ does not divide the order $|G|$ of $G$.*

*Proof.* See Satz V.2.7 in Huppert (1967). Notice that the condition of Proposition 3.7 is always satisfied when $F$ has characteristic zero.

An irreducible module $M$ over an algebra $A$ satisfies $MJ(A) = 0$. So $M$ can naturally be regarded as a module over the factor algebra $A/J(A)$. Obviously $M$ is also irreducible as an $A/J(A)$-module. Evidently $\{0\} = J(A)/J(A)$ is the ideal $J(A/J(A))$ of all elements in $A/J(A)$ annihilating all such irreducible $A/J(A)$-modules $M$. Hence, we have

$$\text{the factor algebra } A/J(A) \text{ is always semi-simple.} \tag{3.8}$$

An algebra $A$ is *simple* if $A \neq \{0\}$ and if $A$ and $\{0\}$ are the only two-sided ideals of $A$. A *semi-simple* algebra can be decomposed into a direct sum of simple ones.

PROPOSITION 3.9. *A semi-simple algebra $A$ has only a finite number $k \geq 0$ of minimal two-sided ideals $A_1, \ldots, A_k$. Each such ideal $A_i$ is a simple subalgebra of $A$, and*

$$A = \oplus \sum_{i=1}^{k} A_i. \tag{3.10}$$

*Any two-sided ideal of $A$ is a direct sum of a subset of the $A_i$'s. Finally, $A_iA_j = \{0\}$, for all $i, j = 1, \ldots, k$ with $i \neq j$.*

*Proof.* See Satz V.3.8 in Huppert (1967).

Notice that the condition $A_iA_j = \{0\}$, for $i \neq j$, implies that algebra multiplication, as well as addition and scalar multiplication, is componentwise in (3.10):

$$(a_1 \oplus \ldots \oplus a_k)(a'_1 \oplus \ldots \oplus a'_k) = (a_1a'_1) \oplus (a_2a'_2) \oplus \ldots \oplus (a_ka'_k),$$
$$\text{if } a_i,\ a'_i \in A_i \quad (i = 1, \ldots, k). \tag{3.11}$$

We indicate this by saying that (3.10) is a *direct sum of algebras* (as well as of vector spaces).

It remains to consider the structure of a simple algebra.

PROPOSITION 3.12. *A simple algebra $A$ has, up to isomorphism, exactly one irreducible module $I$. The commuting ring $D = \mathrm{Hom}_A(I, I)$ is a division algebra over $F$. The natural map of $A$ into $\mathrm{Hom}\,(I, I)$ is an algebra isomorphism of $A$ onto the commuting ring $\mathrm{Hom}_D(I, I)$.*

*Proof.* This follows from lemma I.10.5 and Section V.4 in Huppert (1967).

Of course, a *division algebra* is an algebra in which the non-zero elements form a multiplicative group. Notice that our hypotheses of finite-dimensionality for algebras and modules force $D$ to be finite-dimensional over $F$ and $I$ to be finite-dimensional as a vector space over the skew-field $D$.

The field $F$ is called a *splitting field* for a simple algebra $A$ if the above division algebra $D$ is just $F \cdot 1_D$. Notice that:

PROPOSITION 3.13. *If F is algebraically closed, then it is a splitting field for any simple F-algebra.*

*Proof.* See Lemma V.4.3 in Huppert (1967).

If $A$ is a semi-simple algebra over $F$, and $A_1, \ldots, A_k$ are the simple subalgebras of Proposition 3.9, then $F$ is a *splitting field* for $A$ if and only if it is a splitting field for each $A_i$ $(i = 1, \ldots, k)$.

Fix a semi-simple algebra $A$ over $F$ (which may or may not be a splitting field for $A$). Then $A$ is the direct sum (3.10) of its minimal two-sided ideals $A_1, \ldots, A_k$. There is a corresponding decomposition of $A$-modules as direct sums of $A_i$-modules.

PROPOSITION 3.14. *If M is an A-module, then each $MA_i$ $(i = 1 \ldots, k)$, is an A-submodule of M. The A-module $MA_i$ is "really" an $A_i$-module, in the sense that:*

$$m(a_1 \oplus \ldots \oplus a_t) = ma_i, \quad \textit{for all } m \in MA_i,\ a_1 \in A_1, \ldots, a_k \in A_k. \tag{3.15}$$

*Furthermore,*

$$M = \oplus \sum_{i=1}^{k} MA_i. \tag{3.16}$$

*Proof.* Since $A_i$ is a two-sided ideal of $A$, the product $MA_i$ is an $A$-submodule of $M$. Equation (3.15) comes from the fact that $A_iA_j = \{0\}$, for all $j \neq i$ (see Proposition 3.9). By (3.10) we have

$$M = MA = \sum_{i=1}^{k} MA_i. \tag{3.17}$$

The intersection $L = MA_i \cap \left(\sum_{j \neq i} MA_i\right)$ is "really" an $A_i$-submodule of $MA_i$. Hence it satisfies $L = LA = LA_i$. On the other hand, $LA_i \subseteq \sum_{j \neq i} MA_jA_i = \{0\}$. So $L = \{0\}$ and the sum (3.17) is direct, which finishes the proof of the proposition.

The unique submodule $MA_i$ is called the *$A_i$-primary* (or *$A_i$-homogeneous*) *component* of $M$.

Let $I_i$ be an irreducible $A_i$-module, for each $i = 1, \ldots, k$ (such a module exists by Proposition 3.12). Using (3.15) in reverse, we regard $I_i$ as an obviously irreducible) $A$-module.

PROPOSITION 3.18. *Let M be an A-module. For each $i = 1, \ldots, k$, there is a unique integer $m_i \geq 0$ such that*

$$MA_i \cong m_i \times I_i = \overbrace{I_i \oplus \ldots \oplus I_i}^{m_i\text{-times}} \quad (\textit{as } A\text{- (or as } A_i\text{-) modules}). \tag{3.19}$$

*Proof.* It follows from Proposition 3.12 that the simple algebra $A_i$ is semi-simple. So Satz (V.3.4 of Huppert, 1967) tells us that $MA_i$ is a direct sum of irreducible $A_i$-modules. But $I_i$ is, to within isomorphism, the only irreducible $A_i$-module (by Proposition 3.12). Hence (3.19) holds for some integer $m_i \geq 0$. Clearly $m_i = \dim_F(MA_i)/\dim_F(I_i)$ is unique. So the proposition is proved.

The unique integer $m_i$ of Proposition 3.18 is called the *multiplicity* $m(I_i$ *in* $M)$ of the irreducible $A$-module $I_i$ in $M$. Combining the preceding propositions, we get:

$$M \cong (m(I_1 \text{ in } M) \times I_1) \oplus \ldots \oplus (m(I_k \text{ in } M) \times I_k)$$
$$\textit{(as A-modules), for all A-modules } M. \qquad (3.20)$$

An immediate consequence of this is:

*Any irreducible A-module is isomorphic to exactly one of the modules* $I_1, \ldots, I_k$. (3.21)

**4. Character Identities**

Let $G$ be a finite group. Proposition 3.7 says that the group algebra $FG$ of $G$ over any field $F$ of characteristic zero is semi-simple. If $F$ is algebraically closed, then Proposition 3.13 gives

*F is a splitting field for the group algebra FH of any subgroup H of G.* (4.1)

We assume, from now on, that $F$ is any field of characteristic zero satisfying (4.1). As usual, we identify the ordinary integers with their images in $F$.

It is convenient to extend the definitions of "representation" and "character" from group elements to arbitrary elements of $FG$. Let $M$ be an $FG$-module. Then any element $y$ of $FG$ defines an $F$-linear transformation $R_M(y)$ of the vector space $M$ by

$$mR_M(y) = my, \quad \textit{for all } m \in M. \qquad (4.2)$$

The module identities tell us that the map $R_M$ is an algebra homomorphism of $FG$ into $\operatorname{Hom}_F(M, M)$. It is called the *representation* of $FG$ on $M$ corresponding to the module structure of $M$. It is related to the representation $R$ of $G$ on $M$ given in (3.4) by

$$R_M\Big(\sum_{\sigma \in G} f_\sigma \sigma\Big) = \sum_{\sigma \in G} f_\sigma R(\sigma), \quad \textit{for any } f_\sigma\textit{'s in } F. \qquad (4.3)$$

The *character* $\chi_M$ of the module $M$ is the function from $FG$ to $F$ defined by

$$\chi_M(y) = \operatorname{tr}(R_M(y)), \quad \textit{for all } y \in FG. \qquad (4.4)$$

Since both tr and $R_M$ are $F$-linear, so is $\chi_M$. Evidently, it is related to the character $\chi_R$ of the representation $R$, given in (2.1), by

$$\chi_M\Big(\sum_{\sigma \in G} f_\sigma \sigma\Big) = \sum_{\sigma \in G} f_\sigma \chi_R(\sigma), \quad \textit{for any } f_\sigma\textit{'s in } F. \qquad (4.5)$$

Thus the representation $R_M$ and character $\chi_M$ are just the extensions to $FG$ of the representation $R$ and character $\chi_R$, respectively, by use of $F$-linearity. From now on we shall use these extended definitions of "representation" and "character".

Proposition 3.9 tells us that the semi-simple algebra $FG$ has the unique decomposition

$$FG = A_1 \oplus \ldots \oplus A_k \quad (\textit{as algebras}), \tag{4.6}$$

where the simple subalgebras $A_1, \ldots, A_k$ are the minimal two-sided ideals of $FG$. By Proposition 3.12, each $A_i$ $(i = 1, \ldots, k)$ has, up to isomorphism, a unique irreducible module $I_i$. We regard $I_i$, as usual, as an irreducible $A$-module, and denote the corresponding representation $R_{I_i}$ and character $\chi_{I_i}$ by $R_i$ and $\chi_i$, respectively. The characters $\chi_1, \ldots, \chi_k$ are called the *irreducible characters* of $G$ (or of $FG$).

The definition of the $FG$-module structure of $I_i$ gives

$$R_i(A_j) = \{0\}, \quad \textit{and hence } \chi_i(A_j) = \{0\}, \quad \textit{for all } i, j = 1, \ldots, k \textit{ with } i \neq j. \tag{4.7}$$

Condition (4.1) says that $F$ is a splitting field for each simple algebra $A_i$. This and Proposition 3.12 imply that

$$R_i \textit{ sends the algebra } A_i \textit{ isomorphically onto } \mathrm{Hom}_F(I_i, I_i), \textit{ for all } i = 1, \ldots, k. \tag{4.8}$$

In particular, there exists an element $a_i \in A_i$ such that $\chi_i(a_i) = \mathrm{tr}\,(R_i(a_i)) = 1$, for any $i = 1, \ldots$, k. Since $\chi_j(a_i) = 0$, for $j \neq i$, (by (4.7)), we conclude that

$$\chi_1, \ldots, \chi_k \textit{ are F-linearly independent functions from FG to F.} \tag{4.9}$$

Let $M$ be any $FG$-module. From (2.10), (3.20) and (4.5) we get

$$\chi_M = m(I_1 \text{ in } M)\chi_1 + \ldots + m(I_k \text{ in } M)\chi_k. \tag{4.10}$$

By (4.9) the multiplicities $m(I_1 \text{ in } M), \ldots, m(I_k \text{ in } M)$ are uniquely determined by the character $\chi_M$ via this equation. So (3.20) gives

$$\textit{An FG-module M is determined to within isomorphism by its character } \chi_M. \tag{4.11}$$

The algebra $FG$ is itself an $FG$-module in which the module product $my$ of an element $m$ of the module $FG$ with an element $y$ of the algebra $FG$ is the algebra product of $m$ and $y$ in $FG$. It is clear from (2.6) and the definition of $FG$ that the corresponding representation of $G$ on $FG$ is the regular representation Reg. So (2.8), (4.5) and the identification of arbitrary integers with their images in $F$ give

$$\chi_{FG}(\sigma) = \begin{cases} |G|, & \textit{if } \sigma = 1_G, \\ 0, & \textit{if } \sigma \in G - \{1_G\}. \end{cases} \tag{4.12}$$

We can compute $\chi_{FG}$ another way using (4.10). Clearly $A_i = (FG)A_i$ is the $A_i$-primary component of the module $FG$. So $A_i \simeq \mathrm{m}(I_i \text{ in } FG) \times I_i$ as an $A$-module. The dimension of $I_i$ is $\chi_i(1)$ by (2.3) and (4.5). The dimension of $A_i \simeq \mathrm{Hom}_F(I_i, I_i)$ (by (4.8)) is $(\dim I_i)^2 = \chi_i(1)^2$. Hence,

$$m(I_i \text{ in } FG) = \dim A_i/\dim I_i = \chi_i(1), \quad \textit{for all } i = 1,\ldots, k. \qquad (4.13)$$

This and (4.10) give

$$\chi_{FG} = \chi_1(1)\chi_1 + \ldots + \chi_k(1)\chi_k. \qquad (4.14)$$

For each $i = 1,\ldots, k$, the simple algebra $A_i$ has an identity element $1_{A_i}$. Let $y$ be a general element of $FG$. By (4.6) there are unique elements $y_i \in A_i$ $(i = 1,\ldots, k)$ such that $y = y_1 + \ldots + y_k$. From (3.11) we see that $y_i = y1_{A_i}$ $(i = 1,\ldots, k)$. Hence (4.7) implies

$$\begin{aligned} \chi_j(y1_{A_i}) = \chi_j(y_i) &= 0, \quad \textit{if } j = 1,\ldots, k \textit{ and } j \neq i,\\ &= \chi_i(y), \quad \textit{if } j = i. \end{aligned} \qquad (4.15)$$

In view of (4.14), this gives

$$\chi_{FG}(y1_{A_i}) = \chi_i(1)\chi_i(y), \quad \textit{for all } y \in FG,\ i = 1,\ldots, k. \qquad (4.16)$$

We apply this equation and the other formula (4.12) for $\chi_{FG}$ to compute $1_{A_i}$. As an element of $FG$, the identity $1_{A_i}$ has the form $\sum_{\sigma \in G} f_\sigma \sigma$, for some unique coefficients $f_\sigma \in F$. Evidently $f_\sigma$ is the coefficient of $1_G$ when $\sigma^{-1}1_{A_i}$ is written as a linear combination of elements of $G$:

$$\sigma^{-1}1_{A_i} = f_\sigma 1_G + \ldots$$

Applying $\chi_{FG}$ and using (4.12) we obtain

$$\chi_{FG}(\sigma^{-1}1_{A_i}) = |G| f_\sigma, \quad \textit{for all } \sigma \in G.$$

This and (4.16) give us a formula for $1_{A_i}$ as a function $e(\chi_i)$ of the character $\chi_i$:

$$1_{A_i} = e(\chi_i) = \sum_{\sigma \in G} \frac{\chi_i(1)\chi_i(\sigma^{-1})}{|G|}\sigma, \quad \textit{for } i = 1,\ldots, k. \qquad (4.17)$$

We know from (3.11) that $1_{A_i}1_{A_j} = 0$, if $i \neq j$. So (4.15) implies that $\chi_j(1_{A_i}) = 0$, for $i \neq j$. On the other hand, (4.15) also implies that $\chi_i(1_{A_i}) = \chi_i(1)$, for all $i$. Substituting the expression (4.17) for $1_{A_i}$ in these equations and using the linearity of the characters $\chi_j$, we obtain

$$\begin{aligned} \frac{\chi_i(1)}{|G|} \sum_{\sigma \in G} \chi_i(\sigma^{-1})\chi_j(\sigma) &= 0, \quad \textit{if } i, j = 1,\ldots, k \textit{ and } i \neq j,\\ &= \chi_i(1), \quad \textit{if } i = j = 1,\ldots, k. \end{aligned}$$

By (2.3) the integer $\chi_i(1)$ is strictly positive. So we can divide by it to get:

$$\begin{aligned} \frac{1}{|G|} \sum_{\sigma \in G} \chi_i(\sigma^{-1})\chi_j(\sigma) &= 0, \quad \textit{if } i, j = 1,\ldots, k \textit{ and } i \neq j,\\ &= 1, \quad \textit{if } i = j = 1,\ldots, k. \end{aligned} \qquad (4.18)$$

The identities (4.18) are usually expressed by means of an *inner product*

$(\phi, \psi)_G$ defined for any two linear functions $\phi, \psi$ from $FG$ to $F$ by

$$(\phi, \psi)_G = \frac{1}{|G|} \sum_{\sigma \in G} \phi(\sigma^{-1})\psi(\sigma). \tag{4.19}$$

Since we can replace $\sigma$ by $\sigma^{-1}$ in this summation, $(\cdot, \cdot)_G$ is a symmetric, $F$-bilinear form on the space $(FG)^* = \mathrm{Hom}_F(FG, F)$. Equations (4.18) just say that

*The irreducible characters $\chi_1, \ldots, \chi_k$ are orthonormal with respect to the form $(\cdot, \cdot)_G$.* (4.20)

So far we have no close connections between the irreducible characters $\chi_1, \ldots, \chi_k$ and the structure of the group $G$. One way of forming such connections uses the center $Z(FG)$ of the group algebra $FG$. The center $Z(\mathrm{Hom}_F(I_i, I_i))$ of the algebra of all $F$-linear transformations of $I_i$ is clearly the set $F \cdot 1_{I_i \to I_i}$ of $F$-multiples of the identity transformation. By (4.8) this implies that

$$Z(A_i) = F \cdot 1_{A_i}, \quad (i = 1, \ldots, k). \tag{4.21}$$

This and (4.6) give immediately

$$Z(FG) = Z(A_1) \oplus \ldots \oplus Z(A_k) = F \cdot 1_{A_1} \oplus \ldots \oplus F \cdot 1_{A_k} \quad (\textit{as algebras}). \tag{4.22}$$

There is another way of looking at the center $Z(FG)$. Let $K_1, \ldots, K_c$ be the conjugacy classes of the finite group $G$, and $\tilde{K}_i$ be the *class sum*

$$\tilde{K}_i = \sum_{\sigma \in K_i} \sigma \tag{4.23}$$

in $FG$, for $i = 1, \ldots, c$). Then we have

PROPOSITION 4.24. *The class sums $\tilde{K}_1, \ldots, \tilde{K}_c$ form a basis for the algebra $Z(FG)$.*

*Proof.* Let $y = \sum_{\sigma \in G} f_\sigma \sigma$ be any element of $FG$, where the $f_\sigma$'s all lie in $F$. Evidently $y$ lies in $Z(FG)$ if and only if $y\tau = \tau y$, for all $\tau$ in the basis $G$ of $FG$, i.e. if and only if $\tau^{-1}y\tau = y$, for all $\tau \in G$. Since $\tau^{-1}y\tau = \sum_{\sigma \in G} f_\sigma \sigma^\tau$, this occurs if and only if $f_\sigma = f_{\sigma^\tau}$ for all $\sigma, \tau \in G$, i.e. if and only if $y$ is a linear combination of the class sums $\tilde{K}_1, \ldots, \tilde{K}_c$. Therefore $\tilde{K}_1, \ldots, \tilde{K}_c$ span $Z(FG)$. They are linearly independent, since $K_1, \ldots, K_c$ are pairwise disjoint, non-empty subsets of $G$. So the proposition holds.

Combining this proposition with (4.22) we obtain

PROPOSITION 4.25. *The number $k$ of irreducible characters of $G$ equals the number $c$ of conjugacy classes of $G$.*

*Proof.* By (4.22), $k = \dim_F Z(FG)$. By Proposition 4.24, the latter number is $c$. So the proposition holds.

There is another way to regard Proposition 4.25. A *class function* on the group $G$ is a linear function $\phi : FG \to F$ which is constant on each conjugacy class $K_i$ of $G$. Evidently these functions form an $F$-subspace $CF(G)$ of $(FG)^*$ and $\dim_F(CF(G)) = c$. By Proposition 2.2, each character $\chi_M$ of $G$ is a class function. In view of (4.9) and (4.20), Proposition 4.25 is equivalent to

*The irreducible characters $\chi_1, \ldots, \chi_k$ form an orthonormal basis for $CF(G)$ with respect to the form $(\cdot, \cdot)_G$.* (4.26)

Since $Z(FG)$ is a subalgebra of $FG$, there are unique multiplication coefficients $a_{hij}$, $(h, i, j = 1, \ldots, c)$, such that

$$\tilde{K}_h \tilde{K}_i = \sum_{j=1}^{c} a_{hij} \tilde{K}_j, \quad (h, i = 1, \ldots, c). \tag{4.27}$$

These coefficients have a simple, but important, relationship with multiplication in the group $G$.

PROPOSITION 4.28. *Let $h, i, j = 1, \ldots, c$. If $\rho \in K_j$, then $a_{hij}$ is the number of ordered pairs $(\sigma, \tau)$ of elements of $G$ such that $\sigma \in K_h$, $\tau \in K_i$ and $\sigma\tau = \rho$.*

*Proof.* By (4.27), $a_{hij}$ is the coefficient of $\rho$ in $\tilde{K}_h \tilde{K}_i = \sum_{\sigma \in K_h, \tau \in K_i} \sigma\tau$. This coefficient is obviously the number of the above ordered pairs $(\sigma, \tau)$.

Evidently the decomposition (4.22) gives us $k$ distinct epimorphisms $\theta_1, \ldots, \theta_k$ of the algebra $Z(FG)$ onto $F$ defined by

$$\theta_i(f_1 \cdot 1_{A_1} \oplus \ldots \oplus f_k \cdot 1_{A_k}) = f_i, \quad \textit{for all } i = 1, \ldots, k \textit{ and any } f_j\textit{'s in } F. \tag{4.29}$$

These epimorphisms $\theta_i$ can easily be computed in terms of the irreducible characters $\chi_i$.

If $y \in Z(FG)$, then evidently $R_i(y) = \theta_i(y) \cdot 1_{I_i \to I_i} \in \mathrm{Hom}_F(I_i, I_i)$. Since the trace of the identity transformation $1_{I_i \to I_i}$ is $\dim_F I_i$, which equals $\chi_i(1)$ by (2.3), we have

$$\chi_i(y) = \mathrm{tr}\,(R_i(y)) = \theta_i(y)\,\mathrm{tr}\,(1_{I_i \to I_i}) = \theta_i(y)\chi_i(1).$$

Therefore,

$$\theta_i(y) = \frac{\chi_i(y)}{\chi_i(1)}, \quad \textit{for all } y \in Z(FG) \textit{ and all } i = 1, \ldots, k. \tag{4.30}$$

If we substitute (4.30) in the formula $\theta_i(yz) = \theta_i(y)\theta_i(z)$ (which is valid for all $y, z \in Z(FG)$ since $\theta_i$ is a homomorphism) and then multiply by $\chi_i(1)^2$, we obtain the useful formula:

$$\chi_i(y)\chi_i(z) = \chi_i(1)\chi_i(yz), \quad \textit{for all } y, z \in Z(FG), \text{ and } i = 1, \ldots, k. \tag{4.31}$$

We can use the formulas (4.18) and (4.30) to compute the values of the irreducible characters $\chi_i$ starting from the multiplication table of the group $G$. By Proposition 4.28, we can compute the multiplication coefficients $a_{hij}$ in

(4.27). Since any $\theta_g$ is a homomorphism of $Z(FG)$ into $F$, its value $\theta_g(\tilde{K}_h)$ at a class sum $\tilde{K}_h$ satisfies:

$$\theta_g(\tilde{K}_h)\theta_g(\tilde{K}_i) = \sum_{j=1}^{c} a_{hij}\theta_g(\tilde{K}_j), \quad \textit{for all } i = 1,\ldots, c.$$

Because $\theta_g$ is an epimorphism, not all the $\theta_g(\tilde{K}_i)$ $(i = 1,\ldots, c)$, can be zero. We conclude that $\theta_g(\tilde{K}_h)$ is an eigenvalue of the $c \times c$ matrix $(a_{hij})_{i, j = 1,\cdots, c}$, and that $(\theta_g(\tilde{K}_1),\ldots, \theta_g(\tilde{K})_c)^T$ is a corresponding eigenvector. With this observation one can theoretically (and even practically on a computer) compute the values $\theta_g(\tilde{K}_h)$ of the $k$ distinct epimorphisms $\theta_1,\ldots, \theta_k$ of $Z(FG)$ onto $F$ at the class sums $\tilde{K}_h$, $(h = 1,\ldots, c)$. Applying Proposition 2.2, and (4.30), we get

$$\frac{\chi_g(\sigma)}{\chi_g(1)} = \frac{\theta_g(\tilde{K}_h)}{|K_h|}, \quad \textit{for all } g = 1,\ldots, k \textit{ and } \sigma \in G, \tag{4.32}$$

where $K_h$ is the class $\sigma^G$ of $\sigma$. Using (4.18) we find that

$$\frac{1}{\chi_g(1)^2} = \frac{1}{|G|}\sum_{\sigma \in G} \frac{\chi_g(\sigma^{-1})}{\chi_g(1)} \frac{\chi_g(\sigma)}{\chi_g(1)}, \quad \textit{for all } g = 1,\ldots, k.$$

Therefore $\chi_g(1)^2$ is computable. But $\chi_g(1)$ is a positive integer by (2.3). Hence it is computable. This, together with (4.32), determines the character values $\chi_g(\sigma)$ for all $\sigma \in G$ and all $g = 1,\ldots, k$.

Of course the above program is a bit difficult because of the eigenvalue computation needed in the middle. But it does imply one important fact: *the multiplication coefficients $a_{hij}$ of* (4.27) *determine the irreducible characters $\chi_1,\ldots, \chi_k$ of G; in particular, any information about these coefficients should be reflected in properties of the group characters.*

We illustrate the last principle, which is vital in many proofs, by considering the coefficients of $1_G$ in (4.27). Evidently $\{1_G\}$ is a class, say $K_1$, and $1_G$ is its class sum $\tilde{K}_1$. For any class $K_i$, let $K_i^{-1}$ be the class $\{\sigma^{-1} : \sigma \in K_i\}$ and $\widetilde{K_i^{-1}}$ be its class sum. Using Proposition 4.28 with $\rho = 1_G$, we compute to obtain

$$\begin{aligned} \tilde{K}_i\widetilde{K_j^{-1}} &= 0 \cdot 1_G + \cdots, && \textit{if } i, j = 1,\ldots, c \textit{ and } i \neq j, \\ &= |K_i| \cdot 1_G + \ldots, && \textit{if } i = j = 1,\ldots, c, \end{aligned}$$

where the three dots refer to linear combinations of the other sums $\tilde{K}_h$ $(h > 1)$.

Applying $\chi_{FG}$ to these equations and using (4.12), we obtain

$$\chi_{FG}(\tilde{K}_i\widetilde{K_j^{-1}}) = \begin{cases} 0, & \textit{if } i, j = 1,\ldots, c \textit{ and } i \neq j, \\ |G||K_i|, & \textit{if } i = j = 1,\ldots, c. \end{cases}$$

Next we use (4.14) and (4.31) to get

$$\chi_{FG}(\widetilde{\tilde{K}_i K_j^{-1}}) = \sum_{h=1}^{k} \chi_h(1)\chi_h(\tilde{K}_i K_j^{-1})$$
$$= \sum_{h=1}^{k} \chi_h(\tilde{K}_i)\chi_h(\widetilde{K_j^{-1}}).$$

Choose any elements $\sigma \in K_i$, $\tau \in K_j$. Then Proposition 2.2 implies that $\chi_h(\tilde{K}_i) = |K_i|\chi_h(\sigma)$, and $\chi_h(\widetilde{K_j^{-1}}) = |K_j|\chi_h(\tau^{-1})$, for all $h = 1,\ldots, k$. Hence the above equations become:

$$|K_i||K_j| \sum_{h=1}^{k} \chi_h(\sigma)\chi_h(\tau^{-1}) = \begin{cases} 0, & \text{if } \sigma \underset{G}{\not\sim} \tau, \\ |G||K_i|, & \text{if } \sigma \underset{G}{\sim} \tau, \end{cases}$$

where, of course, $\sigma \underset{G}{\sim} \tau$ means "$\sigma$ is $G$-conjugate to $\tau$", and $\chi \underset{G}{\not\sim} \tau$ means "$\sigma$ is not $G$-conjugate to $\tau$". Dividing by $|K_i||K_j|$, and using the fact that $|G|/|K_j| = |C_G(\tau)|$, we obtain the following identities:

$$\sum_{h=1}^{k} \chi_h(\sigma)\chi_h(\tau^{-1}) = \begin{cases} |C_G(\tau)|, & \text{if } \sigma, \tau \in G \text{ and } \sigma \underset{G}{\sim} \tau, \\ 0, & \text{if } \sigma, \tau \in G \text{ and } \sigma \underset{G}{\not\sim} \tau. \end{cases} \tag{4.33}$$

Besides their relations with the conjugacy classes and the multiplication coefficients $a_{hij}$, the irreducible characters of $G$ are also connected with certain normal subgroups of $G$. To explain this connection, we start with two lemmas.

LEMMA 4.34. *Let $M$ be any $FG$-module and $\sigma$ be any element of $G$. Then there exists a basis $m_1,\ldots, m_n$ for $M$ satisfying:*

$$m_i\sigma = \zeta_i m_i \quad (i = 1,\ldots, n), \tag{4.35}$$

*where $\zeta_1,\ldots, \zeta_n \in F$ are $|G|$th roots of unity (they are even eth roots of unity, where e is the exponent of $G$).*

*Proof.* The cyclic subgroup $\langle\sigma\rangle$ generated by $\sigma$ is abelian, and $F$ is a splitting field for its group algebra $F\langle\sigma\rangle$ by (4.1). It follows from (4.6) and (4.8) that each irreducible $F\langle\sigma\rangle$-module is one-dimensional. Because $F\langle\sigma\rangle$ is semi-simple, (3.20) tells us that $M$, considered as an $F\langle\sigma\rangle$-module, is a direct sum of irreducible $F\langle\sigma\rangle$-submodules $J_1 \oplus \ldots \oplus J_n$. If $m_i$ is a basis element for the one-dimensional module $J_i$ ($i = 1,\ldots$, n), then $m_1,\ldots, m_n$ is a basis for $M$ satisfying (4.35), for some $\zeta_1,\ldots, \zeta_n \in F$. Evidently $\sigma^e = 1$ implies that each $\zeta_i$ is an $e$th root of unity and hence a $|G|$th root of unity. So the lemma holds.

LEMMA 4.36. *Let* $\zeta_1,\ldots,\zeta_n$ $(n \geq 1)$ *be* $|G|$th *roots of unity in* $F$. *If* $\zeta_1+\ldots+\zeta_n = n\zeta$, *where* $\zeta$ *is also a* $|G|$th *root of unity, then* $\zeta_1 = \ldots = \zeta_n = \zeta$.

*Proof.* The $|G|$th roots of unity in $F$ generate a subfield $E$ which is finite-dimensional over the rational subfield $Q$. Evidently it suffices to prove the lemma for $E$ in place of $F$. But $E$ can be isomorphically embedded in the complex numbers $C$. Therefore it suffices to prove the lemma for $F = C$.

In $C$, the absolute value $|\zeta|$ of any $|G|$th root of unity is 1. So, we have

$$|\zeta_1+\ldots+\zeta_n| = |n\zeta| = n = |\zeta_1|+\ldots+|\zeta_n|.$$

This implies that $\zeta_1,\ldots,\zeta_n$ are all positive multiples of each other. Since they all have absolute value 1, they must be equal $\zeta_1 = \ldots = \zeta_n$. Evidently their common value is $(\zeta_1+\ldots+\zeta_n)/n = \zeta$. So the lemma holds.

The normal subgroups associated with the irreducible character $\chi_i$ $(i = 1,\ldots, k)$ are

$$\mathrm{Ker}\,(\chi_i) = \{\sigma\in G : y\sigma = y, \quad \textit{for all } y\in I_i\} \tag{4.37a}$$

$$Z(G \bmod \mathrm{Ker}\,(\chi_i)) = \{\sigma\in G : \sigma\,\mathrm{Ker}\,(\chi_i)\in Z(G/\mathrm{Ker}\,(\chi_i))\}. \tag{4.37b}$$

Their relations with the values of $\chi_i$ are given by

PROPOSITION 4.38. *Fix* $i = 1,\ldots, k$. *If* $\sigma \in G$, *then*

(a) $\sigma \in \mathrm{Ker}\,(\chi_i)$ *if and only if* $\chi_i(\sigma) = \chi_i(1)$,

(b) $\sigma \in Z(G \bmod \mathrm{Ker}\,(\chi_i))$ *if and only if* $\chi_i(\sigma) = \zeta\chi_i(1)$, *for some* $|G|$th *root of unity* $\zeta \in F$.

*Proof.* If $\sigma \in Z(G \bmod \mathrm{Ker}\,(\chi_i))$, then the linear transformation $R_i(\sigma)$ evidently commutes with every $R_i(\tau)$ $(\tau \in G)$. In view of (4.8), $R_i(\sigma)$ must lie in the center of $\mathrm{Hom}_F(I_i, I_i)$. So it has the form $R_i(\sigma) = \zeta\cdot 1_{I_i\to I_i}$, where $\zeta \in F$. This implies that $\chi_i(\sigma) = \zeta \dim_F(I_i) = \zeta\chi_i(1)$, by (2.3). Since $\sigma^{|G|} = 1$, the element $\zeta \in F$ is a $|G|$th root of unity. Obviously $\zeta = 1$ if and only if $\sigma \in \mathrm{Ker}\,(\chi_i)$. So we have proved the "only if" parts of both (a) and (b).

Now suppose that $\sigma$ satisfies $\chi_i(\sigma) = \zeta\chi_i(1)$, for some $|G|$th root of unity $\zeta \in F$ (in the case (a), we take $\zeta = 1$). Lemma 4.34 gives us a basis $m_1,\ldots, m_n$ for $I_i$ and $|G|$th roots of unity $\zeta_1,\ldots,\zeta_n$ satisfying (4.35). The trace $\chi_i(\sigma)$ of $R_i(\sigma)$ is then clearly $\zeta_1+\ldots+\zeta_n = n\zeta$, which implies $\zeta_1 = \ldots = \zeta_n = \zeta$ by Lemma 4.36. So $R_i(\sigma) = \zeta 1_{I_i\to I_i}$ commutes with $R_i(\tau)$, for all $\tau \in G$, which is equivalent to saying that $\sigma \in Z(G \bmod \mathrm{Ker}\,(\chi_i))$. When $\zeta = 1$, we even have $\sigma \in \mathrm{Ker}\,(\chi_i)$. Therefore the proposition is true.

## 5. Induced Characters

We continue to use the notation and hypotheses of Section 4. If $\sigma\in G$ and $\chi \in (FG)^* = \mathrm{Hom}_F(FG, F)$, then the *conjugate function* $\chi^\sigma \in (FG)^*$ is given by:

$$\chi^\sigma(y) = \chi(y^{\sigma^{-1}}) = \chi(\sigma y\sigma^{-1}), \quad \textit{for all } y\in FG. \tag{5.1}$$

This is a linear action of $G$ on $(FG)^*$ contragradient to the conjugation action of $G$ on $FG$. Notice that the form $(\cdot, \cdot)_G$ of (4.19) is invariant under this action:

$$(\xi^\sigma, \chi^\sigma)_G = (\xi, \chi)_G, \quad \textit{for all } \sigma \in G,\ \xi, \chi \in (FG)^*. \tag{5.2}$$

The subspace $CF(G)$ of all class functions on $G$ is evidently given by

$$CF(G) = \{\chi \in (FG)^* : \chi^\sigma = \chi, \quad \textit{for all } \sigma \in G\}. \tag{5.3}$$

Let $H$ be any subgroup of $G$. We extend any function $\phi \in CF(H)$ to a linear function $\hat{\phi} \in (FG)^*$ by setting $\hat{\phi}(\sigma) = 0$, for all $\sigma \in G \backslash H$. So $\hat{\phi}$ is defined by:

$$\hat{\phi}\Big(\sum_{\sigma \in G} f_\sigma \sigma\Big) = \phi\Big(\sum_{\sigma \in H} f_\sigma \sigma\Big), \quad \textit{for any } f_\sigma\textit{'s in } F. \tag{5.4}$$

Since $\phi$ is a class function on $H$, its extension $\hat{\phi}$ satisfies $\hat{\phi}^\sigma = \hat{\phi}$ for all $\sigma \in H$. Hence we can define the *induced function* $\phi^G \in (FG)^*$ to be the "trace from $H$ to $G$" of $\hat{\phi}$:

$$\phi^G = \sum_{\sigma \in \mathrm{rep}\,(G/H)} \hat{\phi}^\sigma, \tag{5.5}$$

where rep $(G/H)$ is any family of representatives for the left cosets $H\sigma$ of $H$ in $G$. Evidently $\phi^G$ is independent of the choice of these representatives, and is fixed under conjugation by any element of $G$. So (5.3) implies

*Induction:* $\phi \to \phi^G$ *is an F-linear map of* $CF(H)$ *into* $CF(G)$. (5.6)

The restriction $\chi_H$ to $FH$ of a class function $\chi$ of $G$ is clearly a class function of $H$. The $F$-linear map $\chi \to \chi_H$ of $CF(G)$ into $CF(H)$ is, in fact, contragradient to induction.

PROPOSITION 5.7 (Frobenius reciprocity law). *If* $\chi \in CF(G)$ *and* $\phi \in CF(H)$, *then*

$$(\chi, \phi^G)_G = (\chi_H, \phi)_H.$$

*Proof.* We know from (5.3) that $\chi = \chi^\sigma$, for all $\sigma \in G$. So (5.5) and (5.2) give:

$$\begin{aligned}(\chi, \phi^G)_G &= \Big(\chi, \sum_{\sigma \in \mathrm{rep}\,(G/H)} \hat{\phi}^\sigma\Big)_G \\ &= \sum_{\sigma \in \mathrm{rep}\,(G/H)} (\chi^\sigma, \hat{\phi}^\sigma)_G = [G:H](\chi, \hat{\phi})_G.\end{aligned}$$

By (4.19) and (5.4), we have:

$$\begin{aligned}(\chi, \hat{\phi})_G &= \frac{1}{|G|} \sum_{\sigma \in G} \chi(\sigma)\hat{\phi}(\sigma^{-1}) = \frac{1}{|G|} \sum_{\sigma \in H} \chi(\sigma)\phi(\sigma^{-1}) \\ &= \frac{1}{[G:H]} (\chi_H, \phi)_H.\end{aligned}$$

The result follows from this and the preceding equation.

The Frobenius reciprocity law has the following important consequence.

PROPOSITION 5.8. *If $\phi$ is a character of $H$, then $\phi^G$ is a character of $G$.*

*Proof.* By (4.26) the class function $\phi^G$ is a unique linear combination $\phi^G = f_1\chi_1 + \ldots + f_k\chi_k$, with coefficients $f_1, \ldots, f_k \in F$, of the irreducible characters $\chi_1, \ldots, \chi_k$ of $G$. In view of (4.10), $\phi^G$ is a character of $G$ if and only if each $f_i$ is a non-negative integer. Using (4.20) and the Frobenius reciprocity law, we have:

$$f_i = (\chi_i, \phi^G)_G = ((\chi_i)_H, \phi)_H.$$

Evidently the restriction $(\chi_i)_H$ is a character of $H$ (the corresponding representation is that of $FH$ on the $FG$-module $I_i$). By definition, $\phi$ is a character of $H$. If $\lambda_1, \ldots, \lambda_l$ are the irreducible characters of $H$, equation (4.10) gives us non-negative integers $n_1, \ldots, n_l, m_1, \ldots, m_l$ such that

$$(\chi_i)_H = n_1\lambda_1 + \ldots + n_l\lambda_l, \qquad \phi = m_1\lambda_1 + \ldots + m_l\lambda_l.$$

But then (4.20) for $H$ implies:

$$f_i = ((\chi_i)_H, \phi)_H = \sum_{i=1}^{l} n_i m_i.$$

The last expression is a non-negative integer, since each $n_i$ or $m_i$ is. This completes the proof of the proposition.

Another consequence of the Frobenius reciprocity law is used often enough to deserve special mention.

PROPOSITION 5.9. *Let $\chi_1, \ldots, \chi_k$ be the irreducible characters of $G$, and $\lambda_1, \ldots, \lambda_l$ be the irreducible characters of $H$, where $\chi_1$, $\lambda_1$ are the trivial characters of $G$, $H$, respectively. If $\phi = \sum_{i=1}^{l} f_i\lambda_i$, with each $f_i \in F$, is any class function on $H$, and $\phi^G = \sum_{j=1}^{k} e_j\chi_j$, with each $e_j \in F$, then $e_1 = f_1$.*

*Proof.* This is shorter to prove than to state. Obviously $(\chi_1)_H = \lambda_1$. So the Frobenius reciprocity law and (4.26) give:

$$e_1 = (\lambda_1, \phi)_H = ((\chi_1)_H, \phi)_H = (\chi_1, \phi^G) = f_1,$$

which proves the result.

A *trivial intersection set* (or *t.i. set*) with $H$ as its normalizer is a non-empty subset $S$ of $H$ satisfying:

$$S^\sigma = S, \quad \text{for all } \sigma \in H, \tag{5.10a}$$

$$S^\sigma \cap S \text{ is empty}, \quad \text{for all } \sigma \in G \backslash H, \tag{5.10b}$$

$$S = S^{-1} = \{\sigma^{-1} : \sigma \in S\}. \tag{5.10c}$$

As an example of a t.i. set, we define a subset $\sqrt{\sigma}$, for any $\sigma \in G$, by:

$$\sqrt{\sigma} = \{\tau \in G : \sigma \in \langle\tau\rangle\}. \tag{5.11}$$

Then we have:

PROPOSITION 5.12. *For any $\sigma \in G$, the set $\sqrt{\sigma}$ is a t.i. set with $N_G(\langle\sigma\rangle)$ as its normalizer.*

*Proof.* Clearly $\sqrt{\sigma}$ is a non-empty subset of $N_G(\langle\sigma\rangle)$. Since $\sigma \in \langle\tau\rangle$ if and only if $\langle\sigma\rangle \leq \langle\tau\rangle$, the subset $\sqrt{\sigma}$ is invariant under conjugation by elements of $N_G(\langle\sigma\rangle)$. If $\rho \in G\backslash N_G(\langle\sigma\rangle)$ and $\tau \in S^\rho \cap S$, then $\langle\tau\rangle$ contains both $\langle\sigma\rangle$ and $\langle\sigma^\rho\rangle = \langle\sigma\rangle^\rho \neq \langle\sigma\rangle$. This is impossible, since the cyclic group $\langle\tau\rangle$ contains only one subgroup of order $|\langle\sigma\rangle| = |\langle\sigma\rangle^\rho|$. Hence $S^\rho \cap S$ is empty. Finally $\langle\tau\rangle = \langle\tau^{-1}\rangle$ implies that $(\sqrt{\sigma})^{-1} = \sqrt{\sigma}$. So the proposition holds.

Let $S$ be any t.i. set with $H$ as its normalizer. We define an $F$-subspace $CF(H|S)$ of $CF(H)$ by

$$CF(H|S) = \{\phi \in CF(H) : \phi(\sigma) = 0, \quad \text{for all } \sigma \in H\backslash S\}. \tag{5.13}$$

The functions in $CF(H|S)$ behave very well under induction.

PROPOSITION 5.14. *If $\phi \in CF(H|S)$ and $\sigma \in G$, then*

$$\phi^G(\sigma) = \begin{cases} 0, & \text{unless } \sigma \underset{G}{\sim} \tau, \quad \text{for some } \tau \in S, \\ \phi(\tau), & \text{if } \sigma \underset{G}{\sim} \tau, \quad \text{for some } \tau \in S. \end{cases}$$

*Proof.* By (5.4) the extension $\hat{\phi}$ is zero on $G\backslash S$. For any $\rho \in \text{rep}\,(G/H)$, the conjugate $\hat{\phi}^\rho$ is zero on $G\backslash S^\rho$. From (5.5) we conclude that $\phi^G$ is zero on $G\backslash \bigcup_\rho S^\rho$, which gives the first of the above equations. For the second, we may assume that $\sigma = \tau \in S$, since $\phi^G$ is a class function on $G$. Then (5.10) implies that $\sigma \notin S^\rho$ for all $\rho \in \text{rep}\,(G/H)$ except the single representative $\rho_0 \in H$. So $\hat{\phi}^\rho(\sigma) = 0$, for all $\rho \neq \rho_0$ in rep $(G/H)$. This and (5.5) imply that $\phi^G(\sigma) = \hat{\phi}^{\rho_0}(\sigma) = \hat{\phi}(\sigma^{\rho_0^{-1}}) = \phi(\sigma^{\rho_0^{-1}}) = \phi(\sigma)$, which completes the proof of the proposition.

From this proposition and the Frobenius reciprocity law, we deduce

PROPOSITION 5.15. *The map $\phi \to \phi^G$ is an isometry of $CF(H|S)$ into $CF(G|S^G)$ $\left(\text{where } S^G = \bigcup_{\sigma \in G} S^\sigma\right)$, i.e.*

$$(\phi^G, \xi^G)_G = (\phi, \xi)_H, \quad \text{for all } \phi, \xi \in CF(H|S). \tag{5.16}$$

*Proof.* By (5.6) and Proposition 5.14 the map $\phi \to \phi^G$ sends $CF(H|S)$ linearly into $CF(G|S^G)$. So we only need to prove (5.16). The Frobenius reciprocity law gives

$$(\phi^G, \xi^G)_G = ((\phi^G)_H, \xi)_H = \frac{1}{|H|} \sum_{\sigma \in H} \phi^G(\sigma)\xi(\sigma^{-1}).$$

Since $\xi \in CF(H|S)$, the value $\xi(\sigma^{-1})$ is zero unless $\sigma^{-1} \in S$. By (5.10c) this occurs if and only if $\sigma \in S$. But then $\phi^G(\sigma) = \phi(\sigma)$ by Proposition 5.14. So we have

$$(\phi^G, \xi^G)_G = \frac{1}{|H|} \sum_{\sigma \in S} \phi^G(\sigma)\xi(\sigma^{-1}) = \frac{1}{|H|} \sum_{\sigma \in S} \phi(\sigma)\xi(\sigma^{-1})$$

$$= \frac{1}{|H|} \sum_{\sigma \in H} \phi(\sigma)\xi(\sigma^{-1}) = (\sigma, \xi)_H,$$

which proves the proposition.

Obviously the identity $1_G$ can lie in $S$ only if $H = G$, which is not a very interesting case. So (2.3) implies that the only character in $CF(H|S)$ is 0, whenever $H < G$. Nevertheless, one can very well have generalized characters of $H$ in $CF(H|S)$. A *generalized character* $\chi$ of the group $G$ is a class function of the form

$$\chi = n_1\chi_1 + \ldots + n_k\chi_k, \quad (n_1, \ldots, n_k \in \mathbf{Z}) \tag{5.17}$$

where, as usual, $\chi_1, \ldots, \chi_k$ are the irreducible characters of $G$. By (4.10) the generalized character $\chi$ is an actual character if and only if each of the coefficients $n_1, \ldots, n_k$ is non-negative. Clearly the generalized characters form the additive subgroup $X(G)$ of $CF(G)$ generated by the characters. From this description and Proposition 5.8 we immediately obtain

*Induction $\phi \to \phi^G$ is an additive homorphism of $X(H)$ into $X(G)$.* (5.18)

We denote by $X(H|S)$ the intersection $X(H) \cap CF(H|S)$, consisting of all generalized characters $\phi$ of $H$ vanishing on $H \backslash S$. Then (5.18) and Proposition 5.15 give

*Induction $\phi \to \phi^G$ is an additive isometry of the subgroup $X(H|S)$ into $X(G|S^G)$.* (5.19)

To obtain the full force of trivial intersection sets, we use them in conjunction with involutions of the group $G$. An *involution* is just an element of order 2. Their most important property is

Proposition 5.20. *If $\iota$, $\iota'$ are involutions in G, then $\sigma = \iota\iota'$ is inverted by both $\iota$ and $\iota'$:*

$$\sigma^{\iota} = \sigma^{\iota'} = \sigma^{-1}. \tag{5.21}$$

*It follows that $\langle \iota, \iota' \rangle = \langle \iota, \sigma \rangle$ is a dihedral subgroup of G, with $\langle \sigma \rangle$ as a normal cyclic subgroup of index* 2.

*Proof.* Since $\iota = \iota^{-1}$ and $\iota' = \iota'^{-1}$, we have:

$$\sigma^{-1} = (\iota')^{-1}\iota^{-1} = \iota'\iota = \iota^{-1}(\iota\iota')\iota = \sigma^{\iota}$$

Similarly we have $\sigma^{-1} = \sigma^{\iota'}$. So (5.21) holds. The rest of the proposition follows easily from this.

COROLLARY 5.22. *If $\sigma \in S$, then $\iota, \iota' \in H$.*

*Proof.* By (5.21) and (5.10c), the intersection $S^{\iota} \cap S$ contains $\sigma^{\iota} = \sigma^{-1}$. So $\iota \in H$ by (5.10b). Similarly $\iota' \in H$.

As a result of the above corollary, we obtain the very useful

PROPOSITION 5.23. *Let $I$, $I'$ be conjugacy classes of involutions in $G$. Denote by $\tilde{I}$, $\tilde{I}'$, $\widetilde{(I \cap H)}$, $\widetilde{(I' \cap H)}$ the sums in $FG$ of the elements of the corresponding sets $I$, $I'$, $I \cap H$, $I' \cap H$, respectively. If $\phi \in CF(H|S)$, then we have*

$$\phi^G(\widetilde{\tilde{I}I'}) = [G : H]\phi(\widetilde{(I \cap H)}\widetilde{(I' \cap H)}). \tag{5.24}$$

*Proof.* Let $\widetilde{\tilde{I}I'} = \sum_{\sigma \in G} f_\sigma \sigma$, for non-negative integers $f_\sigma$. If $\sigma \notin S^G$, then $\phi^G(\sigma) = 0$ by Proposition 5.14. Hence,

$$\phi^G(\widetilde{\tilde{I}I'}) = \sum_{\sigma \in S^G} f_\sigma \phi^G(\sigma).$$

The conditions (5.10) say that $S^G$ is the disjoint union of its conjugate subsets $S^\rho$, for $\rho \in \text{rep}\,(G|H)$. Since $\widetilde{\tilde{I}I'} \in Z(FG)$, the constant $f_{\sigma^\rho}$ equals $f_\sigma$, for all $\sigma \in S$, $\rho \in \text{rep}\,(G|H)$. Applying Proposition 5.14 again, we get

$$\begin{aligned}\phi^G(\widetilde{\tilde{I}I'}) &= \sum_{\rho \in \text{rep}\,(G/H)} f_\sigma \phi^G(\sigma^\rho) \\ &= [G : H] \sum_{\sigma \in S} f_\sigma \phi(\sigma).\end{aligned}$$

If $\sigma \in S$, the coefficient $f_\sigma$ of $\sigma$ in $\widetilde{\tilde{I}I'}$ is just the number of ordered pairs of involutions $(\iota, \iota')$ such that $\iota \in I$, $\iota' \in I'$ and $\iota\iota' = \sigma$. In view of Corollary 5.22, this is just the number $e_\sigma$ of ordered pairs $(\iota, \iota')$, where $\iota \in I \cap H$, $\iota' \in I' \cap H$ and $\iota\iota' = \sigma$. But $e_\sigma$ in turn is the coefficient of $\sigma$ in $\widetilde{(I \cap H)}\widetilde{(I' \cap H)} = \sum_{\sigma \in H} e_\sigma \sigma$. Since $\phi(\sigma) = 0$ for $\sigma \in H \backslash S$, the previous equation becomes

$$\begin{aligned}\phi^G(\widetilde{\tilde{I}I'}) &= [G : H] \sum_{\sigma \in S} e_\sigma \phi(\sigma) = [G : H]\phi\Big(\sum_{\sigma \in H} e_\sigma \sigma\Big) \\ &= [G : H]\phi(\widetilde{(I \cap H)}\widetilde{(I' \cap H)}),\end{aligned}$$

and this proves the proposition.

## 6. Generalized Quaternion Sylow Groups

We apply the ideas of the last section to study the finite groups having generalized quaternion groups as 2-Sylow subgroups.

To construct a generalized quaternion group, we start with a cyclic group

$\langle\tau\rangle$ of order $2^n$, where $n \geq 2$. The automorphism $\sigma \to \sigma^{-1}$ of $\langle\tau\rangle$ then has order 2 and leaves fixed the involution $\iota = \tau^{2^{n-1}}$. It follows that there is a unique group $Q$ containing $\langle\tau\rangle$ as a normal subgroup of index 2, in which some element $\rho \in Q\backslash\langle\tau\rangle$ satisfies

$$\tau^\rho = \tau^{-1}, \qquad \rho^2 = \iota = \tau^{2^{n-1}}. \tag{6.1}$$

This group $Q$ is called the generalized quaternion group of order $2^{n+1}$. If $n = 2$, it is, of course, the ordinary quaternion group.

PROPOSITION 6.2. *Each element $\rho \in Q\backslash\langle\tau\rangle$ satisfies* (6.1). *Hence $\iota$ is the only involution in $Q$.*

*Proof.* Fix an element $\rho_0 \in Q\backslash\langle\tau\rangle$ satisfying (6.1). Since $[Q : \langle\tau\rangle] = 2$, any element $\rho \in Q\backslash\langle\tau\rangle$ has the form $\tau^i\rho_0$, for some integer $i$. From (6.1) for $\rho_0$ we compute that

$$\tau^\rho = \tau^{\tau^i\rho_0} = \tau^{\rho_0} = \tau^{-1},$$

$$\rho^2 = \tau^i\rho_0\tau^i\rho_0 = \tau^i\rho_0^2(\rho_0^{-1}\tau^i\rho_0) = \tau^i\iota\tau^{-i} = \iota.$$

Therefore $\rho$ also satisfies (6.1). In particular, $\rho$ has order 4. So any involution in $Q$ must lie in $\langle\tau\rangle$ and therefore equal $\iota$, which finishes the proof.

We must compute some irreducible characters of $Q$ in a field $F$ of characteristic zero satisfying (4.1) for the group $Q$. Certainly we have the trivial character $1 = 1_{Q\to F}$. Since $Q/\langle\tau\rangle$ is cyclic of order 2, we have a linear character $\lambda$ satisfying

$$\chi = \begin{cases} -1 & \textit{on } Q\backslash\langle\tau\rangle, \\ 1 & \textit{on } \langle\tau\rangle. \end{cases} \tag{6.3}$$

Because $n \geq 2$, there is a linear character $\mu$ of $\langle\tau\rangle$ sending $\tau$ into a primitive fourth root of unity $\sqrt{-1}$. We can find such a root in $F$ by (4.1). So we have two linear characters $\mu$, $\mu^{-1}$ of $\langle\tau\rangle$ given by

$$\mu(\tau^i) = (\sqrt{-1})^i, \qquad \mu^{-1}(\tau^i) = (\sqrt{-1})^{-i} = (-\sqrt{-1})^i, \quad \textit{for all integers } i. \tag{6.4}$$

By (5.5) and (6.1), the induced character $\theta = \mu^Q$ satisfies

$$\theta = \begin{cases} 0 & \textit{on } Q\backslash\langle\tau\rangle, \\ \mu+\mu^{-1} & \textit{on } \langle\tau\rangle. \end{cases} \tag{6.5}$$

Since $\mu$ and $\mu^{-1}$ are distinct irreducible characters of $\langle\tau\rangle$, the Frobenius reciprocity law implies that $(\theta, \theta)_Q = (\theta_{\langle\tau\rangle}, \mu)_{\langle\tau\rangle} = (\mu+\mu^{-1}, \mu)_{\langle\tau\rangle} = 1$. But the character $\theta$ is a linear contribution of the irreducible characters of $Q$ with non-negative integral coefficients. From this and (4.20) we conclude that $\theta$ is an irreducible character of $Q$. Since $\theta(1) = 2 \neq 1 = 1(1) = \lambda(1)$, we have

$$1, \lambda, \theta \textit{ are distinct irreducible characters of } Q. \tag{6.6}$$

From (6.3), (6.4) and (6.5), we compute:

$$1+\lambda-\theta = \begin{cases} 0 & on\ Q\backslash\langle\tau\rangle, \\ 2 & on\ \langle\tau\rangle\backslash\langle\tau^2\rangle, \\ 4 & on\ \langle\tau^2\rangle\backslash\langle\tau^4\rangle, \\ 0 & on\ \langle\tau^4\rangle. \end{cases} \tag{6.7}$$

With this we are ready to prove

THEOREM 6.8 (Brauer–Suzuki). *If $n \geq 3$ and the above generalized quaternion group $Q$ of order $2^{n+1}$ is a 2-Sylow subgroup of a finite group $G$, then $G$ has a proper normal subgroup $N$ containing the involution $\iota$.*

*Proof.* Let $\sigma = \tau^2$. Proposition 5.12 tells us that $\sqrt{\sigma}$ is a t.i. set with $H = N_G(\langle\tau\rangle)$ as its normalizer. We first prove

LEMMA 6.9. *The group $H$ is the semi-direct product $QK$ of $Q$ with a normal subgroup $K$ of odd order centralizing $\sigma$.*

*Proof.* Evidently the characteristic subgroup $\langle\sigma\rangle$ of $\langle\tau\rangle$ is itself normal in $Q$. So $Q \leq H$. Since $Q$ is a 2-Sylow subgroup of $G$, it must also be one of $H$.

Because $n \geq 3$, the element $\sigma$ of order $2^{n-1}$ is not its own inverse. It follows from this and Proposition 6.2 that $Q \cap C_G(\sigma) = \langle\tau\rangle$. Since $C_G(\sigma)$ is normal in $N_G(\langle\sigma\rangle) = H$, this intersection is a 2-Sylow subgroup of $C_G(\sigma)$. It is well known and elementary (see Section I.6 of Huppert, 1967) that the finite group $C_G(\sigma)$ with a cyclic 2-Sylow group $\langle\tau\rangle$ has a unique normal 2-complement $K$, i.e., a normal subgroup $K$ of odd order such that $C_G(\sigma) = \langle\tau\rangle K$.

The factor group $H/C_G(\sigma) = N_G(\langle\sigma\rangle)/C_G(\sigma)$ is isomorphic to a subgroup of the automorphism group of $\langle\sigma\rangle$, and hence is a 2-group. It follows that $[H : K] = [H : C_G(\sigma)]\,[C_G(\sigma) : K]$ is a power of 2. Since $K$ is characteristic in $C_G(\sigma)$, it is normal in $H$. Therefore it is a normal 2-complement in $H$, and the lemma is proved.

Next we must compute the t.i. set $\sqrt{\sigma} \subseteq H$.

LEMMA 6.10. $\sqrt{\sigma} = \langle\tau\rangle K\backslash\langle\tau^4\rangle K$.

*Proof.* Suppose that $\rho\pi \in \sqrt{\sigma}$, for $\rho \in Q$, $\pi \in K$. Passing to the factor group $H/K = QK/K \simeq Q$, we see that $\sigma \in \langle\rho\rangle$. Since $n \geq 3$, the element $\sigma$ is not in $\langle\iota\rangle$. So Proposition 5.2 implies that $\rho \in \langle\tau\rangle$. Obviously $\rho \notin \langle\tau^4\rangle$ since $\sigma = \tau^2 \notin \langle\tau^4\rangle$. Hence $\sqrt{\sigma} \subseteq \langle\tau\rangle K\backslash\langle\tau^4\rangle K$.

Suppose that $\rho \in \langle\tau\rangle\backslash\langle\tau^4\rangle$ and $\pi \in K$. Then either $\langle\rho\rangle = \langle\tau^2\rangle = \langle\sigma\rangle$ or $\langle\rho\rangle = \langle\tau\rangle$. In the first case $\pi$ is an element of odd order commuting with the element $\rho$ of order $2^{n-1}$, by Lemma 6.9. So $\langle\rho\pi\rangle = \langle\rho\rangle\times\langle\pi\rangle \geq \langle\rho\rangle = \langle\sigma\rangle$. In the second case $(\rho\pi)^2 \in \langle\tau^2\rangle K\backslash\langle\tau^4\rangle K$. So $\sigma \in \langle(\rho\pi)^2\rangle \leq \langle\rho\pi\rangle$, by the first case. Hence $\rho\pi \in \sqrt{\sigma}$ in both cases, which completes the proof of the lemma.

Let $\phi$ be the natural epimorphism of $H = QK$ onto $Q$. Then (6.6) implies that the compositions $1\circ\phi$, $\lambda\circ\phi$, $\theta\circ\phi$ are distinct irreducible characters of $H$. From (6.7) and Lemma 6.10 we see that $1\circ\phi+\lambda\circ\phi-\theta\circ\phi = (1+\lambda-\theta)\circ\phi$ is a generalized character of $H$ vanishing outside $\sqrt{\sigma}$, i.e. a member of $X(H|\sqrt{\sigma})$. Since $\sqrt{\sigma}$ is a t.i. set with $H$ as its normalizer, we can apply (5.19) and deduce that $(1\circ\phi+\lambda\circ\phi-\theta\circ\phi)^G \in X(G|(\sqrt{\sigma})^G)$. In view of (5.17) and Proposition 5.9 we have

$$(1\circ\phi+\lambda\circ\phi-\theta\circ\phi)^G = 1+n_2\chi_2+\ldots+n_k\chi_k,$$

where 1 is the trivial character, and $\chi_2,\ldots, \chi_k$ are the other irreducible characters of $G$, and where $n_2,\ldots, n_k$ are arbitrary integers. Using (4.20) and (5.16), we get

$$\begin{aligned}1^2+n_2^2+\ldots+n_k^2 &= ((1\circ\phi+\lambda\circ\phi-\theta\circ\phi)^G,\ \ (1\circ\phi+\lambda\circ\phi-\theta\circ\phi)^G)_G\\ &= (1\circ\phi+\lambda\circ\phi-\theta\circ\phi,\ \ 1\circ\phi+\lambda\circ\phi-\theta\circ\phi)_H\\ &= 3.\end{aligned}$$

Hence exactly two of the $n_i$ are non-zero, and they are both $\pm 1$. Therefore, we have

*There exist distinct non-trivial irreducible characters* $\Lambda$, $\Theta$ *of* $G$ *and integers* $\varepsilon_1$, $\varepsilon_2 = 1$ *such that*: $(1\circ\phi+\lambda\circ\phi-\theta\circ\phi)^G = 1+\varepsilon_1\Lambda-\varepsilon_2\Theta$. (6.11)

Since $\langle\iota\rangle \leq \langle\tau^4\rangle$, it follows from Lemma 6.10 and Proposition 6.2 that $\sqrt{\sigma}$ contains no involution. Hence $(\sqrt{\sigma})^G$ contains neither $\iota$ nor $1_G$. Since $(1\circ\phi+\lambda\circ\phi-\theta\circ\phi)^G$ is zero outside $(\sqrt{\sigma})^G$, this and (6.11) imply that

$$1+\varepsilon_1\Lambda(1)-\varepsilon_2\Theta(1) = 0, \qquad 1+\varepsilon_1\Lambda(\iota)-\varepsilon_2\Theta(\iota) = 0. \tag{6.12}$$

Let $I$ be the conjugacy class of $\iota$ in $G$. From Proposition 6.2 and Lemma 6.9, the intersection $I \cap H$ is contained in $\iota K$. So $(I\cap H)^2 \subseteq (\iota K)^2 = K$. It follows from this and Lemma 6.10 that $(I\cap H)^2 \cap \sqrt{\sigma}$ is empty. Since $1\circ\phi+\lambda\circ\phi-\theta\circ\phi$ is zero on $H\backslash\sqrt{\sigma}$, we conclude that (in the notation of Proposition 5.23)

$$(1\circ\phi+\lambda\circ\phi-\theta\circ\lambda)(\widetilde{(I\cap H)^2}) = 0.$$

Applying (5.24) and (6.11), we get

$$\begin{aligned}0 &= [G : H](1\circ\phi+\lambda\circ\phi-\theta\circ\phi)(\widetilde{(I\cap H)^2})\\ &= (1+\varepsilon_1\Lambda-\varepsilon_2\Theta)(\tilde{I}^2).\end{aligned}$$

In view of (4.31) and Proposition 2.2, this gives

$$0 = \frac{1(\tilde{I})^2}{1} + \varepsilon_1 \frac{\Lambda(\tilde{I})^2}{\Lambda(1)} - \varepsilon_2 \frac{\Theta(\tilde{I})^2}{\Theta(1)}$$
$$= |I|^2 \left(1 + \frac{(\varepsilon_1 \Lambda(\iota))^2}{\varepsilon_1 \Lambda(1)} - \frac{(\varepsilon_2 \Theta(\iota))^2}{\varepsilon_2 \Theta(1)}\right).$$

We can remove the non-zero factor $|I|^2$. Substituting the values of $\varepsilon_2 \Theta(\iota)$, $\varepsilon_2 \Theta(1)$ obtained from (6.12), we have:

$$0 = 1 + \frac{(\varepsilon_1 \Lambda(\iota))^2}{\varepsilon_1 \Lambda(1)} - \frac{(1+\varepsilon_1 \Lambda(\iota))^2}{1+\varepsilon_1 \Lambda(1)}$$
$$= \frac{(\varepsilon_1 \Lambda(\iota) - \varepsilon_1 \Lambda(1))^2}{\varepsilon_1 \Lambda(1)(1+\varepsilon_1 \Lambda(1))}.$$

Therefore $\Lambda(\iota) = \Lambda(1)$. By Proposition 4.38, the involution $\iota$ lies in $N = \text{Ker}(\Lambda)$, which is a proper normal subgroup of $G$, since $\Lambda \neq 1$. This proves the theorem.

## 7. Brauer's Characterization of Generalized Characters

As usual, let $G$ be a finite group and $F$ be a field of characteristic zero satisfying (4.1). A subgroup $E$ of $G$ is called *Brauer elementary* if it is the direct product of a cyclic group and a $p$-group, for some prime $p$. The following theorem, due to Brauer (but whose present proof is due to Brauer and Tate jointly) is difficult but extremely useful for constructing group characters.

THEOREM 7.1. *A class function $\phi$ on G is a generalized character if and only if its restriction $\phi_E$ to each Brauer elementary subgroup E of G is a generalized character of E.*

*Proof.* The "only if" part is trivial, so we need only prove the "if" part. We shall do it in a series of lemmas. First we pass to a different (but actually equivalent) form of the theorem. Let $\mathscr{E}$ be the family of all Brauer elementary subgroups of $G$. For each $E \in \mathscr{E}$, induction maps the additive group $X(E)$ of all generalized characters of $E$ onto a subgroup $X(E)^G$ of $X(G)$ (by (5.18)).

LEMMA 7.2. *Theorem* 7.1 *is implied by the equality:*

$$X(G) = \sum_{E \in \mathscr{E}} X(E)^G. \tag{7.3}$$

*Proof.* Suppose that (7.3) holds. Let $\phi$ be any class function on $G$ whose restriction $\phi_E$ lies in $X(E)$, for any $E \in \mathscr{E}$. Then the Frobenius reciprocity law implies that $(\phi, \xi^G)_G = (\phi_E, \xi)_E \in \mathbf{Z}$, for all $\xi \in X(E)$ and any $E \in \mathscr{E}$. It follows that $(\phi, \psi)_G \in \mathbf{Z}$, for any $\psi$ in the right side of (7.3), i.e., for any $\psi \in X(G)$. If $\chi_1, \ldots, \chi_k \in X(G)$ are the irreducible characters of $G$, we conclude from this, (4.26), and (5.17) that $\phi = (\phi, \chi_1)\chi_1 + \ldots + (\phi, \chi_k)\chi_k$ is a generalized character of $G$. This proves the lemma.

For the rest of this section we regard characters, generalized characters, and class functions as functions from $G$ to $F$, rather than as linear functions from $FG$ to $F$, i.e., we adopt the viewpoint of Section 2 instead of that of Section 4. It follows from (2.14) that the additive group $X(G)$ is closed under multiplication of functions, and hence is a ring of functions from $G$ to $F$. The identity element of this *character ring* $X(G)$ is clearly the trivial character $1 = 1_{G \to F}$ of $G$.

Let $Y \subseteq X(G)$ denote the right side of (7.3).

LEMMA 7.4. *$Y$ is an ideal in the character ring $X(G)$.*

*Proof.* Clearly $Y$ is an additive subgroup of $X(G)$. So we need only show that it is closed under multiplication by an arbitrary element $\psi \in X(G)$. Clearly it suffices to show that $\psi\xi^G \in X(E)^G \subseteq Y$, for any $E \in \mathscr{E}$ and any $\xi \in X(E)$.

As in (5.4), let $\hat{\xi}$ be the extension of $\xi$ to a function from $G$ to $F$ which is zero outside $E$. Then, clearly, we have

$$\psi\hat{\xi} = \widehat{(\psi_E\xi)}$$

Since the restriction $\psi_E$ lies in the ring $X(E)$, so does the product $\psi_E\xi$. Hence $(\psi_E\xi)^G$ lies in $X(E)^G$. Because $\psi$ is a class function on $G$, it equals $\psi^\sigma$, for all $\sigma \in G$. So the above equality and (5.5) imply that

$$\begin{aligned}(\psi_E\xi)^G &= \sum_{\sigma \in \mathrm{rep}(G/E)} \widehat{(\psi_E\xi)}^\sigma = \sum_{\sigma \in \mathrm{rep}(G/E)} (\psi\hat{\xi})^\sigma \\ &= \sum_{\sigma \in \mathrm{rep}(G/E)} \psi^\sigma\hat{\xi}^\sigma = \psi \sum_{\sigma \in \mathrm{rep}(G/E)} \hat{\xi}^\sigma \\ &= \psi\xi^G.\end{aligned}$$

Therefore $\psi\xi^G \in X(E)^G$, which proves the lemma.

The importance of Lemma 7.4 is that any ideal $Y \neq X(G)$ is contained in a maximal ideal of $X(G)$, and that we can say something about these maximal ideals. Before we do so, however, we must make a "ground ring extension".

It follows from (4.1) that $F$ contains a primitive $|\langle\sigma\rangle|$th root of unity (the value $\lambda(\sigma)$ of a suitable linear character $\lambda$ of $\langle\sigma\rangle$), for each $\sigma \in G$. If $e$ denotes the exponent of $G$, we conclude that $F$ contains a primitive $e$th root of unity $\omega$. Let $\mathfrak{O} = \mathbf{Z}[\omega]$ be the subring of $F$ generated by $\omega$.

Lemma 4.34 implies that $\chi(\sigma) \in \mathfrak{O}$, for all $\chi \in X(G)$, $\sigma \in G$. So $X(G)$ is a ring of functions from $G$ to $\mathfrak{O}$. Therefore, so is the family $\mathfrak{O}X(G)$ of all $\mathfrak{O}$-linear combinations of members of $X(G)$:

*$\mathfrak{O}X(G)$ is a ring of functions from $G$ to $\mathfrak{O}$.* (7.5)

Lemma 7.4 and the $F$-linearity of induction imply that the corresponding family $\mathfrak{O}Y$ satisfies

$$\mathfrak{O}Y = \sum_{E \in \mathscr{E}} [\mathfrak{O}X(E)]^G \quad \textit{is an ideal of } \mathfrak{O}X(G). \tag{7.6}$$

We must show that nothing is lost by passing from $\mathbf{Z}$ to $\mathfrak{O}$.

LEMMA 7.7. *If we have*

$$\mathfrak{O}Y = \mathfrak{O}X(G), \tag{7.8}$$

*then* (7.3) *holds.*

*Proof.* By (4.9) and (5.17) the additive group $X(G)$ is a free $\mathbf{Z}$-module with the irreducible characters $\chi_1, \ldots, \chi_k$ of $G$ as a basis. Since $Y \subseteq X(G)$, the elementary divisor theorem gives us a $\mathbf{Z}$-basis $\phi_1, \ldots, \phi_k$ of $X(G)$ and non-negative integers $n_1, \ldots, n_k$ such that $Y = \mathbf{Z}n_1\phi_1 + \ldots + \mathbf{Z}n_k\phi_k$.

Because the $\mathbf{Z}$-basis $\chi_1, \ldots, \chi_k$ of $X(G)$ is $F$-linearly independent (by (4.9)), so is the $\mathbf{Z}$-basis $\phi_1, \ldots, \phi_k$. It follows that $\mathfrak{O}X(G) = \mathfrak{O}\phi_1 \oplus \ldots \oplus \mathfrak{O}\phi_k$ is a free $\mathfrak{O}$-module with $\phi_1, \ldots, \phi_k$ as a basis. If (7.8) holds, then $\mathfrak{O}n_1\phi_1 \oplus \ldots \oplus \mathfrak{O}n_k\phi_k = \mathfrak{O}Y = \mathfrak{O}X(G) = \mathfrak{O}\phi_1 \oplus \ldots \oplus \mathfrak{O}\phi_k$. So,

$$\mathfrak{O}n_i = \mathfrak{O}, \quad (i = 1, \ldots, k).$$

Fix $i = 1, \ldots, k$. The above equation implies that the non-negative integer $n_i$ is positive, and that $\mathfrak{O}$ contains $1/n_i$. So $\mathfrak{O}$ contains the subring $\mathbf{Z}[1/n_i]$ of rational numbers generated by $1/n_i$. Since $\mathfrak{O} = \mathbf{Z}[\omega] = \mathbf{Z}\cdot 1 + \mathbf{Z}\omega + \mathbf{Z}\omega^2 + \ldots + \mathbf{Z}\omega^{e-1}$ is a finitely generated $\mathbf{Z}$-module, so is its submodule $\mathbf{Z}[1/n_i]$. This is only possible if $n_i = 1$

From $n_i = 1$ $(i = 1, \ldots, k)$, we get

$$Y = \mathbf{Z}n_1\phi_1 + \ldots + \mathbf{Z}n_k\phi_k = \mathbf{Z}\phi_1 + \ldots + \mathbf{Z}\phi_k = X(G).$$

This proves the lemma.

Each element $\sigma \in G$ defines an "evaluation" homomorphism $\eta_\sigma$ of the function ring $\mathfrak{O}X(G)$ into $\mathfrak{O}$ given by

$$\eta_\sigma(\psi) = \psi(\sigma), \quad \textit{for all } \psi \in \mathfrak{O}X(G). \tag{7.9}$$

Since $\eta_\sigma(y \cdot 1) = y1(\sigma) = y$, for all $y \in \mathfrak{O}$, we have

*For each $\sigma \in G$, the map $\eta_\sigma$ is an epimorphism of the ring $\mathfrak{O}X(G)$ onto $\mathfrak{O}$.* (7.10)

Now we fix a maximal ideal $M$ in $\mathfrak{O}X(G)$. We shall construct an element of $Y$ not lying in $M$. First we must analyze $M$.

LEMMA 7.11. *There exists an element $\sigma \in G$ and a maximal ideal $P$ of $\mathfrak{O}$ such that $M$ is the inverse image $\eta_\sigma^{-1}(P)$ of $P$ by $\eta_\sigma$.*

*Proof.* The intersection of the kernels $\operatorname{Ker}(\eta_\sigma)$ of the epimorphisms $\eta_\sigma$, $\sigma \in G$, is the set of all functions $\psi \in X(G)$ which vanish at all $\sigma \in G$, i.e., it is

$\{0\}$. Since there are only a finite number of $\sigma \in G$, and the product

$$\prod_{\sigma \in G} \mathrm{Ker}\,(\eta_\sigma) \subseteq \bigcap_{\sigma \in G} \mathrm{Ker}\,(\eta_\sigma) = \{0\}$$

is contained in the maximal ideal $M$ of the ring $\mathfrak{O}X(G)$ with identity, there is some $\sigma \in G$ such that $\mathrm{Ker}\,(\eta_\sigma) \subseteq M$. Evidently this and (7.10) force $M$ to be the inverse image $\eta_\sigma^{-1}(P)$ of a maximal ideal $P$ of $\mathfrak{O}$. So the lemma is true.

We fix $\sigma$ and $P$ satisfying the conditions of Lemma 7.11. Since $P$ is a maximal ideal of $\mathfrak{O}$, the quotient ring $\mathfrak{O}/P$ is a field.

LEMMA 7.12. *The field $\mathfrak{O}/P$ has prime characteristic $p$.*

*Proof.* Suppose not. Then it has characteristic zero. But its additive group is a finitely-generalized **Z**-module, since $\mathfrak{O}$ is. This is impossible, since it contains the additive group of the rationals, which is not finitely-generated. So the lemma holds.

We write the exponent $e$ as a product $e = p^n f$, where $n \geq 0$ and $p$ does not divide the positive integer $f$. Then there exist integers $a$, $b$ such that:

$$ap^n + bf = 1. \tag{7.13}$$

It follows that $\sigma = \sigma^{ap^n}\sigma^{bf}$, where the order of $\sigma^{ap^n}$ divides $f$ and that of $\sigma^{bf}$ is a power of $p$.

LEMMA 7.14. $\psi(\sigma) \equiv \psi(\sigma^{ap^n}) \pmod P$, *for all* $\psi \in \mathfrak{O}X(G)$.

*Proof.* Since both sides of this congruence are $\mathfrak{O}$-linear in $\psi$, it suffices to prove it when $\psi$ is an irreducible character of $G$. Let $I$ be a corresponding irreducible $FG$-module. By Lemma 4.34 there are a basis $y_1, \ldots, y_t$ for $I$ over $F$ and $e$th roots of unity $\zeta_1, \ldots, \zeta_t$ such that $y_i\sigma = \zeta_i y_i$, for all $i = 1, \ldots, t$. This implies that $y_i\sigma^{ap^n} = \zeta_t^{ap^n} y_i$ $(i = 1, \ldots, t)$. Hence,

$$\begin{aligned}\psi(\sigma) - \psi(\sigma^{ap^n}) &= (\zeta_1 + \ldots + \zeta_t) - (\zeta_1^{ap^n} + \ldots + \zeta_t^{ap^n}) \\ &= \zeta_1^{ap^n}(\zeta_1^{bf} - 1) + \ldots + \zeta_t^{ap^n}(\zeta_t^{bf} - 1)\end{aligned}$$

by (7.13). So it suffices to prove that $\zeta_i^{bf} - 1 \in P$, $(i = 1, \ldots, t)$.

Since $\zeta_i$ is an $e$th root of unity, its power $\delta = \zeta_i^{bf}$ is a $p^n$-th root of unity. The image $\bar{\delta}$ of $\delta$ in $\mathfrak{O}/P$ is also a $p^n$-th root of unity. But 1 is the only $p^n$-th root of unity in the field $\mathfrak{O}/P$ of characteristic $p$. Hence $\bar{\delta} = 1$, and $\zeta_i^{bf} - 1 = \delta - 1 \in P$, which finishes the proof of the lemma.

COROLLARY 7.15. *We can assume that the order of $\sigma$ is relatively prime to $p$.*

*Proof.* If not, let $\tau = \sigma^{ap^n}$. The lemma implies that $M = \eta_\sigma^{-1}(P) = \eta_\tau^{-1}(P)$. So we can replace $\sigma$ by $\tau$, whose order divides $f$ and hence is relatively prime to $p$.

Of course, from now on we do assume that $|\langle\sigma\rangle|$ is relatively prime to $p$. Let $S$ be a $p$-Sylow subgroup of $C_G(\sigma)$. Then $E = \langle\sigma\rangle \times S$ is a Brauer elementary subgroup of $G$. Since $\langle\sigma\rangle$ is abelian, each of its irreducible characters $\lambda$ is linear. Using the projection of $E$ on $\langle\sigma\rangle$ (or (2.12)), we obtain a corresponding linear character $\lambda \times 1$ of $E$ satisfying

$$\lambda \times 1(\rho\tau) = \lambda(\rho), \quad \textit{for all } \rho \in \langle\sigma\rangle,\ \tau \in S.$$

Let $\lambda_1, \ldots, \lambda_l$ ($l = |\langle\sigma\rangle|$) be the irreducible characters of $\langle\sigma\rangle$. Then

$$\Lambda = \sum_{i=1}^{l} \lambda_i(\sigma^{-1})(\lambda_i \times 1) \in \mathfrak{O}X(E).$$

Using (4.33) we compute:

$$\Lambda(\rho\tau) = \begin{cases} 0, & \textit{if } \rho \neq \sigma, \\ |\langle\sigma\rangle|, & \textit{if } \rho = \sigma, \end{cases} \quad (\rho \in \langle\sigma\rangle,\ \tau \in S). \tag{7.16}$$

By (7.6) the induced character $\Lambda^G$ lies in $\mathfrak{O}Y$. It satisfies

LEMMA 7.17. $\Lambda^G(\sigma) = [C_G(\sigma) : S] \not\equiv 0 \pmod{P}$.

*Proof.* As in (5.4), we extend $\Lambda$ to a function $\hat{\Lambda}$ from $G$ to $F$ which is zero outside $E$.

Suppose that $\hat{\Lambda}^\tau(\sigma) \neq 0$, for some $\tau \in G$. Then $\sigma^{\tau^{-1}} \in E$ and $\Lambda(\sigma^{\tau^{-1}}) \neq 0$. The order of $\sigma^{\tau^{-1}}$ equals that of $\sigma$. Evidently $\langle\sigma\rangle$ is the set of all elements of $E = \langle\sigma\rangle \times S$ whose orders are not divisible by $p$. Hence $\sigma^{\tau^{-1}} \in \langle\sigma\rangle$. But then $\Lambda(\sigma^{\tau^{-1}}) \neq 0$ implies $\sigma^{\tau^{-1}} = \sigma$, by (7.16). Therefore $\tau \in C_G(\sigma)$.

On the other hand, if $\tau \in C_G(\sigma)$, then (7.16) gives $\hat{\Lambda}^\tau(\sigma) = \hat{\Lambda}(\sigma^{\tau^{-1}}) = \Lambda(\sigma) = |\langle\sigma\rangle| \neq 0$. It follows from this and (5.5) that

$$\Lambda^G(\sigma) = \sum_{\tau \in \mathrm{rep}\,(G/E)} \hat{\Lambda}^\tau(\sigma) = \sum_{\tau \in \mathrm{rep}\,(C_G(\sigma)/E)} |\langle\sigma\rangle| = [C_G(\sigma) : \langle\sigma\rangle \times S]\,|\langle\sigma\rangle|$$
$$= [C_G(\sigma) : S].$$

This is not divisible by $p$, since $S$ is a $p$-Sylow subgroup of $C_G(\sigma)$. So it does not lie in $P$ by Lemma 7.12. This completes the proof of the lemma.

Now we can finish the proof of the theorem. By Lemmas 7.2 and 7.7, we need only show that (7.8) holds. If that is false then $\mathfrak{O}Y$ is an ideal properly contained in $\mathfrak{O}X(G)$, by (7.6). So we can choose our maximal ideal $M$ to contain $\mathfrak{O}Y$. But the above character $\Lambda^G$ lies in $\mathfrak{O}Y$ and not in $M = \eta_\sigma^{-1}(P)$, by (7.9) and Lemmas 7.11 and 7.17. The contradiction proves the theorem.

## 8. *p*-adic Algebras

Fix a prime $p$ in the ring $\mathbf{Z}$ of ordinary integers. From the descending chain $p\mathbf{Z} \supset p^2\mathbf{Z} \supset p^3\mathbf{Z} \supset \ldots$ of ideals of $\mathbf{Z}$ we obtain an infinite chain of natural ring epimorphisms

$$\ldots \to \mathbf{Z}/p^3\mathbf{Z} \to \mathbf{Z}/p^2\mathbf{Z} \to \mathbf{Z}/p\mathbf{Z}. \tag{8.1}$$

The corresponding inverse limit ring $\mathbf{Z}_p = \varprojlim_{n\to\infty} \mathbf{Z}/p^n\mathbf{Z}$ is called the *ring of p-adic integers*.

From its definition $\mathbf{Z}_p$ is provided with a family of natural ring epimorphisms $\mathbf{Z}_p \to \mathbf{Z}/p^n\mathbf{Z}$, for all $n > 0$, commuting with the epimorphisms in (8.1). If $m > n$, then the kernel of the epimorphism $\mathbf{Z}/p^m\mathbf{Z} \to \mathbf{Z}/p^n\mathbf{Z}$ in (8.1) is $p^n\mathbf{Z}/p^m\mathbf{Z}$, which is $p^n(\mathbf{Z}/p^m\mathbf{Z})$ as an additive subgroup of $\mathbf{Z}/p^m\mathbf{Z}$. It follows that the kernel of $\mathbf{Z}_p \to \mathbf{Z}/p^n\mathbf{Z}$ is $p^n\mathbf{Z}_p = \varprojlim_{m\to\infty} p^n(\mathbf{Z}/p^m\mathbf{Z})$. So we have a natural identification of rings:

$$\mathbf{Z}_p/p^n\mathbf{Z}_p = \mathbf{Z}/p^n\mathbf{Z}, \quad \textit{for all } n > 0. \tag{8.2}$$

Under this identification, the chain (8.1) becomes the corresponding chain for $\mathbf{Z}_p$:

$$\ldots \to \mathbf{Z}_p/p^3\mathbf{Z}_p \to \mathbf{Z}_p/p^2\mathbf{Z}_p \to \mathbf{Z}_p/p\mathbf{Z}_p.$$

We conclude that the natural homomorphism of $\mathbf{Z}_p$ into $\varprojlim_{n\to\infty} \mathbf{Z}_p/p^n\mathbf{Z}_p$ is an identity of rings:

$$\mathbf{Z}_p = \varprojlim_{n\to\infty} \mathbf{Z}_p/p^n\mathbf{Z}_p. \tag{8.3}$$

The rings $\mathbf{Z}/p^n\mathbf{Z}$ have identities $1_{\mathbf{Z}/p^n\mathbf{Z}}$ which map onto each other in (8.1). It follows that their inverse $1_{\mathbf{Z}_p} = \varprojlim_{n\to\infty} 1_{\mathbf{Z}/p^n\mathbf{Z}}$ is the identity for $\mathbf{Z}_p$. The unit group $U(\mathbf{Z}_p)$ is easily seen to be the inverse limit

$$U(\mathbf{Z}_p) = \varprojlim_{n\to\infty} U(\mathbf{Z}/p^n\mathbf{Z}),$$

of the unit groups $U(\mathbf{Z}/p^n\mathbf{Z})$ of the rings $\mathbf{Z}/p^n\mathbf{Z}$. But $U(\mathbf{Z}/p^n\mathbf{Z}) = (\mathbf{Z}/p^n\mathbf{Z})\backslash(p\mathbf{Z}/p^n\mathbf{Z})$, for all $n > 0$. Hence, we have

$$U(\mathbf{Z}_p) = \varprojlim_{n\to\infty} ((\mathbf{Z}/p^n\mathbf{Z})\backslash p(\mathbf{Z}/p^n\mathbf{Z})) = \mathbf{Z}_p\backslash p\mathbf{Z}_p. \tag{8.4}$$

From the above information we can deduce the entire ideal structure of $\mathbf{Z}_p$.

PROPOSITION 8.5. *The distinct ideals of* $\mathbf{Z}_p$ *are* $\{0\}$ *and* $p^n\mathbf{Z}_p$ ($n = 0, 1, 2, 3,\ldots$).

*Proof.* By (8.2) we have $\mathbf{Z}_p = p^0\mathbf{Z}_p \supset p\mathbf{Z}_p \supset p^2\mathbf{Z}_p \supset \ldots$ So the ideals $p^n\mathbf{Z}_p$ ($n = 0, 1, 2,\ldots$), are distinct from each other and from $\{0\}$.

Let $I$ be any non-zero ideal of $\mathbf{Z}_p$. By (8.3) the intersection $\bigcap_{n=0}^{\infty} p^n\mathbf{Z}_p = \{0\}$. So $I$ is not contained in every ideal $p^n\mathbf{Z}_p$. On the other hand, $I \subseteq p^0\mathbf{Z}_p = \mathbf{Z}_p$. Hence there exists an integer $n \geq 0$ such that $I \subseteq p^n\mathbf{Z}_p$ but $I \nsubseteq p^{n+1}\mathbf{Z}_p$. Let $y$ be any element of $I$ not in $p^{n+1}\mathbf{Z}_p$. Since $y \in p^n\mathbf{Z}_p$, there is an element $\mu \in \mathbf{Z}_p$ such that $y = p^n\mu$. Since $y \notin p^{n+1}\mathbf{Z}_p$, the element $\mu$ does not lie in $p\mathbf{Z}_p$. By (8.4) $\mu$ is a unit of $\mathbf{Z}_p$. Therefore $I \supseteq y\mathbf{Z}_p = p^n\mu\mathbf{Z}_p = p^n\mathbf{Z}_p$. So $I = p^n\mathbf{Z}_p$, and the proposition is proved.

COROLLARY 8.6. *The ring* $\mathbf{Z}_p$ *is a principal ideal domain of characteristic zero, and* $p\mathbf{Z}_p$ *is its only maximal ideal.*

*Proof.* By the proposition, $\mathbf{Z}_p$ is a commutative ring with identity in which every ideal is principal. The product $p^n\mathbf{Z}_p p^m\mathbf{Z}_p = p^{n+m}\mathbf{Z}_p$ of any two non-zero ideals $p^n\mathbf{Z}_p$, $p^m\mathbf{Z}_p$ of $\mathbf{Z}_p$ is again non-zero. It follows that the product of any two non-zero elements of $\mathbf{Z}_p$ is non-zero. So $\mathbf{Z}_p$ is a principal ideal domain.

The characteristic of the domain $\mathbf{Z}_p$ is either zero or a prime. But $\mathbf{Z}_p$ has a quotient ring $\mathbf{Z}_p/p^2\mathbf{Z}_p$ of characteristic $p^2$, by (8.2). So its characteristic cannot be a prime. Hence it is zero.

The final conclusion of the corollary, that $p\mathbf{Z}_p$ is the only maximal ideal of $\mathbf{Z}_p$, comes directly from the proposition.

We should note that the field of fractions $Q_p$ of the integral domain $\mathbf{Z}_p$ is called the *p-adic number field.*

A *p-adic module M* will be a finitely-generated unitary module over $\mathbf{Z}_p$. Since $\mathbf{Z}_p$ is a principal ideal domain, we can apply the structure theory of finitely-generated modules over such domains (see Section I.13 of Huppert, 1967). In view of Proposition 8.5, we obtain

$$M \simeq \overbrace{\mathbf{Z}_p \oplus \ldots \oplus \mathbf{Z}_p}^{r \text{ terms}} \oplus (\mathbf{Z}_p/p^{n_1}\mathbf{Z}_p) \oplus \ldots \oplus (\mathbf{Z}_p/p^{n_s}\mathbf{Z}_p) \quad \text{(as } \mathbf{Z}_p\text{-modules)}, \tag{8.7}$$

for some unique integers $r, s \geq 0$ and $n_1, \ldots, n_s \geq 1$. This has two important consequences.

In the first place, the factor module $M/pM$ is clearly the direct sum of $r+s$ copies of $\mathbf{Z}_p/p\mathbf{Z}_p$, which is isomorphic to $\mathbf{Z}/p\mathbf{Z}$ by (8.2). Regarding $M/pM$ as a vector space over the field $\mathbf{Z}/p\mathbf{Z}$, we obtain

$$\dim_{\mathbf{Z}/p\mathbf{Z}} (M/pM) = r+s, \textit{ for all p-adic modules } M. \textit{ Furthermore, this dimension is zero if and only if } M = \{0\}. \tag{8.8}$$

The second immediate consequence of (8.7) and (8.3) is that

$$M = \varprojlim_{n\to\infty} M/p^nM, \quad \textit{for all p-adic modules } M, \tag{8.9}$$

in the usual sense that the natural map of the left side into the right is a module isomorphism.

By a *p-adic algebra* we understand a $p$-adic module $A$ together with a $\mathbf{Z}_p$-bilinear, associative product $(a, a') \to aa'$ from $A \times A$ to $A$. Unless otherwise noted we assume that $A$ has an identity $1 = 1_A$ for this multiplication. So $A$ is a ring with identity, and $z \to z1_A$ is an identity-preserving ring homomorphism of $\mathbf{Z}_p$ into the center of $A$.

Obviously the additive subgroup $pA$ is a two-sided ideal of $A$. By (8.8) the factor ring $A/pA$ is finite-dimensional over $\mathbf{Z}/p\mathbf{Z}$. Hence, we have

$$A/pA \quad \textit{is an algebra over } \mathbf{Z}/p\mathbf{Z}. \tag{8.10}$$

We define the *radical* $J(A)$ to be the inverse image in $A$ of the radical $J(A/pA)$

of the algebra $A/pA$. (In fact, one can easily see that $J(A)$ is the Jacobson radical of $A$). Then $J(A)$ is a two-sided ideal of $A$ containing $pA$ and, by (3.8)

$$A/J(A) \simeq [A/pA]/J(A/pA) \textit{ is a semi-simple algebra over } \mathbf{Z}/p\mathbf{Z}. \tag{8.11}$$

By Satz V.2.4 of Huppert (1967), the radical $J(A/pA)$ is a nilpotent ideal in the algebra $A/pA$. It follows that

$$J(A)^d \subseteq pA \subseteq J(A), \quad \textit{for some integer } d > 0. \tag{8.12}$$

This implies that $J(A)^{dn} \subseteq (pA)^n = p^n A \subseteq J(A)^n$, for all integers $n \geq 0$. Hence we have natural ring epimorphisms

$$\varprojlim_{n\to\infty} A/J(A)^{dn} \to \varprojlim_{n\to\infty} A/p^n A \to \varprojlim_{n\to\infty} A/J(A)^n.$$

But the natural epimorphism of the left ring onto the right is clearly an isomorphism. So they are both isomorphic to the center ring. From this and (8.9) we conclude that

$$A = \varprojlim_{n\to\infty} A/p^n A = \varprojlim_{n\to\infty} A/J(A)^n \tag{8.13}$$

in the usual sense that the natural maps among these objects are isomorphisms.

## 9. The Krull–Schmidt Theorem

Let $M$ be a module over a ring $R$. Consider a decomposition

$$M = M_1 \oplus \ldots \oplus M_m, \tag{9.1}$$

where $M_1, \ldots, M_m$ are $R$-submodules of $M$. The corresponding projections $e_i : M \to M_i$ $(i = 1, \ldots, m)$ all lie in the ring $\mathrm{Hom}_R(M, M)$ of $R$-endomorphisms of $M$, and satisfy

$$e_i^2 = e_i \quad (i = 1, \ldots, m), \tag{9.2a}$$

$$e_i e_j = 0 \quad (i, j = 1, \ldots, m \quad \textit{with } i \neq j), \tag{9.2b}$$

$$1 = e_1 + \ldots + e_m. \tag{9.2c}$$

Furthermore, these projections determine (9.1) since $M_i = Me_i$, for $i = 1, \ldots, m$. (Notice that we regard the $R$-endomorphisms as right operators on $M$.)

On the other hand, if $e_1, \ldots, e_m$ are elements of $\mathrm{Hom}_R(M, M)$ satisfying (9.2), then one easily verifies that $M_1 = Me_1, \ldots, M_m = Me_m$ are $R$-submodules of $M$ satisfying (9.1), and that $e_i$ is the corresponding projection of $M$ onto $M_i$ $(i = 1, \ldots, m)$. So there is a natural one to one correspondence between decompositions (9.1) of the $R$-module $M$ and decompositions (9.2c) of the identity into elements $e_1, \ldots, e_m$ of the ring $\mathrm{Hom}_R(M, M)$ satisfying (9.2a, b).

The $R$-module $M$ is *indecomposable* if it is non-zero and cannot be written as the direct sum of two proper $R$-submodules. The former condition is equivalent to $1 \neq 0$ in the ring $\mathrm{Hom}_R(M, M)$. The latter says that there is no decomposition $1 = e_1 + e_2$ in the ring $\mathrm{Hom}_R(M, M)$ satisfying (9.2a, b) for $m = 2$. Since any idempotent $e \neq 0, 1$ in $\mathrm{Hom}_R(M, M)$ gives such a decomposition $1 = e + (1-e)$, we see that

$$M \text{ is indecomposable if and only if } 1 \text{ is the unique non-zero idempotent in } \mathrm{Hom}_R(M, M). \tag{9.3}$$

In the case when $R$ is a $p$-adic algebra $A$ (with identity), we can give another condition for the indecomposability of certain $A$-modules. By a *module $M$ over the $p$-adic algebra $A$* we understand a finitely-generated unitary right $A$-module. Since $A$ is itself a finitely-generated unitary module over $\mathbf{Z}_p$, so is $M$. Hence $M$ is a $p$-adic module in the sense of Section 8.

PROPOSITION 9.4. *For any module $M$ (in the above sense) over the $p$-adic algebra $A$, the ring $\mathrm{Hom}_A(M, M)$ is naturally a $p$-adic algebra.*

*Proof.* Since all multiplications involving $A$ or $M$ are $\mathbf{Z}_p$-bilinear, the ring $\mathrm{Hom}_A(M, M)$ is naturally a unitary $\mathbf{Z}_p$-module, and its multiplication is $\mathbf{Z}_p$-bilinear. So the only problem is to show that $\mathrm{Hom}_A(M, M)$ is a finitely-generated $\mathbf{Z}_p$-module. But it is a $\mathbf{Z}_p$-submodule of $\mathrm{Hom}_{\mathbf{Z}_p}(M, M)$, which is a finitely-generated $\mathbf{Z}_p$-module since $M$ is and since $\mathbf{Z}_p$ is a principal ideal domain. It follows that $\mathrm{Hom}_A(M, M)$ is finitely-generated over $\mathbf{Z}_p$, which proves the proposition.

In view of (9.3) and the above proposition, we must study the $p$-adic algebras for which 1 is the unique non-zero idempotent. To do so, we use

LEMMA 9.5 (Idempotent refinement lemma). *Let $A$ be a $p$-adic algebra and $f$ be an idempotent in $A/J(A)$. Then there exists an idempotent $e$ in $A$ such that $f = e + J(A)$.*

*Proof.* By (8.13) the ring $A$ is the inverse limit of the family of rings and epimorphisms

$$\ldots \to A/J(A)^3 \to A/J(A)^2 \to A/J(A).$$

We shall construct, by induction, idempotents $e_n \in A/J(A)^n$, for $n \geq 1$, satisfying:

$$\ldots \to e_3 \to e_2 \to e_1 = f.$$

Evidently $e = \overset{\lim}{\longleftarrow} e_n$ will be the desired idempotent of $A$.

We start with $e_1 = f$. Suppose that idempotents $e_1 \in A/J(A)$, $e_2 \in A/J(A)^2$, $\ldots$, $e_n \in A/J(A)^n$ have been constructed so that $e_n \to e_{n-1} \to \ldots \to e_1 = f$. Let $g$ be any element of $A/J(A)^{n+1}$ having $e_n$ as image in $A/J(A)^n$. From $e_n^2 = e_n$ we obtain

$$g^2 = g + y,$$

where $y$ lies in the kernel $Y = J(A)^n/J(A)^{n+1}$ of $A/J(A)^{n+1} \to A/J(A)^n$.

Evidently $y = g^2-g$ commutes with $g$. Furthermore $y^2 \in Y^2 = \{0\}$ (since $n \geq 1$). It follows that $e_{n+1} = g+(1-2g)y$ satisfies

$$\begin{aligned} e_{n+1}^2 &= g^2+2g(1-2g)y+(1-2g)^2y^2 \\ &= g+y+(2g-4g^2)y+0 \\ &= g+(1-2g)y+(4g-4g^2)y \\ &= e_{n+1}-4y^2 = e_{n+1}. \end{aligned}$$

Since $e_{n+1} \equiv g \pmod{Y}$, we have found an idempotent $e_{n+1} \in A/J(A)^{n+1}$ satisfying $e_{n+1} \to e_n$. This completes the inductive construction of the $e_n$ and finishes the proof of the lemma.

It is convenient to know the structure of the unit group of $A$.

LEMMA 9.6. *Let $A$ be a p-adic algebra. An element $u$ is a unit in $A$ if and only if its image $u+J(A)$ is a unit in $A/J(A)$.*

*Proof.* If $u$ is a unit in $A$, then $u+J(A)$ is clearly a unit in $A/J(A)$. Suppose, conversely, that $u+J(A)$ is a unit in $A/J(A)$. Then there exists an element $v \in A$ such that $uv \equiv vu \equiv 1 \pmod{J(A)}$. Let $y = 1-uv \in J(A)$. Then $y^n \in J(A)^n$, for all $n \geq 1$. It follows from (8.13) that the sum $1+y+y^2+y^3+\ldots$ "converges" in $\overset{\lim}{\longleftarrow} A/J(A)^n = A$ to a two-sided inverse to $1-y = uv$. Therefore $u$ has the right inverse $v(1-y)^{-1}$. Similarly $u$ has a left inverse. So $u$ is a unit in $A$ and the lemma holds.

As a result of these lemmas, we have:

PROPOSITION 9.7. *The following properties are equivalent for a p-adic algebra $A$:*

(a) 1 *is the unique non-zero idempotent in $A$,*
(b) 1 *is the unique non-zero idempotent in $A/J(A)$,*
(c) *$A/J(A)$ is a division algebra over $\mathbf{Z}/p\mathbf{Z}$,*
(d) *$A\backslash J(A)$ is the unit group of $A$.*

*Proof.* (a) $\Leftrightarrow$ (b). First notice that 0 is the only idempotent in $J(A)$. Indeed, any such idempotent $e$ satisfies $e = e^n \in J(A)^n$, for all $n \geq 1$, and hence $e \in \bigcap_{n \geq 1} J(A)^n = \{0\}$ by (8.13).

Now suppose that (a) holds. Since $1 \neq 0$ in $A$, the idempotent 1 does not lie in $J(A)$. Hence its image 1 is $\neq 0$ in $A/J(A)$. If $f$ is an idempotent of $A/J(A)$ different from 0 and 1, then Lemma 9.5 gives us an idempotent $e \in A$ having $f$ as its image in $A/J(A)$. Clearly $e \neq 0, 1$ in $A$, which is impossible by (a). Therefore (a) implies (b).

Suppose that (b) holds. Then $1 \neq 0$ in $A/J(A)$, which implies $1 \neq 0$ in $A$. If $e$ is an idempotent of $A$ other than 1 or 0, then so is $1-e$ (since $(1-e)^2 = 1-2e+e^2 = 1-2e+e = 1-e$). Both $e$ and $1-e$ must have the image 1 in $A/J(A)$, by (b) and the first paragraph of this proof. That is impossible since their sum 1 also has the image 1 and $1 \neq 0$ in $A/J(A)$. Therefore (b) implies (a).

(b) $\Leftrightarrow$ (c). Suppose that (b) holds. By (8.11), the ring $A/J(A)$ is a semi-simple algebra over $\mathbf{Z}/p\mathbf{Z}$. Proposition 3.9 gives us simple subalgebras $A_1, \ldots, A_k$ such that $A/J(A) = A_1 \oplus \ldots \oplus A_k$. If $k > 1$, then $1_{A_1}$ is an idempotent of $A$ different from 0 and 1, contradicting (b). If $k = 0$, then $1 = 0$ in $A/J(A)$, also contradicting (b). So $k = 1$ and $A/J(A)$ is a simple algebra over $\mathbf{Z}/p\mathbf{Z}$.

Now Proposition 3.12 tells us that $A/J(A) \simeq \mathrm{Hom}_D(I, I)$, where $D$ is a division algebra over $\mathbf{Z}/p\mathbf{Z}$ and $I$ is a finite-dimensional vector space over $D$. Applying (9.3) with $D$, $I$ in place of $R$, $M$, we see from (b) that $I$ is an indecomposable $D$-module. So $I \simeq D$ is one-dimensional over $D$. It follows easily that $A/J(A) \simeq \mathrm{Hom}_D(D, D)$ is a division algebra anti-isomorphic to $D$. So (b) implies (c).

Suppose that (c) holds. Then $1 \neq 0$ in $A/J(A)$. If $f$ is an idempotent in $A/J(A)$ different from 1 and 0, then $f \neq 0$, $1-f \neq 0$, and $f(1-f) = f-f^2 = f-f = 0$, which is impossible in a division algebra. Hence (c) implies (b).

(c) $\Leftrightarrow$ (d). Evidently (c) holds if and only if $[A/J(A)]\backslash\{0\}$ is the unit group of $A/J(A)$. By Lemma 9.6 this is equivalent to (d). So the proposition is proved.

COROLLARY 9.8 (Fitting's Lemma). *The following conditions are equivalent for a module $M$ over a p-adic algebra $A$:*

(a) *$M$ is indecomposable,*
(b) $\mathrm{Hom}_A(M, M)/J(\mathrm{Hom}_A(M, M))$ *is a division algebra over* $\mathbf{Z}/p\mathbf{Z}$,
(c) $\mathrm{Hom}_A(M, M)\backslash J(\mathrm{Hom}_A(M, M))$ *is the group of A-automorphisms of M.*

*Proof.* In view of Proposition 9.4, this follows directly from the above proposition and (9.3).

Since any module over a $p$-adic algebra is, in particular, a $p$-adic module, it satisfies (8.8). This implies immediately that

*A module $M$ over a p-adic algebra $A$ has at least one decomposition $M = M_1 \oplus \ldots \oplus M_m$, where $m \geq 0$ and each $M_i$ is an indecomposable A-submodule of M.* (9.9)

The point of the Krull–Schmidt theorem is that this decomposition is as unique as it can be.

THEOREM 9.10 (Krull–Schmidt Theorem). *Let $M$ be a module over a p-adic algebra $A$. Suppose that $M = M_1 \oplus \ldots \oplus M_m = N_1 \oplus \ldots \oplus N_n$, where $m, n \geq 0$ and each $M_i$ or $N_j$ is an indecomposable A-submodule of M. Then $n = m$ and, after renumbering, $M_i \simeq N_i$ (as A-modules), for $i = 1, \ldots, m$.*

*Proof.* We use induction on $m$. If $m = 0$, then $M = \{0\}$ and the theorem is trivial. So we can assume that $m > 0$, and that the result is true for all smaller values of $m$. Obviously this implies that $M \neq \{0\}$, and hence that $n > 0$.

The decomposition $M = M_1 \oplus \ldots \oplus M_m$ defines projections $e_i$ of $M$ onto $M_i$ $(i = 1, \ldots, m)$. Similarly, the decomposition $M = N_1 \oplus \ldots \oplus N_n$ defines projections $f_j$ of $M$ onto $N_j$ $(j = 1, \ldots, n)$. From (9.2c) we obtain

$$f_1 = 1 f_1 = e_1 f_1 + e_2 f_1 + \ldots + e_m f_1.$$

Let $e_{1i} \in \mathrm{Hom}_A(N_1, M_i)$ be the restriction of $e_i$ to $N_1$, and $f_{i1} \in \mathrm{Hom}_A(M_i, N_1)$ be the restriction of $f_1$ to $M_i$ $(i = 1, \ldots, m)$. Then $e_{1i} f_{i1} \in \mathrm{Hom}_A(N_1, N_1)$ $(i = 1, \ldots, m)$. Since $f_1$ is identity on $N_1$, the above equation implies that

$$1 = e_{1i} f_{i1} + e_{12} f_{21} + \ldots + e_{1m} f_{m1} \quad \textit{in } \mathrm{Hom}_A(N_1, N_1).$$

Because $N_1$ is an indecomposable $A$-module, we may apply Corollary 9.8 to it. It is impossible that every $e_{1i} f_{i1}$ lies in $J(\mathrm{Hom}_A(N_1, N_1)$, $(i = 1, \ldots, m)$, since their sum 1 does not lie in this ideal. After renumbering, we can assume that $e_{11} f_{11} \notin J(\mathrm{Hom}_A(N_1, N_1))$. By Corollary 9.8(c), the product $e_{11} f_{11}$ is an $A$-automorphism of $N_1$. Let $(e_{11} f_{11})^{-1}$ be its inverse in $\mathrm{Hom}_A(N_1, N_1)$. Then $e_{11} \in \mathrm{Hom}_A(N_1, M_1)$ and $f_{11}(e_{11} f_{11})^{-1} \in \mathrm{Hom}_A(M_1, N_1)$ satisfy $e_{11} f_{11}(e_{11} f_{11})^{-1} = 1_{N_1 \to N_1}$. It follows that $e_{11}$ is a monomorphism, and that $M_1 = N_1 e_{11} \oplus \mathrm{Ker}\,(f_{11}(e_{11} f_{11})^{-1})$ (as $A$-modules). Since $M$ is indecomposable and $N_1 \neq \{0\}$, we conclude that $\mathrm{Ker}\,(f_{11} (e_{11} f_{11})^{-1}) = \{0\}$, and that $e_{11}$ is an isomorphism of $N_1$ onto $M_1$.

We now know that the restriction of $e_1$ is an isomorphism of $N_1$ onto $M_1$. It follows that $M = M_1 \oplus M_2 \oplus \ldots \oplus M_m = N_1 \oplus M_2 \oplus \ldots \oplus M_m$. Therefore $M_2 \oplus \ldots \oplus M_m \simeq M/N_1 \simeq N_2 \oplus \ldots \oplus N_n$, (as $A$-modules). By induction $m = n$ and, after renumbering, $M_i \simeq N_i$ (as $A$-modules), for $i = 2, \ldots, m$. Since $M_1 \simeq N_1$ already, this proves the theorem.

Let $R$ be a commutative ring with identity. Then any idempotent $e \in R$ is the identity of the subring $eR = Re$. If $e_1, \ldots, e_m \in R$ satisfy (9.2), then $R$ is the ring direct sum of its subrings $Re_1, \ldots, Re_m$. Conversely, if $R = R_1 \oplus \ldots \oplus R_m$ (as rings), then the identities $e_1, \ldots, e_m$ of $R_1, \ldots, R_m$, respectively, satisfy (9.2) and $R_i = Re_i$ $(i = 1, \ldots, m)$. So our "standard" condition that 1 be the unique non-zero idempotent in $R$ is equivalent to the condition that $R$ be idecomposable as a ring.

We say that an idempotent $e \in R$ is *primitive* if it is the unique non-zero idempotent in the subring $eR$, i.e., if $eR$ is an indecomposable subring of $R$. The importance of decompositions (9.2) in which each $e_i$ is primitive is explained by the

PROPOSITION 9.11. *Let* $e_1, \ldots, e_m$ *be primitive idempotents satisfying* (9.2)

*in a commutative ring R. Then the elements* $e_S = \sum_{i \in S} e_i$, *for* $S \subseteq \{1, \ldots, m\}$, *are precisely the distinct idempotents in R. In particular,* $e_1, \ldots, e_m$ *are the only primitive idempotents in R.*

*Proof.* An elementary calculation shows that $e_s$ is an idempotent, for any $S \subseteq \{1, \ldots, m\}$. Since $e_i e_S = e_i$, if $i \in S$, and is zero if $i \notin S$, $(i = 1, \ldots, m)$, the idempotent $e_S$ determines the set $S$. Therefore distinct subsets $S$ of $\{1, \ldots, m\}$ yield distinct idempotents $e_S$.

Let $f$ be any idempotent of $R$. For $i = 1, \ldots, m$, $fe_i = e_i f$ is an idempotent in $e_i R$. Since $e_i$ is primitive, $fe_i$ is either 0 or $e_i$. Letting $S$ be the set of all $i = 1, \ldots, m$ such that $fe_i = e_i$, we obtain

$$f = f1 = fe_1 + fe_2 + \ldots + fe_m = \sum_{i \in S} e_i = e_S.$$

This proves that the $e_S$ are the only idempotents in $R$.

Since $Re_S$ contains $e_i e_S = e_i$, for all $i \in S$ and any $S \subseteq \{1, \ldots, m\}$, we see immediately that $e_S$ is primitive if and only if $S$ contains exactly one element. So the proposition is proved.

We shall apply the above result to a commutative $p$-adic algebra $A$. In any decomposition $A = A_1 \oplus \ldots \oplus A_m$ (as rings), the subrings $A_i$ are ideals and hence $p$-adic subalgebras of $A$. As in the case of (9.9), statement (8.8) implies that $A$ has such a decomposition in which $m \geq 0$ and each $A_i$ is an indecomposable ring. The equivalence between these decompositions and the decompositions (9.2) in $A$, together with Proposition 9.11, imply that

> *A commutative p-adic algebra A has a finite number of primitive idempotents* $e_1, \ldots, e_m$ $(m \geq 0)$. *These idempotents satisfy* (9.2) *and the elements* $e_S = \sum_{i \in S} e_i$, *for* $S \subseteq \{1, \ldots, m\}$ *are precisely the distinct idempotents in A.* (9.12)

The Idempotent Refinement Lemma 9.5 can be used to give a close connection between idempotents of $A$ and those of $A/J(A)$.

PROPOSITION 9.13. *Let A be a commutative p-adic algebra. The map* $e \to e + J(A)$ *sends the family of all idempotents e of A one to one onto the family of all idempotents of* $A/J(A)$. *Furthermore e is primitive in A if and only if* $e + J(A)$ *is primitive in* $A/J(A)$.

*Proof.* Evidently $e \to e + J(A)$ sends the first family into the second. By Lemma 9.5 the map is onto. So we must prove it to be one to one.

Suppose that $e, f$ are idempotents of $A$ such that $e \equiv f (\text{mod } J(A))$. Then $e \equiv e^2 \equiv ef (\text{mod } J(A))$. Hence $e - ef = e(1-f)$ is an idempotent in $J(A)$. As in the first paragraph of the proof of Proposition 9.7, this implies that $e - ef = 0$, or $e = ef$. Similarly, $ef = f$. Therefore $e \to e + J(A)$ is one to one and the first statement of the proposition is proved.

We know that an idempotent $e$ is non-zero in $A$ if and only if $e+J(A)$ is non-zero in $A/J(A)$. If $f$ is an idempotent in $eA$ different from $e$ and 0, then $f+J(A)$ is an idempotent in $(e+J(A))\,(A/J(A))$ different from $e+J(A)$ and 0. Conversely, if $(e+J(A))\,(A/J(A))$ contains an idempotent $g \neq e+J(A)$, 0, then Lemma 9.5 gives us an idempotent $f$ in $A$ such that $g = f+J(A)$. The product $ef$ is an idempotent in $eA$ whose image in $A/J(A)$ is $(e+J(A))g = g$ (since $g \in (e+J(A))\,(A/J(A))$). Hence $ef \neq 0$, $e$. This proves the second statement and finishes the proof of the proposition.

## 10. Orders

We fix a prime $p$. Recall from Section 8 that the $p$-adic number field $Q_p$ is the field of fractions of the integral domain $\mathbf{Z}_p$ of $p$-adic integers. Let $A$ be an algebra (in the sense of Section 3) over $Q_p$. An *order* (or, more strictly, a $\mathbf{Z}_p$-*order*) in $A$ is a subset $\mathfrak{O}$ satisfying:

$$\mathfrak{O} \textit{ is a subring of } A, \tag{10.1a}$$

$$1_A \in \mathfrak{O}, \tag{10.1b}$$

$$\mathfrak{O} \textit{ is a finitely-generated } \mathbf{Z}_p\textit{-submodule of } A. \tag{10.1c}$$

$$\textit{For each } a \in A, \textit{ there exists } z \neq 0 \textit{ in } \mathbf{Z}_p \textit{ such that } za \in \mathfrak{O}. \tag{10.1d}$$

Evidently (10.1a, b, c) imply that

$$\mathfrak{O} \textit{ is a p-adic algebra.} \tag{10.2}$$

Since $\mathfrak{O}$ is a $\mathbf{Z}_p$-submodule of the vector space $A$ over $Q_p$, it is a torsion-free $\mathbf{Z}_p$-module. Because $\mathbf{Z}_p$ is a principal ideal domain, this and (10.1c) imply that $\mathfrak{O}$ is a free $\mathbf{Z}_p$-module of finite rank $n$. Hence there is a $\mathbf{Z}_p$-basis $a_1,\ldots, a_n$ of $\mathfrak{O}$ such that

$$\mathfrak{O} = \mathbf{Z}_p a_1 \oplus \ldots \oplus \mathbf{Z}_p a_n \simeq \overbrace{\mathbf{Z}_p \oplus \ldots \oplus \mathbf{Z}_p}^{n \text{ times}} \quad (\textit{as } \mathbf{Z}_p\textit{-modules}). \tag{10.3}$$

From (10.1d) it is clear that $a_1,\ldots, a_n$ is also a $Q_p$-basis for the algebra $A$. Therefore, we have

$$\dim_{Q_p} A = n = \operatorname{rank}_{\mathbf{Z}_p} \mathfrak{O}. \tag{10.4}$$

The order $\mathfrak{O}$ determines the algebra $A$ to within isomorphism, since the multiplication coefficients for the basis $a_1,\ldots, a_n$ can be computed in $\mathfrak{O}$.

Of course, orders always exist.

PROPOSITION 10.5. *Any algebra $A$ over $Q_p$ contains at least one order $\mathfrak{O}$.*

*Proof.* Let $b_1,\ldots, b_n$ be any basis for $A$ over $Q_p$. Then there are unique multiplication coefficients $f_{ijk} \in Q_p$, for $i, j, k = 1,\ldots, n$, such that

$$b_i b_j = \sum_{k=1}^{n} f_{ijk} b_k \quad (i, j = 1,\ldots, n).$$

Since $Q_p$ is the field of fractions of $\mathbf{Z}_p$, there is a $z \neq 0$ in $\mathbf{Z}_p$ such that $zf_{ijk} \in \mathbf{Z}_p$ $(i, j, k = 1, \ldots, n)$. The basis $zb_1, \ldots, zb_n$ for $A$ over $Q_p$ then satisfies

$$(zb_i)(zb_j) = \sum_{k=1}^{n} (zf_{ijk})(zb_k) \in \mathbf{Z}_p(zb_1) + \ldots + \mathbf{Z}_p(zb_n) \quad (i, j = 1, \ldots, n).$$

So the $\mathbf{Z}_p$-submodule $\mathfrak{M} = \mathbf{Z}_p(zb_1) + \ldots + \mathbf{Z}_p(zb_n)$ of $A$ is closed under multiplication. It follows that $\mathfrak{O} = \mathbf{Z}_p \cdot 1_A + \mathfrak{M}$ is also closed under multiplication. Therefore $\mathfrak{O}$ satisfies (10.1a). By its construction it satisfies (10.1b, c). It satisfies (10.1d), since $\mathfrak{M}$ does. Hence it is the order we seek.

A general algebra $A$ contains many orders, and it is impossible to single out one of them in any reasonable fashion. However, there is one important exception to this rule when the algebra $A$ is a finite algebraic extension field $F$ of $Q_p$.

PROPOSITION 10.6. *There is a unique maximum order $\mathfrak{O}$ containing all the other orders in $F$.*

*Proof.* Let $\mathfrak{O}_1$ and $\mathfrak{O}_2$ be two orders in $F$. Since $F$ is commutative and $1 \in \mathfrak{O}_1 \cap \mathfrak{O}_2$, the subring generated by $\mathfrak{O}_1$ and $\mathfrak{O}_2$ is their product $\mathfrak{O}_1\mathfrak{O}_2$, the additive subgroup generated by all products $xy$, with $x \in \mathfrak{O}_1$, $y \in \mathfrak{O}_2$. If $a_1, \ldots, a_n$ is a $\mathbf{Z}_p$-basis of $\mathfrak{O}_1$ and $b_1, \ldots, b_n$ is a $\mathbf{Z}_p$-basis of $\mathfrak{O}_2$, then clearly the products $a_ib_j$ $(i, j = 1, \ldots, n)$ generate $\mathfrak{O}_1\mathfrak{O}_2$ as a $\mathbf{Z}_p$-module. Therefore $\mathfrak{O}_1\mathfrak{O}_2$ is an order of $F$ containing both $\mathfrak{O}_1$ and $\mathfrak{O}_2$.

In view of Proposition 10.5, the above argument implies that the union $\mathfrak{O}$ of all the orders in $F$ satisfies (10.1a, b, d). To complete the proof of the proposition, we therefore need only show that $\mathfrak{O}$ is a finitely-generated $\mathbf{Z}_p$-module.

If $f \in F$, then its field trace $\operatorname{tr}(f) \in Q_p$ is the trace of the $Q_p$-linear transformation $L_f : y \to fy$ of $F$. We must prove that

$$\operatorname{tr}(f) \in \mathbf{Z}_p, \quad \textit{for all } f \in \mathfrak{O}. \tag{10.7}$$

Indeed, an element $f \in \mathfrak{O}$ is, by definition, contained in an order $\mathfrak{O}_1$ in $F$. A $\mathbf{Z}_p$-basis $a_1, \ldots, a_n$ for $\mathfrak{O}_1$ is also a $Q_p$-basis for $F$. Since $\mathfrak{O}_1$ is closed under multiplication, there are elements $z_{ij} \in \mathbf{Z}_p$ $(i, j = 1, \ldots, n)$ such that $fa_i = \sum_{j=1}^{n} z_{ij}a_j$, $(i = 1, \ldots, n)$. But, then,

$$\operatorname{tr}(f) = \operatorname{tr}(L_f) = \operatorname{tr}((z_{ij})) = z_{11} + z_{22} + \ldots + z_{nn} \in \mathbf{Z}_p.$$

So (10.7) holds.

Since $\mathbf{Z}_p$ has characteristic zero (by Corollary 8.6), the trace $\operatorname{tr}(1_F) = \dim_{Q_p}(F)$ is non-zero. If $f \neq 0$ in $F$, then $\operatorname{tr}(ff^{-1}) = \operatorname{tr}(1) \neq 0$. It follows that $f, g \to \operatorname{tr}(fg)$ is a non-singular $Q_p$-bilinear form from $F \times F$ into $Q_p$. If $\mathfrak{O}_1$ is an order of $F$ with a $\mathbf{Z}_p$-basis $a_1, \ldots, a_n$, then there exists a dual basis $c_1, \ldots, c_n$ for $F$ over $Q_p$ such that $\operatorname{tr}(a_i c_j)$ is the Kronecker $\delta$-function $\delta_{ij}$ $(i, j = 1, \ldots, n)$. It follows easily that

$$\{f \in F : \operatorname{tr}(fx) \in \mathbf{Z}_p, \quad \textit{for all } x \in \mathfrak{O}_1\} = \mathbf{Z}_p c_1 \oplus \ldots \oplus \mathbf{Z}_p c_n.$$

In view of (10.7), the ring $\mathfrak{O}$ is contained in the left side of this equation. So $\mathfrak{O}$ is a $\mathbf{Z}_p$-submodule of the finitely-generated $\mathbf{Z}_p$-module on the right side. Since $\mathbf{Z}_p$ is a principal ideal domain, this implies that $\mathfrak{O}$ is a finitely-generated $\mathbf{Z}_p$-module, which completes the proof of the proposition.

The above maximum order $\mathfrak{O}$ is usually described differently in the literature. We don't really need the other description, but we put it here anyway out of respect for tradition.

An element $f \in F$ is *integral* over $\mathbf{Z}_p$ if it satisfies an equation of the form:

$$f^m + z_1 f^{m-1} + \ldots + z_m = 0, \quad \textit{for some } m \geq 1 \text{ and } z_1, \ldots, z_m \in \mathbf{Z}_p. \tag{10.8}$$

The set of all such elements $f$ is called the *integral closure* of $\mathbf{Z}_p$ in $F$. In fact, it is simply the maximum order $\mathfrak{O}$.

PROPOSITION 10.9. *The maximum order $\mathfrak{O}$ is the integral closure of $\mathbf{Z}_p$ in $F$.*

*Proof.* Obviously condition (10.8) for an element $f \in F$ implies that $1, f, f^2, \ldots, f^{m-1}$ alone generate the $\mathbf{Z}_p$-module $\mathbf{Z}_p[f]$ generated by all the powers $1, f, f^2, \ldots$ of $f$. On the other hand, if $\mathbf{Z}_p[f]$ is a finitely-generated $\mathbf{Z}_p$-module, then we can find a finite subset $1, f, f^2, \ldots, f^{m-1}$ generating $\mathbf{Z}_p[f]$ in the infinite family $1, f, f^2, \ldots$ of generators. Clearly this implies that (10.8) holds. Hence $f$ is integral over $\mathbf{Z}_p$ if and only if $\mathbf{Z}_p[f]$ is a finitely-generated $\mathbf{Z}_p$-module.

If $f \in \mathfrak{O}$, then $\mathbf{Z}_p[f]$ is a submodule of the finitely-generated $\mathbf{Z}_p$-module $\mathfrak{O}$, and hence is finitely-generated. If $\mathbf{Z}_p[f]$ is a finitely-generated $\mathbf{Z}_p$-module, then $\mathbf{Z}_p[f] \cdot \mathfrak{O}$ is also a finitely-generated $\mathbf{Z}_p$-module. But $\mathbf{Z}_p[f]\mathfrak{O}$ is a subring of $F$ containing $\mathfrak{O}$, and hence is an order in $F$. By Proposition 10.6, $f = f \cdot 1 \in \mathbf{Z}_p[f] \cdot \mathfrak{O} \subseteq \mathfrak{O}$. Therefore $f \in \mathfrak{O}$ if and only if $\mathbf{Z}_p[f]$ is a finitely-generated $\mathbf{Z}_p$-module. Together with the first paragraph, this proves the proposition.

That the order $\mathfrak{O}$ is a local ring is given by

PROPOSITION 10.10. *The radical $J(\mathfrak{O})$ is the unique maximal ideal in the maximum order $\mathfrak{O}$ of $F$.*

*Proof.* Since $F$ is a field, the only idempotents in $F$ are 0 and $1 \neq 0$. It follows that 1 is the unique non-zero idempotent in the $p$-adic algebra $\mathfrak{O}$. By Proposition 9.7 and the commutativity of $\mathfrak{O}$, the factor ring $\mathfrak{O}/J(\mathfrak{O})$ is a field. Hence $J(\mathfrak{O})$ is a maximal ideal in $\mathfrak{O}$. Furthermore, Proposition 9.7 also says that every element of $\mathfrak{O} \backslash J(\mathfrak{O})$ is a unit in $\mathfrak{O}$. Therefore every ideal $\neq \mathfrak{O}$ of $\mathfrak{O}$ is contained in $J(\mathfrak{O})$. So the proposition is true.

We return to an arbitrary algebra $A$ over $Q_p$ and an arbitrary order $\mathfrak{O}$ in $A$. Let $M$ be a module over $A$ in the sense of Section 3. By an *$\mathfrak{O}$-lattice $L$* in $M$ we understand a subset satisfying:

$$L \text{ is a finitely-generated } \mathfrak{O}\text{-submodule of } M, \tag{10.11a}$$

$$\text{For each } y \in M, \text{ there exists } z \neq 0 \text{ in } \mathbf{Z}_p \text{ such that } zy \in L. \tag{10.11b}$$

Evidently $L$ is a module over the $p$-adic algebra $\mathfrak{O}$ in the sense of Section 9, and hence is a $p$-adic module in the sense of Section 8 (see the remarks preceding Proposition 9.4). As in the case of (10.3), this implies that $L$ is a free $\mathbf{Z}_p$-module of finite rank $m$. From (10.11b) we see that any $\mathbf{Z}_p$-basis for $L$ is also a $Q_p$-basis for $M$. Hence,

$$\dim_{Q_p} M = m = \operatorname{rank}_{\mathbf{Z}_p} L. \tag{10.12}$$

As usual, this implies that the $\mathfrak{O}$-module $L$ determines the $A$-module $M$ to within isomorphism.

As in the case of Proposition 10.5, lattices always exist.

PROPOSITION 10.13. *If $\mathfrak{O}$ is an order in an algebra $A$ over $Q_p$, and $M$ is a module over $A$, then there exists at least one $\mathfrak{O}$-lattice in $M$.*

*Proof.* Let $y_1, \ldots, y_m$ be a $Q_p$-basis for $M$. Then $L = \mathfrak{O}y_1 + \ldots + \mathfrak{O}y_m$ is evidently a finitely-generated $\mathfrak{O}$-submodule of $M$ satisfying (10.11b). This is the lattice we are seeking.

## 11. Blocks

Before we can define the blocks of a finite group $G$, we must find a good splitting field.

PROPOSITION 11.1. *If $E$ is a field of characteristic zero, then there is a finite algebraic extension $F$ of $E$ which is a splitting field for the group algebra $FG$.*

*Proof.* In view of Propositions 3.9 and 3.12, an extension $F$ of $E$ is a splitting field for $FG$ if and only if there exist integers $n_1, \ldots, n_k \geq 1$ (for some $k \geq 1$) such that:

$$FG \simeq [F]_{n_1} \oplus \ldots \oplus [F]_{n_k} \quad (\textit{as algebras}),$$

where $[F]_{n_i}$ is the algebra of all $n_i \times n_i$ matrices with entries in $F$. Choosing the usual basis for the matrix algebra $[F]_{n_i}$ (consisting of those matrices with all but one entry equal to zero, and that entry equal to one), we see that this occurs if and only if $FG$ has a basis over $F$ consisting of elements $e_{ij}^{(l)}$, for $l = 1, \ldots, k$ and $i, j = 1, \ldots, n_l$, satisfying

$$e_{ij}^{(l)} e_{i'j'}^{(l')} = \begin{cases} e_{ij'}^{(l)}, & \textit{if } l = l' \textit{ and } j = i', \\ 0, & \textit{otherwise}, \end{cases} \tag{11.2}$$

for all $l, l' = 1, \ldots, k$, all $i, j = 1, \ldots, n_l$ and all $i', j' = 1, \ldots, n_{l'}$.

By Proposition 3.13 the algebraic closure $\hat{E}$ of $E$ is a splitting field for $\hat{E}G$. So $\hat{E}G$ has a basis $\{e_{ij}^{(l)}\}$ of the above form. Since both bases are finite, there are only a finite number of coefficients in the matrix transforming the basis $G$ into the basis $\{e_{ij}^{(l)}\}$. Hence these coefficients generate a finite algebraic extension $F \subseteq \hat{E}$ of $E$ such that $e_{ij}^{(l)}$ lies in $FG \subseteq \hat{E}G$. Evidently the $e_{ij}^{(l)}$ form a basis of $FG$ satisfying (11.2). Hence $F$ is a splitting field for $FG$, and the proposition is proved.

Now let $p$ be any prime. Since the finite group $G$ has only a finite number of subgroups, the above proposition implies the existence of a field $F$ satisfying

*$F$ is a finite algebraic extension of the p-adic number field $Q_p$,* (11.3a)

*$F$ is a splitting field for the group algebra $FH$ of any subgroup $H$ of $G$.* (11.3b)

We fix such a field $F$, and denote by $\mathfrak{O}$ the maximum order in $F$ given by Proposition 10.6.

Because of (11.3a), the group algebra $FG$ is finite-dimensional as a vector space over $Q_p \subseteq F$. Hence it is an algebra over $Q_p$ in the sense of Section 3. Evidently the *group ring* $\mathfrak{O}G$ of $G$ over $\mathfrak{O}$, defined by

$$\mathfrak{O}G = \left\{ \sum_{\sigma \in G} y_\sigma \sigma \in FG : y_\sigma \in \mathfrak{O}, \quad \textit{for all } \sigma \in G \right\}, \tag{11.4}$$

is an order in $FG$.

As in (4.23), let $K_1, \ldots, K_c$ be the conjugacy classes of $G$ and $\tilde{K}_1, \ldots, \tilde{K}_c$ be the corresponding class sums. We can repeat the proof of Proposition 4.24 almost word for word to show that the center $Z(\mathfrak{O}G)$ of the ring $\mathfrak{O}G$ is given by

$$Z(\mathfrak{O}G) = \left\{ \sum_{i=1}^{c} y_i \tilde{K}_i : y_i \in \mathfrak{O} \quad (i = 1, \ldots, c) \right\}. \tag{11.5}$$

In view of that proposition, this implies that $Z(\mathfrak{O}G)$ is an order in the center $Z(FG)$ of the group algebra $FG$. In particular, $Z(\mathfrak{O}G)$ is a commutative $p$-adic algebra. So (9.12) says that the primitive idempotents $e_1, \ldots, e_b \in Z(\mathfrak{O}G)$ satisfy

$$1 = e_1 + \ldots + e_b \textit{ and } e_i e_j = 0, \quad \textit{for all } i, j = 1, \ldots, b \textit{ with } i \neq j. \tag{11.6}$$

The group $G$ has $b$ *blocks* $B_1, \ldots, B_b$ (for the prime $p$) corresponding to the primitive idempotents $e_1, \ldots, e_b$, respectively. The concept "block" is open and not closed. That is, we do not define it once and for all (e.g., by calling $e_i$, which determines $B_i$, *the* block $B_i$). Rather, any time a collection $C$ of objects attached to the group $G$ decomposes naturally into a disjoint union of subsets $C_1, \ldots, C_b$ attached to the idempotents $e_1, \ldots, e_b$, respectively, we put the objects of $C_i$ in the block $B_i$, for each $i = 1, \ldots, b$.

For example, we can put $e_i$ in the block $B_i$, for each $i = 1, \ldots, b$. It is then the unique primitive idempotent of $Z(\mathfrak{O}G)$ lying in $B_i$.

As in (4.6), let $A_1, \ldots, A_k$ be the minimal two-sided ideals of $FG$. From (4.22) we see that their identities $1_{A_1}, \ldots, 1_{A_k}$ are idempotents in $Z(FG)$ satisfying $1 = 1_{A_1} + \ldots + 1_{A_k}$ and $1_{A_i}1_{A_j} = 0$ if $i, j = 1, \ldots, k$ with $i \neq j$. Since $1_{A_i}Z(FG)1_{A_i} = F \cdot 1_{A_i} \simeq F$ is a field, the idempotent $1_{A_i}$ is primitive in $Z(FG)$ ($i = 1, \ldots, k$). Applying Proposition 9.11 to the ring $Z(FG)$, the decomposition $1 = 1_{A_1} + \ldots + 1_{A_k}$, and the idempotents $e_1, \ldots, e_b$ of the subring $Z(\mathfrak{O}G)$, we obtain unique subsets $S_1, \ldots, S_b$ of $\{1, \ldots, k\}$ such that

$$e_i = \sum_{l \in S_i} 1_{A_l} \quad (i = 1, \ldots, b). \tag{11.7}$$

Evidently the condition $e_i e_j = 0$, for $i \neq j$, says that $S_i \cap S_j$ is empty, while the condition $1 = e_1 + \ldots + e_b$ says that $S_1 \cup \ldots \cup S_b = \{1, \ldots, k\}$. Therefore $\{1, \ldots, k\}$ is the disjoint union of its subsets $S_1, \ldots, S_b$. If $l \in S_i$, for some $i = 1, \ldots, b$, we can now put the minimal two-sided ideal $A_l$, its identity $1_{A_l}$, and the corresponding irreducible character $\chi_l$ of $G$ in $B_i$. In view of (4.17), the equation (11.7) becomes

$$e_i = \sum_{\chi_l \in B_i} e(\chi_l) \quad (i = 1, \ldots, b). \tag{11.8}$$

Hence *the block $B_i$ is uniquely determined by its irreducible characters.*

Some blocks only contain one irreducible character.

PROPOSITION 11.9. *Let $p^a$ be the largest power of the prime $p$ dividing $|G|$. If $p^a$ divides $\chi_l(1)$, for some irreducible character $\chi_l$ of $G$, then $\chi_l$ is the only irreducible character in its block.*

*Proof.* In this case (4.17) implies that $e(\chi_l) \in Z(\mathfrak{O}G)$. Since $e(\chi_l)$ is primitive in $Z(FG)$, it is primitive in $Z(\mathfrak{O}G)$. Hence $e(\chi_l) = e_i$, for some $i = 1, \ldots, b$. In view of (11.8), this implies the proposition.

COROLLARY 11.10. *If $p$ does not divide $|G|$, then there is a one to one correspondence between blocks and irreducible characters of $G$, in which each block corresponds to the unique character it contains. After renumbering we then have $e_i = e(\chi_i)$, ($i = 1, \ldots, b$).*

*Proof.* In this case $p^a = 1$ divides $\chi_l(1)$, for all $l = 1, \ldots, k$. So the proposition gives the corollary.

The blocks of the type described in Proposition 11.9 are called *blocks of defect* 0.

Proposition 9.13 says that the images $\bar{e}_1, \ldots, \bar{e}_b$ of $e_1, \ldots, e_b$, respectively, are precisely the primitive idempotents in $Z(\mathfrak{O}G)/J(Z(\mathfrak{O}G))$. Hence we can distribute them among the blocks so that $\bar{e}_i$ is the primitive idempotent of $Z(\mathfrak{O}G)/J(Z(\mathfrak{O}G))$ in $B_i$ ($i = 1, \ldots, b$). Notice that the idempotent $\bar{e}_i$ also determines $B_i$ since, by Proposition 9.13, $e_i$ is the only idempotent of $Z(\mathfrak{O}G)$ having $\bar{e}_i$ as its image in $Z(\mathfrak{O}G)/J(Z(\mathfrak{O}G))$.

Since $Z(\mathfrak{O}G)$ is a commutative $p$-adic algebra, (8.11) and Propositions 3.9 and 3.12 imply that $Z(\mathfrak{O}G)/J(Z(\mathfrak{O}G))$ has a unique decomposition

$$Z(\mathfrak{O}G)/J(Z(\mathfrak{O}G)) = E_1 \oplus \ldots \oplus E_b \quad (\textit{as } \mathbf{Z}/p\mathbf{Z}\textit{-algebras}), \tag{11.11}$$

where $E_1, \ldots, E_b$ are finite-dimensional extension fields of $\mathbf{Z}/p\mathbf{Z}$. Incidentally, we have not made an error in counting the $E_i$'s. Evidently $1_{E_1}, \ldots, 1_{E_b}$ are precisely the primitive idempotents of $Z(\mathfrak{O}G)/J(Z(\mathfrak{O}G))$. So the number of direct summands $E_i$ in (11.11) is in fact the number $b$ of primitive idempotents $\bar{e}_i$, and hence the number of blocks of $G$. We choose the notation so that

$$1_{E_i} = \bar{e}_i, \quad \textit{for } i = 1, \ldots, b. \tag{11.12}$$

Of course we put each $E_i$ in the corresponding block $B_i$.

The natural homomorphism $y \to y1$ of the ring $\mathfrak{O}$ into $Z(\mathfrak{O}G)$ makes the latter ring an algebra (with identity) over the former.

LEMMA 11.13. *For each* $l = 1, \ldots, k$, *the restriction of the epimorphism* $\theta_l : Z(FG) \to F$ *of* (4.29) *is an epimorphism of* $Z(\mathfrak{O}G)$ *onto* $\mathfrak{O}$ *as* $\mathfrak{O}$*-algebras.*

*Proof.* Since $\theta_l$ is a homomorphism of $F$-algebras, its restriction is an $\mathfrak{O}$-algebra epimorphism of $Z(\mathfrak{O}G)$ onto an $\mathfrak{O}$-subalgebra $\theta_l(Z(\mathfrak{O}G))$ of $F$. Evidently $\theta_l(Z(\mathfrak{O}G))$ contains $\mathfrak{O}\theta_l(1) = \mathfrak{O}\cdot 1 = \mathfrak{O}$. On the other hand, the image $\theta_l(Z(\mathfrak{O}G))$ of the order $Z(\mathfrak{O}G)$ of $Z(FG)$ must be an order in $\theta_l(Z(FG)) = F$ (just verify (10.1)!). So $\theta_l(Z(\mathfrak{O}G)) \subseteq \mathfrak{O}$ by Proposition 10.6. That proves the lemma.

We know from (4.29) that $\chi_l(1_{A_j})$ is 1, if $l = j$, and is 0, if $l \neq j$ $(j, l = 1, \ldots, k)$. This and (11.8) imply that

$$\theta_l(e_i) = \begin{cases} 1, & \textit{if } \chi_l \in B_i, \\ 0, & \textit{if } \chi_l \notin B_i, \end{cases} \tag{11.14}$$

for all $l = 1, \ldots, k$ and $i = 1, \ldots, b$.

We denote by $\bar{\theta}_l$ the $\mathfrak{O}$-algebra epimorphism of $Z(\mathfrak{O}G)$ onto $\bar{F} = \mathfrak{O}/J(\mathfrak{O})$ obtained by composing $\theta_l$ with the natural epimorphism of $\mathfrak{O}$ onto $\mathfrak{O}/J(\mathfrak{O})$.

PROPOSITION 11.15. *There are precisely b epimorphisms* $\eta_1, \ldots, \eta_b$ *of* $Z(\mathfrak{O}G)$ *onto* $\bar{F}$ *as* $\mathfrak{O}$*-algebras. For any* $i = 1, \ldots, b$, *the epimorphism* $\eta_i$ *is zero on* $J(Z(\mathfrak{O}G))$, *and the induced epimorphism* $\bar{\eta}_i$ *on* $Z(\mathfrak{O}G)/J(Z(\mathfrak{O}G))$ *is zero on each* $E_h$ $(h \neq i)$ *and an isomorphism of* $E_i$ *onto* $\bar{F}$. *An irreducible character* $\chi_l$ *of* $G$ *lies in a block* $B_i$ *if and only if* $\bar{\theta}_l = \eta_i$.

*Proof.* Let $\eta$ be any $\mathfrak{O}$-algebra epimorphism of $Z(\mathfrak{O}G)$ onto $\bar{F}$. Since $\bar{F}$ is a field of characteristic $p$, the ideal $pZ(\mathfrak{O}G)$ is contained in the kernel Ker $(\eta)$ of $\eta$. In view of (8.12), the image $\eta(J(Z(\mathfrak{O}G)))$ is nilpotent. Because $\bar{F}$ is a field, this image is zero. Therefore $\eta$ induces an $\mathfrak{O}$-algebra epimorphism $\bar{\eta}$ of $Z(\mathfrak{O}G)/J(Z(\mathfrak{O}G))$ onto $\bar{F}$.

Since $\bar{F}$ is a field, and each field $E_i$ $(i = 1,\ldots, b)$ is an $\mathfrak{O}$-subalgebra of $Z(\mathfrak{O}G)/J(Z(\mathfrak{O}G))$, it is clear from (11.11) that the epimorphism $\bar{\eta}$ must be zero on all but one of the $E_i$ and an isomorphism on that one. Because $\bar{F} = \mathfrak{O}\cdot 1_F$, the exceptional $E_i$ satisfies $E_i = \mathfrak{O}\cdot 1_{E_i}$. Hence it has exactly one $\mathfrak{O}$-isomorphism onto $\bar{F}$. Therefore $\eta = \eta_i$ is uniquely determined by this value of $i$.

Now let $i$ be any of the integers $1,\ldots, b$ and $\chi_l$ be any irreducible character of $G$ in the block $B_i$. Then $\bar{\theta}_l$ is an $\mathfrak{O}$-algebra epimorphism of $Z(\mathfrak{O}G)$ onto $\bar{F}$. By the above argument, $\bar{\theta}_l$ induces an $\mathfrak{O}$-algebra epimorphism $\bar{\bar{\theta}}_l$ of $Z(\mathfrak{O}G)/J(Z(\mathfrak{O}G))$ onto $\bar{F}$. From (11.14) we see that $\bar{\bar{\theta}}_l(\bar{e}_i) = \bar{\theta}_l(e_i) = 1$. Hence $\bar{\bar{\theta}}_l$ is not zero on $E_i$. The above argument implies that $\bar{\theta}_l$ is the unique $\mathfrak{O}$-epimorphism $\eta_i$ associated with this value of $i$. This completes the proof of the proposition.

COROLLARY 11.16. *Two irreducible characters $\chi_j$, $\chi_l$ of $G$ belong to the same block if and only if*

$$\frac{\chi_j(\tilde{K}_i)}{\chi_j(1)} \equiv \frac{\chi_l(\tilde{K}_i)}{\chi_l(1)} \pmod{J(\mathfrak{O})}, \tag{11.17}$$

*for each class sum $\tilde{K}_i$ $(i = 1,\ldots, c)$ of $G$.*

*Proof.* This follows from the proposition, (11.5), and (4.30).

As usual, we put $\eta_i$ in the block $B_i$, for each $i = 1,\ldots, b$. Evidently $\eta_i$ determines $e_i$ (and hence $B_i$) by the condition

$$\eta_i(e_h) = \begin{cases} 1_F, & \text{if } i = h, \\ 0, & \text{if } i \neq h, \end{cases} \qquad (i,\ h = 1,\ldots, b). \tag{11.18}$$

As a final example of objects which can be put in blocks, we consider the indecomposable $\mathfrak{O}G$-modules. If $L$ is any $\mathfrak{O}G$-module (in the sense of Section 9) then conditions (11.6) and the fact that the idempotents $e_i$ all belong to the center of $\mathfrak{O}G$ imply that

$$L = Le_1 \oplus \ldots \oplus Le_b \quad (\textit{as } \mathfrak{O}G\textit{-modules}). \tag{11.19}$$

In particular, if $L$ is indecomposable, then exactly one of the $\mathfrak{O}G$-submodules $Le_i$ is non-zero, and that one equals $L$. Obviously we put $L$ in the corresponding block. Hence,

*An indecomposable $\mathfrak{O}G$-module $L$ lies in the block $B_i$*
*(where $i = 1,\ldots, b$) if and only if $L = Le_i$.* (11.20)

One relation between indecomposable lattices and irreducible characters in a block is very useful.

PROPOSITION 11.21. *Let $M$ be an $FG$-module* (*in the sense of Section* 3) *and $L$ be an $\mathfrak{O}G$-lattice in $M$. If $L$ is indecomposable and lies in a block $B_i$ of $G$,* *then the character* $\chi_M$ *of* $M$ *(defined by* (4.4)) *has the form*

$$\chi_M = \sum_{\chi_l \in B_i} c_l \chi_l, \tag{11.22}$$

*for some integers* $c_l \geq 0$.

*Proof.* By (11.20) multiplication by $e_i$ is identity on $L$. It follows that $e_i$ also acts as identity on $M$. In view of (11.8) this implies that

$$M = \oplus \sum_{\chi_l \in B_i} Me(\chi_l) \quad (\textit{as } FG\textit{-modules})$$

From Proposition 3.18 we see that each $Me(\chi_l) = M1_{A_l} = MA_l$ is a direct sum of $c_l$ copies of the irreducible $FG$-module $I_l$ corresponding to the character $\chi_l$, for some integer $c_l \geq 0$. The proposition results directly from this, the preceding equation, and (2.10).

We close this section with a useful example of groups for which the blocks are highly non-trivial.

PROPOSITION 11.23. *Suppose that G is a p-group. Then we have*

$$J(\mathfrak{O}G) = J(\mathfrak{O})1 + \sum_{\sigma \in G,\, \sigma \neq 1} \mathfrak{O}(\sigma - 1). \tag{11.24}$$

*Therefore:*

$$\mathfrak{O}G/J(\mathfrak{O}G) \simeq \mathfrak{O}/J(\mathfrak{O}) \simeq \bar{F}. \tag{11.25}$$

*Proof.* By definition $J(\mathfrak{O}G)$ is the inverse image in $\mathfrak{O}G$ of $J(\mathfrak{O}G/p\mathfrak{O}G)$. Let $I$ be an irreducible $\mathfrak{O}G/p\mathfrak{O}G$-module. Since $I$ has finite dimension over $\mathbf{Z}_p$, its additive group is a finite $p$-group. Using the operation of $G$ on $I$, we form the semi-direct product $GI$, which is also a finite $p$-group having $I$ as a non-trivial normal subgroup. It follows (see Satz III.7.2 in Huppert, 1967) that $I \cap Z(GI)$ is non-trivial. Hence there is an element $y \neq 0$ in $I$ such that $y\sigma = y$, for all $\sigma \in G$. Because $I$ is irreducible, we must have $I = z(\mathfrak{O}/p\mathfrak{O})$ Furthermore, $I$ must be irreducible as an $\mathfrak{O}/p\mathfrak{O}$-module. Hence $yJ(\mathfrak{O}/p\mathfrak{O}) = 0$. It follows that *the only irreducible* $\mathfrak{O}G/p\mathfrak{O}G$*-module is* $\bar{F} = \mathfrak{O}/J(\mathfrak{O})$ *with trivial action of* $G$. Equation (11.24) follows directly from this and (3.6), while (11.25) follows from (11.24).

COROLLARY 11.26. *The p-group G has just one block containing every irreducible character of G.*

*Proof.* The proposition and Proposition 9.7 imply that 1 is the unique non-zero idempotent in $\mathfrak{O}G$. Hence 1 is also the unique non-zero idempotent in the subring $Z(\mathfrak{O}G)$. This implies the corollary.

## 12. Orthogonality Relations

We continue to use the notation and hypotheses of the last section.

Let $L$ be an $\mathfrak{O}G$-module, and $K$ be an $\mathfrak{O}H$-submodule of $L$, for some su bgroup $H$ of $G$. We say that $L$ is *induced* from $K$ (and write $L = K^G$) if

$$L = \oplus \sum_{\sigma \in \operatorname{rep}(G/H)} K\sigma \quad (\textit{as } \mathfrak{O}\textit{-modules}) \tag{12.1}$$

where rep $(G/H)$ is, as in (5.5), a family of representatives for the left cosets $H\sigma$ of $H$ in $G$. For each $\sigma \in \operatorname{rep}(G/H)$ and $\tau \in G$, there are unique $\sigma' \in \operatorname{rep}(G/H)$ and $\rho \in H$ such that $\sigma\tau = \rho\sigma'$. Since $K$ is an $\mathfrak{O}H$-submodule of $L$, it follows that:

$$(k\sigma)\tau = (k\rho)\sigma' \in K\sigma', \quad \textit{for all } k \in K. \tag{12.2}$$

Evidently $k \to k\sigma$ is an $\mathfrak{O}$-isomorphism of $K$ on to $K\sigma$. So this and (12.1) imply that the $\mathfrak{O}G$-module structure of $L$ is completely determined by the $\mathfrak{O}H$-module structure of $K$. Furthermore it is evident that we can start with an arbitrary $\mathfrak{O}H$-module $K$ and construct via (12.1) and (12.2) an $\mathfrak{O}G$-module $L$ satisfying $L = K^G$ (to get $K$ to be a submodule of $L$, pick rep $(G/H)$ to contain 1 and identify $K$ with the summand $K1$ in (12.1)).

There is a simple connection between induced modules and the induced characters of Section 5. To express it, it is convenient to write $\chi_L$ for the character $\chi_M$ of an $FG$-module $M$ having $L$ as an $\mathfrak{O}G$-lattice.

PROPOSITION 12.3. *Let $M$ be an $FG$-module, $L$ be an $\mathfrak{O}G$-lattice in $M$, and $K$ be an $\mathfrak{O}H$-submodule such that $L = K^G$. Then $K$ is an $\mathfrak{O}H$-lattice in the $F$-subspace $N$ of $M$ which it spans. Furthermore,*

$$\chi_L = \chi_{K^G} = (\chi_K)^G, \tag{12.4}$$

*where $\chi_K$ is the $H$-character of $K$.*

*Proof.* Clearly $N$ is an $FH$-submodule of $M$. A glance at (10.11) shows that K is an $\mathfrak{O}H$-lattice in $N$. From (12.1) we easily obtain the equation

$$M = \oplus \sum_{\sigma \in \operatorname{rep}(G/H)} N\sigma \quad (\textit{as } F\textit{-spaces})$$

Let $\tau$ be any element of $G$ and $T$ be the linear transformation $m \to m\tau$ of $M$. The above decomposition gives us unique $F$-linear maps $T_{\sigma,\pi} : N\sigma \to N\pi$, for $\sigma, \pi \in \operatorname{rep}(G/H)$, such that

$$T(m) = \oplus \sum_{\pi \in \operatorname{rep}(G/H)} T_{\sigma,\pi}(m), \quad \textit{for all } \sigma \in \operatorname{rep}(G/H),\ m \in N\sigma.$$

Furthermore,

$$\chi_L(\tau) = \chi_M(\tau) = \operatorname{tr}(T) = \sum_{\sigma \in \operatorname{rep}(G/H)} \operatorname{tr}(T_{\sigma,\sigma}).$$

Let $\sigma$ be any element of rep $(G/H)$, and $\sigma' \in \operatorname{rep}(G/H)$, $\rho \in H$ be the unique elements such that $\sigma\tau = \rho\sigma'$. Then $T(N\sigma) = N\sigma\tau = N\rho\sigma' = N\sigma'$. Hence $T_{\sigma,\pi} = 0$ for $\pi \neq \sigma'$. In particular $T_{\sigma,\sigma} = 0$ unless $\sigma = \sigma'$, which occurs if and only if $\tau^{\sigma^{-1}} = \rho \in H$. In that case the $F$-isomorphism $n \to n\sigma$ of $N$ onto $N\sigma$ defines an equivalence between $T_{\sigma,\sigma}$ and the linear transformation $n \to n\tau^{\sigma^{-1}}$ of $N$. Therefore tr $(T_{\sigma,\sigma})$ equals the trace $\chi_N(\tau^{\sigma^{-1}})$ of this last transformation. Using (5.1), (5.4) and (5.5), we conclude that

$$\chi_L(\tau) = \sum_{\sigma \in \mathrm{rep}\,(G/H),\ \tau^{\sigma^{-1}} \in H} \chi\,(\tau^{\sigma^{-1}}) = (\chi_N)^G(\tau) = (\chi_K)^G(\tau).$$

So the proposition holds.

We shall need some lemmas to aid us to compute endomorphism rings of induced modules.

LEMMA 12.5. *Let $L$ be an $\mathfrak{O}G$-module and $K$ be an $\mathfrak{O}H$-submodule such that $L = K^G$. If $\phi \in \mathrm{Hom}_{\mathfrak{O}G}(L, L)$ then the restriction $\phi_K$ to $K$ lies in $\mathrm{Hom}_{\mathfrak{O}H}(K, L)$. Furthermore the map $\phi \to \phi_K$ is an $\mathfrak{O}$-isomorphism of $\mathrm{Hom}_{\mathfrak{O}G}(L, L)$ onto $\mathrm{Hom}_{\mathfrak{O}H}(K, L)$.*

*Proof.* Clearly $\phi_K \in \mathrm{Hom}_{\mathfrak{O}H}(K, L)$, for all $\phi \in \mathrm{Hom}_{\mathfrak{O}G}(L, L)$. Furthermore $\phi_K$ determines $\phi$ because of (12.1) and the equation

$$\phi(k\sigma) = \phi(k)\sigma = \phi_K(k)\sigma, \quad \text{for all } k \in K, \sigma \in \mathrm{rep}\,(G/H). \tag{12.6}$$

Hence the map $\phi \to \phi_K$ is an $\mathfrak{O}$-monomorphism of $\mathrm{Hom}_{\mathfrak{O}G}(L, L)$ into $\mathrm{Hom}_{\mathfrak{O}H}(K, L)$.

Suppose we are given $\phi_K \in \mathrm{Hom}_{\mathfrak{O}H}(K, L)$. By (12.1) there is a unique $\mathfrak{O}$-linear map $\phi$ of $L$ into $L$ satisfying (12.6). Choosing $\sigma \in H$, we see that $\phi_K$ is indeed the restriction of $\phi$ to $K$. From (12.2) we compute easily that $\phi \in \mathrm{Hom}_{\mathfrak{O}G}(L, L)$. So $\phi \to \phi_K$ is an epimorphism, and the lemma is proved.

COROLLARY 12.7. *The inverse of the above isomorphism $\phi \to \phi_K$ sends $\mathrm{Hom}_{\mathfrak{O}H}(K, K) \subseteq \mathrm{Hom}_{\mathfrak{O}H}(K, L)$ monomorphically into $\mathrm{Hom}_{\mathfrak{O}G}(L, L)$ as $\mathfrak{O}$-algebras. This monomorphism carries the identity $1_{K \to K}$ of the first algebra into the identity $1_{L \to L}$ of the second.*

*Proof.* If $\phi, \psi \in \mathrm{Hom}_{\mathfrak{O}G}(L, L)$ satisfy $\phi_K, \psi_K \in \mathrm{Hom}_{\mathfrak{O}H}(K, K)$, then clearly $(\phi\psi)_K = \phi_K\chi_K$. The first statement of the corollary follows from this and the lemma. The second comes from the remark that $1_{K \to K}$ is obviously the restriction to $K$ of $1_{L \to L}$.

We shall use the above lemma in a very special case. Let $\langle\pi\rangle$ be a cyclic $p$-group of order $p^d > 1$. We consider an $\mathfrak{O}\langle\pi\rangle$-module $L$ and an $\mathfrak{O}\langle\pi^p\rangle$-submodule $K$ such that $L = K^{\langle\pi\rangle}$. Since $\langle\pi\rangle$ is commutative, the map $\Pi : l \to l\pi$ is a central element of $\mathrm{Hom}_{\mathfrak{O}\langle\pi\rangle}(L, L)$. We identify $\mathrm{Hom}_{\mathfrak{O}\langle\pi^p\rangle}(K, K)$ with its image in $\mathrm{Hom}_{\mathfrak{O}\langle\pi\rangle}(L, L)$ via the algebra monomorphism of Corollary 12.7. Then we have

LEMMA 12.8. *The power $\Pi^p$ is the central unit $\psi : k \to k\pi^p$ of $\mathrm{Hom}_{\mathfrak{O}\langle\pi^p\rangle}(K, K)$. Furthermore,*

$$\mathrm{Hom}_{\mathfrak{O}\langle\pi\rangle}(L, L) =$$
$$\mathrm{Hom}_{\mathfrak{O}\langle\pi^p\rangle}(K, K) \oplus \mathrm{Hom}_{\mathfrak{O}\langle\pi^p\rangle}(K, K)\Pi \oplus \ldots \oplus \mathrm{Hom}_{\mathfrak{O}\langle\pi^p\rangle}(K, K)\Pi^{p-1}$$
$$\textit{(as } \mathfrak{O}\textit{-modules)} \tag{12.9}$$

*Proof.* The first statement is clear from the definition of $\Pi$ and Lemma 12.5. For the second, notice that $\Pi^i : k \to k\pi^i$ is an $\mathfrak{O}\langle\pi^p\rangle$-isomorphism of $K$ onto $K\pi^i$ $(i = 0, 1, \ldots, p-1)$. Hence $\mathrm{Hom}_{\mathfrak{O}\langle\pi^p\rangle}(K, K)\Pi^i$ is isomorphic to $\mathrm{Hom}_{\mathfrak{O}\langle\pi^p\rangle}(K, K\pi^i)$ $(i = 0, 1, \ldots, p-1)$. Since $L = K \oplus K \oplus \ldots \oplus K\pi^{p-1}$ (as $\mathfrak{O}\langle\pi^p\rangle$-modules), this and Lemma 12.5 prove (12.9). So the lemma holds.

From this knowledge of the structure of $\mathrm{Hom}_{\mathfrak{O}\langle\pi\rangle}(L, L)$ we easily prove

PROPOSITION 12.10 (Green). *If, in the situation of Lemma* 12.8, *the* $\mathfrak{O}\langle\pi^p\rangle$-*submodule K is indecomposable, then so is the* $\mathfrak{O}\langle\pi\rangle$-*module L.*

*Proof.* Let $A = \mathrm{Hom}_{\mathfrak{O}\langle\pi\rangle}(L, L)$ and $B$ be its subalgebra $\mathrm{Hom}_{\mathfrak{O}\langle\pi^p\rangle}(K, K)$. By Lemma 12.8, $A$ is obtained from $B$ by adjoining the central element $\Pi : A = B[\Pi]$. If $I$ is an irreducible $A/pA$-module, we conclude that $IJ(B)$ is an $A/pA$-submodule of $I$. By (8.12), the ideal $J(B)$ is nilpotent modulo $pB \subseteq pA$. So $IJ(B) = I$ would imply that $I = IJ(B) = IJ(B)^2 = \ldots = 0$, which is impossible. Therefore $IJ(B) = 0$, and $I$ is really an irreducible module over the ring $\bar{A} = A/J(B)A$. It follows that $J(A)$ is the inverse image in $A$ of $J(\bar{A})$.

The ring $\bar{A}$ is generated over its subring $\bar{B} = B/(B \cap J(B)A)$ by the image $\bar{\Pi}$ of $\Pi$. From the definition of $\Pi$ it is clear that $\Pi^{p^d} = 1$ in $A$. Hence $\bar{\Pi}^{p^d} = 1$ in $\bar{A}$. But $\bar{A}$ is a ring of characteristic $p$. Therefore $0 = \bar{\Pi}^{p^d} - 1 = (\bar{\Pi} - 1)^{p^d}$. It follows that the central element $\bar{\Pi} - 1$ generates a nilpotent two-sided ideal of $\bar{A}$. As usual, this ideal is contained in $J(\bar{A})$. Hence $J(A)$ is the inverse image in $A$ of the radical of $\bar{A}/(\bar{\Pi}-1)\bar{A} \simeq \bar{B}/(\bar{B} \cap (\bar{\Pi}-1)\bar{A})$.

The indecomposability of $K$ and Corollary 9.8 tell us that $B/J(B)$ is a division algebra over $\mathbf{Z}/p\mathbf{Z}$. Since $B \cap J(B)A \supseteq J(B)$, we conclude that $\bar{B} = B/(B \cap J(B)A) \neq 0$ is also a division algebra over $\mathbf{Z}/p\mathbf{Z}$. Hence so is its epimorphic image $\bar{B}/(\bar{B} \cap (\bar{\Pi}-1)\bar{A}) \simeq \bar{A}/(\bar{\Pi}-1)\bar{A}$. We conclude that $A/J(A) \simeq \bar{A}/(\bar{\Pi}-1)\bar{A}$ is a division algebra over $\mathbf{Z}/p\mathbf{Z}$. By Corollary 9.8 again, this implies that $L$ is indecomposable. So the proposition is proved.

We combine Propositions 12.3 and 12.10 to obtain an extremely useful criterion for the vanishing of certain characters.

PROPOSITION 12.11. *Let* $\langle\sigma\rangle$ *be a cyclic group whose order is divisible by p, and* $\langle\pi\rangle$ *be the p-Sylow subgroup of* $\langle\sigma\rangle$. *Suppose that M is an* $F\langle\sigma\rangle$-*module, L is an* $\mathfrak{O}\langle\sigma\rangle$-*lattice in M, and K is an* $\mathfrak{O}\langle\pi^p\rangle$-*submodule of L such that* $L = K^{\langle\pi\rangle}$ *as an* $\mathfrak{O}\langle\pi\rangle$-*module. If L′ is any* $\mathfrak{O}\langle\sigma\rangle$-*direct summand of L, then L′ is an* $\mathfrak{O}\langle\sigma\rangle$-*lattice and*

$$\chi_{L'}(\tau) = 0, \quad \textit{for all } \tau \in \langle\sigma\rangle \backslash \langle\sigma^p\rangle. \qquad (12.12)$$

*Proof.* Of course, we assume implicitly that the conditions of Section 11, in particular (11.3), are satisfied by the group $\langle\sigma\rangle$ and field $F$. Obviously $L'$ is an $\mathfrak{O}\langle\sigma\rangle$-lattice in the $F$-subspace of $M$ which it generates. So the problem is to prove (12.12). In doing so, we can assume that $L'$ is an indecomposable $\mathfrak{O}\langle\sigma\rangle$-module, since a decomposition $L' = L'_1 \oplus L'_2$ (as $\mathfrak{O}\langle\sigma\rangle$-modules) implies that $\chi_{L'} = \chi_{L'_1} + \chi_{L'_2}$.

We can find a subgroup $\langle\rho\rangle$ of the cyclic group such that $p$ does not divide the order of $\langle\rho\rangle$ and $\langle\sigma\rangle = \langle\rho\rangle \times \langle\pi\rangle$. Let $\lambda_1, \ldots, \lambda_k$ be the irreducible, and hence linear, characters of $\langle\rho\rangle$. By Corollary 11.10 the corresponding idempotents $e(\lambda_1), \ldots, e(\lambda_k)$ of $F\langle\rho\rangle$ all lie in $\mathfrak{O}\langle\rho\rangle \subseteq \mathfrak{O}\langle\sigma\rangle$. Since $\mathfrak{O}\langle\sigma\rangle$ is abelian, this implies that

$$L' = L'e(\lambda_1) \oplus \ldots \oplus L'e(\lambda_k) \quad (\textit{as } \mathfrak{O}\langle\sigma\rangle\textit{-modules}).$$

Because $L'$ is an indecomposable $\mathfrak{O}\langle\sigma\rangle$-module, we conclude that $L' = L'e(\lambda_i)$, for some $i = 1, \ldots, k$. It follows that $l'\rho^j = \lambda_i(\rho^j)l'$, for all $l' \in L'$ and $\rho^j \in \langle\rho\rangle$. Since all the values of the linear character $\lambda_i$ lie in $\mathfrak{O}$ (by (11.3)), this implies that any $\mathfrak{O}$-submodule of $L'$ is an $\mathfrak{O}\langle\rho\rangle$-submodule of $L'$. Hence any $\mathfrak{O}\langle\pi\rangle$-submodule of $L'$ is an $\mathfrak{O}(\langle\rho\rangle \times \langle\pi\rangle) = \mathfrak{O}\langle\sigma\rangle$-submodule. In particular, $L'$ is indecomposable as an $\mathfrak{O}\langle\pi\rangle$-module.

Choose indecomposable $\mathfrak{O}\langle\pi^p\rangle$-submodules $K_1, \ldots, K_t$ of $K$ so that $K = K_1 \oplus \ldots \oplus K_t$. From (12.1) it is clear that

$$L = K^{\langle\pi\rangle} = K_1^{\langle\pi\rangle} \oplus \ldots \oplus K_t^{\langle\pi\rangle} \quad (\textit{as } \mathfrak{O}\langle\pi\rangle\textit{-modules}).$$

Proposition 12.10 tells us that each $K_j^{\langle\pi\rangle}$ is an indecomposable $\mathfrak{O}\langle\pi\rangle$-submodule of $L$. Since $L'$ is an indecomposable $\mathfrak{O}\langle\pi\rangle$-direct summand of $L$, the Krull–Schmidt Theorem 9.10 implies that $L' \simeq K_j^{\pi}$ (as $\mathfrak{O}\langle\pi\rangle$-modules), for some $j = 1, \ldots, t$. Hence there is an $\mathfrak{O}\langle\pi^p\rangle$-submodule $K'$ of $L'$ such that $L' = (K')^{\langle\pi\rangle}$ (as $\mathfrak{O}\langle\pi\rangle$-modules). The $\mathfrak{O}$-submodule $K'$ is invariant under $\langle\rho\rangle$, and hence is an $\mathfrak{O}(\langle\rho\rangle \times \langle\pi^p\rangle) = \mathfrak{O}\langle\sigma^p\rangle$-submodule of $L'$. From this and (12.1) we see easily that $L' = (K')^{\langle\sigma\rangle}$ (as $\mathfrak{O}\langle\sigma\rangle$-modules). Now Proposition 12.3 tells us that $\chi_{L'}$ is induced from the character $\chi_{K'}$ on $\langle\sigma^p\rangle$. Using (5.1), (5.4) and (5.5) we compute directly that

$$\chi_{L'}(\tau) = (\chi_{K'})^{\langle\sigma\rangle}(\tau) = \begin{cases} 0, & \textit{if } \tau \in \langle\sigma\rangle \backslash \langle\sigma^p\rangle, \\ p\chi_{K'}(\tau), & \textit{if } \tau \in \langle\sigma^p\rangle. \end{cases}$$

In particular, (12.12) holds. This proves the proposition.

As an application of the above proposition, we prove an orthogonality relation for the characters in a single block which should be compared with the identity (4.33) for all the characters.

For any element $\sigma$ of a finite group $G$, we define the *p-part* $\sigma_p$ and the *p'-part* $\sigma_{p'}$ of $\sigma$ to be the unique elements of $\langle\sigma\rangle$ whose orders are, respectively, a power of $p$ and relatively prime to $p$, and which satisfy $\sigma = \sigma_p\sigma_{p'} = \sigma_{p'}\sigma_p$.

THEOREM 12.13. *Let $B_i$ be any block of the finite group $G$, and $\sigma$, $\tau$ be any two elements of $G$ whose p-parts $\sigma_p$, $\tau_p$ are not G-conjugate. Then*

$$\sum_{\chi_j \in B_i} \chi_j(\sigma)\chi_j(\tau^{-1}) = 0. \tag{12.14}$$

*Proof*. The cyclic subgroup $\langle\tau\times\sigma\rangle$ of $G\times G$ acts naturally on the set $G$ so that

$$\rho(\tau\times\sigma)=\tau^{-1}\rho\sigma \quad (\textit{for all } \rho\in G). \tag{12.15}$$

Evidently this operation makes $\mathfrak{O}G$ an $\mathfrak{O}\langle\tau\times\sigma\rangle$-lattice in the $F\langle\tau\times\sigma\rangle$-module $FG$.

Suppose that $\pi=(\tau\times\sigma)_p=\tau_p\times\sigma_p$ fixes an element $\rho\in G$. Then $\tau_p^{-1}\rho\sigma_p=\rho$, which implies that $\sigma_p=\rho^{-1}\tau_p\rho$. This is impossible since $\sigma_p$ and $\tau_p$ are not $G$-conjugate. Therefore each $\langle\pi\rangle$-orbit $R$ of $G$ has length $p^d>1$. It follows that $R$ is the disjoint union of $\langle\pi^p\rangle$-orbits of the form $R=S\cup S\pi\cup\ldots\cup S\pi^{p-1}$. Hence the submodules $\mathfrak{O}R$, $\mathfrak{O}S$ generated by $R$, $S$, respectively, in $\mathfrak{O}G$, satisfy

$$\mathfrak{O}R=\mathfrak{O}S\oplus\mathfrak{O}S\pi\oplus\ldots\oplus\mathfrak{O}S\pi^{p-1} \quad (\textit{as } \mathfrak{O}\textit{-modules}).$$

Comparing with (12.1), we see that the $\mathfrak{O}\langle\pi\rangle$-module $\mathfrak{O}R$ is induced from its $\mathfrak{O}\langle\pi^p\rangle$-submodule $\mathfrak{O}S$. Since $\mathfrak{O}G$ is the direct sum of the $\mathfrak{O}R$ (as $\mathfrak{O}\langle\pi\rangle$-modules), where $R$ runs over all the $\langle\pi\rangle$-orbits of $G$, we conclude that $\mathfrak{O}G$ is induced, as an $\mathfrak{O}\langle\pi\rangle$-module, from an $\mathfrak{O}\langle\pi^p\rangle$-submodule.

Because the primitive idempotents $e_1,\ldots,e_b$ all lie in the center of $Z(\mathfrak{O}G)$, we have

$$\mathfrak{O}G=e_1\mathfrak{O}G\oplus\ldots\oplus e_b\mathfrak{O}G \quad (\textit{as two-sided } \mathfrak{O}G\textit{-modules}).$$

From (12.15) it is clear that this is also a decomposition as $\mathfrak{O}\langle\tau\times\sigma\rangle$-modules. Now all the conditions of Proposition 12.11 are satisfied with $\langle\tau\times\sigma\rangle$ as the cyclic group, $\langle\pi\rangle$ as its $p$-Sylow subgroup, $\mathfrak{O}G$ as the $\mathfrak{O}\langle\tau\times\sigma\rangle$-lattice $L$, and $e_i\mathfrak{O}G$ as its $\mathfrak{O}\langle\tau\times\sigma\rangle$-direct summand $L'$. Since $\tau\times\sigma\notin\langle\tau^p\times\sigma^p\rangle$, equation (12.12) implies that

$$\chi_{e_i\mathfrak{O}G}(\tau\times\sigma)=0.$$

Evidently $e_i\mathfrak{O}G$ is an $\mathfrak{O}\langle\tau\times\sigma\rangle$-lattice in $e_iFG$. Using (11.8), (4.6) and (4.7) we see that

$$e_iFG=\bigoplus_{\chi_j\in B_i}e(\chi_j)FG=\bigoplus_{\chi_j\in B_i}A_j \quad (\textit{as two-sided } FG\textit{-modules}).$$

It follows that

$$0=\chi_{e_i\mathfrak{O}G}(\tau\times\sigma)=\chi_{e_iFG}(\tau\times\sigma)=\sum_{\chi_j\in B_i}\chi_{A_j}(\tau\times\sigma).$$

Fix $j=1,\ldots,k$ so that $\chi_j\in B_i$. From (4.8) and (12.15) we see that the representation of $\langle\tau\times\sigma\rangle$ on $A_j\simeq\mathrm{Hom}_F(I_j,I_j)$ is obtained by restriction from the representation $\mathrm{Hom}\,(R_j^{-1},R_j)$ of $G\times G$ on $\mathrm{Hom}_F(I_j,I_j)$ defined by (2.15). So (2.16) tells us that $\chi_{A_j}(\tau\times\sigma)=\chi_j(\tau^{-1})\chi_j(\sigma)$. Substituting this in the preceding equation, we obtain (12.14). Hence the theorem is proved.

Let $T$ be any family of $p$-elements of $G$ closed under inverses, i.e., $T^{-1}=T$. The *p-section* $S_p(T)$ is defined by

$$S_p(T)=\{\sigma\in G:(\sigma_p)^\tau\in T, \textit{ for some } \tau\in G\}. \tag{12.16}$$

Clearly $S_p(T)$ is a union of conjugacy classes of $G$ which is also closed under inverses.

As in (5.13), we denote by $CF(G|S_p(T))$ the $F$-vector space of all class functions from $G$ to $F$ which vanish outside $S_p(T)$. For each block $B_i$ of $G$, let $CF(G|S_p(T), B_i)$ be the subspace of all $F$-linear combinations $\phi$ of the irreducible characters $\chi_j \in B_i$ such that $\phi = 0$ on $G \backslash S_p(T)$.

PROPOSITION 12.17. *The inner product* $(\cdot, \cdot)_G$ *of* (4.19) *is non-singular on the subspace* $CF(G|S_p(T))$ *of* $CF(G)$. *With respect to this inner product,* $CF(G|S_p(T))$ *is the perpendicular direct sum of its subspaces* $CF(G|S_p(T), B_i)$, *i.e.*

$$CF(G|S_p(T)) = \sum_{i=1}^{b} CF(G|S_p(T), B_i). \tag{12.18}$$

*Proof.* Let $K_1, \ldots, K_l$ be the conjugacy classes of $G$ contained in $S_p(T)$. Since $S_p(T)$ is closed under inverses, there is an involutory permutation $\pi$ of $1, \ldots, l$ such that $K_i^{-1} = K_{\pi(i)}$, for each $i = 1, \ldots, l$. The characteristic functions $\phi_1, \ldots, \phi_l$ of the classes $K_1, \ldots, K_l$, respectively, form a basis for $CF(G|S_p(T))$. In view of (4.19) we have

$$(\phi_i, \phi_{\pi(j)})_G = \begin{cases} \dfrac{|K_i|}{|G|} & \text{if } i = j, \\ 0 & \text{if } i \neq j, \end{cases}$$

for all $i, j = 1, \ldots, l$. It follows that $(\cdot, \cdot)_G$ is non-singular on $CF(G|S_p(T))$.

For any $\rho \in S_p(T)$ and any block $B_i$ of $G$, we define a class function $\psi_{\rho, B_i}$ by

$$\psi_{\rho, B_i} = \sum_{\chi_j \in B_i} \chi_j(\rho^{-1})\chi_j.$$

From (12.16), we see that $\rho_p$ is not conjugate to $\tau_p$, for any $\tau \in G \backslash S_p(T)$. Therefore $\psi_{\rho, B_i}$ vanishes at all such $\tau$ by Theorem 12.13. Hence $\psi_{\rho, B_i} \in CF(G|S_p(T), B_i)$. By (4.33) the sum

$$\sum_{i=1}^{b} \psi_{\rho, B_i} = \sum_{j=1}^{k} \chi_j(\rho^{-1})\chi_j$$

is $|C_G(\rho)|$ times the characteristic function $\phi_h$ of the class $K_h$ containing $\rho$. Since every $\phi_h$ can be obtained in this fashion, this proves that

$$CF(G|S_p(T)) = \sum_{i=1}^{b} CF(G|S_p(T), B_i).$$

Suppose that $\theta \in CF(G|S_p(T), B_i)$ and $\theta' \in CF(G|S_p(T), B_{i'})$, for two distinct blocks $B_i, B_{i'}$ of $G$. Then $\theta$ is a linear combination of the $\chi_j \in B_i$ and $\theta'$ is a linear combination of the $\chi_{j'} \in B_{i'}$. Hence $(\theta, \theta')_G$ is a linear combination of the $(\chi_j, \chi_{j'})_G$, for $\chi_j \in B_i$, $\chi_{j'} \in B_{i'}$. But all these $(\chi_j, \chi_{j'})_G$ are zero by (4.20) since $B_i \neq B_{i'}$. Therefore $CF(G|S_p(T), B_i)$ is perpendicular to

$CF(G|S_p(T), B_{i'})$, for all $i, i' = 1, \ldots, b$ with $i \neq i'$. Since $(\cdot, \cdot)_G$ is non-singular on $CF(G|S_p(T)) = \sum_{i=1}^{b} CF(G|S_p(T), B_i)$, this is enough to prove the proposition.

We use the above result to prove another orthogonality relation similar to (4.18). For any class function $\phi$ on $G$, we denote by $\phi|_{S_p(T)}$ the class function which equals $\phi$ on $S_p(T)$ and is zero on $G \backslash S_p(T)$.

THEOREM 12.19. *If the irreducible character $\chi_j$ belongs to the block $B_i$ of $G$, then $\chi_j|_{S_p(T)} \in CF(G|S_p(T), B_i)$. If $\chi_{j'}$ is another irreducible character belonging to a different block $B_{i'}$ of $G$, then*

$$0 = (\chi_j|_{S_p(T)}, \chi_{j'}|_{S_p(T)})_G = \frac{1}{|G|} \sum_{\sigma \in S_p(T)} \chi_j(\sigma^{-1})\chi_{j'}(\sigma). \qquad (12.20)$$

*Proof.* Clearly $\chi_j|_{S_p(T)} \in CF(G|S_p(T))$. If $B_{i'}$ is any block of $G$ different from $B_i$, and $\theta \in CF(G|S_p(T), B_{i'})$, then $(\chi_j, \theta)_G = 0$ by (4.20), since $\theta$ is a linear combination of the irreducible characters $\chi_{j'} \in B_{i'}$, all of which are different from $\chi_j$. Since $\theta = 0$ on $G \backslash S_p(T)$, this and (4.19) give

$$0 = (\chi_j, \theta)_G = \frac{1}{|G|} \sum_{\sigma \in S_p(G)} \chi_j(\sigma^{-1})\theta(\sigma)$$
$$= (\chi_j|_{S_p(T)}, \theta)_G.$$

Therefore $\chi_j|_{S_p(T)}$ is perpendicular to $CF(G|S_p(T), B_{i'})$, for every $i' \neq i$. In view of (12.18), this implies that $\chi_j|_{S_p(T)} \in CF(G|S_p(T), B_i)$. The rest of the theorem follows directly from this and (12.18).

## 13. Some Brauer Main Theorems

As in the last section, we continue to use the hypotheses and notations of Section 11.

Let $P$ be a $p$-subgroup of $G$. We define an $\mathfrak{O}$-linear map $S = S_{1 \to P}$ of $Z(\mathfrak{O}G)$ into $\mathfrak{O}G$ by

$$S(\tilde{K}_i) = \sum_{\sigma \in K_i \cap C_G(P)} \sigma, \textit{ for each class sum } \tilde{K}_i \ (i = 1, \ldots, c) \textit{ of } G, \qquad (13.1)$$

where an empty sum is understood to be zero.

PROPOSITION 13.2. *If $H$ is a subgroup of $G$ satisfying $C_G(P) \leq H \leq N_G(P)$, then $S(Z(\mathfrak{O}G)) \subseteq Z(\mathfrak{O}H)$. Furthermore,*

$$S(\tilde{K}_i)S(\tilde{K}_j) \equiv S(\tilde{K}_i\tilde{K}_j) \pmod{pZ(\mathfrak{O}H)} \quad (i, j = 1, \ldots, c). \qquad (13.3)$$

*Proof.* Since $C_G(P)$ is a normal subgroup of $H$ and $K_i$ is invariant under $H$-conjugation, the intersection $K_i \cap C_G(P)$ is $H$-invariant, $(i = 1, \ldots, c)$.

Hence its sum $S(\tilde{K}_i)$ lies in $Z(\mathfrak{O}H)$. This proves the first statement of the proposition.

For the second, let $\rho$ be any element of $H$. We must prove that the coefficient $x$ of $\rho$ in the product $S(\tilde{K}_i)S(\tilde{K}_j)$ is congruent to the coefficient $y$ of $\rho$ in $S(\tilde{K}_i\tilde{K}_j)$ modulo $p$ when both products are written as linear combinations of elements of $H$. Since $S(\tilde{K}_i)$ and $S(\tilde{K}_j)$ both lie in $\mathfrak{O}C_G(P)$, so does $S(\tilde{K}_i)S(\tilde{K}_j)$. Therefore $x = 0$ for $\rho \notin C_G(P)$. In that case $y$ is also zero by (13.1). Hence the result is true for $\rho \notin C_G(P)$.

Now suppose that $\rho \in C_G(P)$. From Proposition 4.28 and (13.1) it is clear that $y$ is the number of elements in the set $T$ of all ordered pairs $(\sigma, \tau)$ such that $\sigma \in K_i$, $\tau \in K_j$ and $\sigma\tau = \rho$. Because $P$ centralizes $\rho$, it operates naturally on $T$ by conjugation

$$(\sigma, \tau) \in T, \pi \in P \to (\sigma, \tau)^\pi = (\sigma^\pi, \tau^\pi) \in T.$$

Let $T_1$ be the subset of all elements of $T$ fixed by $P$. Evidently $(\sigma, \tau) \in T_1$ if and only if $\sigma \in K_i \cap C_G(P)$, $\tau \in K_j \cap C_G(P)$ and $\sigma\tau = \rho$. In view of (13.1), the coefficient $x$ of $\rho$ in $S(\tilde{K}_i)S(\tilde{K}_j)$ is precisely the order of $T_1$. But every $P$-orbit of $T \backslash T_1$ has length divisible by $p$, since $P$ is a $p$-group. Therefore $x = |T_1| \equiv |T| = y \pmod p$, and the proposition is proved.

COROLLARY 13.4. *The map $S$ induces an identity-preserving $\mathfrak{O}$-algebra homomorphism of $Z(\mathfrak{O}G)/pZ(\mathfrak{O}G)$ into $Z(\mathfrak{O}H)/pZ(\mathfrak{O}H)$.*

*Proof.* This follows directly from the proposition and the observation that $S(1_G) = 1_H$.

Let $\hat{B}$ be a block of $H$. By Proposition 11.15 there is a unique $\mathfrak{O}$-algebra epimorphism $\eta_{\hat{B}} : Z(\mathfrak{O}H) \to \bar{F}$ lying in $B$. Clearly $\eta_{\hat{B}}$ is zero on $pZ(\mathfrak{O}H)$. So Corollary 13.4 implies that the composition $\eta_{\hat{B}} \circ S$ is an $\mathfrak{O}$-algebra epimorphism of $Z(\mathfrak{O}G)$ onto $\bar{F} = \mathfrak{O} \cdot 1_{\bar{F}}$. We denote by $\hat{B}^G$ the unique block of $G$ having $\eta_{\hat{B}} \circ S$ as its $\mathfrak{O}$-algebra epimorphism $Z(\mathfrak{O}G) \to \bar{F}$, i.e., satisfying

$$\eta_{\hat{B}^G} = \eta_{\hat{B}} \circ S. \tag{13.5}$$

There is another way of considering the relation between $\hat{B}$ and $\hat{B}^G$. For any block $\hat{B}$ of $H$, let $e_{\hat{B}}$ be the corresponding primitive idempotent of $Z(\mathfrak{O}H)$. Define $e_B \in Z(\mathfrak{O}G)$ similarly, for any block $B$ of $G$. Then we have

PROPOSITION 13.6. *If $H$ is any subgroup of $G$ satisfying $C_G(P) \le H \le N_G(P)$ and if $B$ is any block of $G$, then*

$$S(e_B) \equiv \sum e_{\hat{B}} \pmod{pZ(\mathfrak{O}H)}, \tag{13.7}$$

*summed over all blocks $\hat{B}$ of $H$ such that $\hat{B}^G = B$.*

*Proof.* Since (8.11) holds for $Z(\mathfrak{O}H)$, Proposition 9.13 implies that the primitive idempotents of $Z(\mathfrak{O}H)/pZ(\mathfrak{O}H)$ are precisely the images of the primitive idempotents $e_{\hat{B}}$ of $Z(\mathfrak{O}H)$. By Corollary 13.4 the image of $S(e_B)$ is an idempotent in $Z(\mathfrak{O}H)/pZ(\mathfrak{O}H)$. So Proposition 9.11 gives us a unique set $T$ of blocks of $H$ such that

$$S(e_B) \equiv \sum_{\hat{B} \in T} e_{\hat{B}} \pmod{pZ(\mathfrak{O}H)}.$$

Suppose that $\hat{B} \in T$. From (11.8) for $H$ we see that $\eta_{\hat{B}} \circ S(e_B) = 1_F$. Hence $\eta_{\hat{B}} \circ S = \eta_B$, by (11.18) for $G$, and $B = \hat{B}^G$. If $\hat{B}$ is a block of $H$ not in $T$, then $\eta_{\hat{B}} \circ S(e_B) = 0$ by (11.18) for $H$. Therefore $\eta_{\hat{B}} \circ S \neq \eta_B$, by (11.18) for $G$. So $T$ is precisely the set of all blocks $\hat{B}$ of $H$ such that $B = \hat{B}^G$, and the proposition is proved.

For Brauer's second main theorem we need a simple criterion (due to D. Higman) telling us when we can apply Proposition 12.11.

LEMMA 13.8. *Let $\langle \pi \rangle$ be a non-trivial cyclic p-group, and $L$ be an $\mathfrak{O}\langle \pi \rangle$-module. Suppose there is an $\mathfrak{O}\langle \pi^p \rangle$-endomorphism $\phi : l \to l\phi$ of $L$ satisfying*

$$1_{L \to L} = \phi + \pi^{-1}\phi\pi + \ldots + \pi^{-(p-1)}\phi\pi^{(p-1)}. \tag{13.9}$$

*Then there is an $\mathfrak{O}\langle \pi^p \rangle$-submodule $K$ such that $L = K^{\langle \pi \rangle}$.*

*Proof.* Form an $\mathfrak{O}\langle \pi \rangle$-module $L^*$ having $L$ as an $\mathfrak{O}\langle \pi^p \rangle$- (but *not* $\mathfrak{O}\langle \pi \rangle$-) submodule so that $L^* = L^{\langle \pi \rangle}$. We write the module product in $L^*$ with $*$ to distinguish it from that in $L$. Then (12.1) gives

$$L^* = L \oplus L * \pi \oplus \ldots \oplus L * \pi^{p-1} \quad (\textit{as } \mathfrak{O}\textit{-modules}).$$

Let $\psi : L \to L^*$ be the map:

$$l \to l\psi = l\phi \oplus l\pi^{-1}\phi * \pi \oplus \ldots \oplus l\pi^{-(p-1)}\phi * \pi^{p-1}.$$

Using (12.2), we easily compute that $\psi$ is an $\mathfrak{O}\langle \pi \rangle$-homomorphism of $L$ into $L^*$. Similarly the map $\xi : L^* \to L$ defined by

$$l_0 \oplus l_1 * \pi \oplus \ldots \oplus l_{p-1} * \pi^{p-1} \to l_0 + l_1\pi + \ldots + l_{p-1}\pi^{p-1},$$
$$\textit{for all } l_0, l_1, \ldots, l_{p-1} \in L,$$

is an $\mathfrak{O}\langle \pi \rangle$-homomorphism of $L^*$ into $L$. Condition (13.9) says that $\psi\xi$ is the identity on $L$. Hence $\psi$ is an $\mathfrak{O}\langle \pi \rangle$-monomorphism and $L^* = \psi(L) \oplus \operatorname{Ker} \xi$ (as $\mathfrak{O}\langle \pi \rangle$-modules).

Because the $\mathfrak{O}\langle \pi \rangle$-module $L^*$ is induced from its $\mathfrak{O}\langle \pi^p \rangle$-submodule $L$, Proposition 12.10 and the Krull–Schmidt Theorem 9.10 imply that every $\mathfrak{O}\langle \pi \rangle$-direct summand of $L$ is induced from one of its $\mathfrak{O}\langle \pi^p \rangle$-submodules. We have just seen that $L$ is $\mathfrak{O}\langle \pi \rangle$-isomorphic to such an $\mathfrak{O}\langle \pi \rangle$-direct summand $\psi(L)$. Therefore the lemma holds.

To state the following theorem, we choose $P$ to be a non-trivial cyclic $p$-subgroup $\langle \pi \rangle$ of $G$. If $\chi_j$ is any irreducible character of $G$, then its restriction $(\chi_j)_{C_G(\pi)}$ is a character of $C_G(\pi) = C_G(P)$, and hence a linear combination of the irreducible characters of that group. Since these are partitioned among the blocks $\hat{B}$ of $C_G(\pi)$, we have a unique decomposition

$$(\chi_j)_{C_G(\pi)} = \sum_{\hat{B}} \chi_{j,\hat{B}}, \tag{13.10}$$

where each $\chi_{j,\hat{B}}$ is a linear combination of the irreducible characters in the block $\hat{B}$ of $C_G(\pi)$.

With this notation we have

THEOREM 13.11 (Brauer's second main theorem). *Let the irreducible character $\chi_j$ lie in the block $B$ of $G$. If $\rho$ is a $p'$-element of $C_G(\pi)$, then:*

$$\chi_j(\pi\rho) = \sum_{\hat{B}^G = B} \chi_{j,\hat{B}}(\pi\rho). \tag{13.12}$$

*Proof.* (Nagao). As in Proposition 13.6, let $e_B$ be the primitive central idempotent of $Z(\mathfrak{O}G)$ lying in $B$. Let $e$ be the idempotent $\sum_{\hat{B}^G = B} e_{\hat{B}}$ of $Z(\mathfrak{O}C_G(\pi))$. Choose any $\mathfrak{O}G$-lattice $L$ in the irreducible $FG$-module $I_j$ corresponding to $\chi_j$. Then $L = L(1-e) \oplus Le$ (as $\mathfrak{O}C_G(\pi)$-modules). Evidently $\sum_{\hat{B}^G = B} \chi_{j,\hat{B}}$ is the character $\chi_{Le}$ of $Le$. So (13.10) implies that (13.12) is equivalent to

$$\chi_{L(1-e)}(\pi\rho) = 0. \tag{13.13}$$

Let $K_i$ be any class of $G$. Then $K_i \backslash (K_i \cap C_G(\pi))$ is a union of $\langle\pi\rangle$-orbits, each of which has length $p^n$, for some $n > 0$. It follows that there is a $\langle\pi^p\rangle$-invariant subset $H_i$ of $K_i \backslash (K_i \cap C_G(\pi))$, such that the latter set is the disjoint union of the conjugates $H_i, H_i^{\pi}, \ldots, H_i^{\pi^{p-1}}$ of $H_i$. Writing $\tilde{H}_i$ for the sum of the elements of $H_i$, we obtain

$$\tilde{K}_i - S(\tilde{K}_i) = \tilde{H}_i + \pi^{-1}\tilde{H}_i\pi + \ldots + \pi^{-(p-1)}\tilde{H}_i\pi^{p-1} \quad (i = 1, \ldots, c).$$

Write $e_B = \sum_{i=1}^{c} a_i K_i$, with coefficients $a_i \in \mathfrak{O}$. Then $S(e_B) = \sum_{i=1}^{c} a_i S(\tilde{K}_i)$. By (13.7) the difference $S(e_B) - e$ has the form

$$S(e_B) - e = px = x + \pi^{-1}x\pi + \ldots + \pi^{-(p-1)}x\pi^{p-1},$$

for some $x \in Z(\mathfrak{O}C_G(\pi))$. We conclude that

$$\begin{aligned} e_B &= (e_B - S(e_B)) + (S(e_B) - e) + e \\ &= \sum_{i=1}^{c} a_i(\tilde{K}_i - S(\tilde{K}_i)) + px + e \\ &= y + \pi^{-1}y\pi + \ldots + \pi^{-(p-1)}y\pi^{p-1} + e, \end{aligned}$$

where

$$y = \left(\sum_{i=1}^{c} a_i \tilde{H}_i\right) + x$$

is an element of $\mathfrak{O}G$ commuting with $\pi^p$. Since $e$ is an idempotent commuting with $\pi$, this implies that

$$e_B(1-e) = y(1-e) + \pi^{-1}y(1-e)\pi + \ldots + \pi^{-(p-1)}y(1-e)\pi^{p-1}.$$

Because $\chi_j$ lies in the block $B$, the idempotent $e_B$ acts as identity on $I_j$. It follows that $e_B(1-e)$ acts as identity on $L(1-e)$. Since $L$ is an $\mathfrak{O}G$-module and $y(1-e) \in \mathfrak{O}G$ commutes with $\pi^p$, the map $\phi : l \to \phi = ly(1-e)$ is an $\mathfrak{O}\langle\pi^p\rangle$-endomorphism of $L(1-e)$. The above equation now tells us that $\langle\pi\rangle$, $L(1-e)$, and $\phi$ satisfy the hypotheses of Lemma 13.8. Hence there is an $\mathfrak{O}\langle\pi^p\rangle$-submodule $K$ such that $L(1-e) = K^{\langle\pi\rangle}$.

Since $\rho$ is a $p'$-element centralizing $\pi$, the group $\langle\pi\rangle$ is a $p$-Sylow subgroup of the cyclic group $\langle\pi\rho\rangle$. Now we can apply Proposition 12.11 to $\langle\pi\rho\rangle$, $\langle\pi\rangle$, $I_j(1-e)$, $L(1-e)$ and $K$, with $L(1-e)$ in the role of both $L$ and $L'$. From (12.12) we conclude that (13.13) holds. So the theorem is proved.

To make the most effective use of Brauer's second main theorem we need Brauer's "third main theorem" which tells us that prinicipal blocks correspond only to principal blocks in the relation $\hat{B}^G = B$. The *principal block* of a group $G$ is the block containing the trivial character 1 of $G$. We shall denote this block by $B_0(G)$. In view of Proposition 11.15 and equation (4.30), the corresponding $\mathfrak{O}$-epimorphism $\eta_{B_0(G)}$ of $Z(\mathfrak{O}G)$ onto $\bar{F}$ is given by

$$\eta_{B_0(G)}(\tilde{K}_i) = |K_i| \cdot 1_{\bar{F}}, \quad \textit{for all classes } K_i \textit{ of } G. \tag{13.14}$$

As you might expect, we have

PROPOSITION 13.15. *If $P$ is a $p$-subgroup of $G$ and $C_G(P) \leq H \leq N_G(P)$ then $B_0(H)^G = B_0(G)$.*

*Proof.* Let $K_i$ be any class of $G$. Letting $P$ operate by conjugation on $K_i$, we see that the non-trivial $P$-orbits form the subset $K_i \backslash (K_i \cap C_G(P))$, whose order is divisible by $p$. From (13.1), (13.5) and (13.14) we obtain

$$\begin{aligned} \eta_{B_0(H)^G}(\tilde{K}_i) &= \eta_{B_0(H)}(S(\tilde{K}_i)) = |K_i \cap C_G(P)| \cdot 1_{\bar{F}} \\ &= |K_i| \cdot 1_{\bar{F}} = \eta_{B_0(G)}(\tilde{K}_i) \quad (i = 1 \ldots, c), \end{aligned}$$

since $\bar{F}$ has characteristic $p$. This proves the proposition.

The difficult thing is to show that the converse to Proposition 13.15 holds whenever $H$ contains $PC_G(P)$. In that case any block $\hat{B}$ of $H$ satisfying $\hat{B}^G = B_0(G)$ must be equal to $B_0(H)$. The proof of this, which is Brauer's third main theorem, is rather roundabout.

We start with the study of defect groups. A *defect group* $D(K_i) = D_G(K_i)$ of a conjugacy class $K_i$ of $G$ is any $p$-Sylow subgroup of the centralizer $C_G(\sigma)$ of any element $\sigma \in K_i$. Evidently $D(K_i)$ is a $p$-subgroup of $G$ determined up to $G$-conjugation by the class $K_i$.

In order to talk about conjugacy classes of subgroups of $G$, we adopt the notation $D_1 \lesssim D_2$ (or $D_1 \lesssim_G D_2$) to mean that $D_1$ and $D_2$ are subgroups of $G$ and that $D_1$ is $G$-conjugate to a subgroup of $D_2$. Evidently $\lesssim$ is a partial ordering on the subgroups of $G$. Two subgroups $D_1$, $D_2$ are equivalent for this partial ordering if and only if they are conjugate in $G$, in which case we write $D_1 \sim D_2$ (or $D_1 \sim_G D_2$).

LEMMA 13.16. *Let $a_{ijh}$ be the integers satisfying*

$$\tilde{K}_i\tilde{K}_j = \sum_{h=1}^{c} a_{ijh}\tilde{K}_h, \quad \textit{for all } i, j = 1, \ldots, c.$$

*If $a_{ijh} \not\equiv 0 \pmod p$, for some $i, j, h$, then $D(K_h) \lesssim D(K_i)$ and $D(K_h) \lesssim D(K_j)$.*

*Proof.* Pick $\rho \in K_h$. We know from Proposition 4.28 that $a_{ijh}$ is the number of elements in the set $T = \{(\sigma, \tau) : \sigma \in K_i, \tau \in K_j, \sigma\tau = \rho\}$. Choose the defect group $D(K_h)$ to be a $p$-Sylow subgroup of $C_G(\rho)$. Then $D(K_h)$ operates by conjugation on the set $T$ of ordered pairs $(\sigma, \tau)$. If $a_{ijh} = |T| \not\equiv 0 \pmod p$, then there must be a $(\sigma, \tau) \in T$ fixed by the $p$-group $D(K_h)$. Then $D(K_h)$ is a $p$-subgroup of both $C_G(\sigma)$ and $C_G(\tau)$, and hence is contained in $p$-Sylow subgroups of these two groups, which we may take to be $D(K_i)$ and $D(K_j)$, respectively. So the lemma holds.

With the aid of the above lemma we can define defect groups of blocks.

PROPOSITION 13.17. *Let $B$ be a block of $G$, and $\eta$ be the corresponding $\mathfrak{O}$-algebra epimorphism of $Z(\mathfrak{O}G)$ onto $\bar{F}$. Then there is a unique $G$-conjugacy class of $p$-subgroups $D(B)$ of $G$ satisfying*

*There exists a conjugacy class $K_i$ such that $D(B) \sim D(K_i)$ and $\eta_B(\tilde{K}_i) \neq 0$.* (13.18a)

*If $K_j$ is any conjugacy class of $G$ such that $\eta_B(\tilde{K}_j) \neq 0$, then $D(B) \lesssim D(K_j)$.* (13.18b)

*Proof.* Choose $D(B)$ among the minimal elements for the partial ordering $\lesssim$ on the set of all defect groups of all classes $K_j$ such $\eta_B(\tilde{K}_j) \neq 0$. Then there exists a class $K_i$ such that $D(B) \sim D(K_i)$ and $\eta_B(\tilde{K}_i) \neq 0$. If $K_j$ is any class of $G$ such that $\eta_B(\tilde{K}_j) \neq 0$, then $\eta_B(\tilde{K}_i\tilde{K}_j) = \eta_B(\tilde{K}_i)\eta_B(\tilde{K}_j) \neq 0$ in the field $\bar{F}$. In the notation of Lemma 13.16 we have

$$0 \neq \eta_B(\tilde{K}_i\tilde{K}_j) = \sum_{h=1}^{c} a_{ijh}\eta_B(\tilde{K}_h).$$

So there must exist an $h = 1, \ldots, c$ such that $\eta_B(\tilde{K}_h) \neq 0$ and $a_{ijh} \not\equiv 0 \pmod p$. From Lemma 13.16 we obtain $D(K_h) \lesssim D(K_i) \sim D(B)$. The minimality of $D(B)$ forces $D(K_h) \sim D(B)$. But then Lemma 13.16 again tells us that $D(B) \sim D(K_h) \lesssim D(K_j)$. Therefore $D(B)$ satisfies (13.18). Evidently the properties (13.18) determine $D(B)$ to within conjugation. So the proposition is proved.

The groups $D(B)$ are called the *defect groups* of the block $B$. When $\eta_B$ is known, they are, of course, very easy to calculate. For example, (13.14) implies that $\eta_{B_0(G)}(\tilde{K}_i) \neq 0$ if and only if $p$ does not divide $|K_i|$, i.e., if and only if $D(K_i)$ is a $p$-Sylow subgroup of $G$. Hence,

$$D(B_0(G)) \text{ is a } p\text{-Sylow subgroup of } G. \tag{13.19}$$

To obtain a different characterization of the defect groups of a block we shall use

LEMMA 13.20. *Let $A$ be an $\mathfrak{O}$-sub algebra (with or without identity) of $Z(\mathfrak{O}G)$, and $\lambda$ be an $\mathfrak{O}$-algebra epimorphism of $A$ onto $\bar{F}$. Then there is an $\mathfrak{O}$-algebra epimorphism $\eta$ of $Z(\mathfrak{O}G)$ onto $\bar{F}$ whose restriction to $A$ is $\lambda$.*

*Proof.* By Proposition 11.15 and (11.11) the $\mathfrak{O}$-algebra epimorphisms $\eta_1, \ldots, \eta_b$ of $Z(\mathfrak{O}G)$ onto $\bar{F}$ satisfy $\operatorname{Ker}\eta_1 \cap \ldots \cap \operatorname{Ker}\eta_b = J(Z(\mathfrak{O}G))$. By (8.12) there is an integer $t > 0$ such that $J(Z(\mathfrak{O}G))^t \subseteq pZ(\mathfrak{O}G)$. Since $A$ is a $\mathbf{Z}_p$-submodule of the finitely-generated $\mathbf{Z}_p$-module $Z(\mathfrak{O}G)$, there is an integer $s > 0$ such that $p^s Z(\mathfrak{O}G) \cap A \subseteq pA \subseteq \operatorname{Ker}\lambda$. It follows that $A \cap \operatorname{Ker}\eta_1, \ldots, A \cap \operatorname{Ker}\eta_b$ are ideals of $A$ satisfying

$$\begin{aligned}
[(A \cap \operatorname{Ker}\eta_1)\ldots(A \cap \operatorname{Ker}\eta_b)]^{st} &\subseteq A \cap [(\operatorname{Ker}\eta_1)\ldots(\operatorname{Ker}\eta_b)]^{st} \\
&\subseteq A \cap [(\operatorname{Ker}\eta_1 \cap \ldots \cap \operatorname{Ker}\eta_b)]^{st} \\
&= A \cap J(Z(\mathfrak{O}G))^{st} \\
&\subseteq A \cap p^s Z(\mathfrak{O}G) \subseteq pA \subseteq \operatorname{Ker}\lambda.
\end{aligned}$$

Since $\lambda$ is an epimorphism of $A$ onto a field $\bar{F}$, we conclude that $A \cap \operatorname{Ker}\eta_i \subseteq \operatorname{Ker}\lambda$, for some $i = 1, \ldots, b$. But then $\lambda$ is the restriction to $A$ of an $\mathfrak{O}$-algebra epimorphism $\eta$ of $A + \operatorname{Ker}\eta_i$ onto $\bar{F}$ such that $\operatorname{Ker}\eta_i \subseteq \operatorname{Ker}\eta$. Since

$$\operatorname{Ker}\eta_i \subseteq \operatorname{Ker}\eta \subset A + \operatorname{Ker}\eta_i \subseteq Z(\mathfrak{O}G),$$

and $Z(\mathfrak{O}G)/\operatorname{Ker}\eta_i \simeq \bar{F}$ (as $\mathfrak{O}$-algebras), we must have $\operatorname{Ker}\eta = \operatorname{Ker}\eta_i$, $A + \operatorname{Ker}\eta_i = Z(\mathfrak{O}G)$, and $\eta = \eta_i$. Therefore the lemma holds.

The other characterization of the defect groups of a block $B$ is as the largest groups $P$, in the partial ordering $\lesssim$, from which $B$ can be obtained via (13.5).

PROPOSITION 13.21. *Let $P$ be a $p$-subgroup of $G$, $H$ be any subgroup satisfying $C_G(P) \le H \le N_G(P)$, and $B$ be a block of $G$. Then there exists a block $\hat{B}$ of $H$ such that $\hat{B}^G = B$ if and only if $P \lesssim D(B)$.*

*Proof.* Suppose such a block $\hat{B}$ exists. Then the $\mathfrak{O}$-algebra epimorphism $\eta_{\hat{B}}$ of $Z(\mathfrak{O}H)$ onto $\bar{F}$ lying in $\hat{B}$ defines the corresponding epimorphism $\eta_B : Z(\mathfrak{O}G) \to \bar{F}$ by (13.5). From (13.1) it is clear that $S(\tilde{K}_i) = 0$, for all classes $K_i$ of $G$ such that $P \not\lesssim D(K_i)$. Hence $\eta_B(\tilde{K}_i) = \eta_{\hat{B}}(S(\tilde{K}_i)) = 0$, for all such $K_i$. In view of (13.18*a*), this implies that $P \lesssim D(B)$.

Now suppose that $P \lesssim D(B)$. From (13.18*b*) we see that $\eta_B(K_i) = 0$, for all classes $K_i$ of $G$ such that $P \not\lesssim D(K_i)$. But the kernel of $S : Z(\mathfrak{O}G) \to Z(\mathfrak{O}H)$

is precisely the $\mathfrak{D}$-linear span of the sums $\tilde{K}_i$ of these classes (by 13.1). Hence there is a unique $\mathfrak{D}$-linear map $\lambda_1$ of the image $S(Z(\mathfrak{D}G))$ into $\bar{F}$ such that $\eta_B = \lambda_1 \circ S$. It is clear from (13.1) that $S(pZ(\mathfrak{D}G)) = pS(Z(\mathfrak{D}G)) = pZ(\mathfrak{D}H) \cap S(Z(\mathfrak{D}G))$. Since $\eta_B$ is zero on $pZ(\mathfrak{D}G)$, the map $\lambda_1$ is zero on $pS(Z(\mathfrak{D}G))$. So it has a unique extension to an $\mathfrak{D}$-linear map $\lambda$ of $A = pZ(\mathfrak{D}H) + S(Z(\mathfrak{D}G))$ onto $\bar{F}$ such that $\lambda(pZ(\mathfrak{D}H)) = 0$. In view of Corollary 13.4, $A$ is an $\mathfrak{D}$-subalgebra of $Z(\mathfrak{D}H)$ and $\lambda$ is an $\mathfrak{D}$-algebra epimorphism. Lemma 13.20 now gives us an $\mathfrak{D}$-algebra epimorphism $\eta : Z(\mathfrak{D}H) \to \bar{F}$ whose restriction to $A$ is $\lambda$. Then $\eta \circ S = \lambda \circ S = \lambda_1 \circ S = \eta_B$, and $\hat{B}^G = B$, where $\hat{B}$ is the block of $H$ corresponding to $\eta$ in Proposition 11.15. This proves the proposition.

We need some information about the blocks of a group $H$ having a normal $p$-subgroup $P$. Let $\phi$ be the $\mathfrak{D}$-algebra epimorphism of $\mathfrak{D}H$ onto $\mathfrak{D}(H/P)$ induced by the natural group epimorphism $H \to H/P$. Then the restriction of $\phi$ is an $\mathfrak{D}$-algebra homomorphism of $Z(\mathfrak{D}H)$ into $Z(\mathfrak{D}(H/P))$.

LEMMA 13.22. *If $\hat{B}$ is a block of $H$, then there exists a block $B$ of $H/P$ such that $\eta_{\hat{B}} = \eta_B \circ \phi$.*

*Proof.* Let Aug ($\mathfrak{D}P$) be the *augmentation ideal* of $\mathfrak{D}P$, having the elements $\sigma - 1$, for $\sigma \in P - \{1\}$, as $\mathfrak{D}$-basis. Proposition 11.23 implies that Aug $(\mathfrak{D}P) \subseteq J(\mathfrak{D}P)$. So (8.12) gives us an integer $d > 0$ such that $[\text{Aug}(\mathfrak{D}P)]^d \subseteq p\mathfrak{D}P$. Evidently the kernel of $\phi$ is $\text{Ker}\,\phi = \text{Aug}(\mathfrak{D}P)\cdot\mathfrak{D}H = \mathfrak{D}H\cdot\text{Aug}(\mathfrak{D}P)$. Hence,

$$(\text{Ker}\,\phi)^d = [\text{Aug}(\mathfrak{D}P)]^d \cdot \mathfrak{D}H \subseteq p\mathfrak{D}H.$$

It follows that $\text{Ker}\,\phi \cap Z(\mathfrak{D}H)$, the kernel of the restriction of $\phi$ to $Z(\mathfrak{D}H)$, is an ideal of $Z(\mathfrak{D}H)$ which is nilpotent modulo $pZ(\mathfrak{D}H)$. This implies that $\text{Ker}\,\phi \cap Z(\mathfrak{D}H) \subseteq \text{Ker}\,\eta_{\hat{B}}$. So there is a unique $\mathfrak{D}$-algebra epimorphism $\lambda$ of $A = \phi(Z(\mathfrak{D}H))$ onto $\bar{F}$ such that $\eta_{\hat{B}} = \lambda \circ \phi$. Applying Lemma 13.20, we obtain an $\mathfrak{D}$-algebra epimorphism $\eta$ of $Z(\mathfrak{D}(H/P))$ onto $\bar{F}$ whose restriction to $A$ is $\lambda$. By Proposition 11.15, $\eta = \eta_B$, for a block $B$ of $H/P$. We have proved the lemma.

COROLLARY 13.23. *If $K$ is a conjugacy class of $H$ lying in $H\backslash C_H(P)$ then $\eta_{\hat{B}}$ vanishes on the corresponding class sum $\tilde{K}$. Hence $P \leq D(\hat{B})$.*

*Proof.* Let $\sigma$ be an element of $K$, and $C_H(\phi(\sigma))$ be the inverse image in $H$ of $C_{H/P}(\phi(\sigma))$. Then the image in $H/P$ of the class $K$ of $\sigma$ is the class $L$ of $\phi(\sigma)$. But the corresponding class sums $\tilde{K}$ and $\tilde{L}$ are related by

$$\phi(\tilde{K}) = [C_H(\phi(\sigma)) : C_H(\sigma)]\tilde{L},$$

since $[C_H(\phi(\sigma)) : C_H(\sigma)]$ members of $K$ map onto each member of $L$. Evidently $C_H(\phi(\sigma))$ contains $PC_H(\sigma)$. Therefore $[C_H(\phi(\sigma)) : C_H(\sigma)]$ is divisible by $[PC_H(\sigma) : C_H(\sigma)] = [P : C_P(\sigma)] = p^s$, for some $s \geq 1$, since $P$ does not centralize $\sigma$. It follows that $\phi(\tilde{K}) \in pZ(\mathfrak{O}(H/P))$, and hence that $\eta_{\hat{B}}(\tilde{K}) = \eta_B \circ \phi(\tilde{K}) = 0$ in $\bar{F}$, which is the first conclusion of the corollary. The second conclusion comes from the first and the definition (13.18) of $D(\hat{B})$.

Since $P$ is a normal subgroup of $H$, so is $C_H(P)$. Hence $H$ acts by conjugation as $\mathfrak{O}$-algebra automorphisms of $Z(\mathfrak{O}C_H(P))$. It follows that $H$ permutes the $\mathfrak{O}$-algebra epimorphisms $\eta : Z(\mathfrak{O}C_H(P)) \to \bar{F}$ among themselves by conjugation:

$$\eta^\sigma(y) = \eta(y^{\sigma^{-1}}), \quad \textit{for all } \sigma \in H,\ y \in Z(\mathfrak{O}C_H(P)). \tag{13.24}$$

Applying Proposition 11.15, we obtain a natural action of $H$ on the corresponding blocks of $C_H(P)$.

PROPOSITION 13.25. *There is a one-to-one correspondence between blocks $\hat{B}$ of $H$ and $H$-conjugacy classes of blocks $\bar{B}$ of $C_H(P)$. The block $\hat{B}$ corresponds to the class of $\bar{B}$ if and only if the corresponding $\mathfrak{O}$-algebra epimorphisms $\eta_B : Z(\mathfrak{O}H) \to \bar{F}$ and $\eta_{\bar{B}} : Z(\mathfrak{O}C_H(P)) \to \bar{F}$ have the same restriction to $Z(\mathfrak{O}H) \cap Z(\mathfrak{O}C_H(P))$.*

*Proof.* For each block $\hat{B}$ of $H$, let $T(\hat{B})$ be the family of all blocks $\bar{B}$ of $C_H(P)$ such that $\eta_{\hat{B}}$ and $\eta_{\bar{B}}$ have the same restriction to $A = Z(\mathfrak{O}H) \cap Z(\mathfrak{O}C_H(P))$. Since $1 \in A$, the restriction $\lambda$ of $\eta_{\hat{B}}$ is an $\mathfrak{O}$-algebra epimorphism of $A$ onto $\bar{F}$. So Lemma 13.20 and Proposition 11.15 imply that $T(\hat{B})$ is not empty. Because each element of $A$ is fixed under conjugation by elements of $H$, it follows from (13.24) that $T(\hat{B})$ is an $H$-invariant set of blocks of $C_H(P)$.

Let $\bar{B}_1, \ldots, \bar{B}_s$ be an $H$-orbit in $T(\hat{B})$, and $e_1, \ldots, e_s$ be the corresponding primitive idempotents of $Z(\mathfrak{O}C_H(P))$. Evidently $e_1, \ldots, e_s$ is an $H$-conjugacy class of primitive idempotents of $Z(\mathfrak{O}C_H(P))$. So $e_1 + \ldots + e_s$ is invariant under $H$, and hence lies in $A$. From (11.18) we see that $\lambda(e_1 + \ldots + e_s) = \eta_{\bar{B}_1}(e_1 + \ldots + e_s) = 1$. If $\bar{B}$ is any other block of $C_H(P)$, then (11.18) gives $\eta_{\bar{B}}(e_1 + \ldots + e_s) = 0 \neq \lambda(e_1 + \ldots + e_s)$. Therefore $\bar{B}$ does not lie in $T(\hat{B})$. Hence $T(\hat{B}) = \{\bar{B}_1 \ldots, \bar{B}_s\}$ is an $H$-conjugacy class of blocks of $C_H(P)$.

If $\bar{B}$ is any block of $C_H(P)$, then the restriction $\lambda$ of $\eta_{\bar{B}}$ to $A$ is an $\mathfrak{O}$-algebra epimorphism of $A$ onto $\bar{F}$, and hence is the restriction to $A$ of $\eta_{\hat{B}}$, for some block $\hat{B}$ of $H$, by Lemma 13.20 and Proposition 11.15. Therefore $\bar{B} \in T(\hat{B})$, for some block $\hat{B}$ of $H$.

Corollary 13.23 implies that $\eta_{\hat{B}}$ is determined by its restriction to $Z(\mathfrak{O}H) \cap \mathfrak{O}C_H(P) = A$, for any block $\hat{B}$ of $H$. Hence $\hat{B}$ is determined by $T(\hat{B})$. We have shown that $\hat{B} \leftrightarrow T(\hat{B})$ is a one to one correspondence between blocks of $H$ and $H$-conjugacy classes of blocks of $C_H(P)$. That is the proposition.

We need one more lemma.

LEMMA 13.26. *Let $P$ be a $p$-subgroup of the group $G$. Then the map $K_i \to K_i \cap C_G(P)$ sends the family of all conjugacy classes $K_i$ of $G$ such that $P \sim D_G(K_i)$ one to one onto the family of all conjugacy classes $L_j$ of $N_G(P)$ such that $P = D_{N_G(P)}(L_j)$.*

*Proof.* If $P$ is a defect group of a conjugacy class $K_i$ of $G$, then $K_i$ contains some element $\sigma$ having $P$ as a Sylow subgroup of its centralizer $C_G(\sigma)$. Hence $\sigma \in K_i \cap C_G(P)$. Therefore $K_i \cap C_G(P)$ is not empty.

If $\tau$ is another element of $K_i \cap C_G(P)$, then $\tau = \sigma^\rho$, for some $\rho \in G$. Evidently both $P$ and $P^\rho$ are $p$-Sylow subgroups of $C_G(\tau)$. Hence there exists $\pi \in C_G(\tau)$ such that $P^{\rho\pi} = P$. Now $\rho\pi$ is an element of $N_G(P)$ satisfying $\sigma^{\rho\pi} = \tau^\pi = \tau$. We conclude that $K_i \cap C_G(P)$ is a conjugacy class of $N_G(P)$. Because $P$ is a $p$-Sylow subgroup of $C_{N_G(P)}(\sigma)$, the defect group $D_{N_G(P)}(K_i \cap C_G(P))$ is precisely $P$.

Now let $L_j$ be any conjugacy class of $N_G(P)$ such that $P = D_{N_G(P)}(L_j)$, and let $K_i$ be the conjugacy class of $G$ containing $L_j$. Evidently $L_j \subseteq K_i \cap C_G(P)$. Pick an element $\sigma \in L_j$. Let $D$ be a $p$-Sylow subgroup of $C_G(\sigma)$ containing $P$. If $P < D$, then $P < N_D(P) \leq C_{N_G(P)}(\sigma)$, which is impossible since $P$ is a $p$-Sylow subgroup of $C_{N_G(P)}(\sigma)$. Hence $P = D \sim D(K_i)$. Because $K_i \cap C_G(P)$ is a class of $N_G(P)$, we have $L_j = K_i \cap C_G(P)$. The map $K_i \to K_i \cap C_G(P)$ being clearly one-to-one, this is enough to prove the lemma.

At last we have

THEOREM 13.27 (Brauer's third main theorem). *Suppose that $P$ is a $p$-subgroup of $G$, and that $H$ is a subgroup satisfying $PC_G(P) \leq H \leq N_G(P)$. If $\hat{B}$ is a block of $H$, then $\hat{B}^G = B_0(G)$, if and only if $\hat{B} = B_0(H)$.*

*Proof.* We already know that $\hat{B} = B_0(H)$ implies $\hat{B}^G = B_0(G)$, by Proposition 13.15. So it is only necessary to prove the converse. Choose a counterexample $P$, $H$, $\hat{B}$ with $|P|$ maximal. We must first show that we can assume $H = N_G(P)$.

Since $P$ is a normal subgroup of $H$ and $C_G(P) = C_H(P)$, Proposition 13.25 tells us that $\hat{B}$ and $B_0(H) \neq \hat{B}$ correspond to distinct $H$-conjugacy classes of blocks of $C_G(P)$. It is evident from (13.14) that $B_0(H)$ corresponds to the $H$-conjugacy class containing $B_0(C_G(P))$. Let $\bar{B} \neq B_0(C_G(P))$ be a block of $C_G(P)$ in the $H$-conjugacy class corresponding to $\hat{B}$. It follows from (13.14) that $B_0(C_G(P))$ is $N_G(P)$-invariant. So the block $\hat{B}_1$ of $N_G(P)$ corresponding to the $N_G(P)$-class of $\bar{B}$ in Proposition 13.25 is different from $B_0(N_G(P))$. But $\eta_{\hat{B}_1}$, $\eta_{\bar{B}}$ and $\eta_{\hat{B}}$ have the same restriction to

$$S(Z(\mathfrak{O}G)) \subseteq Z(\mathfrak{O}C_G(P)) \cap Z(\mathfrak{O}N_G(P)) \subseteq Z(\mathfrak{O}C_G(P)) \cap Z(\mathfrak{O}H).$$

Therefore $\eta_{\hat{B}_1} \circ S = \eta_{\hat{B}} \circ S$. Hence $\hat{B}_1^G = \hat{B}^G = B_0(G)$, and $P$, $N_G(P)$, $\hat{B}_1$ is a counterexample to the theorem.

Now assume that $H = N_G(P)$. By Corollary 13.23, the defect group $D(\hat{B})$ contains $P$. Suppose that $D(\hat{B}) = P$. Then $\eta_{\hat{B}}(\tilde{L}_j) \neq 0$, for some sum $\tilde{L}_j$ of some class $L_j$ of $N_G(P)$ having $D(L_j) = P$. Lemma 13.26 gives us a conjugacy class $K_i$ of $G$ satisfying $D(K_i) \sim P$ and $K_i \cap C_G(P) = L_j$. By (13.1), $S(\tilde{K}_i) = \tilde{L}_j$. Therefore $\eta_{B_0(G)}(\tilde{K}_i) = \eta_{\hat{B}} \circ S(\tilde{K}_i) = \eta_{\hat{B}}(\tilde{L}_j) \neq 0$. From (13.18b) and (13.19) we conclude that $P$ is a $p$-Sylow subgroup of $G$. Now every class $L_j$ of $N_G(P)$ contained in $C_G(P)$ has $P$ as its defect group. Since $\hat{B} \neq B_0(N_G(B))$, Corollary 13.23 implies the existence of such a class $L_j$ such that $\eta_{\hat{B}}(\tilde{L}_j) \neq \eta_{B_0}(\tilde{L}_j)$. But then the corresponding class $K_i$ of $G$ satisfies

$$\eta_{\hat{B}} \circ S(\tilde{K}_i) = \eta_{\hat{B}}(\tilde{L}_j) \neq \eta_{B_0}(\tilde{L}_j) = \eta_{B_0} \circ S(\tilde{K}_v),$$

which is impossible, since $\hat{B}^G = B_0^G = B_0(G)$ (by Proposition 13.15). We conclude that $D(\hat{B})$ strictly contains $P$.

From $P < D(\hat{B})$ we obtain $C_G(D(\hat{B})) \leq C_G(P) \leq N_G(P)$. Therefore $D(\hat{B})C_G(D(\hat{B})) \leq N_G(P)$. Proposition 13.21 gives us a block $\hat{\hat{B}}$ of $D(\hat{B})C_G(D(\hat{B}))$ such that $\hat{\hat{B}}_{N_G(P)} = \hat{B}$. Since $\hat{B} \neq B_0(N_G(P))$, Proposition 13.15 implies that $\hat{\hat{B}}$ is not $B_0(D(\hat{B})C_G(D(\hat{B})))$. From (13.1) we compute immediately that $\hat{\hat{B}}^G$ (defined with respect to the $p$-group $D(\hat{B})$) is $(\hat{\hat{B}}^{N_G(P)})^G = \hat{B}^G = B_0(G)$. Therefore $D(\hat{B})$, $D(\hat{B})C_G(D(\hat{B}))$, $\hat{\hat{B}}$ is a counter-example with $|D(\hat{B})| > |P|$. This contradicts the maximality of $|P|$. The final contradiction proves the theorem.

We denote by $O_{p'}(G)$ the largest normal $p'$-subgroup of $G$, i.e., the largest normal subgroup whose order is not divisible by $p$. The following relation between $O_{p'}(G)$ and $B_0(G)$ is very useful in applications.

PROPOSITION 13.28. $O_{p'}(G) = \bigcap_{\chi_j \in B_0(G)} \operatorname{Ker}(\chi_j)$.

*Proof.* Let $f = |O_{p'}(G)|^{-1} \sum_{\sigma \in O_{p'}(G)} \sigma$ be the primitive idempotent of $Z(FO_{p'}(G))$ corresponding to the trivial character of $O_{p'}(G)$. By Corollary 11.10, $f$ is an idempotent in $\mathfrak{O}G$. Clearly it lies in $Z(\mathfrak{O}G)$. So Proposition 9.11 gives a unique decomposition $f = e_1 + \ldots + e_t$, where $e_1, \ldots, e_t$ are primitive idempotents of $Z(\mathfrak{O}G)$ corresponding to blocks $B_1, \ldots, B_t$, respectively, of $G$. Using (11.8), we see that the unique decomposition of $f$ as a sum of primitive idempotents in $Z(FG)$ is

$$f = \sum_{i=1}^{t} \sum_{\chi_j \in B_i} e(\chi_j) \tag{13.29}$$

Evidently the idempotent $e(1) = |G|^{-1} \sum_{\sigma \in G} \sigma$ of $Z(FG)$ corresponding to the trivial character of $G$ satisfies $fe(1) = e(1)$. So it appears in the decomposition (13.29) of $f$. Hence the block $B_0(G)$ containing $e(1)$ appears in the list $B_1, \ldots, B_t$, say as $B_1$. If $\chi_j$ is any character in $B_0(G)$, then (13.29) implies that $e(\chi_j)f = e(\chi_j)$. If $I_j$ is any irreducible $FG$-module with character $\chi_j$, then $I_j = I_j e(\chi_j)$ and $e(\chi_j)f = e(\chi_j)$ imply that $f$ acts as identity on $I_j$. Since $f\sigma = f$, for all $\sigma \in O_{p'}(G)$, we conclude that $O_{p'}(G)$ acts trivially on $I_j$. Therefore $O_{p'}(G) \leq \operatorname{Ker}(\chi_j)$, by (4.37a). This proves that

$$O_{p'}(G) \subseteq \bigcap_{\chi_j \in B_0(G)} \operatorname{Ker}(\chi_j).$$

Evidently $N = \bigcap_{\chi_j \in B_0(G)} \operatorname{Ker}(\chi_j)$ is a normal subgroup of $G$. Suppose that $N \nsubseteq O_{p'}(G)$. Then $p$ must divide $|N|$. So there is a $p$-element $\pi \neq 1$ in $N$. Applying Theorem 12.13, we obtain

$$0 = \sum_{\chi_j \in B_0(G)} \chi_j(\pi)\chi_j(1).$$

But this sum is

$$\sum_{\chi_j \in B_0(G)} \chi_j(1)^2 > 0,$$

since $\pi \in \operatorname{Ker}(\chi_j)$, for all $\chi_j \in B_0(G)$ (see Proposition 4.38(a)). The contradiction shows that $N \leq O_{p'}(G)$, and finishes the proof of the proposition.

## 14. Quaternion Sylow Groups

In addition to the hypotheses of the last three sections, we suppose that $G$ has a t.i. set $S$ satisfying

$$\textit{An element } \sigma \in G \textit{ lies in } S \textit{ if and only if } \sigma_p \in S. \tag{14.1}$$

We denote by $H$ the normalizer of $S$ in $G$.

Let $T$ be the subset of all $p$-elements in $S$. By (5.10a, c), the subset $T$ is closed under inverses and invariant under $H$-conjugation. From (14.1) it is clear that $S$ is the $p$-section of $H$ defined by $T$ via (12.16). We write this section as $S_p(T \text{ in } H)$, instead of $S_p(T)$, to indicate the group $H$. Then $S^G$ is the $p$-section $S_p(T \text{ in } G)$.

We use the notation of Proposition 12.17 and Theorem 12.19 with respect to $CF(H|S, \hat{B})$, or $CF(G|S^G, B)$, for blocks $\hat{B}$, $B$ of $H$, $G$ respectively.

PROPOSITION 14.2. *Induction:* $\phi \to \phi^G$ *is an isometry of* $CF(H|S, B_0(H))$ *on to* $CF(G|S^G, B_0(G))$.

*Proof.* Proposition 5.15 says that $\phi \to \phi^G$ is an isometry of $CF(H|S)$ into $CF(G|S^G)$. If $\psi$ is any class function on $G$, let $\psi|_S$ be the class function on $H$ which equals $\psi$ on $S$ and is zero on $H \backslash S$. From Proposition 5.14 it is clear that $(\psi|_S)^G$ equals $\psi$ on $S^G$ and is zero on $G \backslash S^G$. It follows that $\phi \to \phi^G$ is an isometry of $CF(H|S)$ onto $CF(G|S^G)$, with $\psi \to \psi|_S$ as its inverse.

Let $\pi$ be any $p$-element of $S$. Since $S$ is a t.i. set, the centralizer $C = C_G(\pi)$ is contained in $H$, and hence equals $C_H(\pi)$. Let $R = R(\pi) = S_p(\{\pi, \pi^{-1}\}$ in $C)$. Evidently $R$ is the set of all $\sigma \in C$ such that $\sigma_p = \pi$ or $\pi^{-1}$. By (14.1), $S$ is the union of its subsets $R(\pi)$, where $\pi$ runs over all elements of $T$.

Suppose that $\psi$ is a class function on some group containing $C$. For each block $\bar{B}$ of $C$ we denote by $\psi_{\bar{B}}$ the unique $F$-linear combination of the irreducible characters in $\bar{B}$ such that the restriction $\psi_C$ of $\psi$ to $C$ satisfies

$$\psi_C = \sum_{\bar{B}} \psi_{\bar{B}}.$$

As usual, we denote by $\psi|_R$ the class function on $C$ which equals $\psi$ on $R$ and is zero on $C\backslash R$. Theorem 12.19 implies that $\psi_{\bar{B}}|_R \in CF(C|R, \bar{B})$, for all blocks $\bar{B}$ of $C$. Hence,

$$\psi|_R = \psi_C|_R = \sum_{\bar{B}} \psi_{\bar{B}}|_R \tag{14.3}$$

is the unique decomposition of $\psi|_R \in CF(C|R)$ as a sum of elements $\psi_{\bar{B}}|_R \in CF(C|R, \bar{B})$, given by (12.18).

Now let $\psi$ be an element of $CF(G|S^G, B)$, for some block $B$ of $G$. Brauer's second main theorem, applied to both $\pi$ and $\pi^{-1}$, tells us that

$$\psi|_R = \sum_{\bar{B}^G = B} \psi_{\bar{B}}|_R.$$

Since the decomposition (14.3) is unique, we conclude that

$$\psi_{\bar{B}}|_R = 0, \quad \textit{for all } \bar{B} \textit{ such that } \bar{B}^G \neq B. \tag{14.4}$$

We decompose $\psi|_S$ in the form

$$\psi|_S = \sum_{\hat{B}} \psi_{\hat{B}},$$

where $\psi^{\hat{B}} \in CF(H|S, \hat{B})$, for each block $\hat{B}$ of $H$. As in (14.4), we have

$$(\psi_{\hat{B}})_{\bar{B}}|_R = 0, \quad \textit{for all } \bar{B}, \hat{B} \textit{ such that } \bar{B}^H \neq \hat{B}. \tag{14.5}$$

Evidently $R \subseteq S$ implies that

$$\psi|_R = (\psi|_S)|_R = \sum_{\hat{B}} \psi_{\bar{B}}|_R = \sum_{\hat{B}, \bar{B}} (\psi_{\hat{B}})_{\bar{B}}|_R.$$

The unicity of (14.3) and (14.5) give

$$\psi_{\bar{B}}|_R = \sum_{\hat{B}} (\psi_{\hat{B}})_{\bar{B}}|_R = (\psi_{\bar{B}^H})_{\bar{B}}|_R, \quad \textit{for all } \bar{B}. \tag{14.6}$$

Suppose that $B = B_0(G)$. Brauer's third main theorem and (14.4) tell us that $\psi_{\bar{B}}|_R = 0$, for all $\bar{B} \neq B_0(C)$. If $\hat{B} \neq B_0(H)$, then (14.5), (14.6) and Brauer's third main theorem now imply that $(\psi_{\hat{B}})_{\bar{B}}|_R = 0$, for all $\bar{B}$. Hence $\psi_{\hat{B}}|_R = 0$ by (14.3). Since $\pi$ is an arbitrary element of $T$, this says that $\psi_{\hat{B}}$ vanishes on the union $S$ of the $R(\pi)$, $\pi \in T$. Therefore $\psi_{\hat{B}} = 0$. We conclude that $\psi|_S = \psi_{B_0(H)}$. So $\psi \to \psi|_S$ maps $CF(G|S^G, B_0(G))$ into $CF(H|S, B_0(H))$.

Suppose that $B \neq B_0(G)$. Brauer's third main theorem and (14.4) tell us that $\psi_{B_0(C)}|_R = 0$. So (14.6), (14.5), and Brauer's third main theorem imply that $(\psi_{B_0(H)})_{\bar{B}}|_R = 0$, for all $\bar{B}$. Therefore $\psi_{B_0(H)}|_R = 0$. As above, we conclude that $\psi_{B_0(H)} = 0$. So $\psi \to \psi|_S$ maps $\sum_{B \neq B_0(G)} CF(G|S^G, B)$ into $\sum_{\hat{B} \neq B_0(H)} CF(H|S, \hat{B})$. Since $\psi \to \psi|_S$ sends $CF(G|S^G)$ onto $CF(H|S)$, this and (12.18) are enough to prove the proposition.

COROLLARY 14.7. *The inverse map to $\phi \to \phi^G$ sends $\psi \in CF(G|S^G, B_0(G))$ into $\psi|_S \in CF(H|S, B_0(H))$.*

*Proof.* This was shown in the first paragraph of the above proof.

The above proposition gives us a "method" for finding those irreducible characters $\chi_j \in B_0(G)$ which do not vanish on $S^G$. By Theorem 12.19 and (12.18), an irreducible character $\chi_j$ of $G$ satisfies these conditions if and only if $\chi_j|_{S^G}$ is a non-zero element of $CF(G|S^G, B_0(G))$, hence if and only if there exists $\psi \in CF(G|S^G, B_0(G))$ such that $(\psi, \chi_j)_G = (\psi, \chi_j|_{S^G})_G \neq 0$. By the proposition this occurs if and only if $(\phi^G, \chi_j)_G \neq 0$, for some $\phi \in CF(H|S, B_0(H))$. Therefore we can find all these $\chi_j$ by taking all the characters $\phi \in CF(H|S, B_0(H))$ (actually, a basis will do), inducing them to $G$, writing the resulting characters $\phi^G$ as linear combinations of irreducible characters $\chi_j$ of $G$, and taking all the $\chi_j$ having a non-zero coefficient in one of these decompositions. This doesn't *sound* too practical, but, as we shall see later, it can be made to work in certain cases.

The best of all cases is the *coherent case*. To define this, let $\phi_1, \ldots, \phi_c$ be the irreducible characters in $B_0(H)$ which do not vanish on $S$. The coherent case occurs when we can find irreducible characters $\chi_1, \ldots, \chi_c$ of $G$ and signs $\varepsilon_1, \ldots, \varepsilon_c = \pm 1$ satisfying

If $f_1, \ldots, f_c \in F$ and $f_1\phi_1 + \ldots + f_c\phi_c \in CF(H|S, B_0(H))$, *then*

$$(f_1\phi_1 + \ldots + f_c\phi_c)^G = f_1\varepsilon_1\chi_1 + \ldots + f_c\varepsilon_c\chi_c. \tag{14.8}$$

In the coherent case we have

PROPOSITION 14.9. *The characters $\chi_1, \ldots, \chi_c$ are precisely the irreducible characters in $B_0(G)$ which do not vanish on $S^G$. Furthermore, they satisfy:*

$$\chi_i|_S = \varepsilon_i\phi_i|_S \quad (i = 1, \ldots, c). \tag{14.10}$$

*Proof.* Evidently every $\phi \in CF(H|S, B_0(H))$ has the form $f_1\phi_1 + \ldots + f_c\phi_c$, for some $f_1, \ldots, f_c \in F$. We have seen in the above discussion that the irreducible characters in $B_0(G)$ which do not vanish on $S^G$ are precisely those which occur with a non-zero coefficient in some such $\phi^G$. So these characters are among $\chi_1, \ldots, \chi_c$ by (14.8).

Fix $i = 1, \ldots, c$. By Theorem 12.19, the class function $\phi_i|_S$ is a non-zero member of $CF(H|S, B_0(H))$. So Proposition 12.17 gives us an element $\phi \in CF(H|S, B_0(H))$ such that $(\phi_i|_S, \phi)_H \neq 0$. Evidently $\phi = f_1\phi_1 + \ldots + f_c\phi_c$, for some $f_1, \ldots, f_c \in F$, and

$$f_i = (\phi_i, \phi)_H = (\phi_i|_S, \phi)_G \neq 0.$$

From (14.8) we see that $\chi_i$ has a non-zero coefficient $f_i\varepsilon_i$ in $\phi^G$. Therefore $\chi_i \in B_0(G)$ and $\chi_i|_{S^G} \neq 0$. This proves the first statement of the proposition.

For the second statement, notice that condition (14.8) implies that

$$(\chi_i, \phi^G)_G = (\varepsilon_i\phi_i, \phi)_H, \quad \textit{for all } \phi \in CF(H|S, B_0(H)).$$

Using the Frobenius reciprocity law, this becomes

$$(\chi_i|_S, \phi)_H = ((\chi_i)_H, \phi)_H = (\chi_i, \phi^G)_G = (\varepsilon_i\phi_i|_S, \phi)_H,$$
$$\text{for all } \phi \in CF(H|S, B_0(H)).$$

By Corollary 14.7 and Theorem 12.19, both $\chi_i|_S$ and $\varepsilon_i\phi_i|_S$ lie in $CF(H|S, B_0(H))$. So Proposition 12.17 and the above equation imply (14.10), which finishes the proof of the proposition.

We shall apply the above ideas in the case $p = 2$ to prove

THEOREM 14.11 (Brauer–Suzuki). *Let $G$ have a quaternion group $Q$ (of order 8) as a 2-Sylow subgroup. Then $G$ has a normal subgroup $N \neq G$ containing the involution $\iota$ of $Q$.*

To show this we shall prove a series of lemmas, all based on the hypothesis that $G$ is a counterexample. We begin with

LEMMA 14.12. *All the elements of $Q\backslash\langle\iota\rangle$ are conjugate to each other in $G$.*

*Proof.* Let $\rho$, $\tau$ be two elements of $Q\backslash\langle\iota\rangle$ which are not $G$-conjugate to each other. Since $\rho$ is $Q$-conjugate to $\rho^{-1}$, the cosets $\rho\langle\iota\rangle, \tau\langle\iota\rangle, \rho\tau\langle\iota\rangle$ are the three involutions in the four-group $Q/\langle\iota\rangle$. One of $\rho$ and $\tau$, say $\rho$, is not $G$-conjugate to $\rho\tau$. Then the two cosets $\langle\rho\rangle = \{1, \rho, \iota, \rho^{-1}\}$ and $\tau\langle\rho\rangle = \{\tau, \rho\tau, \tau^{-1}, (\rho\tau)^{-1}\}$ of $\langle\rho\rangle$ in $Q$ have the property that no element in one is $G$-conjugate to any element in the other.

We define a function $\lambda : G \to F$ by

$$\lambda(\sigma) = \begin{cases} +1, & \textit{if } \sigma_2 \textit{ is } G\textit{-conjugate to an element of } \langle\rho\rangle, \\ -1, & \textit{if } \sigma_2 \textit{ is } G\textit{-conjugate to an element of } \tau\langle\rho\rangle \end{cases}$$

for any $\sigma \in G$. Evidently $\lambda$ is a well-defined class-function on $G$ whose restriction to $Q$ is a linear character. If $E$ is any nilpotent subgroup of $G$, and $E = D \times A$, where $D \leq Q$ and $A$ is a subgroup of odd order, then the restriction of $\lambda$ to $E$ is clearly a linear character. Since every nilpotent subgroup of $G$ is conjugate to such an $E$, we conclude from Theorem 7.1 that $\lambda$ is a generalized character of $G$.

By definition $\lambda(\sigma) = \pm 1$, for all $\sigma \in G$. Furthermore, $\lambda(\sigma^{-1}) = \lambda(\sigma)$. Hence,

$$(\lambda, \lambda)_G = \frac{1}{|G|} \sum_{\sigma \in G} \lambda(\sigma^{-1})\lambda(\sigma) = \frac{1}{|G|} \sum_{\sigma \in G} (\pm 1)^2 = 1.$$

But $\lambda = \sum_{j=1}^{k} a_j\chi_j$, where $a_1, \ldots, a_k \in \mathbf{Z}$ and $\chi_1, \ldots, \chi_k$ are the irreducible characters of $G$. From (4.20) we get $1 = (\lambda, \lambda)_G = \sum_{j=1}^{k} a_j^2$, which implies that all the $a_j$ except one are zero and that that one is $\pm 1$. Hence $\pm\lambda$ is an irreducible character of $G$. Since $\lambda(1) = 1 > 0$, we conclude that $\lambda$ is a linear character of $G$. Now $N = \text{Ker}\,\lambda$ is a normal subgroup of $G$ containing $\iota$ and not containing $\rho$. This contradicts the fact that $G$ is a counterexample to Theorem 14.11. The contradiction proves the lemma.

Fix an element $\rho \in Q \backslash \langle \iota \rangle$. The above lemma implies immediately that

COROLLARY 14.13. *The elements* 1, $\iota$, $\rho$ *are representatives for the distinct conjugacy classes of* 2*-elements of* $G$.

In fact, we can say more. Let $H$ be the normalizer in $G$ of $\langle \iota \rangle$. Of course, $H = C_G(\iota)$, and $Q$ is a 2-Sylow subgroup of $H$.

LEMMA 14.14. *The elements* 1, $\iota$, $\rho$ *are representatives for the distinct conjugacy classes of* 2*-elements of* $H$.

*Proof.* Let $\tau$ be any element of $Q \backslash \langle \iota \rangle$. By Lemma 14.12 there exists $\pi \in G$ such that $\rho^\pi = \tau$. But then $\iota^\pi = (\rho^2)^\pi = \tau^2 = \iota$. Hence $\pi \in H = C_G(\iota)$. Therefore all elements of $Q \backslash \langle \iota \rangle$ are $H$-conjugate to $\rho$. The lemma follows from this.

Evidently $H$ contains the group $L = N_G(\langle \rho \rangle)$. Hence $L = N_H(\langle \rho \rangle)$. The structure of $L$ is quite simple.

LEMMA 14.15. *The group* $L$ *is the semi-direct product* $QK$ *of* $Q$ *with an odd normal subgroup* $K$ *centralizing* $\rho$.

*Proof.* Since $\langle \rho \rangle$ is normal in $Q$, and $Q$ is a 2-Sylow subgroup of $G$, the group $Q$ is a 2-Sylow subgroup of $L$. Inversion $\rho \to \rho^{-1}$, is the only non-trivial automorphism of the cyclic group $\langle \rho \rangle$ of order 4, and any element of $Q \backslash \langle \rho \rangle$ inverts $\langle \rho \rangle$. It follows that $L = QC$, where $C = C_G(\rho)$, and that $\langle \rho \rangle = Q \cap C$ is a cyclic 2-Sylow subgroup of $C$. This implies that $C = \langle \rho \rangle K$, where $K = 0_{2'}(C)$ (see Section I.6 of Huppert, 1967). Since $K$ is characteristic in $C$, it is normal in $L$, and $L = QK$ is the semi-direct product of $Q$ and $K$. This proves the lemma.

The group $Q$ has four linear characters 1, $\lambda_1$, $\lambda_2$, $\lambda_3$, coming from the four-group $Q/\langle \iota \rangle$, and the irreducible character $\theta$ defined by (6.5). If $\tau \in Q \backslash \langle \rho \rangle$, then 1, $\iota$, $\rho$, $\tau$, $\rho\tau$ are representatives for the five $Q$-conjugacy classes of $Q$, and we have (after renumbering) the character table:

| | 1 | $\iota$ | $\rho$ | $\tau$ | $\rho\tau$ |
|---|---|---|---|---|---|
| 1 | 1 | 1 | 1 | 1 | 1 |
| $\lambda_1$ | 1 | 1 | 1 | $-1$ | $-1$ |
| $\lambda_2$ | 1 | 1 | $-1$ | 1 | $-1$ |
| $\lambda_3$ | 1 | 1 | $-1$ | $-1$ | 1 |
| $\theta$ | 2 | $-2$ | 0 | 0 | 0 |

(14.16)

By Proposition 4.25, $1, \lambda_1 \lambda_2, \lambda_3, \theta$ are all the irreducible characters of $Q$. We denote by $\tilde{1}, \tilde{\lambda}_1, \tilde{\lambda}_2, \tilde{\lambda}_3, \tilde{\theta}$ the corresponding characters of $L = QK$ obtained from $1, \lambda_1, \lambda_2, \lambda_3, \theta$, respectively, by composition with the natural epimorphism of $QK$ onto $Q = QK/K$.

LEMMA 14.17. *The characters* $\tilde{1}, \tilde{\lambda}_1, \tilde{\lambda}_2, \tilde{\lambda}_3, \tilde{\theta}$ *are precisely the irreducible characters in* $B_0(L)$.

*Proof.* By Proposition 13.28 any irreducible character in $B_0(L)$ has $K = 0_{2'}(L)$ in its kernel. Therefore it must be among the characters $\tilde{1}, \tilde{\lambda}_1, \tilde{\lambda}_2, \tilde{\lambda}_3, \tilde{\theta}$ which come from those of $Q = L/K$. On the other hand, one verifies easily from (4.30), (13.14) and Proposition 11.15 that each character on our list lies in $B_0(K)$. This proves the lemma.

By Proposition 5.12, the set $\sqrt{\rho}$ is a t.i. set in both $H$ and $G$ with $L$ as its normalizer. Since $\rho$ is a 2-element, this t.i. set clearly satisfies (14.1) (for both $H$ and $G$). From Lemma 14.15 it is evident that $\sqrt{\rho} = \rho K \cup \rho^{-1}K$. Now the table (14.16) and Lemma 14.17 imply that

$$CF(L|\sqrt{\rho}, B_0(L)) = F(\tilde{1}+\tilde{\lambda}_1-\tilde{\lambda}_2-\tilde{\lambda}_3). \tag{14.18}$$

In view of (5.19), the induced character $(\tilde{1}+\tilde{\lambda}_1-\tilde{\lambda}_2-\tilde{\lambda}_3)^H = \Phi$ is a generalized character of $H$ satisfying:

$$(\Phi, \Phi)_H = (\tilde{1}+\tilde{\lambda}_1-\tilde{\lambda}_2-\tilde{\lambda}_3, \tilde{1}+\tilde{\lambda}_1-\tilde{\lambda}_2-\tilde{\lambda}_3)_L = 4.$$

Proposition 5.9 implies that $\Phi = 1+\sum_{j=1}^{d} a_j\phi_j$, where 1 is the trivial character and $\phi_1,\ldots, \phi_d$ are the non-trivial irreducible characters of $H$, and where $a_1,\ldots, a_d$ are integers. Since $4 = 1^2+\sum_{j=1}^{d} a_j^2$, all the $a_j$ but three must be zero, and those three must be $\pm 1$. Hence there are three distinct non-trivial irreducible characters, say $\phi_1, \phi_2, \phi_3$ of $H$, and three signs $a_1, a_2, a_3 = \pm 1$ such that

$$(\tilde{1}+\tilde{\lambda}_1-\tilde{\lambda}_2-\tilde{\lambda}_3)^H = 1+a_1\phi_1+a_2\phi_2+a_3\phi_3.$$

By (14.18) and (14.8) we are in the coherent case. So Proposition 14.9 tells us that

$$1, \phi_1, \phi_2, \phi_3 \textit{ are precisely the irreducible characters in } B_0(H) \textit{ which do not vanish on } (\sqrt{\rho})^H. \tag{14.19}$$

Furthermore, (14.10) and (14.16) give

$$a_1\phi_1 = a_2\phi_2 = a_3\phi_3 = 1 \quad on\ (\sqrt{\rho})^H. \tag{14.20}$$

To compute the rest of the characters in $B_0(H)$ we shall use a process of modification of generalized characters based on Theorem 7.1. Let $H_{2'}$ be the subset of all elements of odd order in $H$. Since $\iota$ is central in $H$, it is clear that $\langle\iota\rangle H_{2'} = H_{2'} \cup \iota H_{2'}$ is the 2-section of all $\sigma \in H$ such that $\sigma_2 = 1$ or $\iota$. We define $X(H|\langle\iota\rangle H_{2'}, B_0(H))$ to be the intersection $X(H) \cap CF(H|\langle\iota\rangle H_{2'}, B_0(H))$, i.e., the additive group of all **Z**-linear combinations $\phi$ of the irreducible characters $\phi_j \in B_0(H)$ such that $\phi = 0$ on $H \backslash \langle\iota\rangle H_{2'}$.

Let $X_+$, $X_-$ be the two additive subgroups of $X(H|\langle\iota\rangle H_{2'}, B_0(H))$ defined by

$$X_+ = \{\phi \in X(H|\langle\iota\rangle H_{2'}, B_0(H)) : \phi(\sigma\iota) = \phi(\sigma), \quad \textit{for all } \sigma \in H\}, \qquad (14.21a)$$

$$X_- = \{\phi \in X(H|\langle\iota\rangle H_{2'}, B_0(H)) : \phi(\sigma\iota) = \phi(\sigma), \quad \textit{for all } \sigma \in H\}. \qquad (14.21b)$$

For any $\phi \in X_+ \cup X_-$ we define a class function $\phi^*$ on $H$ by

$$\phi^*(\sigma) = \begin{cases} \phi(\sigma), & \textit{for } \sigma \in H_{2'}, \\ -\phi(\sigma), & \textit{for } \sigma \in \iota H_{2'}, \\ 0, & \textit{for } \sigma \in H \backslash \langle\iota\rangle H_{2'}. \end{cases} \qquad (14.22)$$

Then we have

LEMMA 14.23. *The map $\phi \to \phi^*$ sends $X_+$ isometrically into $X_-$ and $2X_-$ isometrically into $X_+$.*

*Proof.* Let $\phi$ be any element of $X_+ \cup 2X_-$. We first show that $\phi^* \in X(H)$. By Theorem 7.1 it suffices to prove that the restriction $\phi^*_E$ lies in $X(E)$, for any nilpotent subgroup $E$ of $H$. Since we can pass to any $H$-conjugate of $E$, Lemma 14.14 implies that we can assume $E$ to have one of the four forms $T$, $\langle\iota\rangle \times T$, $\langle\rho\rangle \times T$, or $Q \times T$, where $|T|$ is odd.

If $E = T$, then $\phi^*_E \in X(E)$, by (14.22). If $E = \langle\iota\rangle \times T$, then (14.21) implies that $\phi_E = \lambda \times \psi$, where $\psi \in X(T)$ and $\lambda$ is one of the two linear characters of $\langle\iota\rangle$. Evidently $\phi^*_E$ is then $\lambda' \times \psi$, where $\lambda'$ is the other linear character of $\langle\iota\rangle$. So $\phi^*_E \in X(E)$.

If $E = \langle\rho\rangle \times T$, then the fact that $\phi$ is zero on $(\langle\rho\rangle \backslash \langle\iota\rangle) \times T \subseteq H \backslash \langle\iota\rangle H_{2'}$, together with (14.21), implies that $\phi_E = \lambda^{\langle\rho\rangle} \times \psi$, where $\psi \in X(T)$ and $\lambda$ is one of the linear characters of $\langle\iota\rangle$. Then $\phi^*_E = (\lambda')^{\langle\rho\rangle} \times \psi \in X(E)$, where $\lambda'$ is the other linear character of $\langle\iota\rangle$.

There remains the case in which $E = Q \times T$. Suppose that $\phi \in X_+$. Since $\phi = 0$ on $(Q \backslash \langle\iota\rangle) \times T$, it follows from (14.16) and (14.21a) that $\phi_E = (1 + \lambda_1 + \lambda_2 + \lambda_3) \times \psi$, for some $\psi \in X(T)$. But then $\phi^*_E = 2\theta \times \psi \in X(E)$. On the other hand, if $\phi \in 2X_-$, we see from (14.16) and (14.21b) that $(\frac{1}{2}\phi)_E = \theta \times \psi$, for some $\psi \in X(T)$. Hence $\phi_E = 2\theta \times \psi$ and $\phi^*_E = (1 + \lambda_1 + \lambda_2 + \lambda_3) \times \psi \in X(E)$. Therefore $\phi^*_E \in X(E)$ in all cases, which proves that $\phi^* \in X(H)$.

It is evident from (14.22) that $\phi^*$ lies in $X(H|\langle\iota\rangle H_{2'})$. We must show that it lies in $X(H|\langle\iota\rangle H_{2'}, B_0(H))$. Let $\phi_j$ be an irreducible character of $H$ such that $(\phi^*, \phi_j)_H \neq 0$. It suffices to prove that $\phi_j \in B_0(H)$.

Since $\phi$ is a linear combination of the irreducible characters in $B_0(H)$,

Theorem 12.19 implies that $\phi|_{H_{2'}} \in CF(H|H_{2'}, B_0(H))$ and $\phi|_{\iota H_{2'}} \in CF(H|\iota H_{2'}, B_0(H))$. From (14.22) we get

$$0 \neq (\phi^*, \phi_j)_H = (\phi|_{H_{2'}} - \phi|_{\iota H_{2'}}, \phi_j)_H = (\phi|_{H_{2'}}, \phi_j|_{H_{2'}})_H - (\phi|_{\iota H_{2'}}, \phi_j|_{\iota H_{2'}})_H$$

So one of $(\phi|_{H_{2'}}, \phi_j|_{H_{2'}})_H$, $(\phi|_{\iota H_{2'}}, \phi_j|_{\iota H_{2'}})_H$ is non-zero. In view of Theorem 12.19 and (12.18), this implies that $\phi_j \in B_0(H)$.

We now know that $\phi^* \in X(H|\langle\iota\rangle H_{2'}, B_0(H))$. It is clear from (14.21) and (14.22) that $\phi^* \in X_-$ if $\phi \in X_+$ and $\phi^* \in X_+$ if $\phi \in 2X_-$. Since the map $\phi \to \phi^*$ is obviously an isometry (with respect to $(\cdot, \cdot)_H$), this proves the lemma.

Because the involution $\iota$ lies in the center of $H$, it acts on any irreducible $FH$-module $I$ either as multiplication by 1 or as multiplication by $-1$. So the corresponding irreducible character $\phi_j$ of $H$ satisfies either

$$\phi_j(\sigma\iota) = \phi_j(\sigma), \quad \textit{for all } \sigma \in H, \textit{ or} \tag{14.24a}$$

$$\phi_j(\sigma\iota) = -\phi_j(\sigma), \quad \textit{for all } \sigma \in H, \tag{14.24b}$$

respectively.

LEMMA 14.25. *The characters* 1, $\phi_1$, $\phi_2$, $\phi_3$ *are precisely the irreducible characters* $\phi_j$ *in* $B_0(H)$ *satisfying* (14.24a).

*Proof.* Since $\rho$ is conjugate to $\rho\iota = \rho^{-1}$, any irreducible character $\phi_j$ satisfying (14.24b) is zero on $\rho$. This and (14.20) imply that each of 1, $\phi_1$, $\phi_2$, $\phi_3$ satisfies (14.24a).

Now let $\phi_j$ be any irreducible character in $B_0(H)$ satisfying (14.24a). Assume that $\phi_j$ is not 1, $\phi_1$, $\phi_2$, or $\phi_3$. Then $\phi_j = 0$ on $(\sqrt{\rho})^H$ by (14.19). Lemma 14.14 implies that $(\sqrt{\rho})^H = H\backslash\langle\iota\rangle H_{2'}$. Hence $\phi_j \in X_+$. Applying Lemma 14.23, we see that $\phi_j^* \in X_-$ and $(\phi_j^*, \phi_j^*)_H = (\phi_j, \phi_j)_H = 1$. Therefore $\pm\phi_j^*$ is an irreducible character of $H$. Since $\phi_j^*(1) = \phi_j(1) > 0$, we conclude that $\phi_j$, $\phi_j^*$ are (obviously distinct) irreducible characters in $B_0(H)$.

Because $\phi_j$ is zero on $Q\backslash\langle\iota\rangle$ and satisfies (14.24a), its restriction to $Q$ is an integral multiple of $(1+\lambda_j+\lambda_2+\lambda_3)$. Hence 4 divides $\phi_j(1)$. Now (4.17) and (14.22) imply that the primitive idempotents $f_j$, $f_j^*$ of $Z(FH)$ corresponding to $\phi_j$, $\phi_j^*$, respectively, satisfy

$$f_j + f_j^* = \sum_{\sigma \in H} \frac{\phi_j(1)\,[\phi_j(\sigma^{-1}) + \phi_j^*(\sigma^{-1})]\sigma}{|H|}$$

$$= \sum_{\sigma \in H_{2'}} \frac{2\phi_j(1)\phi_j(\sigma^{-1})\sigma}{|H_1|}$$

Since 8, which is the highest power of 2 dividing $|H|$, divides $2\phi_j(1)$, we conclude that $f_j + f_j^* \in Z(\mathfrak{O}H)$. This is impossible by (11.8), since $B_0(H)$ contains other characters besides $\phi_j$ and $\phi_j^*$. Therefore $\phi_j$ does not exist and the lemma is proved.

COROLLARY 14.26. $X_+ = \{c_0 1 + c_1 a_1 \phi_1 + c_2 a_2 \phi_2 + c_3 a_3 \phi_3 :$
$c_0, c_1, c_2, c_3 \in \mathbf{Z}, c_0 + c_1 + c_2 + c_3 = 0\}$

*Proof.* It is evident from (14.21a) and (14.24) that any $\phi \in X_+$ is a **Z**-linear combination of those irreducible characters $\phi_j \in B_0(H)$ satisfying 14.24a, i.e., of $1, \phi_1, \phi_2, \phi_3$. So $X_+$ consists precisely of all $\phi = c_0 1 + c_1 a_1 \phi_1 + c_2 a_2 \phi_2 + c_3 a_3 \phi_3$, with $c_0, c_1, c_2, c_3 \in \mathbf{Z}$, such that $\phi = 0$ on $H \backslash \langle \iota \rangle H_{2'}$. But $H \backslash \langle \iota \rangle H_{2'} = (\sqrt{\rho})^H$ by Lemma 14.14. By (14.20), $\phi = 0$ on $(\sqrt{\rho})^H$ if and only if $c_0 + c_1 + c_2 + c_3 = 0$. This proves the corollary.

COROLLARY 14.27. *$B_0(H)$ contains precisely three irreducible characters $\phi_4, \phi_5, \phi_6$ satisfying* (14.24b). *Furthermore* $X_- = \{c_4 \phi_4 + c_5 \phi_5 + c_6 \phi_6 : c_4, c_5, c_6 \in \mathbf{Z}\}$.

*Proof.* Any irreducible character $\phi_j \in B_0(H)$ satisfying (14.24b) is certainly not one of $1, \phi_1, \phi_2, \phi_3$. So $\phi_j = 0$ on $(\sqrt{\rho})^H = H \backslash \langle \iota \rangle H_{2'}$ by (14.9). Hence $\phi_j \in X_-$. It follows that these $\phi_j$ form a **Z**-basis for $X_-$. Since Lemma 14.23 implies that the **Z**-rank of $X_-$ is the same as that of $X_+$, which is 3 by Corollary 14.26, this completes the proof of the corollary.

Actually, we can compute $\phi_4, \phi_5, \phi_6$ quite explicitly.

LEMMA 14.28. *There exist signs $a_4, a_5, a_6 = \pm 1$, such that (after renumbering)*

$$\left.\begin{aligned} 2a_4 \phi_4^* &= 1 + a_1 \phi_1 - a_2 \phi_2 - a_3 \phi_3 \\ 2a_5 \phi_5^* &= 1 - a_1 \phi_1 + a_2 \phi_2 - a_3 \phi_3 \\ 2a_6 \phi_6^* &= 1 - a_1 \phi_1 - a_2 \phi_2 + a_3 \phi_3 \end{aligned}\right\} \tag{14.29}$$

*Proof.* If $j = 4$, 5 or 6, then Lemma 14.23 implies that $(2\phi_j)^*$ is an element of $X_+$ satisfying $((2\phi_j)^*, (2\phi_j)^*)_H = 4$. From the description of $X_+$ in Corollary 14.26 it is clear that $(2\phi_j)^*$ has the form $\pm 1 \pm a_1 \phi_1 \pm a_2 \phi_2 \pm a_3 \phi_3$ where the sum of the four $\pm$'s is zero. So, with a suitable choice of $a_j = \pm 1$, the character $2a_j \phi_j^*$ has one of the three forms on the right of (14.29). Since there are three distinct $\phi_j$, they must exhaust these three forms. This proves the lemma.

Now we can compute the module $X(H|\sqrt{\iota}, B_0(H))$ of all **Z**-linear combinations $\phi$ of irreducible characters in $B_0(H)$ such that $\phi = 0$ on $H \backslash \sqrt{\iota} = H_{2'}$.

LEMMA 14.30. *The **Z**-module $X(H|\sqrt{\iota}, B_0(H))$ has the **Z**-basis*

$$1 + a_1 \phi_1 + a_2 \phi_2 + a_3 \phi_3, \quad a_2 \phi_2 + a_3 \phi_3 + a_4 \phi_4, \quad a_1 \phi_1 + a_3 \phi_3 + a_5 \phi_5,$$
$$a_1 \phi_1 + a_2 \phi_2 + a_6 \phi_6. \tag{14.31}$$

*Proof.* We know from its definition that $1 + a_1 \phi_1 + a_2 \phi_2 + a_3 \phi_3$ vanishes outside $(\sqrt{\rho})^H$. In particular, it is zero on $H_{2'}$. So $1 + a_1 \phi_1 + a_2 \phi_2 + a_3 \phi_3$

$\in X(H|\sqrt{\iota}, B_0(H))$. From this, (14.22) and (14.29) we see that

$$2a_4\phi_4|_{H_2'} = 1+a_1\phi_1|_{H_2'}-a_2\phi_2|_{H_2'}-a_3\phi_3|_{H_2'}$$
$$= -2(a_2\phi_2|_{H_2'}+a_3\phi_3|_{H_2'}).$$

It follows that $a_2\phi_2+a_3\phi_3+a_4\phi_4 \in X(H|\sqrt{\iota}, B_0(H))$. Similarly $a_1\phi_1+a_3\phi_3+a_5\phi_5$, $a_1\phi_1+a_2\phi_2+a_6\phi_6 \in X(H|\sqrt{\iota}, B_0(H))$.

Since $\phi_4$, $\phi_5$, $\phi_6$ are $F$-linearly independent on $H$, vanish on $(\sqrt{\rho})^H = H\backslash\langle\iota\rangle H_2'$, and satisfy (14.24b), their restrictions to $H_{2'}$ are $F$-linearly independent. It follows that the module $X(H|B_0(H))$ of all **Z**-linear combinations of 1, $\phi_1,\ldots,\phi_6$ maps by restriction to $H_{2'}$ onto a **Z**-module isomorphic to $X(H|B_0(H))/X(H|\sqrt{\iota}, B_0(H))$ of rank at least 3. But the elements (14.31) generate a submodule $M$ of rank 4 in $X(H|\sqrt{\iota}, B_0(H))$. Hence $X(H|\sqrt{\iota}, B_0(H))$ has rank 4. Since $X(H|B_0(H))/L$ is clearly **Z**-torsion free, $M$ equals $X(H|\sqrt{\iota}, B_0(H))$, and the lemma is proved.

By Proposition 5.12 the set $\sqrt{\iota}$ is a t.i. set in $G$ with normalizer $H$.

LEMMA 14.32. *There exist distinct non-trivial irreducible characters* $\chi_1,\ldots,\chi_6$ *of G and signs* $\varepsilon_1,\ldots,\varepsilon_6 = \pm 1$ *such that*

$$(1+a_1\phi_1+a_2\phi_2+a_3\phi_3)^G = 1+\varepsilon_1\chi_1+\varepsilon_2\chi_2+\varepsilon_3\chi_3. \qquad (14.33a)$$
$$(a_2\phi_2+a_3\phi_3+a_4\phi_4)^G = \varepsilon_2\chi_2+\varepsilon_3\chi_3+\varepsilon_4\chi_4 \qquad (14.33b)$$
$$(a_1\phi_1+a_3\phi_3+a_5\phi_5)^G = \varepsilon_1\chi_1+\varepsilon_3\chi_3+\varepsilon_5\chi_5 \qquad (14.33c)$$
$$(a_1\phi_1+a_2\phi_2+a_6\phi_6)^G = \varepsilon_1\chi_1+\varepsilon_2\chi_2+\varepsilon_6\chi_6 \qquad (14.33d)$$

*Proof.* It follows as usual from Lemma 14.30, Proposition 5.9 and (5.19) that there exist three distinct non-trivial irreducible characters $\chi_1$, $\chi_2$, $\chi_3$ of $G$ and three signs $\varepsilon_1$, $\varepsilon_2$, $\varepsilon_3 = \pm 1$ such that (14.33a) holds. (Compare the argument preceding (14.19)). Since (5.19) and Lemma 14.30 imply that $\Psi = (a_2\phi_2+a_3\phi_3+a_4\phi_4)^G$ is a generalized character of $G$ satisfying $(\Psi, \Psi)_G = a_2^2+a_3^2+a_4^2 = 3$, there exist three distinct irreducible characters $\psi_1$, $\psi_2$, $\psi_3$ of $G$ and three signs $d_1$, $d_2$, $d_3 = \pm 1$ such that $\Psi = d_1\psi_1+d_2\psi_2+d_3\psi_3$. By Proposition 5.9, none of $\psi_1$, $\psi_2$, $\psi_3$ is 1. We also know from (5.19) that

$$(1+\varepsilon_1\chi_1+\varepsilon_2\chi_2+\varepsilon_3\chi_3,\ d_1\psi_1+d_2\psi_2+d_3\psi_3)_G = a_2^2+a_3^2 = 2.$$

It follows that two of the $\psi_i$ are equal to two of the $\chi_j$, say $\psi_1 = \chi_2$, $\psi_2 = \chi_3$ that $d_1 = \varepsilon_2$, $d_2 = \varepsilon_3$, and that $\psi_3$ is distinct from 1, $\chi_1$, $\chi_2$, $\chi_3$. Hence (14.33b) holds with $\chi_4 = \psi_3$ and $\varepsilon_4 = d_3$.

Similarly $(a_1\phi_1+a_3\phi_3+a_5\phi_5)^G = \varepsilon_i\chi_i+\varepsilon_j\chi_j+\varepsilon_5\chi_5$, where $i, j = 1, 2, 3$, $i \neq j$, $\varepsilon_5 = \pm 1$ and $\chi_5$ is an irreducible character of $G$ distinct from 1, $\chi_1$, $\chi_2$, $\chi_3$. If $i, j = 2, 3$, then (5.19) gives

$$1 = a_3^2 = (\varepsilon_2\chi_2+\varepsilon_3\chi_3+\varepsilon_4\chi_4,\ \varepsilon_2\chi_2+\varepsilon_3\chi_3+\varepsilon_5\chi_5)_G = 2+\varepsilon_4\varepsilon_5(\chi_4, \chi_5)_G.$$

Hence $\chi_4 = \chi_5$ and $\varepsilon_4 = -\varepsilon_5$. But both $\varepsilon_2\chi_2+\varepsilon_3\chi_3+\varepsilon_4\chi_4 = (a_2\phi_2+a_3\phi_3 a_4\phi_4)^G$ and $\varepsilon_2\chi_2+\varepsilon_3\chi_3-\varepsilon_4\chi_4 = (a_1\phi_1+a_3\phi_3+a_5\phi_5)^G$ vanish at 1. Hence $\chi_4(1) = 0$, which is impossible. Therefore one of $i, j$ is equal to 1 and the other is equal to 2 or 3, say 3. Now (14.33c) holds. Since

$$1 = (\varepsilon_2\chi_2+\varepsilon_3\chi_3+\varepsilon_4\chi_4,\ \varepsilon_1\chi_1+\varepsilon_3\chi_3+\varepsilon_5\chi_5)_G = 1+\varepsilon_1\varepsilon_5(\chi_4, \chi_5)_G,$$

the characters $1, \chi_1, \chi_2, \chi_3, \chi_4, \chi_5$ are all distinct.

A similar argument shows that (14.33d) holds for an irreducible character $\chi_6$ distinct from $1, \chi_1, \ldots, \chi_5$ and for an $\varepsilon_6 = \pm 1$. So the lemma is proved.

COROLLARY 14.34. $\varepsilon_i\chi_i|\sqrt{\iota} = a_i\phi_i|\sqrt{\iota}$ $(i = 1, \ldots, 6)$.

*Proof.* This follows directly from the lemma, Proposition 14.9, and Lemma 14.30.

At last we can prove Theorem 14.11. Let $I$ be the conjugacy class of $\iota$ in $G$. Evidently $\iota$ is the only involution in $H$, and $\iota^2 = 1 \notin \sqrt{\iota}$. So Corollary 5.22 implies that $I^2 \cap \sqrt{\iota}$ is empty. It follows from this and Proposition 5.23 that $\phi^G(\tilde{I}^2) = 0$, for any $\phi \in X(H|\sqrt{\iota})$. Applying this to $\phi = a_2\phi_2+a_3\phi_3+a_4\phi_4$, and using (14.33b), we get

$$\begin{aligned} 0 &= \varepsilon_2\chi_2(\tilde{I}^2)+\varepsilon_3\chi_3(\tilde{I}^2)+\varepsilon_4\chi_4(\tilde{I}^2) \\ &= \frac{[\varepsilon_2\chi_2(\tilde{I})]^2}{\varepsilon_2\chi_2(1)}+\frac{[\varepsilon_3\chi_3(\tilde{I})]^2}{\varepsilon_3\chi_3(1)}+\frac{[\varepsilon_4\chi_4(\tilde{I})]^2}{\varepsilon_4\chi_4(1)} \quad \textit{(by (4.31))} \\ &= |I|^2\left\{\frac{[\varepsilon_2\chi_2(\iota)]^2}{\varepsilon_2\chi_2(1)}+\frac{[\varepsilon_3\chi_3(\iota)]^2}{\varepsilon_3\chi_3(1)}+\frac{[\varepsilon_4\chi_4(\iota)]^2}{\varepsilon_4\chi_4(1)}\right\} \quad \text{(by Proposition 2.2)} \end{aligned}$$

Since $1+a_1\phi_1+a_2\phi_2+a_3\phi_3 = (1+\lambda_1-\lambda_2-\lambda_3)^H$ vanishes at $\iota$, we compute from (14.29) and (14.22) that

$$2a_4\phi_4(\iota) = -1-a_1\phi_1(\iota)+a_2\phi_2(\iota)+a_3\phi_3(\iota) = 2a_2\phi_2(\iota)+2a_3\phi_3(\iota)$$

Using Corollary 14.34, we see that this implies

$$\varepsilon_4\chi_4(\iota) = \varepsilon_2\chi_2(\iota)+\varepsilon_3\chi_3(\iota).$$

Because $\phi^G = \varepsilon_2\chi_2+\varepsilon_3\chi_3+\varepsilon_4\chi_4$ vanishes at 1, we have

$$\varepsilon_4\chi_4(1) = -\varepsilon_2\chi_2(1)-\varepsilon_3\chi_3(1).$$

Substituting these in the previous equation we obtain:

$$\begin{aligned} 0 &= \frac{[\varepsilon_2\chi_2(\iota)]^2}{\varepsilon_2\chi_2(1)}+\frac{[\varepsilon_3\chi_3(\iota)]^2}{\varepsilon_3\chi_3(1)}-\frac{[\varepsilon_2\chi_2(\iota)+\varepsilon_3\chi_3(\iota)]^2}{\varepsilon_2\chi_2(1)+\varepsilon_3\chi_3(1)} \\ &= \frac{[\varepsilon_2\chi_2(\iota)\varepsilon_3\chi_3(1)-\varepsilon_3\chi_3(\iota)\varepsilon_2\chi_2(1)]^2}{\varepsilon_2\chi_2(1)\varepsilon_3\chi_3(1)(\varepsilon_2\chi_2(1)+\varepsilon_3\chi_3(1))} \end{aligned}$$

Hence,

$$\chi_2(\iota)\chi_3(1)-\chi_3(\iota)\chi_2(1) = 0.$$

or

$$\frac{\chi_2(\iota)}{\chi_2(1)} = \frac{\chi_3(\iota)}{\chi_3(1)}.$$

By symmetry we also have $\dfrac{\chi_2(\iota)}{\chi_2(1)} = \dfrac{\chi_1(\iota)}{\chi_3(1)}$. So there is a constant $d$ such that $\chi_i(\iota) = d\chi_i(1)$, for $i = 1, 2, 3$. Since $1+\varepsilon_1\chi_1+\varepsilon_2\chi_2+\varepsilon_3\chi_3 = (1+a_1\phi_1 + a_2\phi_2+a_3\phi_3)^G = (1+\lambda_1-\lambda_2-\lambda_3)^G$ vanishes outside $(\sqrt{\rho})^G$, it is zero at 1 and at $\iota$. This gives

$$0 = 1+\varepsilon_1\chi_1(1)+\varepsilon_2\chi_2(1)+\varepsilon_3\chi_3(1)$$
$$0 = 1+\varepsilon_1\chi_1(\iota)+\varepsilon_2\chi_2(\iota)+\varepsilon_3\chi_3(\iota)$$
$$= 1+d(\varepsilon_1\chi_1(1)+\varepsilon_2\chi_2(1)+\varepsilon_3\chi_3(1)).$$

Therefore $d = 1$, and $\chi_i(\iota) = \chi_i(1)$, for $i = 1, 2, 3$. Proposition 4.38 now tells us that $N = \operatorname{Ker}(\chi_1)$ is a normal subgroup of $G$ containing $\iota$. Since $\chi_1 \neq 1$, its kernel $N$ is not the whole of $G$. Therefore $G$ is not a counterexample to the theorem. This contradiction proves the theorem.

## 15. Glauberman's Theorem

We have now reached the goal of these lectures, which is the proof of

THEOREM 15.1 (Glauberman). *Let $G$ be a finite group, and $T$ be a 2-Sylow subgroup of $G$. Suppose that $T$ contains an involution $\iota$ satisfying*

$$\text{if } \iota^\sigma \in T, \quad \text{for some } \sigma \in G, \quad \text{then } \iota^\sigma = \iota. \tag{15.2}$$

*Then $\iota$ lies in $Z(G \bmod O_{2'}(G))$. In particular, if $G$ is simple, then $G = \langle \iota \rangle$.*

*Proof.* We use induction on the order of $G$. The result is obvious if $|G| = 2$, so we can assume it to be true for all groups of smaller order than $|G|$.

Assume that $G$ contains a normal subgroup $N$ with $\iota \in N < G$. Then $T \cap N$ is a 2-Sylow subgroup of $N$ containing $\iota$, and $N$, $T \cap N$, $\iota$ satisfy the hypotheses of the theorem. So the image of $\iota$ is central in $N/O_{2'}(N)$ by induction. Since $O_{2'}(N)$ is characteristic in $N$, it is normal in $G$. It follows that $O_{2'}(N) = N \cap O_{2'}(G)$. Therefore the image $\bar{\iota}$ of $\iota$ in $\bar{G} = G/O_{2'}(G)$ is central in the image $\bar{N}$ of $N$.

Suppose that $\bar{\iota}$ is not central in $\bar{G}$. Then there exists some element $\sigma \in G$ such that $\bar{\iota}$ is different from the image $\overline{(\iota^\sigma)}$ of $\iota^\sigma$. Both $\iota$ and its $\bar{G}$-conjugate $\overline{(\iota^\sigma)}$ lie in the 2-Sylow subgroup of $Z(\bar{N})$, which is characteristic in $\bar{N}$ and hence normal in $\bar{G}$. So they both lie in the image $\bar{T}$ of $T$. Regarding things in the inverse image $TO_{2'}(G)$ of $\bar{T}$, we see that $\iota^{\sigma\tau} \in T$, for some $\tau \in O_{2'}(G)$. By (15.2), this implies that $\iota^{\sigma\tau} = \iota$. But then the image $\overline{(\iota^{\sigma\tau})} = \overline{(\iota^\sigma)}$ of $\iota^{\sigma\tau}$ equals $\bar{\iota}$. This contradiction forces $\bar{\iota}$ to be central in $\bar{G}$, which is the theorem in this case. Therefore we can assume from now on that

$$\textit{The only normal subgroup of } G \textit{ containing } \iota \textit{ is } G \textit{ itself} \tag{15.3}$$

One case in which (15.2) is satisfied is that in which $\iota$ is the only involution in $T$. To handle this case, we shall use the following old result about 2-groups.

LEMMA 15.4. *Let $T$ be a finite 2-group containing exactly one involution $\iota$. Then $T$ is either generalized quaternion of order $2^{n+1}$, for some $n \geq 2$, or else cyclic.*

*Proof.* Choose $A$ maximal among the normal abelian subgroups of $T$. We must show that

$$A = C_T(A). \tag{15.5}$$

Because $A$ is an abelian normal subgroup of $T$, its centralizer $C_T(A)$ is a normal subgroup containing $A$. If $C_T(A) > A$, let $B$ be minimal among all the normal subgroups of $T$ satisfying $A < B \leq C_T(A)$. Then $[B : A] = 2$, and $B = \langle \beta, A \rangle$, for any $\beta \in B \backslash A$. Since $A$ is abelian and $\beta \in C_T(A)$ centralizes $A$, the normal subgroup $B$ of $T$ is also abelian. This contradicts the maximality of $A$. Therefore (15.5) holds.

Evidently $\iota$ is the only involution in the abelian subgroup $A$. So $A = \langle \tau \rangle$ is cyclic of some order $2^n > 1$. In view of (15.5), conjugation in $T$ defines an isomorphism of $T/A$ onto a subgroup $\bar{T}$ of the automorphism group of $A$. If $\bar{T} = 1$, then $T = A$ is cyclic, and the proof is finished. So we can assume that $\bar{T} \neq 1$. Then there exists an element $\rho \in T$ whose image $\bar{\rho}$ is an involution in $\bar{T}$. Let $k$ be an integer such that $\tau^\rho = \tau^k$. Because $\tau^\rho$ also generates $A = \langle \tau \rangle$, the integer $k$ is a unit modulo $2^n$. This, $\bar{\rho} \neq 1$, and $\bar{\rho}^2 = 1$, imply that

$$k \equiv 1 \pmod 2, \qquad k \not\equiv 1 \pmod{2^n}, \qquad k^2 \equiv 1 \pmod{2^n}. \tag{15.6}$$

Obviously $n \geq 2$. If $n = 2$, the only solution to (15.6) is $k \equiv -1 \pmod{2^n}$. If $n > 2$, there are two other solutions: $k \equiv 1+2^{n-1}$ and $k \equiv -1+2^{n-1} \pmod{2^n}$ which we must eliminate.

Suppose that $n > 2$ and that $k \equiv 1+2^{n-1} \pmod{2^n}$. Since $\bar{\rho}^2 = 1$, the element $\rho^2$ lies in $A$. Hence it lies in $C_A(\rho) = \langle \tau^2 \rangle$. So $\rho^2 = \tau^{2l}$, for some integer $l$. If $h$ is any integer, then

$$(\rho\tau^h)^2 = \rho\tau^h\rho\tau^h = \rho^2(\rho^{-1}\tau^h\rho)\tau^h = \tau^{2l}\tau^{(1+2^{n-1})h}\tau^h = \tau^{2(l+(1+2^{n-2})h)}.$$

Since $n > 2$, the integer $1+2^{n-2}$ is a unit modulo $2^n$. Hence we can choose $h$ so that $(1+2^{n-2})h \equiv -l \pmod{2^n}$. Then $(\rho\tau^h)^2 = 1$. So $\rho\tau^h$ is an involution in $T$, which is impossible since $\rho\tau^h \notin A$ and the only involution $\iota$ lies in $A$.

Suppose that $n > 2$ and that $k \equiv -1+2^{n-1} \pmod{2^n}$. Then $\rho^2 \in C_A(\rho) = \langle \tau^{2^{n-1}} \rangle = \langle \iota \rangle$. Since $\rho$ is not an involution, we must have $\rho^2 = \tau^{2^{n-1}}$. But then

$$(\rho\tau)^2 = \rho\tau\rho\tau = \rho^2(\rho^{-1}\tau\rho)\tau = \tau^{2^{n-1}}\tau^{-1+2^{n-1}}\tau = \tau^{2^n} = 1.$$

So $\rho\tau$ is an involution not in $A$, which is impossible.

In view of the above arguments, we must have $n \geq 2$ and $k \equiv -1 \pmod{2^n}$. Since $\rho^2 \in C_A(\rho) = \langle \tau^{2^{n-1}} \rangle$ and $\rho^2 \neq 1$, we have $\rho^2 = \tau^{2^{n-1}} = \iota$. Therefore $\langle \rho, A \rangle = \langle \rho, \tau \rangle$ is a generalized quaternion group of order $2^{n+1}$ (by (6.1)).

If $\langle \rho, A \rangle < T$, then $\langle \bar{\rho} \rangle < \bar{T}$. The above arguments show that $\bar{\rho}$ is the only involution in $\bar{T}$. So there must exist a $\sigma \in T$ whose image $\bar{\sigma} \in \bar{T}$ satisfies $\bar{\sigma}^2 = \bar{\rho}$. Then $\tau^\sigma = \tau^m$ for some integer $m$ satisfying $m^2 \equiv -1 \pmod{2^n}$. Since $n \geq 2$, this implies $m^2 \equiv -1 \pmod 4$, which is impossible. Therefore $\langle \rho, A \rangle = T$, and the lemma is proved.

Now assume that $\iota$ is the only involution in $T$ (our Sylow subgroup). By the above lemma, $T$ is either cyclic or generalized quaternion. If $T$ is cyclic, then $G = T\mathrm{O}_{2'}(G)$ (see Section I.6 of Huppert, 1967). So the image of $\iota$ is central in $G/\mathrm{O}_{2'}(G) \simeq T$, which is the theorem in this case. If $T$ is generalized quaternion of order $2^{n+1}$, for $n \geq 2$, then Theorem 6.8 (for $n \geq 3$) or Theorem 14.11 (for $n = 2$) give us a normal subgroup $N < G$ containing $\iota$. This contradicts (15.3). Therefore we can assume from now on that

$$T \text{ contains an involution other than } \iota. \tag{15.7}$$

Another "trivial case" is that in which some involution $\iota_1$ is central in $G$. If $\iota_1 = \iota$, the theorem is clearly true. If $\iota_1 \neq \iota$, then the image $\bar{\iota}$ of $\iota$ in $\bar{G} = G/\langle \iota_1 \rangle$ satisfies (15.2) with respect to the image $\bar{T}$ of $T$, which is a 2-Sylow subgroup of $\bar{G}$. Indeed, if $\bar{\iota}^{\bar{\sigma}} \in \bar{T}$, for some $\bar{\sigma} \in \bar{G}$, and $\sigma \in G$ has $\bar{\sigma}$ as its image in $\bar{G}$, then $\iota^\sigma$ lies in the inverse image $T$ of $\bar{T}$. So $\iota^\sigma = \iota$ (by (15.2)) and $\bar{\iota}^{\bar{\sigma}} = \bar{\iota}$. Since $|\bar{G}| < |G|$, induction tells us that $\bar{\iota} \in Z(\bar{G} \bmod \mathrm{O}_{2'}(\bar{G}))$. It follows that the inverse image in $G$ of $\langle \iota \rangle \mathrm{O}_{2'}(\bar{G})$ is a normal subgroup containing $\iota$. By (15.3), it must be $G$ itself. The inverse image $N$ of $\mathrm{O}_{2'}(\bar{G})$ is then a normal subgroup of $G$ with $G = \langle \iota \rangle N$. The 2-Sylow subgroup $\langle \iota_1 \rangle$ of $N$ is cyclic and centralizes $N$. We conclude that $N = \langle \iota_1 \rangle \times \mathrm{O}_{2'}(N)$. Now $\mathrm{O}_{2'}(N) = \mathrm{O}_{2'}(G)$. Since $G/\mathrm{O}_{2'}(G)$ has order 4, it is abelian. So the theorem is true in this case. Therefore we can assume from now on that

$$C_G(\iota_1) < G, \quad \text{for all involutions } \iota_1 \in G. \tag{15.8}$$

One almost obvious remark is

LEMMA 15.9. *If two $G$-conjugates $\iota^\sigma$, $\iota^\tau$ of $\iota$ lie in the same 2-subgroup $P$ of $G$, then $\iota^\sigma = \iota^\tau$.*

*Proof.* The 2-subgroup $P$ is contained in a conjugate $T^\pi$ of the 2-Sylow subgroup $T$. So $\iota^{\sigma\pi^{-1}}$, $\iota^{\tau\pi^{-1}}$ both lie in $T$. By (15.2), $\iota^{\sigma\pi^{-1}} = \iota = \iota^{\tau\pi^{-1}}$. Hence $\iota^\sigma = \iota^\tau$.

COROLLARY 15.10. *If $H$ is a subgroup of $G$ such that $\iota \in H < G$, then $\iota \in Z(H \bmod \mathrm{O}_{2'}(H))$.*

*Proof.* Applying the lemma to a 2-Sylow subgroup $P$ of $H$ containing $\iota$, we conclude that $H$, $P$ and $\iota$ satisfy the hypotheses of the theorem. Since $|H| < |G|$, induction gives this corollary.

Now we come to the heart of the matter.

LEMMA 15.11. *Let $\iota_2$ be any involution in $T$ other than $\iota$. If $\chi$ is an irreducible character in $B_0(G)$ and $\sigma \in G$, then $\chi(\iota\iota_2^\sigma) = \chi(\iota\iota_2)$.*

*Proof.* First assume that $\iota\iota_2^\sigma$ is an involution. Then $\iota$ centralizes $\iota_2^\sigma$. We can therefore find an element $\pi \in C_G(\iota_2^\sigma)$ such that $\iota$ and $\iota^{\sigma\pi}$ lie in the same 2-Sylow subgroup $P$ of $C_G(\iota_2^\sigma)$. By Lemma 15.9 we have $\iota = \iota^{\sigma\pi}$. Hence $\iota\iota_2^\sigma = \iota^{\sigma\pi}\iota_2^{\sigma\pi}$ is $G$-conjugate to $\iota\iota_2$. Therefore $\chi(\iota\iota_2^\sigma) = \chi(\iota\iota_2)$ in this case.

Now let $\tau = \iota\iota_2^\sigma$ be arbitrary. By Proposition 5.20, the group $D = \langle \iota, \iota_2^\sigma \rangle$ is dihedral, with $\langle \tau \rangle$ as a cyclic normal subgroup of index 2 inverted by both $\iota$ and $\iota_2^\sigma$. For any integer $i$ we have

$$(\iota_2^\sigma)^{\tau^i} = \tau^{-i}\iota_2^\sigma\tau^i = \tau^{-i}(\iota_2^\sigma)^{-1}\tau^i(\iota_2^\sigma)(\iota_2^\sigma) = \tau^{-2i}\iota_2^\sigma, \tag{15.12}$$

since $\iota_2^\sigma$ is an involution. If $|\langle \tau \rangle|$ is odd, we can choose $i$ so that $\tau^{-2i}\iota_2^\sigma = \tau\iota_2^\sigma = \iota$. This is impossible, since $\iota_2$ is not $G$-conjugate to $\iota$ (by (15.2)). Therefore $|\langle \tau \rangle|$ is even, and $\langle \tau \rangle$ contains an involution $\iota_1$ centralized by all of $D$.

In view of (15.8) the group $H = C_G(\iota_1)$ is properly contained in $G$. This group contains $D$ and hence $\iota$. By Corollary 15.10, $\iota$ lies in $Z(H \bmod O_{2'}(H))$. Therefore $\tau = \iota\iota_2^\sigma \equiv \iota_2^\sigma\iota \pmod{O_{2'}(H)}$. It follows that $\tau^2 \in O_{2'}(H)$. Hence $\langle \tau^2 \rangle$ has odd order, and we are in the following situation:

$$\langle \tau \rangle = \langle \iota_1 \rangle \times \langle \tau^2 \rangle, \qquad \langle \tau^2 \rangle = \langle \tau \rangle \cap O_{2'}(H). \tag{15.13}$$

The Brauer second and third main theorems tell us that there exist integers $a_j$, one for each irreducible character $\phi_j \in B_0(H)$, such that

$$\chi(\iota_1\rho) = \sum_{\phi_j \in B_0(H)} a_j\phi_j(\iota_1\rho),$$

for all elements $\rho$ of odd order in $H$. From (15.13) we see that $\tau = \iota_1\rho$, where $\rho \in \langle \tau^2 \rangle \leq O_{2'}(H)$ has odd order. Proposition 13.28 then implies that $\phi_j(\iota_1\rho) = \phi_j(\iota_1)$, for all $\phi_j \in B_0(H)$. So, we have

$$\chi(\iota\iota_2^\sigma) = \chi(\tau) = \chi(\iota_1\rho) = \sum_{\phi_j \in B_0(H)} a_j\phi_j(\iota_1\rho) = \sum_{\phi_j \in B_0(H)} a_j\phi_j(\iota_1) = \chi(\iota_1).$$

But (15.12) and (15.13) imply that $\iota_1 = \iota(\iota_2^\sigma)^{\tau^i}$, for some integer $i$. Since $\iota_1$ is an involution, we have already seen that this implies $\chi(\iota_1) = \chi(\iota\iota_2)$. Therefore $\chi(\iota\iota_2^\sigma) = \chi(\iota_1) = \chi(\iota\iota_2)$, and the lemma is proved.

Let $I$ be the conjugacy class of $\iota$ in $G$, and $\tilde{I}$ be the corresponding class sum. Define $I_2$, ${}_2$ similarly for any involution $\iota_2 \neq \iota$ in $T$. If $\chi$ is any irreducible character in $B_0(G)$, then (4.31) gives

$$\chi(\tilde{I})\chi(\tilde{I}_2) = \chi(1)\chi(\tilde{I}\tilde{I}_2).$$

Evidently $\tilde{I}\tilde{I}_2$ is a sum of $|I||I_2|$ elements of the form $\iota^\sigma \iota_2^\tau = (\iota\iota_2^{\tau\sigma^{-1}})^\sigma$, for $\sigma$, $\tau \in G$. By Lemma 15.11, the value of $\chi$ at any such element is $\chi(\iota\iota_2)$. Therefore the above equation becomes

$$|I|\chi(\iota)|I_2|\chi(\iota_2) = \chi(1)|I||I_2|\chi(\iota\iota_2)$$

or

$$\chi(\iota)\chi(\iota_2) = \chi(1)\chi(\iota\iota_2).$$

Since $\iota$ is central in $T$ (by (15.2)), the product $_2$ $\iota\iota$is also an involution different from $\iota$ in $T$. Applying the above equation to $\iota\iota_2$ in place of $\iota_2$, we obtain

$$\chi(\iota)\chi(\iota\iota_2) = \chi(1)\chi(\iota_2).$$

Hence,

$$\chi(\iota)^2\chi(\iota_2) = \chi(\iota)\chi(1)\chi(\iota\iota_2) = \chi(1)^2\chi(\iota_2).$$

If $\chi(\iota_2) \neq 0$, we conclude that $\chi(\iota) = \pm\chi(1)$, and hence that $\iota \in Z(G \text{ mod } \text{Ker}(\chi))$ (by Proposition 4.38(b)). So we have

*If $\chi$ is any irreducible character in $B_0(G)$ such that $\chi(\iota_2) \neq 0$, for some involution $\iota_2 \in T\backslash\langle\iota\rangle$, then $\iota \in Z(G$ mod Ker $(\chi))$.* (15.14)

Let $N$ be the intersection of the kernels, Ker $(\chi)$, of all the irreducible characters $\chi$ of $G$ satisfying the hypotheses of (15.14). Then $N$ is a normal subgroup of $G$ and $\iota \in Z(G \text{ mod } N)$. From Proposition 13.28 we see that $N \geq O_{2'}(G)$. To show that $N \leq O_{2'}(G)$, which will finish the proof of the theorem, it suffices to prove that no involution of $T$ lies in $N$.

Suppose that $N$ contains an involution $\iota_2$ of $T$ other than $\iota$. Then the definition of $N$ and Proposition 4.38(a) imply that the value $\chi(\iota_2)$ is either 0 or $\chi(1)$, for any irreducible character $\chi \in B_0(G)$. Now Theorem 12.13 tells us that

$$0 = \sum_{\chi \in B_0(G)} \chi(\iota_2)\chi(1) = \sum_{\chi \in B_0(G),\ \chi(\iota_2) \neq 0} \chi(1)^2$$

Since at least one $\chi$, the trivial character, satisfies $\chi(\iota_2) \neq 0$, and since each $\chi(1)$ is a positive integer, this is impossible. So $\iota_2 \notin N$.

Suppose that $\iota \in N$. Then $N = G$ by (15.3). So $N$ contains an involution $\iota_2 \in T$ other than $\iota$, by (15.7). We have just seen that this is impossible. Therefore $N$ contains no involutions, $N = O_{2'}(G)$, and the theorem is proved.

## Bibliography

Curtis, C. W. and Reiner, I. (1962). "Representation Theory of Finite Groups and Associative Algebras". Interscience. New York.

Feit, W. (1969). "Representations of Finite Groups". Department of Mathematics, Yale University.

Huppert, B. (1967). "Endliche Gruppen I". Springer–Verlag, Heidelberg.